Elementary Modern Physics

Third Edition

Richard T. Weidner

Rutgers University
New Brunswick, New Jersey

Robert L. Sells

State University College of Arts and Sciences
Geneseo, New York

Allyn and Bacon, Inc.
Boston London Sydney Toronto

To Jean and Pat

Library of Congress Cataloging in Publication Data

Weidner, Richard T
Elementary modern physics.

Bibliography: p.
Includes index.
1. Physics. I. Sells, Robert L., joint author.
II. Title.
QC23.W4 1980 539 79-20503
ISBN 0-205-06559-7
ISBN 0-205-06609-7 International ed.

Printed in the United States of America.

10 9 8 7 6 5 4 3 2 85 84 83 82 81

Production Editor: David Dahlbacka
Art Editor: Leora Haywood
Preparation Buyer: Patricia Hart

Cover note: The cover displays the electronic density inside a gallium arsenide crystal. The circles (copper) represent the atomic cores; copper lines are drawn along the bonds. The electron density is represented by contour lines. The density piles up along the bonds (covalent bonding), but the peak is shifted toward the more positive arsenic atom (ionic bonding). These results were obtained using a pseudo-potential method by Professor Marvin L. Cohen and Dr. Jisoon Ihm at the University of California at Berkeley.

Contents

Preface

In this third edition of *Elementary Modern Physics,* our aim remains that of treating the fundamentals of twentieth-century physics fairly rigorously, but at an elementary level. The textbook is intended primarily for an introductory course in modern physics. The prerequisites are merely an elementary knowledge of classical physics and introductory calculus.

Although this revision is extensive, the basic organization of the text has not been changed. The overall length has been reduced. This is not a smorgasbord, with sketchy remarks on a whole variety of contemporary developments; rather, the book concentrates on a limited number of truly fundamental principles and treats them in reasonable depth, while quite deliberately maintaining a modest level of mathematical sophistication. After brief preliminaries, we begin with special relativity and use it extensively in what follows. The quantum theory is introduced through the photon-electron interactions and is followed by the wave properties of particles. Then come atomic physics, solid-state physics, nuclear physics, and the characteristics of elementary particles. Throughout we attempt to capitalize on what students already know according to classical physics, and have given prominent attention to the general correspondence principle—that classical physics is the correspondence limit of relativity and quantum physics.

The principal changes in organization and content are these:

- The detailed derivation of the Lorentz transformation relations has been moved from the chapter on relativistic kinematics to Appendix I. This makes the chapter more concise and leaves the instructor more options as to how and whether the detailed derivation will be used.
- Solid-state physics has been expanded to two chapters, one on fundamental principles and the second on applications. New topics treated are superconductivity, solid-state detectors, the solar cell, and holography.
- Material on nuclear physics has been condensed into two chapters.
- Elementary-particle physics has been substantially revised and updated. An elementary, but meaningful, treatment of the quark model is included.
- There are many more worked examples.
- Almost all problems are new, and answers are given in the back of the book for all of them. In addition, an *Instructor's Solution Manual,* with detailed solutions to all 480 problems, is available to course instructors from the publisher. Problems are identified by section number and topic. Problems are also identified according to level of difficulty—short and easy (identified by ●), of average difficulty (identified by problem number only), and challenging (identified by ■).
- All references, grouped according to the relevant text chapters and with comments on each cited work, have been placed in Appendix III.

To instructors who use the text for a relatively short course we suggest that the principal parts of chapters 1 through 6 are basic. The remaining groups of chapters—7 (many-electron atoms), 8 and 9 (fundamentals and applications of solid-state physics), 10 and 11 (nuclear physics), and 12 (elementary particles)—are relatively independent of one another and may be included or skipped at the instructor's option.

We are grateful to many users of past editions who have been good enough to communicate their suggestions and criticisms to us or the publishers. The present revision was, in fact, undertaken only after a comprehensive review of user opinions by the publisher. Their comments and suggestions for changes have been invaluable. Some suggestions were in mutual opposition; obviously, we have not been able to accommodate them all. We assume responsibility, of course, for any residual errors and would appreciate having them pointed out.

Our special thanks go to Judy MacDonald for preparation of the typescript, to Janet Wright for her excellent copyediting of the typescript, and to the editorial and production staffs of Allyn and Bacon who have given support to our efforts over many years—particularly to our Production Editor, David Dahlbacka.

R. T. W.
R. L. S.

1
Some Preliminaries

Albert Einstein (1879, Austria–1955). Giant of 20th-century physics. Doctorate, Polytechnic Academy, Zurich, 1905; also year of four epochal papers–special theory of relativity, equivalence of mass and energy, theory of Brownian motion, photon theory of light. 1916, general theory of relativity. 1921, Nobel Prize, photoelectric effect. 1933, joined Institute for Advanced Studies, Princeton, which was founded for Einstein. 1939, wrote to President Franklin D. Roosevelt urging the development of nuclear energy from uranium. (Photo courtesy of American Institute of Physics, Niels Bohr Library.)

What is modern physics? How does it differ from classical physics, and how is it similar? Which central ideas of classical physics can be carried over unchanged into modern physics, and which must be modified or replaced? These questions and other important ones are dealt with in this introductory chapter.

1-1 The Program of Physics

The program of physics is to devise concepts and laws that can help us to understand the universe. Physical laws are constructions of the human mind, subject to all the limitations of human understanding. They are not necessarily immune to change, and nature is not compelled to obey them.

A law in physics is a statement, usually in the succinct and precise language of mathematics, of a relation that has been found by repeated experiment to hold among physical quantities and that reflects persistent regularities in the behavior of the physical world. A "good" physical law has the greatest possible generality, simplicity, and precision. The final criterion of a successful law of physics is how accurately it predicts experimental results. On the other hand, extrapolating any law beyond its range of tested validity may predict results inconsistent with later experiments. One famous example of this was the Michelson-Morley experiment (see Chapter 2), which refuted the 19th-century conception of the ether as the medium for the propagation of electromagnetic waves. Such contradictions of theory are an important part of the evolution of physics. Early theories and laws that prove inadequate are supplanted by more general, comprehensive theories and laws that describe phenomena in the new, as well as the old, regions of investigation. Figure 1–1 shows the regions in which *classical physics, relativity physics, quantum physics,* and *relativistic quantum physics* apply.

Classical physics is the physics of ordinary-sized objects moving at ordinary speeds; it embraces Newtonian mechanics and electromagnetism. For speeds approaching the speed of light, classical physics must be supplanted by relativity physics; for sizes of about 10^{-10} m (approximately

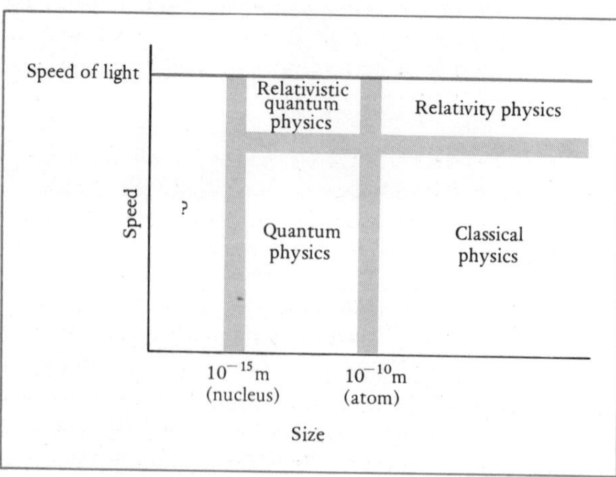

FIGURE 1–1. *Regions of applicability of various physical theories.*

the size of an atom), classical physics must be supplanted by quantum physics. For subatomic dimensions and speeds approaching the speed of light, only relativistic quantum physics is adequate. The limits of the several physical theories are not sharply defined; in fact, they overlap. Relativistic quantum physics is the most comprehensive and complete theoretical structure in present-day physics. At dimensions of about 10^{-15} m (the approximate size of the atomic nucleus) perplexing phenomena appear, at present only partly understood. Similarly, important cosmological questions remain unresolved in the domain of the very large (of the order of 10^{25} m).

Our understanding of atomic and nuclear structure is grounded in the two great ideas of modern physics, relativity theory and quantum theory. Both originated early in this century, when improved experimental techniques first allowed physicists to study phenomena at small enough dimensions and high enough speeds and energies. Indeed, by *modern* physics we mean the physics of the twentieth century.

After reviewing some crucial aspects of classical physics, we shall study relativity theory and quantum theory and use them to analyze atomic and nuclear structure. We shall deal with situations in which some familiar notions in physics may be inapplicable—situations in which classical physics is downright wrong. Does this mean, then, that all the time and effort spent in studying elementary classical physics is wasted, that one might better begin with relativity and quantum theory? Not at all! All results of experiment, however remote from our ordinary experience, must ultimately be expressed in classical terms—that is, in the classical concepts of momentum, energy, position, and time. Furthermore, we shall see that many of the concepts and laws of classical physics carry over into the new physics.

1-2 The Correspondence Principle

As we mentioned before, any theory or law in physics is more or less tentative and approximate; extrapolation to untested situations may show that it is incomplete or incorrect. If a new, more general theory is proposed, there is a completely reliable guide for relating the new theory to the older, more restricted one. This guide, the *correspondence principle,* was first proposed by the Danish physicist Niels Bohr in 1923 and applied to the theory of atomic structure. We shall find it helpful to apply this principle in a broader sense to both relativity physics and quantum physics.

> *The Correspondence Principle: Any new theory in physics, whatever its character or details, must reduce to the well-established classical theory to which it corresponds when it is applied under the circumstances in which the less general theory is known to hold.*

For example, when we are analyzing the motion of a projectile with a comparatively small range, we make the following assumptions: (1) The weight of the projectile is constant in magnitude and is given by the mass times a gravitational acceleration that is constant in magnitude; (2) the earth is represented by a plane surface, and (3) the weight of the projectile is constant in direction, vertically downward. With these assumptions, the theory predicts a parabolic path—in excellent agreement with experiment, provided that the projectile motion extends over only relatively short distances. However, if we try to describe the motion of an earth satellite on the same assumptions, *very* serious errors will be made. To discuss the satellite motion we must instead assume that (1) the weight of the body is *not* constant in magnitude but varies inversely with the square of its distance from the earth's center; (2) the earth's surface is spherical, not flat; and (3) the direction of the weight is *not* constant but always points toward the earth's center. With these assumptions, the theory predicts an elliptical path and describes satellite motion properly. Now, if we apply the second, more general, theory to the motion of a body traveling a distance that is small compared with the earth's radius at the surface of the earth, notice what happens. The weight appears to be constant in both magnitude and direction, the earth appears flat, and the elliptical path becomes parabolic. This is precisely what the correspondence principle requires!

The correspondence principle asserts that when the conditions of the new and old theories correspond, the predictions will also correspond; that is, a new (general) theory will yield the old (restricted) theory as a special approximation. We have, then, an infallible guide when testing a new theory or law: The new theory must reduce to the theory it supplants. Any new theory that fails in this respect is so fundamentally defective that it cannot possibly be accepted. Therefore, we know that the relativity and quantum theories *must* yield classical physics when applied to large-scale objects moving at speeds much lower than the speed of light. In the next section, we shall see another familiar example of the correspondence principle.

1-3 Ray Optics and Wave Optics

There are two means of describing the propagation of light: ray (geometrical) optics and wave (physical) optics. Only wave optics is capable of explaining such phenomena as interference and diffraction; however, ray optics can satisfactorily describe such phenomena as the rectilinear propagation, reflection, and refraction of light. Wave optics can, of course, also account for these phenomena. Therefore, wave optics is a comprehensive theory of light, while ray optics is an adequate theory only in certain restricted situations. Even the wave theory can't account for all the effects of light; it must thus be supplanted by a quantum theory of electromagnetic radiation (Chapter 4).

The correspondence principle requires that the comprehensive theory reduce to the restricted theory in the correspondence limit. Thus, wave optics must become, in effect, ray optics in those conditions in which such distinctive wave phenomena as diffraction and interference are unimportant. We know that interference and diffraction are clearly discernible only if the dimensions d of the obstacles or apertures that the light encounters are comparable to the wavelength λ of the light. When $\lambda \ll d$, the wave treatment gives the same results as the ray treatment. Symbolically, we can write

$$\text{Limit}_{\lambda/d \to 0} \text{ (wave optics)} = \text{ray optics}$$

Figure 1–2 illustrates the transition from conditions in which wave

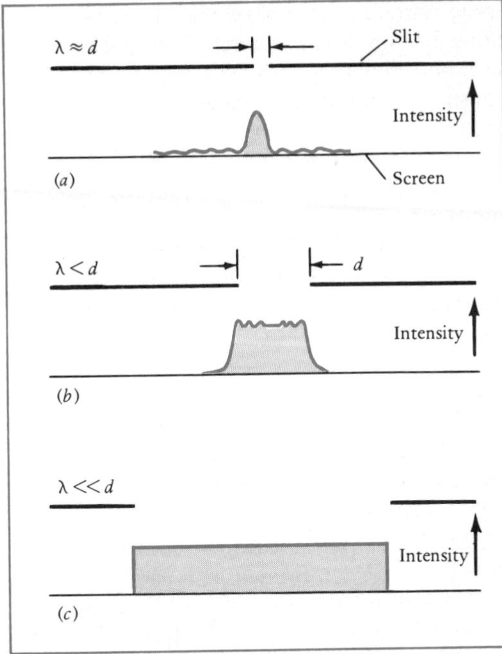

FIGURE 1–2. *Distribution of intensity of monochromatic light passing through single slits of increasing width. [Alternatively, the screen is far from the slit in (a) but close to the screen in (b) and (c)].*

optics is required to the simpler conditions in which wave and ray optics yield identical results. The figure shows the diffraction pattern of monochromatic light passing through a single parallel-edged slit for (a) a wavelength λ comparable to the slit width d, (b) a wavelength less than the slit width, and (c) a wavelength *much* less than the slit width. In the first diagram, the wave disturbance is spread far into the geometrical shadow and has the characteristic alternating light and dark diffraction bands. In the second diagram, the diffraction is less pronounced, and the light is concentrated mainly within the area bounded by the geometrical shadow. In the third diagram, where the wavelength is much smaller than the width of the slit opening, the intensity pattern is indistinguishable from that predicted by ray optics.†

Ray optics is concerned solely with the path of light, which may be represented as rays in the direction of light propagation. This suggests a model for describing the character of light, the *particle model.* In such a model, light is assumed to consist of small, essentially weightless particles, or corpuscles. The particle model is consistent with the observed facts that (1) in free space, light follows a straight-line path like that of a stream of particles; (2) when being reflected, light behaves like particles making elastic collisions with the surface; (3) when being refracted in a transparent material, such as glass, the light behaves as though the direction of the particles abruptly changed at the interface; and (4) the intensity from a point source varies inversely with the square of the distance from it.

The most celebrated advocate of the particle theory of light was Sir

† Between (a) and (b) in Figure 1–2, the diffraction patterns may be quite complicated; some even show a dark band at the center.

Isaac Newton. He showed that according to the particle concept, the speed of light in a refracting medium should be greater than the speed of light in a vacuum. Foucault, however, experimentally found that the speed of light through water was less than that through air; the wave theory, of course, predicts a lower speed in a refracting medium. The Foucault experiment, with the earlier work of Young and Fresnel on the interference and diffraction of light, convinced physicists that light consisted of waves, as Huygens first proposed.

Although physicists knew, well before 1864, that light consisted of waves and they thus could describe interference and diffraction, they did not know "what was waving" before Maxwell proposed his electromagnetic wave theory in that year.

1-4 The Particle and Wave Descriptions in Classical Physics

The ideas of particle and wave play a central role in both classical physics and modern physics. Here, we will briefly summarize their characteristics.

An ideal particle can be localized completely. Its mass and electric charge can be determined with such high precision that the particle can be considered to be a mass point. Although, in nature, all particles have finite sizes, we can regard them as mass points under the appropriate circumstances. For example, in the kinetic theory molecules are considered point particles, although their size is finite and they have internal structures; similarly, stars are considered particles when we discuss the behavior of galaxies. In short, an object is effectively a particle whenever its dimensions are very small relative to the dimensions of the system of which it is a part, and when its internal structure is unimportant to the problem under consideration. Newtonian mechanics deals with ideal particles; given the initial position and velocity of a particle and a knowledge of the forces acting on it, we can predict in detail its future position and velocity.

The simplest type of wave is strictly sinusoidal and is distinguished by its frequency or by its wavelength. Consider the electric field E of a perfectly monochromatic electromagnetic wave; its amplitude is E_0, its frequency ν, and its wavelength λ. It travels in the positive x-direction with a speed $v = \nu\lambda$. Then

$$E = E_0 \sin 2\pi\left(\frac{x}{\lambda} - \nu t\right) = -E_0 \sin(\omega t - kx) \tag{1-1}$$

where $\omega = 2\pi\nu$ and $k = 2\pi/\lambda$. Such a wave shows a sinusoidal variation of E in space for any fixed time; conversely, it shows a sinusoidal variation of E in time for any fixed point x in space. Equation (1–1) implies that the electric disturbance extends over *all* possible values of x for all possible instants of time t.

An ideal wave, one whose wavelength and frequency can be known with infinite precision, cannot be confined to any restricted region of space; rather, it must have an infinite extension along the direction in which it is propagated. It is simple to show by a hypothetical (thought) experiment that if we are to measure and thereby know the frequency of a wave with no uncertainty the wave must be infinite.

Suppose we have a standard clock for measuring how many wave crests pass a fixed point per unit time. For simplicity, we imagine that the standard clock is an oscillator producing waves whose frequency is to be compared with that of some incoming wave. How can we state with complete assurance that the frequency of the incoming wave is precisely the same as the frequency of the wave generated by our standard clock?

We shall allow the two waves to interfere with one another, so as to produce beats. The number of beats per unit time equals the difference between the frequencies of the two waves. If the two waves are of precisely the same frequency, then we will detect no beats whatsoever. If we observe the resultant amplitude of the two interfering waves over some limited period, we may find no appreciable change in this amplitude, but we cannot assert that there is none on the basis of such a measurement. If we had waited longer, we might have found that the combined amplitude of the two waves was decreasing or increasing (see Figure 1–3); this would indicate an incipient beat, or a difference between the frequencies. To be absolutely sure that no beats occur, we must wait for an infinite period. If we do wait for an infinite time, then the wave we measure will have traveled for an infinite time and have an infinite extension in space.

We now wish to determine the uncertainty of the result if we measure the beats ν_2 produced with the standard clock ν_1 over a *finite* time Δt. On a conservative basis, we can be confident of observing a complete beat in time

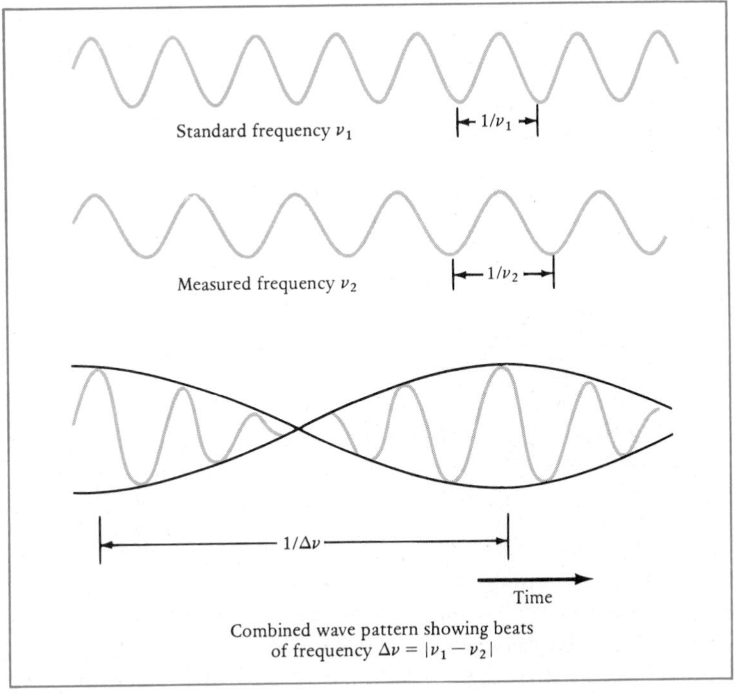

FIGURE 1–3. *Beat pattern resulting from the superposition of waves of frequencies ν_1 and ν_2.*

Δt only if the beat has a period of Δt or smaller. This corresponds to $1/\Delta t$ or more beats per unit time, or a minimum observable frequency difference of $\Delta\nu$:

$$\frac{1}{\Delta t} \leq \Delta\nu$$

or

$$\Delta t \Delta\nu \geq 1 \tag{1–2}$$

It follows from Equation 1–2 that the uncertainty is large if the frequency is measured over a very short time interval and that if $\Delta\nu$ is to be zero, Δt must be infinite.

When a musical tone, for example, produces 2 beats per second when sounded with a standard 440-Hz tuning fork, the frequency of the tone is either 438 Hz or 442 Hz. It takes at least $\Delta t \geq 1/\Delta\nu = 1/(2\text{ Hz}) = 0.5$ s to hear one beat.

A relation giving the corresponding uncertainty in wavelength can easily be deduced from Equation 1–2. Suppose the wave has been observed only over the finite time interval Δt; then during this time the wave will have traveled a distance $\Delta x = v\,\Delta t$, where v is the speed of the wave. Therefore, it will have been observed only over the distance Δx,

$$\Delta x = v\,\Delta t$$

and substituting from Equation 1–2,

$$\Delta x \geq \frac{v}{\Delta\nu} \tag{1–3}$$

But since

$$\nu = \frac{v}{\lambda}$$

we have

$$\Delta\nu = \frac{v\,\Delta\lambda}{\lambda^2} \tag{1–4}$$

(We have omitted the minus sign since we are concerned with magnitudes only.) Substituting Equation 1–4 in Equation 1–3 yields

$$\Delta x\,\Delta\lambda \geq \lambda^2 \tag{1–5}$$

If the extension of a wave in space is uncertain by an amount Δx, its wavelength will be uncertain by an amount $\Delta\lambda \geq \lambda^2/\Delta x$. As Equation 1–5 shows, $\Delta\lambda = 0$ only if $\Delta x = \infty$.

Our discussion on waves up to this point has been concerned only with monochromatic sinusoidal waves. Wavepulses, which are wave disturbances confined at a given time to some limited region of space, also can be propagated. Any pulse can be shown to be equivalent mathematically to a number of superimposed sinusoidal waves of different frequencies. If we compute the number of waves of different frequencies that must be added together to give a completely sharp pulse, we find that *all* frequencies from zero to infinity must be included (see the analysis below). This agrees completely with what we have already found. If a wavepulse is confined to an infinitesimally small region of space, then we cannot determine its wavelength. Actually, we cannot speak of a single "frequency" for a pulse.

WAVE PACKETS. A monochromatic wave, traveling along the x-axis with a velocity $v = \nu\lambda$ is represented by

$$A = A_0 \cos 2\pi \left(\frac{x}{\lambda} - \nu t\right) \tag{1–6}$$

Here λ represents the wavelength and ν the frequency, respectively. The wave disturbance A is given as a function of both position x and time t; its maximum value, A_0, is the *amplitude* of the wave. Variable A can stand for many quantities, depending on the wave in question: electric or magnetic field for an electromagnetic wave, air pressure for a sound wave in air, transverse displacement for a transverse wave on a string. With the definition

$$k = \frac{2\pi}{\lambda} \tag{1–7}$$

where k is the *wavenumber*, Equation 1–6 can be written

$$A = A_0 \cos k(x - vt) \tag{1–8}$$

Figure 1–4 shows the amplitude of a single monochromatic wave having a wavenumber k, or wavelength $2\pi/k$, and Figure 1–5 shows A as a function of x at the particular time $t = 0$.

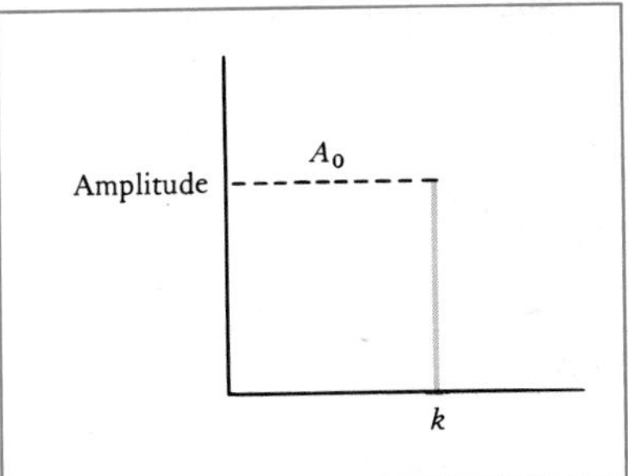

FIGURE 1–4. *Frequency spectrum of a monochromatic wave.*

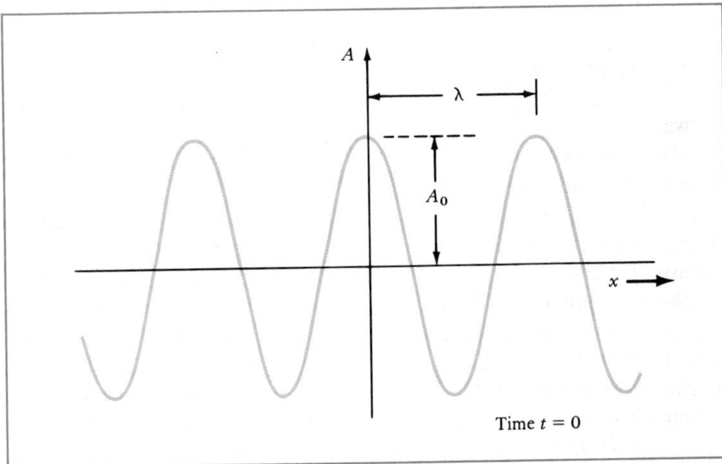

FIGURE 1–5. *Spatial variation of a monochromatic wave.*

Let us now consider a collection, or packet, of monochromatic waves, all traveling at the same speed v (showing no dispersion) in the $+x$-direction. For convenience, we imagine that all the waves have the same amplitude A_0 and that the wave packet includes all wavenumbers running from $k - \Delta k/2$ to $k + \Delta k/2$. Therefore, all waves lie within the band of width Δk, as shown in Figure 1–6. If $\Delta k = 0$, the band of waves becomes the single, monochromatic wave of Figure 1–4. Figure 1–7, showing the spatial extent of the wave packet at time $t = 0$, corresponds to Figure 1–5.

At the origin ($x = 0$), all the individual waves are in phase and add constructively, giving a large resultant amplitude. As we leave the origin in either direction, the waves become increasingly out of phase, and the algebraic addition of the individual waves gives a resultant amplitude A, which rapidly approaches zero.

We now compute for any point x and any time t the resultant amplitude A, composed of contributions of all monochromatic waves within the band Δk. We sum the individual contributions $A_k\, dk$ from $k - \Delta k/2$ to $k + \Delta k/2$. It is convenient to let $x - vt = x'$. Then Equation 1–8 becomes

$$A_k = A_0 \cos kx'$$

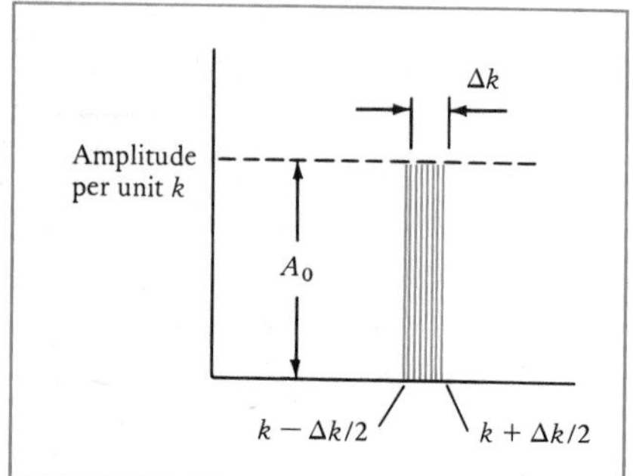

FIGURE 1–6. *Frequency spectrum of a packet of waves.*

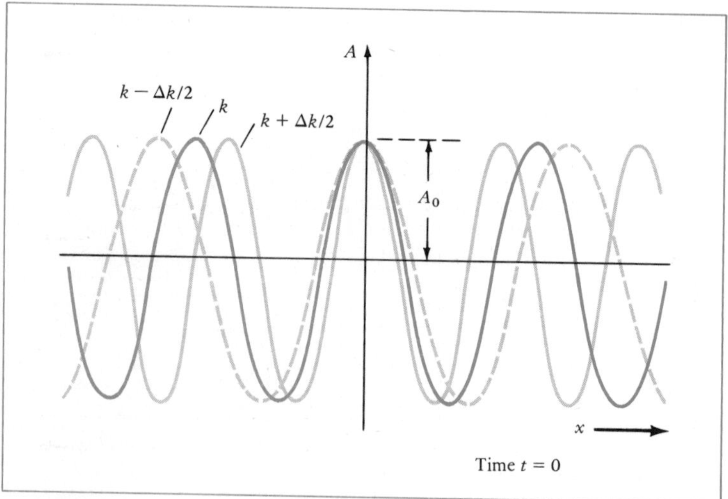

FIGURE 1–7. *Spatial variation of a packet of monochromatic waves.*

where A_0 is now the amplitude per unit k. The resultant displacement is given by

$$A = \int_{k-\Delta k/2}^{k+\Delta k/2} A_k\, dk = A_0 \int_{k-\Delta k/2}^{k+\Delta k/2} \cos kx'\, dk = \left(\frac{A_0}{x'}\right) \sin kx' \Big|_{k-\Delta k/2}^{k+\Delta k/2}$$

$$= \frac{A_0}{x'}\left[\sin x'\left(k + \frac{\Delta k}{2}\right) - \sin x'\left(k - \frac{\Delta k}{2}\right)\right]$$

We can simplify this using the trigonometric identity

$$\sin(a + b) - \sin(a - b) = 2 \sin b \cos a$$

so that it becomes

$$A = \left(\frac{2A_0}{x'}\right) \sin\left(\frac{x'\,\Delta k}{2}\right) \cos x'k \qquad (1\text{–}9)$$

The first three diagrams of Figure 1–8 show the separate factors of

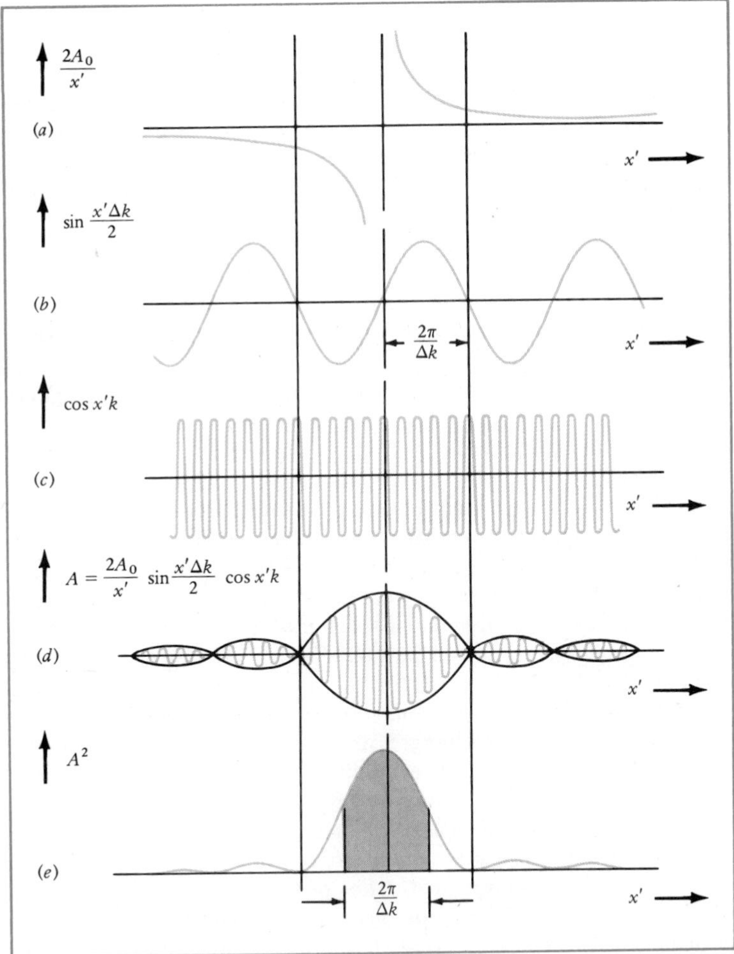

FIGURE 1–8. *Spatial variation of the factors appearing in Equation 1–9 for a wave packet [parts (a), (b), and (c)]; the resultant wave [part (d)]; and the envelope of A^2 [part (e)]. All these appear as functions of x'.*

Equation 1–9 plotted against x'; the last two show the resultant wave A and the envelope of A^2 as a function of x'. Since x' is equal to $x - vt$ (that is, the "viewpoint" of x' travels with the pulse), Figure 1–8*e* is also a snapshot of the pulse intensity, which is proportional to the square of the amplitude.

Region $2\pi/\Delta k$ (the shaded area in Figure 1–8*e*) is about three-quarters of the total area under the curve, so clearly more than half of the total energy of the packet is in this region. At any instant, the uncertainty Δx in the width of the wave packet is at least $2\pi/\Delta k$:

$$\Delta x \geq \frac{2\pi}{\Delta k} \tag{1–10}$$

We have introduced wavenumber k merely for convenience in evaluating the integrals. To rewrite Equation 1–10 in terms of $\lambda = 2\pi/k$, we start with

$$|\Delta k| = \frac{2\pi\,\Delta\lambda}{\lambda^2}$$

and Equation 1–10 becomes

$$\Delta x\,\Delta\lambda \geq \lambda^2 \tag{1–11}$$

agreeing with the simple arguments that lead to Equation 1–5. Therefore, if $\Delta\lambda$ is small (that is, if the monochromatic waves making up the packet have similar wavelengths), the spatial extent of the packet Δx will be very large. Likewise, if the packet is to be confined to a small region of space, $\Delta\lambda$ must be very large.

Waves and particles are important in physics because they represent the only modes of energy transport (communication) between two points. We can be warmed by infrared waves from a blazing fire or be knocked over by a flying particle such as a baseball; we can signal another person with a thrown rock (a particle), a shout (sound waves), a gesture (light waves), a telephone call (electric waves in conductors), or a radio message (electromagnetic waves in space).

Interactions, or transfers of energy, take place only between particles and particles or between waves and particles. A particle-particle interaction takes place when two or more particles collide, either elastically or inelastically. A wave-particle interaction takes place when a particle gives up all or part of its energy to generate a wave, or when all or part of the energy carried by a wave is absorbed by a nearby particle. A wood chip dropped into water, or an electric charge under acceleration, generates waves; these waves carry away energy that can be absorbed by other wood chips or electric charges.

Two waves, however, do *not* interact. If they meet, they pass through each other essentially unchanged, and their respective effects at every point in space simply add together according to the *principle of superposition* to form a resultant at that point. Everyone has seen this effect. Two water waves will travel toward each other, meet in a mottled interference pattern, and then travel onward as if oblivious to each other's existence. This behavior of waves contrasts sharply with that of two small, impenetrable particles, which cannot occupy the same place at the same time. The superposition principle is the basis for treating all problems in interference and diffraction.

1–5 Phase and Group Velocities

The distinction we make between *phase velocity* and *group velocity* will be important in our later consideration of the quantum wave properties of matter (Section 5–7).

When two sinusoidal waves of different frequency travel through a medium in the same direction at the *same* speed, the energy transported by the resultant wave travels at the same speed as the individual components. But when waves of different frequency travel through the same medium at *different* wavespeeds, the energy is transported at a speed—the group velocity —that differs from the phase velocity of either of the component waves.

Consider first sinusoidal waves of frequencies ν_1 and ν_2, traveling through a medium at the same speed. For simplicity, assume that amplitude A is the same for the two waves. The resultant waveform (at one instant) is found by superposing the component waves as shown in Figure 1–9. (Note that Figure 1–9 is a snapshot of the component waves and their resultant as a function of displacement along the direction of wave propagation. It looks like, but is not the same as, Figure 1–3, which shows the component and resultant oscillations at one point in space as a function of time.) The alternating constructive and destructive interference of the individual waveforms produces a slowly varying envelope. Since the energy of a simple harmonic oscillator is proportional to the square of the amplitude of oscillation, the energy carried by the resultant wave is concentrated in regions in which the amplitude of the envelope is large. Thus, the speed with which the waves' energy is transported through the medium is the speed with which the envelope advances. For equal-component wavespeeds, the speed of the envelope—the so-called group velocity—is the same as the *phase velocity* v of either component wave. By phase velocity is meant the speed with which

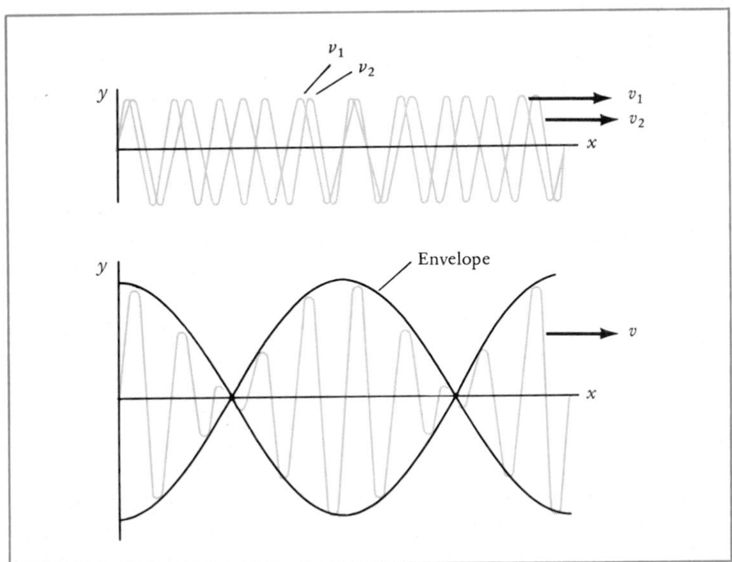

FIGURE 1–9. *Two monochromatic waves of differing frequency and their resultant.*

a point of constant phase on the wave, such as a crest, travels along the propagation direction. By definition,

$$v = \nu\lambda = \frac{\omega}{k}$$

for a frequency ν, wavelength λ, angular frequency $\omega = 2\pi\nu$, and wavenumber $k = 2\pi/\lambda$. When waves of all frequencies have the same phase velocity, riding with the crest of one wave is like riding with the crest of any other wave, or with their resultant.

Now consider two sinusoidal waves of slightly different frequencies that travel through a medium in the same direction but with different phase velocities: $v_1 = \nu_1\lambda_1 = \omega_1/k_1$ and $v_2 = \nu_2\lambda_2 = \omega_2/k_2$. Such a medium is said to exhibit *dispersion.* A simple form of dispersion occurs when polychromatic white light passes through a refracting medium. Violet, blue, and green light travel through a glass prism at lower speeds than yellow, orange, or red light, because the refractive index of glass is greater for violet light than for blue, for blue than for green, and so on. Consequently, white light passing through a glass prism is dispersed into a spectrum with violet at one end and red at the other.

It is easy to see that the group velocity v_{gr} differs from the phase velocity in a dispersive medium. Because one of the two sets of waves of different frequency travels faster than the other, riding on the crest of one wave is not the same as riding on the crest of the other wave, and a region of strong interference shifts as one wave gains on the other. The resultant wave envelope no longer remains locked to either component waveform. As Figure 1–10 shows, in a time t a crest of the wave of frequency ν_1 (or any other point of constant phase) advances a distance v_1t, and a crest of the other wave advances a different distance v_2t. During the same time the envelope shifts by an amount $v_{gr}t$.

A single sinusoidal wave traveling along the positive x-axis can be represented by $A \sin 2\pi(\nu t - x/\lambda) = A \sin(\omega t - kx)$. Then the resultant displacement y of two waves described above is given as a function of x and t by

$$y = A \sin(\omega_1 t - k_1 x) + A \sin(\omega_2 t - k_2 x) \tag{1–12}$$

Using the trigonometric identity

$$\sin\alpha + \sin\beta = 2\cos\frac{\alpha - \beta}{2}\sin\frac{\alpha + \beta}{2}$$

we can rewrite the resultant wave in the form

$$y = 2A\cos\left(\frac{\omega_1 - \omega_2}{2}t - \frac{k_1 - k_2}{2}x\right)\sin\left(\frac{\omega_1 + \omega_2}{2}t - \frac{k_1 + k_2}{2}x\right) \tag{1–13}$$

If the component waves differ only slightly in angular frequency and wavenumber, we can write the sums as

$$\omega = \frac{\omega_1 + \omega_2}{2} \quad \text{and} \quad k = \frac{k_1 + k_2}{2} \tag{1–14}$$

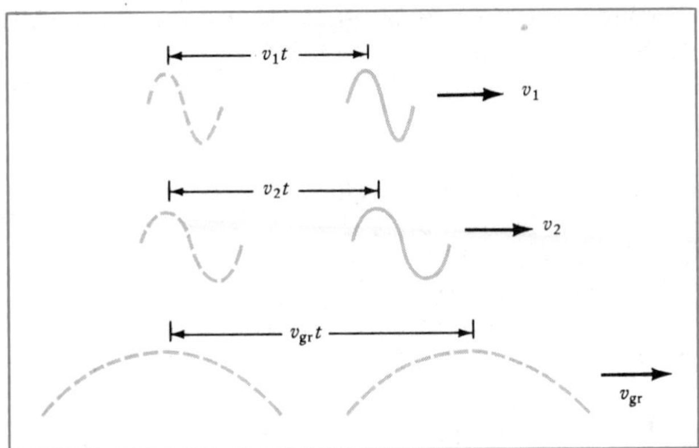

FIGURE 1–10. *One wave advances at the phase velocity v_1, a second monochromatic wave of different frequency advances at the phase velocity v_2, and the envelope of their resultant advances at the group velocity v_{gr}.*

where ω and k now represent average values. We can also write the differences as

$$d\omega = \omega_1 - \omega_2 \qquad \text{and} \qquad dk = k_1 - k_2 \tag{1–15}$$

Using Equations 1–14 and 1–15, we write Equation 1–13 in the simpler form:

$$y = \left[2A \cos\left(\frac{d\omega}{2}t - \frac{dk}{2}x\right)\right] \sin(\omega t - kx) \tag{1–16}$$

The equation for the resultant wave comprises two factors. The first (in brackets) represents the envelope, and the second represents an "average" component wave traveling with phase velocity, $v_{ph} = \omega/k$. The group velocity is the velocity of the envelope; we find it, as in the case of the phase velocity, by taking the ratio of the coefficients of t and of x. Thus,

$$\begin{aligned} v_{gr} &= \frac{d\omega}{dk} \\ v_{ph} &= \frac{\omega}{k} \end{aligned} \tag{1–17}$$

The group velocity is the *derivative* of ω with respect to k, whereas the phase velocity is the *ratio* of ω to k. Writing ω as $v_{ph}k$, we have

$$v_{gr} = \frac{d(v_{ph}k)}{dk} = v_{ph} + k\frac{dv_{ph}}{dk} \tag{1–18}$$

If the phase speed is the same for all frequencies, and $dv_{ph}/dk = 0$, the phase and group velocities are the same. But for a dispersive medium with a frequency-dependent phase velocity, the group velocity is greater than the phase velocity when $dv_{ph}/dk > 0$ and less than the phase velocity when $dv_{ph}/dk < 0$.

2
Relativistic Kinematics: Space and Time

Albert A. Michelson (1852, Prussia—1931). Graduate U.S. Naval Academy. Developed precision optical instruments, including the interferometer that was used especially for measuring the speed of light, a lifetime passion of Michelson. 1887, carried out first ether-drift experiment (with Edward W. Morley). The Michelson-Morley experiment is the most significant negative experiment in the history of science. (Photo, Michelson Museum—courtesy of the American Institute of Physics, Niels Bohr Library.)

The theory of special relativity, set forth by Albert Einstein in 1905, is one of the greatest achievements of the human intellect. Often regarded as esoteric and recondite, its principal features arise naturally from the two fundamental postulates of relativity. Postulate 1, *the principle of relativity,* is basic also to classical (Newtonian) mechanics. Postulate 2, *the constancy of the speed of light,* is at variance with classical mechanics and also with Postulate 1, if the classical concepts of space and time are adhered to. The brilliant work of Einstein reconciled the two postulates in a self-consistent theory of the physical universe quite different from that presented in classical physics. However, the theory of special relativity is *not* hypothetical or conjectural; a variety of experiments have firmly established its essential correctness.

Relativity theory is very important to atomic and nuclear physics and we shall examine it closely, first considering relativistic kinematics, or the relativity of space and time, and subsequently (Chapter 3), relativistic dynamics, or the relativity of momentum and energy.

2-1 The Principle of Relativity

First, we will explore the meaning and consequences of Postulate 1 in classical physics.

Postulate 1: The laws of physics are the same, or covariant, in all inertial systems—that is, the mathematical form of a physical law remains the same.

An *inertial system* is defined as a coordinate frame of reference within which the law of inertia, Newton's first law, obtains. An object subject to no net external force moves with a constant velocity when observed in an inertial system. A simple test of whether an observer is attached to an inertial system is to have him or her throw an object and then notice whether this object travels in an undeviating path at a constant speed. It does so only in a truly inertial system. Such a system can exist, strictly speaking, only in empty space, far from any mass. However, a reference system attached to the earth's surface may be regarded as an approximate inertial system, since it has only a small acceleration with respect to true inertial systems.† Thus, an object sliding on a frictionless horizontal plane on the earth has a zero net force acting on it and will move in a nearly straight line with a nearly constant speed. The first postulate of relativity physics implies that all inertial systems are equivalent in that no one inertial system can be distinguished by any experiment in physics from any other inertial system, since any law of physics is valid in all inertial systems.

To determine the full significance of Postulate 1, we must first find the relations between the spatial and temporal coordinates of one inertial system and the spatial and temporal coordinates of a second inertial system moving relative to the first.

† This is due mostly to the earth's rotation about its own axis.

2-2 Galilean Transformations

In classical physics, a set of equations called the *Galilean* (or *Newtonian*) *transformations* relates the space and time coordinates of one coordinate system (call it S) to the coordinates of another system (call it S') moving at constant velocity $\mathbf{v}$ relative to the first. Assume that S' is moving with velocity $\mathbf{v}$ to the *right* as viewed from S. (Viewed from S', of course, S will be moving to the left with velocity $-\mathbf{v}$. We can speak only of the *relative* motions of S and S'.) We can then choose the x- and x'-axes coincident and parallel with $\mathbf{v}$ so that the positive x-direction is along the motion of S' relative to S (see Figure 2–1). System S is imagined to contain an infinite number of observers at rest relative to each other, one at every point in space. The observers are equipped with identical metersticks and synchronized clocks. (That is, when a clock at the origin of S reads time t, all other clocks in S read the same time t.) For simplicity, all the many observers at rest in S will be referred to as observer S. Similarly, all the many observers at rest in S' will be called observer S'.

By specifying the location and time of some physical phenomenon, such as the explosion of a firecracker, an observer describes an *event*. The space and time coordinates of an event E (Figure 2–1), as described by observer S, are (x, y, z, t); the coordinates of the *same* event, as described by observer S' are (x', y', z', t'). The space coordinates x, y, and z give the displacements of this event from the origin in the x-, y-, and z-directions as measured by the meterstick of observer S; the time coordinate t gives the instant at which this event takes place as measured by the clock of observer S.

Suppose that observers S and S' are synchronizing their clocks and comparing their metersticks while temporarily at rest relative to one another. They then set their clocks so that when S' is set in motion with respect to S, both clocks will read zero when the origin of S' passes the origin of S. (That is, when $t = 0$, then $t' = 0$, and at that instant $x = x'$.) The y- and z-axes of the two coordinate systems are always parallel.

From Figure 2–1, we can immediately write down the Galilean coordinate transformations expressing the space and time coordinates of an event as measured by observer S' in terms of the coordinates as measured by observer S for the same event:

Galilean coordinate transformations:

$$\begin{aligned} x' &= x - vt \\ y' &= y \\ z' &= z \\ t' &= t \end{aligned} \tag{2–1}$$

That $y' = y$ and $z' = z$ follows, since the relative motion between systems S' and S is at right angles to these coordinates. To get the coordinates of system S in terms of system S', we merely interchange primes and unprimes and change v to $-v$; this is proper, since the unprimed and primed systems are

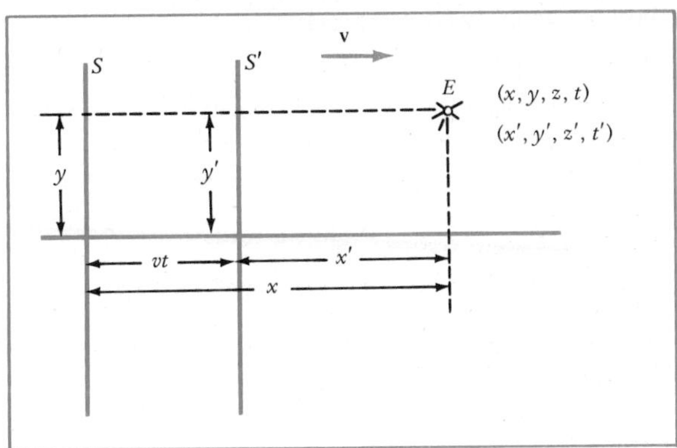

FIGURE 2–1. *Space and time coordinates of an event E as measured by two observers moving at a constant relative velocity* **v**.

purely arbitrary. Saying that S' moves with velocity $\mathbf{v}$ with respect to S is equivalent to saying that S moves with a velocity $-\mathbf{v}$ with respect to S'.

These classical transformation equations may seem completely axiomatic and self-evident, but it is very important that we appreciate the profound assumptions implicit in them. *Space is absolute* and *time is absolute* in the following sense: the space interval (separation distance) between any two events is the same for all observers, and the time interval between them is also the same for all observers. If the metersticks of observers S and S' are compared at some time and found to be the same length, they will always be the same length, regardless of their relative motion. Likewise, if their clocks are synchronized and calibrated initially, they will always thereafter agree, regardless of their relative motion. Our everyday, commonsense ideas of space and time are contained in the Galilean transformation equations and expressed formally by them.

Velocity and acceleration transformation relations follow directly from Equations (2–1) by differentiation with respect to time. The x-component of the velocity measured by observer S is defined as dx/dt. For convenience, we designate it $\dot{x}$; the dot above the coordinate means the first derivative with respect to time. Similarly, we write the y and z velocity components $\dot{y} = dy/dt$ and $\dot{z} = dz/dt$. The velocity components in system S' are $\dot{x}' = dx'/dt'$, and so on. Acceleration is given by $\ddot{x}' = d^2x'/dt'^2, \ldots$ for observer S'. It is important to appreciate the exact meaning of the concept of velocity. We define dx/dt as the limit of the distance traversed in the x-direction, or dx, measured by the meterstick of observer S, divided by the time interval dt, measured by the clock of the same observer S as the time interval approaches zero. It is meaningless to speak of $dx/dt', \ldots$, since the two measurements must be made with respect to a *single* coordinate system. For the Galilean transformations, this careful definition of velocity may not appear important, since time is regarded as absolute, $dt = dt'$, and therefore $dx/dt' = dx/dt$. We shall find later, however, that coordinate transformations satisfying the postulates of the theory of special relativity do not have such simplicity.

By differentiating Equations 2–1, we immediately get the velocity transformations, and the derivatives of these give in turn the acceleration transformations:

Galilean velocity transformations:

$$\dot{x}' = \dot{x} - v$$
$$\dot{y}' = \dot{y} \qquad (2\text{–}2)$$
$$\dot{z}' = \dot{z}$$

Galilean acceleration transformations:

$$\ddot{x}' = \ddot{x}$$
$$\ddot{y}' = \ddot{y} \qquad (2\text{–}3)$$
$$\ddot{z}' = \ddot{z}$$

Equations 2–2 show that the velocity of a particle as measured in system S' equals the velocity of the same particle as measured in S, minus the velocity $\mathbf{v}$ of S' relative to S. Thus, velocities can be combined using the standard rules of vector addition. From Equations 2–3, we see that corresponding acceleration components are equal in any two inertial systems moving at a constant velocity with respect to one another.

2-3 Covariance of Classical Mechanics under Galilean Transformations

To see the significance of Postulate 1, the principle of relativity, consider two well-known physical laws of mechanics under a Galilean transformation: the conservation of linear momentum and the conservation of energy.

CONSERVATION OF LINEAR MOMENTUM. Suppose an observer in system S' watches a head-on collision between two particles of respective masses m and M (see Figure 2–2*a*). Figure 2–2*b* shows the same collision, but now as seen by an observer in S. As before, system S' moves to the right with a velocity $\mathbf{v}$ with respect to system S, and the velocities measured by the two observers are related by the Galilean velocity transformations, Equations 2–2. In Figure 2–2, the small letters and large letters refer to two particles of masses m and M; the primed and unprimed velocities refer to the two observers S' and S; and the subscripts b and a refer to velocities measured before and after the collision.

We ask, "Is the law of momentum conservation a good physical law? That is, does it satisfy Postulate 1 of relativity theory and remain covariant under a Galilean transformation?" To answer this question, we test whether observers S and S' will find the *same mathematical form* for the statement of the momentum-conservation law as each watches the same head-on collision.

For the observer in the inertial system S', the momentum-conservation law is written

Momentum before collision = momentum after collision

Therefore, (2–4)

$$m\dot{x}'_{\mathrm{b}} + M\dot{X}'_{\mathrm{b}} = m\dot{x}'_{\mathrm{a}} + M\dot{X}'_{\mathrm{a}}$$

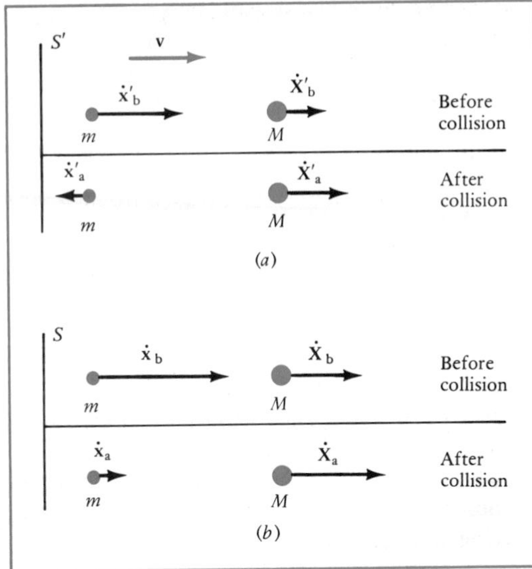

FIGURE 2–2. *Collision of two particles as viewed by two observers moving at a constant relative velocity* **v**.

Using the Galilean velocity transformations, we can rewrite this equation in terms of the velocities measured by the observer in inertial system S,

$$m(\dot{x}_b - v) + M(\dot{X}_b - v) = m(\dot{x}_a - v) + M(\dot{X}_a - v)$$

which reduces to

$$m\dot{x}_b + M\dot{X}_b = m\dot{x}_a + M\dot{X}_a \tag{2–5}$$

Equations 2–4 and 2–5 are of identical mathematical form; that is, they differ only in the primes and unprimes. Therefore, an observer S in one inertial system and an observer S' in *any* other system moving with a constant velocity with respect to S would both agree that *momentum is conserved.* In short, the conservation law for momentum *is* a "good" law of classical mechanics. The total linear momentum $m\dot{x} + M\dot{X}$ before (or after) the collision as observed in S is *not* the same as the total linear momentum $m\dot{x}' + M\dot{X}'$ before (or after) the collision as observed in S'; the total momentum is greater in S in this example.

COVARIANCE OF NEWTON'S SECOND LAW. It is easy to show that Newton's second law of motion is covariant under a Galilean transformation. Suppose two bodies, of masses m and M, interact through some force such as the gravitational force. If no net external force is applied to the system (the two bodies), the system is isolated, and its total linear momentum must remain constant. For simplicity, suppose both masses lie on the x-axis and move along it. For observer S, momentum conservation implies that

$$m\dot{x} + M\dot{X} = \text{constant}$$

Taking the time derivative of this equation gives

$$\frac{d}{dt}m\dot{x} = -\frac{d}{dt}M\dot{X} \tag{2–6}$$

Since force is the rate of change of momentum, the left-hand side of this equation is the force f on m caused by M (as measured in S), and the right-hand side is the force F on M caused by m (also as measured in S). Therefore,

$$f = -F$$

which is Newton's third law.

Considering the force acting on mass m alone, we have

$$f = \frac{d}{dt} m\dot{x} = m\ddot{x} \tag{2-7}$$

where it is assumed that the mass m has the same value in all inertial systems. Likewise, observer S' would write

$$f' = \frac{d}{dt'} m\dot{x}' = m\ddot{x}' \tag{2-8}$$

Equations 2–3 show that $\ddot{x} = \ddot{x}'$, and therefore from Equations 2–7 and 2–8 it is seen that $f = f'$. Newton's second law is covariant. Since the forces and acceleration are unchanged, it follows that if S is an inertial system, then *any* coordinate system S' moving at a constant velocity $\mathbf{v}$ with respect to S is also an inertial system. From Newton's second law, all inertial systems, of which there are an infinite number, are equivalent and indistinguishable. On the other hand, any coordinate frame of reference S_F that is *accelerated* with respect to some inertial system cannot itself be an inertial system, because no longer does $\ddot{x} = \ddot{x}_F$, where $\ddot{x}_F$ is the acceleration of mass m as measured in the accelerated frame S_F.

We shall restrict our considerations to the special case of inertial systems moving with a constant velocity with respect to one another. We shall not discuss the more general case, in which one system may be accelerated with respect to another. Thus our discussion will be confined to the theory of *special relativity;* the more general case of accelerated systems is treated in the theory of *general relativity.*

CONSERVATION OF ENERGY. To examine the covariance of the energy conservation law under a Galilean transformation, consider again the collision in Figure 2–2, taking it to be perfectly *elastic.* (Although this might seem to rule out inelastic collisions, *all* collisions are elastic collisions at the subatomic level.)

From the viewpoint of the observer S', the energy conservation law is written

$$\text{Kinetic energy before collision} = \text{kinetic energy after collision}$$

$$\tfrac{1}{2}m(\dot{x}'_{\mathrm{b}})^2 + \tfrac{1}{2}M(\dot{X}'_{\mathrm{b}})^2 = \tfrac{1}{2}m(\dot{x}'_{\mathrm{a}})^2 + \tfrac{1}{2}M(\dot{X}'_{\mathrm{a}})^2 \tag{2-9}$$

This equation can be rewritten from the viewpoint of observer S by using Equations 2–2, the Galilean velocity transformations:

$$\tfrac{1}{2}m(\dot{x}_{\mathrm{b}})^2 - m\dot{x}_{\mathrm{b}}v + \tfrac{1}{2}mv^2 + \tfrac{1}{2}M(\dot{X}_{\mathrm{b}})^2 - M\dot{X}_{\mathrm{b}}v + \tfrac{1}{2}Mv^2$$
$$= \tfrac{1}{2}m(\dot{x}_{\mathrm{a}})^2 - m\dot{x}_{\mathrm{a}}v + \tfrac{1}{2}mv^2 + \tfrac{1}{2}M(\dot{X}_{\mathrm{a}})^2 - M\dot{X}_{\mathrm{a}}v + \tfrac{1}{2}Mv^2$$

Using Equation 2–5, the covariance of momentum conservation, we can cancel the terms involving v. Because the terms in v^2 also cancel, there remains merely

$$\tfrac{1}{2}m(\dot{x}_b)^2 + \tfrac{1}{2}M(\dot{X}'_b)^2 = \tfrac{1}{2}m(\dot{x}_a)^2 + \tfrac{1}{2}M(\dot{X}_a)^2 \qquad (2\text{–}10)$$

Equations 2–9 and 2–10 are of identical mathematical form, differing only in the unprimes and primes. Therefore, the energy conservation law is valid for all inertial systems.

To summarize, we have found that *the laws of classical mechanics* (momentum conservation, Newton's law of motion, and energy conservation) *are all covariant under a Galilean transformation. Thus, all inertial systems are equivalent in classical mechanics, and it is impossible by means of any experiment in mechanics to distinguish one inertial system from any other.* The covariance of the laws of mechanics, which has been formally proved here, is implicitly assumed in all elementary physics. We are confident, for example, that a Ping-Pong game played on a steadily moving train will follow the same physical laws for an observer fixed on the ground as for an observer traveling with the train.

In our analysis, we have implicitly assumed that the ratio of momentum to speed (in classical physics, the mass m of the particle) is also a constant, independent of the particle's motion with respect to an observer. We can then conclude (in line with day-to-day experience) that length, time, and mass, the three basic quantities in physical measurements, are all independent of the relative motion of an observer. As we shall see, Einstein's relativity physics drastically revises this notion.

2–4 Failure of Galilean Transformations

One might well ask whether the other laws of physics, not merely Newtonian mechanics, are covariant under a Galilean transformation. Inasmuch as Postulate 1 requires that *all* the laws of physics be covariant, the Galilean transformations are universally valid only if all laws in physics can likewise be shown to be covariant.

A discussion of the propagation of electromagnetic waves alone will enable us to analyze the covariance of classical electromagnetism. First, however, we consider an analogous situation in mechanics.

Suppose that a *sound* pulse is traveling to the right with respect to the medium transmitting it. The medium is at rest in system S, and the velocity of the pulse, as measured in S, is $\dot{\mathbf{x}}$. (By *velocity,* we mean the pulse velocity in the medium in which it is propagated—in this example, air.) For observer S', moving at a velocity $\mathbf{v}$ with respect to S, the measured (apparent) velocity is $\dot{\mathbf{x}}' = \dot{\mathbf{x}} - \mathbf{v}$ (Figure 2–3 or Equation 2–2). Observer S', measuring the velocity of the sound pulse, would get a different value from what Observer S would measure.

Suppose, for example, a cannon is fired in still air on the ground, producing a sound pulse that travels at 1100 ft/s ($\dot{x}$) with respect to the ground S. An observer in an airplane S' moving away from the cannon at 400 ft/s (v) measures the pulse velocity from his inertial frame as $\dot{x}' = \dot{x} - v = 1100 - 400 = 700$ ft/s. On the other hand, an observer in another airplane (S'')

moving *toward* the cannon at the same speed ($v = -400$ ft/s) measures the pulse velocity from her inertial frame as $\dot{x}'' = 1100 - (-400) = 1500$ ft/s. It follows that in general, the measured velocity of a sound pulse depends on the observer's velocity relative to the medium through which the pulse travels. Only when the observer is at rest with respect to the medium (here, air) will the pulse velocity be the same in all directions. This result is confirmed by experiments with sound waves. Note also that the measured velocity of the sound pulse in S, the system in which the air is at rest, does *not* depend on the velocity of the source of the sound. If a source generates sinusoidal variations in air pressure rather than a pulse, the frequency and the wavelength of the sound will depend on the relative motion of the source and the medium (the *Doppler effect*), but the velocity of propagation will still be independent of the relative motion.

Consider now the analogous case for light in terms of the Galilean transformations. A pulse of light is traveling to the right at a speed $\dot{x} \equiv c$ (Figure 2–3) with respect to a medium through which it is propagated. An observer at rest in S will measure the speed of the pulse as c in the x-direction and also in *any other direction* in which the pulse might travel. An observer at rest in another system S' will measure the speed as $\dot{x}' = \dot{x} - v = c - v$ when S' travels to the right; when S' moves to the left, he will measure a different speed, $c + v$. This implies that the speed of light measured by any observer except S depends on the velocity of the coordinate system with respect to the medium through which the pulse of light is propagated. Therefore, the speed of light is certainly *not* invariant under Galilean transformations. In fact, if these transformations are to apply to light, then there must exist in nature a unique inertial system in which the medium is at rest; in this system, and in this system alone, will the measured speed of light be exactly c.

An invisible, impalpable substance called *ether* was proposed in the nineteenth century as a medium through which a light pulse could be propagated. Physicists of that era, firmly convinced that all physical phenomena were ultimately mechanical in origin, found it unthinkable that an electromagnetic "wave" could be propagated through empty space. The only conspicuous property of ether was that it "carried" electromagnetic disturbances; by its existence, however, it implied also an inertial system in which it was at rest, and thus in which the speed of light would equal c.

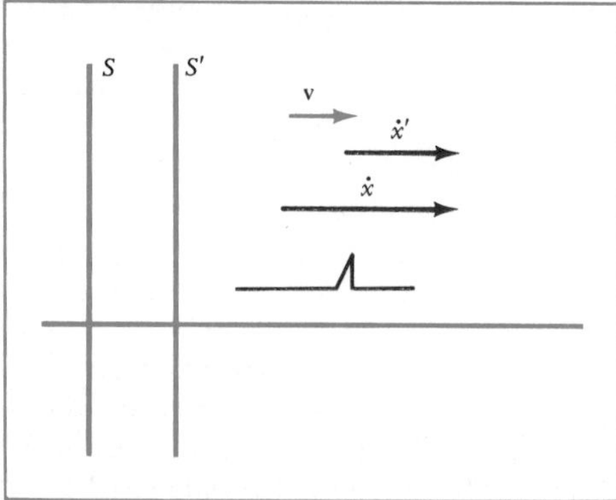

FIGURE 2–3. *A pulse as viewed by two observers moving at a constant relative velocity* **v**.

The essence of an experiment to confirm the existence of the ether is simple. Measure the speed of light in a variety of inertial systems and note whether the measured speed is different in the different systems. Note especially whether there is evidence of a single, unique inertial system in which the speed of light is c. Performing such an experiment, however, is far more difficult than the corresponding experiment with sound waves, because the speed of light is so very high. In 1887, in one of the most celebrated physics experiments of all time, Michelson and Morley tried to find a change in the measured speed of light as the earth's rotation around its axis and revolution around the sun moved their apparatus through the ether.

Let us analyze how such experiments as the Michelson-Morley experiment (and its present-day analogs using microwaves or laser beams) try to find the unique inertial system S. The experimenter cannot know whether the equipment is at rest in S when it is used to measure c. Therefore, he or she must assume that it is in *any* system S' moving at velocity $\mathbf{v}$ with respect to S. If at some moment the equipment is at rest in S, then S' is S, and the speed of light will be measured as c. Six months later, the earth will be moving in the opposite direction in its orbit around the sun; then presumably, the equipment will be in motion with respect to S, and the speed of light will be measured as different from c.

The speed of light c is extremely large compared with the earth's orbital speed. Therefore, for practical reasons, c must be determined by measuring the time required for a light beam to travel a known distance to a mirror and back. In determining the round-trip time, one is not measuring the speed in a single direction but the *average* speed in opposite directions along a single line. The time for a trip from A to B and back to A will be the same as the time for a trip from B to A and then back to B. Therefore, one must compare round-trip times along two *nonparallel* lines. The difference in the round-trip times is at its maximum when one compares time intervals for lines parallel to the ether flow with intervals for lines perpendicular to the flow.

Consider a cylinder of length l at rest in S' and aligned along the x-axis, which is the direction of the relative motion of S and S' (Figures 2–4d to f). As before, S' moves to the right with a speed v relative to system S. While a light pulse is traveling to the right, an observer in S' measures the speed of the pulse as $c - v$, and the time required for the pulse to reach the right-end plate is $l/(c - v)$. After being reflected, the pulse travels to the left with a speed $c + v$ relative to S' and reaches the left-end plate in time $l/(c + v)$. The time interval Δt_x for the light pulse to travel a complete round trip is

$$\Delta t_x = \frac{l}{c - v} + \frac{l}{c + v} = \frac{2l/c}{1 - (v/c)^2} \tag{2–11}$$

The sequence of events as seen by an observer in S is illustrated in Figures 2–4a to c.

Now consider the situation in which observer S' aligns the same cylinder along his y-axis. The time required for the pulse to make a round trip between what are now the bottom-end and top-end plates is designated Δt_y. The sequence of events, as seen by observer S, is shown in Figures 2–5a to c; and as seen by observer S', in Figures 2–5d to f. From the point of view of observer S, only a light pulse (necessarily traveling at speed c) that leaves the origin of system S in direction θ will reach the center of the top-end plate at

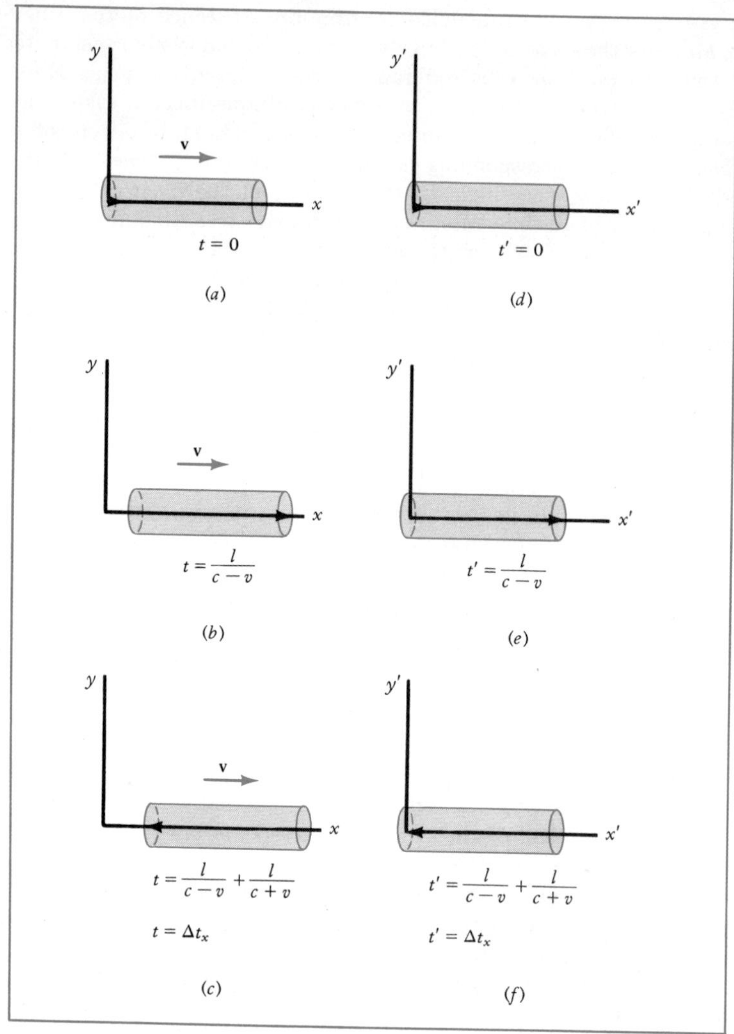

FIGURE 2–4. *Time of flight of a light pulse as measured by two observers traveling at a constant relative velocity* **v**. *The pulse moves parallel to* **v**.

A, as shown in Figure 2–5b. The pulse goes from O to A at a speed c in a time $\Delta t_y/2$ while the cylinder goes at a speed v from O to B in the same time. Therefore,

$$OA = \frac{c\,\Delta t_y}{2} \qquad OB = \frac{v\,\Delta t_y}{2} \qquad \text{and} \qquad AB = l$$

But

$$OA^2 = OB^2 + AB^2$$

By substitution

$$\left(\frac{c\,\Delta t_y}{2}\right)^2 = \left(\frac{v\,\Delta t_y}{2}\right)^2 + l^2$$

Solving for Δt_y gives, finally,

$$\Delta t_y = \frac{2l/c}{\sqrt{1-(v/c)^2}} \tag{2–12}$$

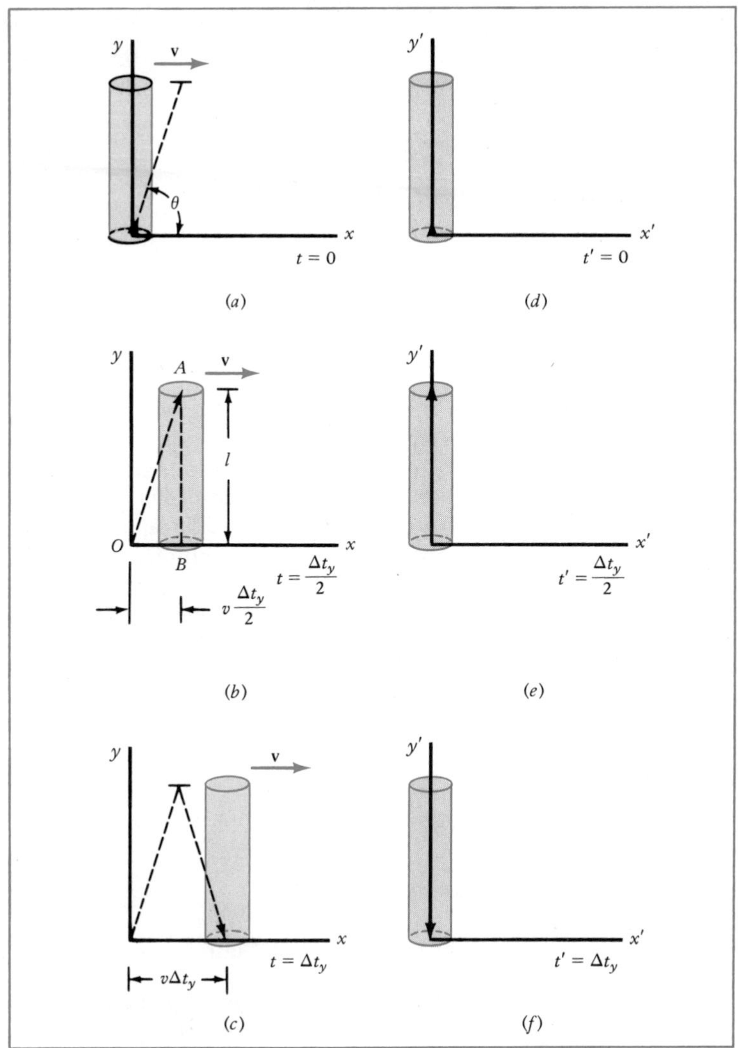

FIGURE 2–5. *Time of flight of a light pulse as measured by two observers traveling at a constant relative velocity* **v**. *The pulse moves perpendicular to* **v**, *according to observer S′.*

Comparing Equation 2–11 with Equation 2–12 shows that $\Delta t_x \neq \Delta t_y$; that is, the time taken by the light pulse in a round trip is not the same for the two perpendicular orientations. If the cylinder were at rest in system S, then v would be zero, and $\Delta t_x = \Delta t_y = 2l/c$. Therefore, $\Delta t_x - \Delta t_y = 0$ when $v = 0$. In general, however, with system S' in motion relative to system S,

$$\Delta t_x - \Delta t_y = \frac{2l}{c}\{[1 - (v/c)^2]^{-1} - [1 - (v/c)^2]^{-1/2}\}$$

Since $v/c \ll 1$, the binomial expansion can be used (with higher-order terms neglected) to yield

$$\Delta t_x - \Delta t_y = \frac{2l}{c}\left[1 + (v/c)^2 - 1 - \frac{1}{2}(v/c)^2\right] = \frac{2l}{c}\frac{v^2}{2c^2} \qquad (2\text{–}13)$$

The time for a round trip is approximately $\Delta t_x = 2l/c$, according to Equations 2–11 and 2–12. Therefore, the maximum fractional change in the round-trip time interval for orientation at 90° is, according to Equation 2–13:

$$\frac{\Delta t_x - \Delta t_y}{\Delta t_x} = \frac{v^2}{2c^2} \tag{2–14}$$

The maximum speed v attainable on the earth is the orbital speed of the earth around the sun, 3×10^4 m/s. Substituting this value in Equation 2–14 shows that the fractional change in the round-trip time interval might be as large as 5×10^{-9}, or five parts in 1 billion! But Michelson and Morley were confident of detecting a change 100 times smaller than this, five parts in 100 billion.

The device they used to measure Δt_x and Δt_y was a precision optical instrument, the *interferometer,* developed a few years earlier by Michelson. In this device, a light beam was divided into two separate beams traveling at right angles to each other. The beams bounced off mirrors at the ends of their respective paths, returned to the center of the apparatus, and finally recombined to form an interference pattern. Any difference between Δt_x and Δt_y would have appeared as a shift in the interference pattern.

Michelson and Morley performed this experiment many times, at various times of the year and in various locations. Every time, it produced the same result: $\Delta t_x - \Delta t_y = 0$. *The result was always null.*†

This can have only one meaning: *Any* inertial system S' behaves as if it were the "unique" inertial system S—or the speed of light as measured in *every* inertial system is the same, namely c, for all directions and all observers. This fundamental assertion is supported not only by the Michelson-Morley experiment but by a variety of other experiments, as we shall see.

2–5 The Second Postulate and the Lorentz Transformations

The second postulate of relativity theory—actually, an experimental fact—is

> *Postulate 2: The speed of light in a vacuum is a constant, independent of the inertial system, the source, and the observer.*

This postulate is obviously incompatible with the Galilean transformations, which imply that the measured speed of light depends on the motion of the observer. Einstein noted the inconsistency between Postulate 2 and the Galilean transformations. Postulate 2 could not be relinquished, for it was an experimental fact; therefore the Galilean transformations had to give way, despite their success in classical mechanics and their obvious appeal to common experience. How drastic such a change would be is demonstrated by the fact that our very ideas of space and time and their apparently absolute character are contained in the Galilean transformations.

† Many workers have conducted similar experiments. In one such experiment, two identical infrared lasers were oriented with axes perpendicular to each other, and the resonant frequencies of the two lasers were compared at various points in the earth's orbit. [See T. S. Jaseja, A. Javin, and C. H. Townes, *Phys. Rev., 133,* A1221 (1964).] This and other recent experiments greatly exceed in precision the original Michelson-Morley experiment.

Two criteria had to be met by the new coordinate transformations. First, Postulate 1: physical laws had to be covariant in all inertial systems. Then, Postulate 2: the speed of light had to be invariant in all inertial systems. The correspondence principle requires, of course, that the new transformations reduce to the Galilean transformations under appropriate limiting conditions.

In 1905, Einstein found transformation equations satisfying both of these criteria; the equations were the *Lorentz coordinate relations,* which H. A. Lorentz had derived two years earlier to describe electromagnetism. Like the Galilean transformations, these relations derive from the following assumptions: (1) The xyz- and $x'y'z'$-axes are mutually parallel, with the x- and x'-axes coincident. (2) The velocity $\mathbf{v}$ of S' relative to S is along the positive x- (or x'-) axis. (3) When the two origins coincide, clocks in both S and S' read zero (that is, when $x = x' = 0$, then $t = t' = 0$).

The additional, distinguishing, assumption is that the speed of light is invariant in all inertial systems. From these assumptions, a detailed, but straightforward, analysis yields the Lorentz transformations†:

$$\begin{aligned} x' &= \frac{x - vt}{\sqrt{1 - (v/c)^2}} \\ y' &= y \\ z' &= z \\ t' &= \frac{t - (v/c^2)x}{\sqrt{1 - (v/c)^2}} \end{aligned} \tag{2–15}$$

These equations give the four primed coordinates on the left in terms of the four unprimed coordinates on the right corresponding to the same event. The equations are *linear*—that is, the variables appear as first-power terms only. This means that in the mathematical model, as in the physical situation, a single event (x, y, z, t) in S will always correspond to only *one* event (x', y', z', t') in S', and conversely. To find the inverse equations with the unprimed quantities on the left, one could solve Equations 2–15 algebraically for x, y, z, and t (see Problem 2–7). A consideration used earlier with the Galilean transformations will yield the inverse equations immediately, however. If the velocity of S' relative to S is $\mathbf{v}$, the velocity of S relative to S' is $-\mathbf{v}$. Therefore, the inverse equations can be produced by simply interchanging primed and unprimed coordinates in Equations 2–15 and simultaneously replacing v wherever it appears by $-v$. The first of the four relations in Equations 2–15, for example, becomes

$$x = \frac{x' + vt'}{\sqrt{1 - (v/c)^2}}$$

The Lorentz transformations are the unique transformations that meet the requirements of the two relativity postulates and therefore supplant the Galilean transformations. By the correspondence principle (Section 1–2), we know that the Lorentz transformations must reduce to the Galilean transformations in that range of speeds in which the latter are known to be

† Appendix 1 gives a derivation of the Lorentz transformations from fundamental principles.

essentially correct. By comparing Equations 2–15 and 2–1, we see that the two sets of transformations become identical as $v/c \to 0$. Thus, when v is much less than c, the Galilean transformations are an excellent approximation to the universally valid Lorentz transformations. Mathematically, making $v/c \ll 1$ is equivalent to imagining $c \to \infty$; therefore, we may regard the Galilean transformations as the correct coordinate transformations in a hypothetical universe in which the speed of light is infinite. We may write symbolically

$$\underset{c \to \infty}{\text{Limit}} \text{ (Lorentz tranformations)} = \text{Galilean transformations}$$

When the speed of an object is close to the speed of light, only the Lorentz transformations apply and neither the time nor the space coordinates are absolute. Time coordinate t' is no longer independent of space coordinate x, and the clocks and metersticks of observers S and S' no longer agree ($t \neq t'$, $x \neq x'$). (We will discuss this more fully in Example 2–1 and in Section 2–7.)

For ordinary objects, it is obvious from the Lorentz transformations that the relative speed v between any two inertial systems cannot exceed c. The quantity $\sqrt{1 - (v/c)^2}$ in Equations 2–15 is a real number only if $v < c$. If $v > c$, the quantity becomes an *imaginary* number (a real number times the square root of -1) and a "real" event in S becomes an "imaginary," hence unobservable, event in S'.† This would violate Postulate 1, which requires that a single observable event in one inertial system must correspond to a single observable event in any other inertial system.

Example 2–1. A star is moving away from the earth at a speed of $0.60c$ as measured in inertial system S, at rest with respect to the earth. At time $t = 0$, when the star is 1.0×10^9 light-years (ly) away, the following two events occur simultaneously as observed in S:‡ (1) a bomb explodes nearby on the earth; (2) a bomb explodes on the moving star.

Give the space and time coordinates for events 1 and 2 as observed in inertial system S', which is at rest with respect to the star, according to (*a*) the Galilean transformations; (*b*) the Lorentz transformations.

(*a*) For convenience, we locate the earth at the origin. The Galilean transformations, Equations 2–1, give

	S	S'
EVENT 1 (BOMB EXPLODES ON EARTH)	$x_1 = 0$ $t_1 = 0$	$x'_1 = 0$ $t'_1 = 0$
EVENT 2 (BOMB EXPLODES ON STAR)	$x_2 = 1.0 \times 10^9$ ly $t_2 = 0$	$x'_2 = 1.0 \times 10^9$ ly $t'_2 = 0$

Under the Galilean transformations, the two explosions are simultaneous in both systems ($t_2 - t_1 = t'_2 - t'_1 = 0$), and the distance

† Physicists have, however, conjectured that particles exist whose speed always exceeds c. Although the theory of relativity does not preclude such particles, no experimental evidence of their existence has yet been produced.

‡ One light-year equals the distance a pulse of light travels in one year. Convenient units of distance and time in astronomy and in relativity are the light-year (abbreviated "ly") and the year (abbreviated "yr"). The speed of light in these units is simply $c = (1 \text{ ly})/(1 \text{ yr}) = 1$ ly/yr.

between the two events is also the same ($x_2 - x_1 = x_2' - x_1' = 1.0 \times 10^9$ ly).

(*b*) With the earth again at the origin, events 1 and 2 have the same space and time coordinates in system S as in part (*a*). To get the corresponding coordinates in system S', we use the Lorentz transformations, Equations 2–15, with $v = 0.60c$,

$$\sqrt{1 - (v/c)^2} = \sqrt{1 - (0.60)^2} = 0.80$$

The space-and-time coordinates of the two explosions in S' follow directly from Equations 2–15. For example, the time coordinate of event 2 in the S' system is

$$t_2' = \frac{t_2 - vx_2/c^2}{\sqrt{1 - (v/c)^2}}$$

$$= \frac{0 - [(0.60 \text{ ly/yr})(1.0 \times 10^9 \text{ ly})/(1 \text{ ly/yr})^2]}{0.80}$$

$$= -0.75 \times 10^9 \text{ yr}$$

	S	S'
EVENT 1 (BOMB EXPLODES ON EARTH)	$x_1 = 0$ $t_1 = 0$	$x_1' = 0$ $t_1' = 0$
EVENT 2 (BOMB EXPLODES ON STAR)	$x_2 = 1.0 \times 10^9$ ly $t_2 = 0$	$x_2' = 1.25 \times 10^9$ ly $t_2' = -0.75 \times 10^9$ yr

The two events are *not* simultaneous under the Lorentz transformations; $t_2 - t_1 = 0$, whereas $t_2' - t_1' = -0.75 \times 10^9$ yr, with event 2 *preceding* event 1 in the star system S'. The distances between the two events are also different; $x_2 - x_1 = 1.0 \times 10^9$ ly, whereas $x_2' - x_1' = 1.25 \times 10^9$ yr.

2–6 The Relativistic Velocity Relations

The Lorentz transformations for velocity components are obtained in the same way as the Galilean velocity transformations, Equations 2–2. In any inertial system S, the x-component of velocity $\dot{x}$ is defined as the ratio of the differential displacement dx to the corresponding time interval dt, both as measured by observers at rest in system S. Thus, the x-component of the velocity observed in reference frame S is $\dot{x} = dx/dt$; and the x-component observed in S' is $\dot{x}' = dx'/dt'$.

Taking the differential of each coordinate in Equations 2–15 gives

$$dx' = \frac{dx - v\,dt}{\sqrt{1 - (v/c)^2}}$$

$$dy' = dy$$

$$dz' = dz$$

$$dt' = \frac{dt - (v/c^2)\,dx}{\sqrt{1 - (v/c)^2}}$$

Dividing dx', dy', and dz' in turn by dt' gives the velocity components in S'

$$\dot{x}' \equiv \frac{dx'}{dt'} = \frac{dx - v\,dt}{dt - (v/c^2)\,dx}$$

and

$$\dot{y}' \equiv \frac{dy'}{dt'} = \frac{dy\sqrt{1-(v/c)^2}}{dt - (v/c^2)\,dx}$$

The relation for $\dot{z}'$ is similar to the relation for $\dot{y}'$. Dividing the numerator and denominator above by dt gives

$$\dot{x}' = \frac{\dot{x} - v}{1 - (v/c^2)\dot{x}}$$

$$\dot{y}' = \frac{\dot{y}\sqrt{1-(v/c)^2}}{1 - (v/c^2)\dot{x}} \qquad (2\text{–}16)$$

$$\dot{z}' = \frac{\dot{z}\sqrt{1-(v/c)^2}}{1 - (v/c^2)\dot{x}}$$

One surprising result of the Lorentz velocity transformations is that the y- and z-components of the velocity of a particle as measured in S' depend on the x-component as measured in S! As before, to get the velocity components as measured in S in terms of those as measured in S', simply interchange primes and unprimes and replace v by $-v$ in Equations 2–16. Applying the correspondence principle to the Lorentz velocity transformations, we get the Galilean velocity transformations, Equations 2–2

$$\lim_{v/c \to 0} \text{(Lorentz velocity transformations)}$$

$$= \text{(Galilean velocity transformations)}$$

Example 2–2. Show that the Lorentz velocity transformation relations yield, as they must, the result that all observers measure the speed of light to be c.

For simplicity, suppose that a beam of light is directed along the x-axis; therefore, in reference frame S the velocity components are

$$\dot{x} = c \qquad \dot{y} = 0 \qquad \dot{z} = 0$$

From Equations 2–16 we then have

$$\dot{x}' = \frac{\dot{x} - v}{1 - (v/c^2)\dot{x}} = \frac{c - v}{1 - (v/c^2)c} = \frac{c - v}{c - v}c = c$$

$$\dot{y}' = 0$$

$$\dot{z}' = 0$$

Observer S', traveling at *any* velocity v (less than c), also measures the speed of light to be c.

Example 2–3. A particle moves north at speed $0.80c$ relative to the earth (system S). What is the velocity of this particle as measured by an observer in a spaceship (system S') traveling east relative to the earth at speed $0.98c$?

Choose the positive xx'-axes pointing east and the positive yy'-axes pointing north. We then have

Velocity of S' relative to S $\quad v = 0.98c$ (east)

Velocity of particle relative to S $\quad \dot{x} = 0$

$\dot{y} = 0.80c$

Substituting these values in the relativistic velocity relations (Equations 2–16) gives the velocity components of the particle relative to the spaceship S':

$$\dot{x}' = \frac{\dot{x} - v}{1 - (v\dot{x}/c^2)} = \frac{(0 - 0.98)c}{1 - (.98)(0)} = -0.98c$$

$$\dot{y}' = \frac{\dot{y}\sqrt{1 - (v/c)^2}}{1 - (v\dot{x}/c^2)} = \frac{0.80c\sqrt{1 - (0.98)^2}}{1} = 0.16c$$

To compute the velocity of the particle relative to reference frame S', we must remember that the velocity within any inertial system is obtained by adding velocity vector components in that system. Therefore, the particle's speed v' in the S' system is given by

$$v' = \sqrt{(\dot{x}')^2 + (\dot{y}')^2} = \sqrt{(-0.98)^2 + (0.16)^2}\,c = 0.993c$$

and the angle θ' between the direction of the particle velocity and the negative x'-axis is determined by

$$\tan\theta' = \frac{-\dot{y}'}{\dot{x}'} = \frac{0.16c}{0.98c}$$

or

$$\theta' = 9.3^\circ \text{ north of west}$$

Note that the classical velocity-combination rule (Galilean velocity transformations, Equations 2–2) would have yielded a speed $v' = \sqrt{(-0.98)^2 + (0.80)^2}\,c = 1.3c$, a speed in *excess* of the speed of light, whereas the relativistic relations ensure that the particle speed is less than c (here, $0.993c$). In addition, the Galilean relations predict an angle $\theta' = \arctan(0.80/0.98) = 39^\circ$ north of west, in contrast to the much smaller angle of 9.3° found above from the relativistic relations.

2–7 Length and Time Intervals in Relativity Physics

SPACE CONTRACTION. In classical physics, the length of an object is the same for all observers, whatever their velocities with respect to it. Let us now examine the meaning of length in the Lorentz transformations.

Suppose that observers S and S' are at rest with respect to each other, compare their respective metersticks, and agree that both have the same length L_0. Then system S' is set in motion to the right with a speed v with respect to system S. Observer S aligns his meterstick along the x-axis, its left end at x_l and its right end at x_r; then, from his point of view, $L_0 = x_r - x_l$. Similarly, observer S' aligns her meterstick along the x-axis, the left and right ends at the points x_l' and x_r'; then, from her point of view, $L_0 = x_r' - x_l'$. Each observer sees *his or her own* meterstick as having length L_0; this, of course, is because all inertial systems are equivalent and indistinguishable.

Now we ask, "What is the length of a moving meterstick, say S''s meterstick, as measured by S?" First, we must recognize that when measuring the length of an object—even in classical physics—one must mark both ends of the object simultaneously. If one marked one end of a moving meterstick at one time and the other end at a later time, the distance between the two marks would not correspond to the length of the moving object; this distance could assume any value, depending on the time interval between the two marking operations. Since observer S must mark both ends of a moving object simultaneously, we must choose the two events, (x_l, t_l) and (x_r, t_r), representing the left and right ends of S''s meterstick as measured by S, with $t_l = t_r$. Equations 2–15 give the space-time coordinates (x_l', t_l') and (x_r', t_r') of these same two events, as measured by S', in terms of the events as measured by S. Because the meterstick is always at rest in S''s system, the space coordinates x_l' and x_r' are independent of the times t_l' and t_r' at which S' measures them. Thus, it is not essential that the spatial measurements be made simultaneously in the system in which the object is at rest.

Using Equations 2–15, we have

$$x_r' - x_l' = \frac{(x_r - x_l) - v(t_r - t_l)}{\sqrt{1 - (v/c)^2}} \tag{2–17}$$

with $t_r = t_l$. The length of S''s meterstick, as measured by S', is $x_r' - x_l' = L_0$; the length of this same meterstick, as measured by S, is $x_r - x_l = L$. Therefore, Equation 2–17 becomes

$$L = L_0\sqrt{1 - (v/c)^2} \tag{2–18}$$

This relation describes *space contraction*. The length of S''s moving meterstick (as measured by S) is shorter by a factor of $\sqrt{1 - (v/c)^2}$ along the direction of motion. This contraction is reciprocal; that is, S's moving meterstick (as measured by S') is also shorter by the same factor. (Note, however, that when S' marks the two ends of her meterstick simultaneously, these two events are not simultaneous as observed by S.) Because the contraction is not a result of any physical disturbance (such as cooling or compression) but rather reflects the properties of space and time, the phenomenon is known as *space contraction*. Since $y = y'$ and $z = z'$, there is no space contraction perpendicular to the relative motion.

Although a moving object no longer has the same length when measured by all inertial observers, one length is invariant: the *proper length,* or the length of an object at rest with respect to an observer. Thus, all inertial observers would agree that L_0 is the length of a given meterstick when it is at rest in their respective inertial systems. We are accustomed to events in which the velocity v of S' is much less than c and the length L is essentially equal to L_0.

TIME DILATION. Relativity physics shows that time intervals, like space intervals, are not absolute but, rather, depend on the relative motion of the observer.

Two observers S and S' synchronize their respective clocks while at rest with respect to one another, and agree that the time interval between two events—for example, the interval between successive ticks of a clock—is T_0. Now we imagine that system S' is moving at a speed v to the right with respect to system S. Observer S keeps his clock at rest at a particular point x in his system and measures time interval T_0 as elapsing between instants t_0 and t. Similarly, S' keeps her clock at a fixed position x' and measures the interval T_0 as elapsing between instants t_0' and t'. Each observer measures the time interval on his or her own clock to be T_0. We wish to find out how the intervals registered on the two clocks compare when *both* intervals are measured *by observer* S'. From Equations 2–15 we see that the time interval $t' - t_0'$ between the two events, as measured by observer S', is related to the space interval $(x - x_0)$ and to the time interval $(t - t_0)$ between the same two events, as measured by observer S, by

$$t' - t_0' = \frac{(t - t_0) - (v/c^2)(x - x_0)}{\sqrt{1 - (v/c)^2}} \tag{2–19}$$

Now, observer S' wishes to measure the time interval on a clock that is at rest in inertial system S. Because all clocks at rest in system S' have been synchronized, it is not important at what location in S' the time of an event is measured. Therefore, we choose a clock at rest in S at the point $x_0 = x$ and consider the events (x_0, t_0) and (x_0, t). In the equation above the time interval $t' - t_0'$ between the two events, as measured by observer S' is, then,

$$t' - t_0' = \frac{t - t_0}{\sqrt{1 - (v/c)^2}}$$

Letting the time interval that observer S measures with a clock at rest in his system be $T_0 = t - t_0$ and the time interval, between the same events, that S' measures for the moving clock be $T = t' - t_0'$, we obtain

$$T = \frac{T_0}{\sqrt{1 - (v/c)^2}} \tag{2–20}$$

This relation illustrates *time dilation*. For an example of this, suppose that $v = 0.98c$; then $\sqrt{1 - (v/c)^2} = \frac{1}{5}$, and $T' = 5T_0$. Therefore, if the time interval between consecutive ticks on each of two identical clocks is T_0 (*the proper time*) when both are at rest with respect to an observer, then, when one clock is in motion at $0.98c$ with respect to the other clock, the time interval between consecutive ticks of the moving clock will be $5T_0$ as measured on the resting clock. That is, the resting clock makes five ticks for every tick of the moving clock. We say that moving clocks run slow, or "live longer." Again, this effect is reciprocal, in that each observer finds the other moving observer's clock to run slow.

It should be emphasized that by *clock* we mean *any* device that repeats itself at regular time intervals, whether it is an ordinary mechanical clock producing ticks, a quartz crystal undergoing oscillations in a digital clock, vibrating atoms in an atomic clock, or even the repetitive heartbeats of a human being.

For $v \ll c$, the relative time of relativity physics reduces to the absolute time of classical physics and our everyday experience. Although time dilation may seem bizarre, inescapable evidence for its existence has been found in the decay of high-speed unstable particles. The following examples show the unusual behavior of space-time at relativistic velocities.†

† For another interesting example of relativistic behavior, see the discussion of the "twin paradox" in Section 2–8.

Example 2–4. An interesting confirmation of the time-dilation phenomenon (Equation 2–20) is found in the decay of high-energy (therefore, high-speed) *muons*. These particles are produced in the decay of other unstable particles, called π mesons, which in turn are created in high-energy collisions of nuclei. (These particles will be discussed in Section 12–2.) The muon is like the electron in its properties except that its mass is approximately 207 times greater and it is unstable, quickly decaying into other particles. The measured half-life of muons that are at rest with respect to an observer is 1.52×10^{-6} s; that is, if 10,000 muons are at rest with respect to an observer, the observer will find only an average of 5000 muons surviving after 1.52×10^{-6} s has elapsed. If on the other hand, the muons are moving with respect to an observer, their lifetime will be dilated and they will appear to this observer to live longer. According to Equation 2–20, this observer will measure a half-life T given by

$$T = \frac{T_0}{\sqrt{1-(v/c)^2}}$$

where $T_0 = 1.52 \times 10^{-6}$ s is the half-life in the inertial system of the muons, and v is the relative velocity between the mesons and the observer.

In the high-energy collision between cosmic ray particles from outer space (mostly protons) and nuclei in the atmosphere, muons are produced; and some approach the earth's surface with speeds very nearly c. For such particles, time dilation will therefore be appreciable and readily observed.

Consider 10,000 muons approaching the earth's surface at speed $0.98c$ from an altitude $L_0 = 2.23$ km, this height measured by an observer fixed on the earth. We may think of this distance as registered on a long, vertical post *fixed* to the earth; see Figure 2–6. As measured by the observer on the earth, the muons' half-life is

$$T = \frac{T_0}{\sqrt{1-(v/c)^2}} = 5T_0 = 7.60 \times 10^{-6}\ \text{s}$$

and their flight time is

$$\frac{L_0}{0.98c} = \frac{2.23 \times 10^3\ \text{m}}{0.98 \times 3.0 \times 10^8\ \text{m/s}} = 7.60 \times 10^{-6}\ \text{s}$$

The two are equal only because of our convenient choice of L_0. Therefore, of the original 10,000 muons at 2.23-km altitude, only 5000 will have survived decay on reaching the earth's surface.

How do these decay events appear to an observer moving with the decaying muons? He finds that half the muons decay in a time $T_0 = 1.52 \times 10^{-6}$ s. Moreover, he sees the earth and vertical post approach him at a speed of $0.98c$. Because of the space-contraction phenomenon, the earth's distance from him is contracted (that is, the vertical post fixed to the earth is contracted). At the time that he counts 10,000 muons, the earth's distance from him by his measurements is only $(2.23\ \text{km})\sqrt{1-(0.98)^2} = 0.446$ km. Thus, the time of flight, according to him, is

$$\frac{446\ \text{m}}{0.98 \times 3 \times 10^8\ \text{m/s}} = 1.52 \times 10^{-6}\ \text{s}$$

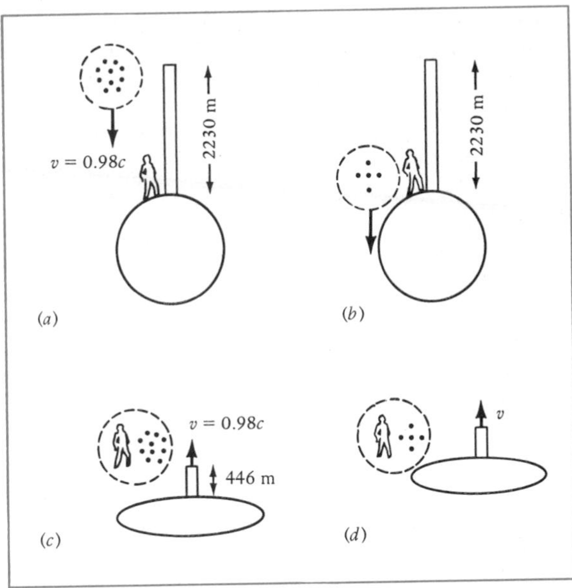

FIGURE 2–6. *Muons and Earth approaching each other at 0.98c. As seen by observer fixed on the earth: (a) 10 000 muons approaching Earth; (b) 5 000 undecayed muons arriving at Earth. As seen by observer traveling with the muons: (c) Earth approaching 10 000 muons; (d) Earth arriving at 5 000 undecayed muons.*

But this is just the decay half-life in the rest system of the muons; therefore, an observer in that system also will find 5000 of the original 10,000 muons surviving when the earth reaches the muons. Although the two observers, one on earth and one with the muons, disagree on the measurements of time and of length intervals, both agree that 5000 muons survive when the earth and muons meet.

If our classical notions of space and time were valid and there were no dilation of time, the observer on the earth would measure the half-life of the moving muons to be 1.52×10^{-6} s, and the flight time would be $(7.60 \times 10^{-6})/(1.52 \times 10^{-6}) = 5$ times the half-life. Since only half the muons survive each half-life, the observer on the earth would then predict that $(\frac{1}{2})^5 = \frac{1}{32}$ of the original 10,000 would survive to the earth's surface. Without relativistic effects, he would find only 310 muons surviving, in disagreement with experiment. Time-dilation effects observed with high-speed particles give emphatic confirmation to relativistic physics.

Example 2–5. We here consider some specific events–the births and deaths of three people. So that we can use the Lorentz coordinate transformations, Equations 2–15, we consider two inertial systems S and S', with axes aligned as in Figure 2–1; S' moves along the $+x$-axis at a speed of $0.98c$ with respect to S. We think of S as an inertial system in which the earth is at rest and of S' as one in which a spaceship is at rest.

At time $t = t' = 0$, observers in S record the simultaneous births of the three people: Jack, born at the origin; Astrid, born at $x = +10$ light-years (see Example 2–1 for the definition of the light-year); and Jill, born at the origin. See Figure 2–7a. Both Jack and Astrid are always at rest in S throughout their 70-year life span, according to observers in S. Jill, on the other hand, is born and remains at rest in the spaceship, which is at rest in S'. With respect to S', Jill also lives 70 years.

(*a*) According to observers in S (the earth at rest), how long does Jill live?
(*b*) According to observers in S' (spaceship at rest), how long do Jack and Astrid live?
(*c*) According to observers in S', how far apart are Jack and Astrid?

Equations 2–15 relate the space-time coordinates of any event as measured from two inertial systems S and S'. In our example, there are

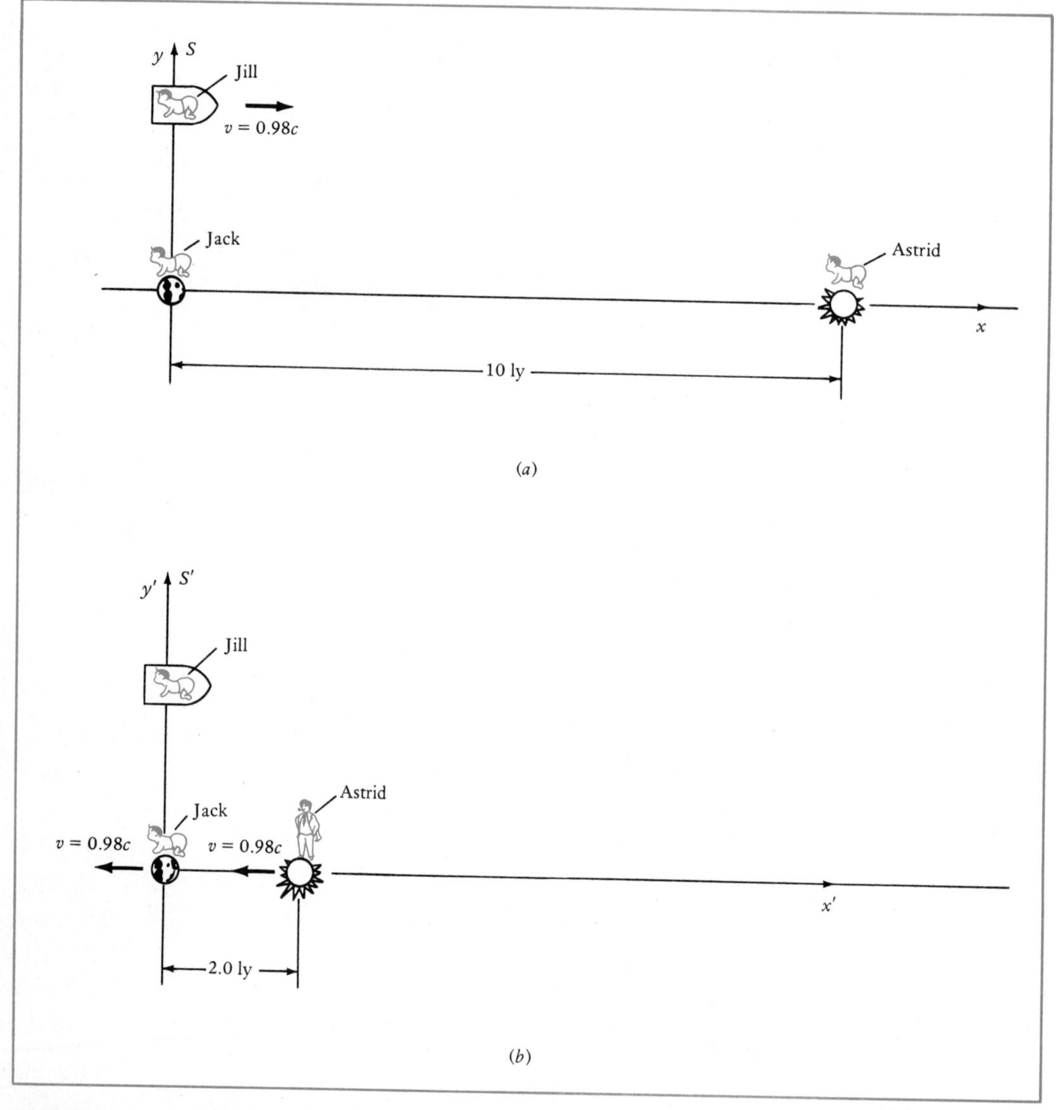

FIGURE 2–7. *The simultaneous births (a) of Jack, Jill, and Astrid, as observed in S, and (b) of Jack and Jill, as observed in S′. In (b), observers in S′ find that Astrid is 49 years old and 2.0 light-years from Jack and Jill. See Example 2–5.*

six relevant events: the birth and death of Jack, the birth and death of Astrid, and the birth and death of Jill. If we know the space-time coordinates of an event as observed in one of the systems, we can use Equations 2–15 to get the coordinates of the same event as observed in the other system. The table below lists the space-time coordinates of the six events as observed in the two system S and S' and indicates which are *given* and which are *computed* by Equations 2–15. A sample calculation follows; it shows how the specific values of the coordinates in one system are deduced from the given values in the other system.

	SPACE-TIME COORDINATES	
EVENT	$S(x, t)$	$S'(x', t')$ $v/c = 0.98$
Birth of Jack	(0 , 0) given	(0 , 0) computed
Death of Jack	(0 , 70 yr) given	(−343 ly, 350 yr) computed
Birth of Astrid	(10 ly, 0) given	(50 ly, −49 yr) computed
Death of Astrid	(10 ly, 70 yr) given	(−293 ly, 301 yr) computed
Birth of Jill	(0 , 0) given	(0 , 0) computed
Death of Jill	(343 ly, 350 yr) computed	(0 , 70 yr) given

Consider the birth of Astrid. Observers at rest in the earth system S record the birthplace of Astrid as $x = +10$ light-years and the time of birth as $t = 0$. Observers at rest in the spaceship system S' record the location and time of Astrid's birth by Equation 2–1:

$$x' = \frac{x - vt}{\sqrt{1 - (v/c)^2}} = \frac{10\text{ ly} - 0}{\sqrt{1 - (0.98)^2}} = 5(10\text{ ly}) = 50\text{ ly}$$

$$t' = \frac{t - (v/c^2)x}{\sqrt{1 - (v/c)^2}} = 5\left(0 - 0.98\frac{10\text{ ly}}{1\text{ ly/yr}}\right) = -49\text{ yr}$$

Observers in S' record Astrid's birth as 50 light-years from the birthplace of Jack and Jill and 49 years before the births of Jack and Jill.

The death of Astrid, according to observers in S, takes place at $x =$ 10 light-years and $t = 70$ years. According to observers in S' by Equations 2–15, the event is

$$x' = \frac{x - vt}{\sqrt{1 - (v/c)^2}} = 5\left(10\text{ ly} - \frac{0.98\text{ ly}}{\text{yr}}70\text{ yr}\right) = -293\text{ ly}$$

$$t' = \frac{t - (v/c^2)x}{\sqrt{1 - (v/c)^2}} = 5\left\{70\text{ yr} - \left(\frac{0.98\text{ ly}}{\text{yr}}\right)\left[\frac{10\text{ ly}}{(1\text{ ly/yr})^2}\right]\right\} = 301\text{ y}$$

The other space-time coordinates are calculated in the same way.

Using entries in the above table, you can now easily answer the questions asked at the beginning of this example.

(*a*) According to observers S, how long does Jill live? From the table we have
Life span of Jill according to observers S:

$$t(\text{death}) - t(\text{birth}) = 350\text{ yr} - 0\text{ yr} = 350\text{ yr}$$

This agrees with the time dilation of moving clocks, Equation 2–20. For this example, the lifetime interval of the moving woman is increased by a factor of 5.

(*b*) According to spaceship observers S', how long do Jack and Astrid live?

Life span of Jack according to observers S':

$$t'(\text{death}) - t'(\text{birth}) = 350 \text{ yr} - 0 \text{ yr} = 350 \text{ yr}$$

Life span of Astrid according to observers S':

$$t'(\text{death}) - t'(\text{birth}) = 301 \text{ yr} - (-49 \text{ yr}) = 350 \text{ yr}$$

This, too, agrees with Equation 2–20.

(*c*) According to the observers stationed in S', how far apart are Jack and Astrid at any instant? The locations of Jack and Astrid must be determined at some common time t' in S'. Since we know that Astrid travels at a velocity of 0.98 light-years per year relative to S', let us for convenience find her location at $t' = 0$, at which time we already know that Jack is at the origin. From the table we see that Astrid is born at $x' = +50$ light-years at time $t' = -49$ years. Then, 49 years later, Astrid will have moved to the left a distance $vt = (0.98 \text{ ly/yr})(49 \text{ yr}) = 48 \text{ ly}$. Therefore, at time $t' = 0$ Astrid will be at the location

$$50 \text{ ly} - 48 \text{ ly} = +2 \text{ ly from } S' \text{ origin}$$

At this time Jack is at S''s origin (and is just being born). Observed from S', Astrid and Jack are therefore 2.0 light-years apart at time $t' = 0$. This is different from their separation, 10 light-years, observed in the reference frame S in which they are at rest. These distances agree with the phenomenon of space contraction, Equation 2–18. Notice that at time $t' = t = 0$, observers in S' find that Astrid is 49 years old when Jack and Jill are just being born; see Figure 2–7. On the other hand, observers in S find Astrid, Jack, and Jill all born simultaneously.

2–8 The Twin Paradox

One of the most startling predictions of Einstein's theory of relativity is the famous twin paradox. It involves the difference in rates of aging of two biological clocks, twins, in relative motion. Stated simply, the situation posed is this. Two clocks, both at rest in an inertial frame S, are synchronized. One clock is then set in motion; it travels outward and later returns to the other clock fixed in system S. The time intervals of the two clocks are then compared. According to Equation 2–20, the traveling clock should always be behind the fixed clock in time—a traveling clock (our traveling twin) ages less rapidly than one at rest.

The paradox is apparent if we consider an observer riding with the "moving" clock. From that person's point of view, the clock fixed in S moves out and back, and on return, should be behind in time! Which clock lags, then, in time behind the other?

Resolving the paradox is straightforward if we recall that measurements and comparisons of time intervals should always be made in a *single* inertial system. The clock in system S is initially at rest and remains at rest in a single inertial system; measurements made in this single frame correctly predict that the second clock ages less rapidly. On the other hand, the second clock does not remain at rest in a *single* inertial system throughout the trip. Moving out in the first segment of the trip, it is in one system; and moving back in the second segment, it is in another system. Applying the time-dilation relation (Equation 2–20) in a frame that has the spaceship clock always at rest throughout the trip is therefore improper, inasmuch as Equation 2–20 relates time intervals of a clock at rest and a clock in motion only from the viewpoint of a *single inertial system*. Example 2–6 shows the difference in aging of twins, one a traveler, the other at rest on Earth.

Example 2–6. Consider twins, Jack and Jill, initially at rest on Earth. Jill departs on a spaceship that travels outward at speed $0.98c$. It quickly stops and returns at the same speed, $0.98c$. According to Earth clocks, the trip takes five years. During this trip how many birthdays does each twin celebrate?

This example can be discussed more easily in time rates rather than in time intervals; the time rate is by definition just the reciprocal of the corresponding time interval (just as frequency is related to period):

$$R = \frac{1}{T}$$

Equation 2–20 can therefore be written in time rates as follows:

$$T = \frac{T_0}{\sqrt{1-(v/c)^2}} \qquad (2\text{–}20)$$

Therefore $$R = R_0\sqrt{1-(v/c)^2}$$

where R_0 is the time rate of a series of periodic events (for example, birthdays) taking place at a fixed point in an inertial frame, and R is the time rate of a corresponding series of events taking place in a reference frame moving at speed v.

Jack, who stays at home on Earth, remains at rest in a *single* inertial system. Time rates for each of the twins can correctly be determined in this inertial system. If the period T_0 between Jack's birthdays is one year, then the time rate of his birthdays is

$$R_0 = \frac{1}{T_0} = \frac{1 \text{ birthday}}{\text{yr}}$$

and since the trip takes five years, the number of birthdays Jack celebrates is

$$(R_0)(\text{time of trip}) = (1 \text{ birthday/yr})(5 \text{ yr}) = 5 \text{ birthdays}$$

Because Jill is always moving at speed $v = 0.98c$ relative to Earth's inertial frame, the time rate of her birthdays is

$$R = R_0\sqrt{1-(0.98)^2} = \tfrac{1}{5}R_0 = (\tfrac{1}{5})(1 \text{ birthday/yr})$$
$$= (\tfrac{1}{5}) \text{ birthday/yr}$$

and the number of birthdays Jill celebrates during the five-year trip is

$$(R)(\text{time of trip}) = (\tfrac{1}{5}\ \text{birthday/yr})(5\ \text{yr}) = 1\ \text{birthday}$$

Thus, during the round trip, Jill ages one year, whereas Jack ages five years! The moving twin ages less rapidly than the fixed twin.

It is possible to measure this difference in aging of the twins from any inertial system. One could choose, for example, an inertial system S' that always moves away from the earth at speed $0.98c$ and that has Jill at rest during the first outgoing segment of the trip. Relative to this inertial system, Jack always moves at $0.98c$ and therefore celebrates birthdays at the rate of $\frac{1}{5}$ birthday/yr. Jill, on the other hand, is at rest during the first segment of the trip and celebrates birthdays at the rate of 1 birthday/yr during this segment. During the return segment of the trip, however, she travels at a speed greater than $0.98c$ in system S' and eventually catches up with Jack; Jill celebrates birthdays during this segment at a rate even lower than that for Jack, which is $\frac{1}{5}$ birthday/yr. Although the calculations as viewed in the S' inertial frame are more complicated, one again would find that when the twins reunite, Jack has celebrated five birthdays and Jill only one.

We can see that twins Jack and Jill are not equivalent in still another way. Suppose, for the sake of argument, that they spend their lives each inside a spaceship. Jack's spaceship remains on Earth, and his experiences in it are uneventful. Jill, whose spaceship travels to a distant planet and back, has different experiences. At takeoff, under the spaceship's acceleration, Jill and all unattached objects within the spaceship are thrown against its wall during the change in inertial systems. She has a similar experience when the spaceship arrives at the distant planet, slows down, and then speeds up in the opposite direction in still another inertial frame. Finally, when Jill arrives back at the earth, and her spaceship again decelerates, she experiences a shock for the third time. Thus, when the twins are reunited and compare life histories, they find not only that their ages are different but also that Jack's existence has been quiet while Jill's has been punctuated by three accelerations.

2-9 Space-Time Events and the Light Cone

Returning to the sequence of space-time events in Example 2–5, let us examine the six events shown in the table of Example 2–5. Figure 2–8 shows graphically the events as observed in S and S', with time t as ordinate and displacement x as abscissa. (In the usual graph, the displacement is the ordinate and the time the abscissa.) In Figure 2–8(a) it is obvious that all events occurring along the same horizontal line are simultaneous in system S. Thus, all three births occur at the same time, $t = 0$; and the deaths of Jack and Astrid are simultaneous events in S, occurring at time $t = 70$ yr. Similarly, events along the same vertical line occur at the same location in space. The births of Jack and Jill, for example, and the death of Jack all occur at the same space point, $x = 0$ light-years, and the birth and death of Astrid occur at the same point, $x = 10$ light-years.

The "path" representing the time sequence of events of a single object on a space-time diagram specifies the location x of the object as a function of time t. Since the velocity of the object is defined as $\Delta x/\Delta t$, the slope, *measured against the t-axis,* of the space-time path gives the velocity of the object. Figure 2–8*a* shows that the paths of Jack and Astrid are both vertical and the slopes zero; therefore, the velocities of both are zero. On the other hand, the path of Jill is a straight line from the origin to the point $x = 343$ ly, $t = 350$ yr. The slope, or velocity, is

$$v\,(\text{Jill}) = \frac{\Delta x}{\Delta t} = \frac{343\text{ ly}}{350\text{ yr}} = \frac{0.98\text{ ly}}{\text{yr}}$$

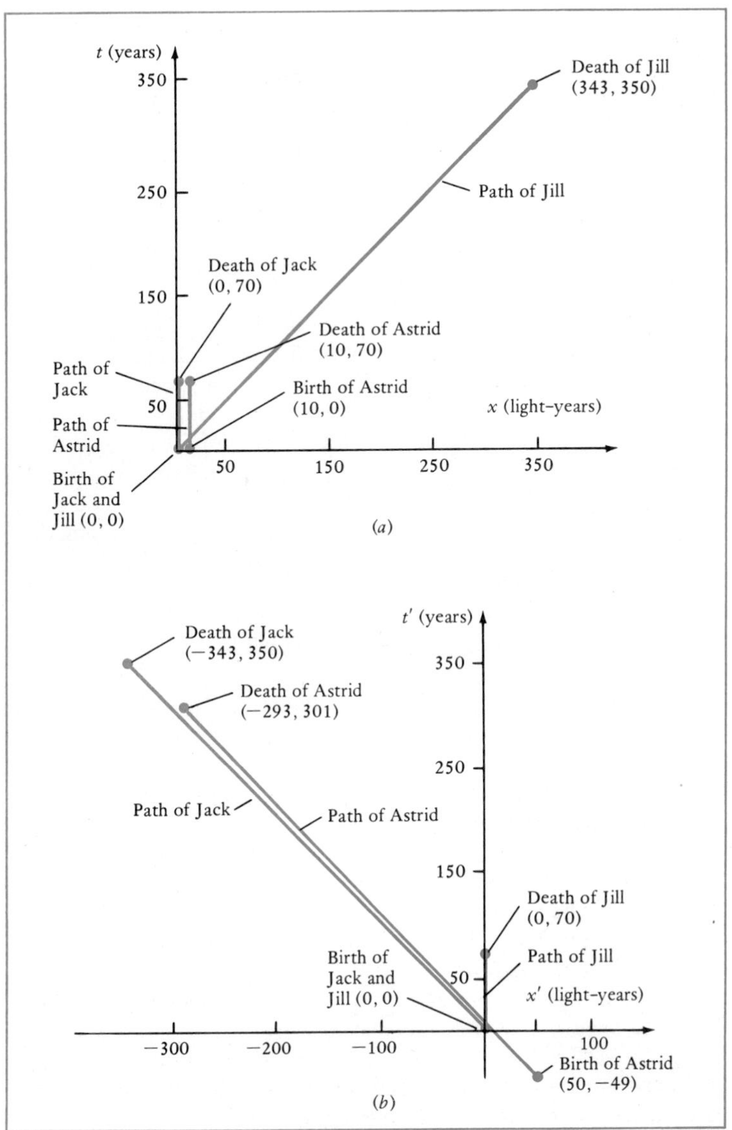

FIGURE 2–8. *Space-time graph of events in Example 2–5 as observed (a) in system S and (b) in system S′.*

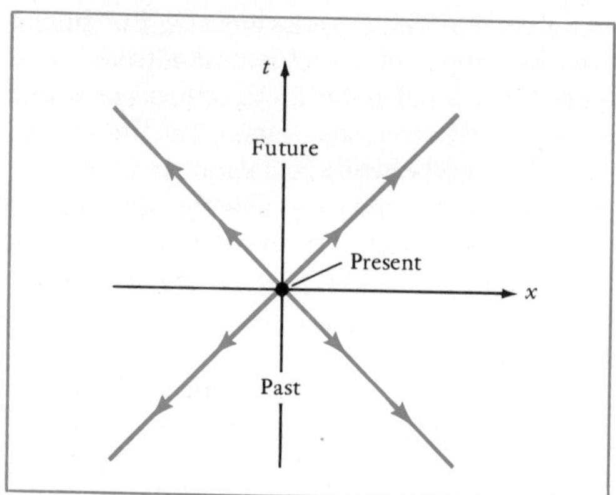

FIGURE 2–9. *Space-time paths of two light pulses traveling on the x-axis in opposite directions.*

Figure 2–8(b) is a space-time diagram of the same six events, but now as observed in system S'. From the point of view of observers in S' only two events are simultaneous—the birth of Jack and the birth of Jill. The space-time path of Jill is now vertical, showing that Jill remains at rest in S'. The space-time paths of both Jack and Astrid have the same slope against the t' axis, corresponding to the velocity -0.98 light-years per year.

All the earlier conclusions about events can be read directly from these two figures.

Is there a limit to the magnitude of the slope of a space-time path? There is. We know that the speed of light is the *same* in all inertial systems and moreover that this speed is the *maximum* speed. Therefore, in any inertial system no space-time path can have a slope greater than c. If we choose units of length and time so that the magnitude of c equals 1, then the maximum slope is 1. Thus, the space-time path of a light pulse makes an angle of 45° with the time axis. Figure 2–9 shows the space-time paths of two light pulses, one traveling along the $+x$-axis and the other along the $-x$-axis. At time $t = 0$, both pulses are at the origin, at $x = 0$. The slope of each light pulse is $c = 1$ ly/yr. The slope of the space-time path of any particle can never exceed this. If we set the y-axis at right angles to the plane of the paper, the light paths now generate a *light cone* whose axis is the time axis (Figure 2–10).

Any event, say the event (0, 0), is limited in its ability to be influenced

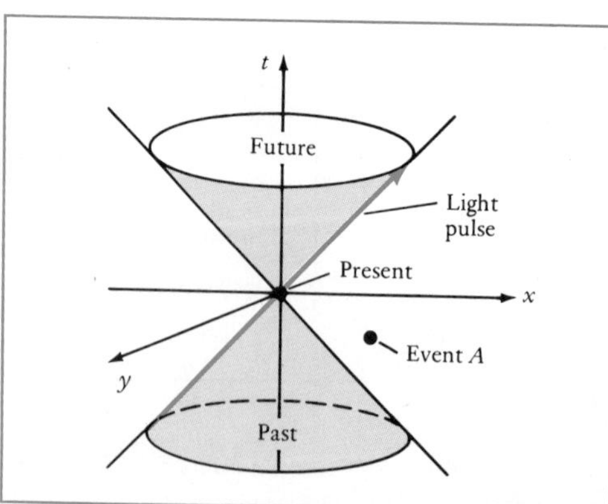

FIGURE 2–10. *Light cone with time t along the axis of the cone. Axis y is perpendicular to the plane of the paper, that is, perpendicular to axes t and x.*

by something in the past ($t < 0$) or to influence events in the future ($t > 0$). In Figure 2–10, only those events within the cone marked "past" can influence the present at the space-time origin; similarly, a present event can influence only those events within the "future" cone. Events falling outside the light cone of Figure 2–10, such as event A, cannot influence an event at the apex of the cone. For A to influence an event, the signal from A to the apex would have to travel faster than c. In short, all events within the light cone can be causally related to an event at the apex; all events outside cannot.

Let us examine causally related events by again looking at the six events of Example 2–5. We choose as our apex event the simultaneous births of Jack and Jill, first as observed in S and then as observed in S'. From Figure 2–8*a* we can see that all the events except the birth of Astrid could be causally related. There is no question that the birth of Jack influences his own death; likewise for the birth and death of Jill. Since the death of Astrid falls within the future part of the apex event's cone, it too could be influenced by the event at the apex.

In Figure 2–8*b*, in which we observe events from the point of view of observer S', we again see that Jack and Jill's births can be causally related to all the other events except Astrid's birth. Even though Astrid is born 49 years before Jack and Jill, she is 50 light-years away from them, and the fastest signal possible, light, would take 50 years to reach Jack's birthplace.

A related question pertains to the sequence of events. If an observer S sees an event A and then an event B, is it possible that in some other inertial system B precedes A? We shall see that it is indeed possible in some cases, but not in others. Does this mean that relativity theory does not preserve the causality of two related events for all observers? We can use the Lorentz transformations to answer these questions.

Consider any two events in four-dimensional space-time. S observes the two events shown in Figure 2–11*a*. We can always choose our space-time coordinates so that one of the events, say A, occurs at the origin (0, 0, 0, 0). The coordinates of event B we label (x, y, z, t).

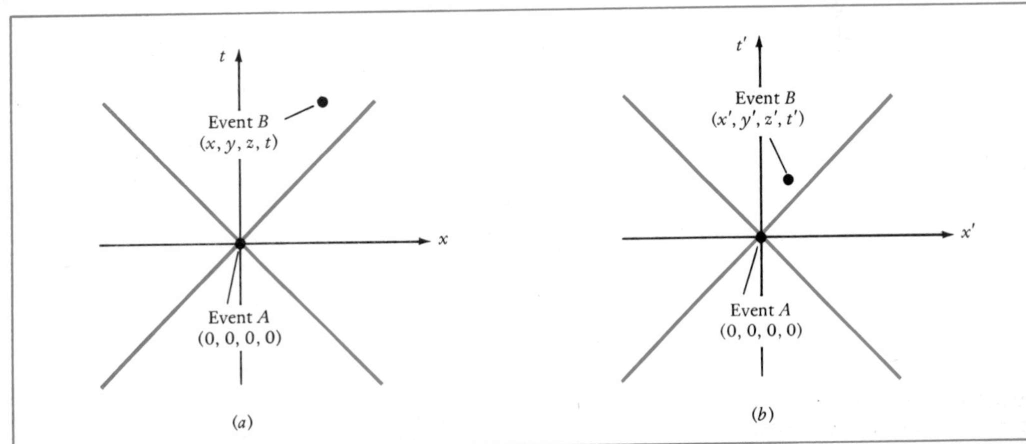

FIGURE 2–11. *Two space-time events observed (a) from system S and (b) from system S′.*

The space-time coordinates of the two events A and B as measured by a second observer S' are given by the Lorentz coordinate transformations, Equations 2–15. For simplicity we let

$$\frac{1}{\sqrt{1-(v/c)^2}} \equiv \gamma$$

Then the coordinates observed by S' are as follows.

Event A:

$$x' = \gamma(x - vt) = 0$$

$$y' = y = 0$$

$$z' = z = 0$$

$$t' = \gamma\left(t - \frac{v}{c^2}x\right) = 0$$

Event B:

$$x' = \gamma(x - vt)$$

$$y' = y$$

$$z' = z$$

$$t' = \gamma\left(t - \frac{v}{c^2}x\right)$$

Figure 2–11*b* shows events A and B as observed in system S', which moves along the $+x$-axis at speed v. Event A is common to all inertial systems, having the coordinates (0, 0, 0, 0). Therefore, space and time intervals between the two events A and B will be those of coordinates B alone. It is obvious that the spatial separations between the events are not the same, or invariant, in the two systems. Calling these L and L', we have

$$L^2 = (x^2 + y^2 + z^2)$$

$$L'^2 = (x'^2 + y'^2 + z'^2) \neq L^2$$

Nor are the time intervals the same:

$$t' \neq t$$

Is there some interval, analogous to the separate space intervals and time intervals of Galilean physics, that is invariant under a Lorentz transformation? There is such an interval, and we can easily find it by the Lorentz transformations. We recall that the Lorentz transformations were developed on the postulate that all observers measure the same constant speed c of a light signal. Therefore, if event A were to represent the start of a light pulse and event B the arrival of the pulse at the point (x, y, z) in space, then both observers S and S' must get for light signals

$$x^2 + y^2 + z^2 = c^2t^2$$

$$x'^2 + y'^2 + z'^2 = c^2t'^2$$

or

$$c^2t^2 - (x^2 + y^2 + z^2) = 0$$

$$c^2t'^2 - (x'^2 + y'^2 + z'^2) = 0$$

All inertial observers measure the *same* value for the quantity $c^2t^2 - (x^2 + y^2 + z^2)$ for the events connected by light signals; it is zero. Since this quantity is a combination of the space interval and the time interval between two events, we might wonder whether it is an invariant quantity between any two events. Let us find out. For system S' the interval is $c^2t'^2 - (x'^2 + y'^2 + z'^2)$. Using the Lorentz transformations, we can express all coordinates by the coordinates in system S:

$$c^2t'^2 - (x'^2 + y'^2 + z'^2) = c^2\gamma^2\left(t - \frac{v}{c^2}x\right)^2 - [\gamma^2(x - vt)^2 + y^2 + z^2]$$

Multiplying out the right-hand side, collecting terms, and recalling that $\gamma = 1/[1 - (v/c)^2]^{1/2}$, we get

$$\Delta S^2 = c^2t'^2 - (x'^2 + y'^2 + z'^2) = c^2t^2 - (x^2 + y^2 + z^2) \tag{2–21}$$

Thus the quantity ΔS^2, the square of the *space-time interval,* is an invariant quantity, having the *same value in any inertial system.* As derived here, it is the space-time interval between any space-time event (x, y, z, t) and the space-time event $(0, 0, 0, 0)$. There are three ranges of values for this quantity:

$\Delta S^2 = 0$. In this case $c^2t^2 = x^2 + y^2 + z^2$ and the events are connected by light signals. All space-time points connected to the origin with $\Delta S^2 = 0$ generate the light cone. See Figure 2–12.

$\Delta S^2 < 0$. In this case $c^2t^2 < x^2 + y^2 + z^2$ and the region is said to be *spacelike.* Events in this region are not causally related to the event $(0, 0, 0, 0)$, since signals would have to travel faster than c to make them so.

$\Delta S^2 > 0$. In this case $c^2t^2 > x^2 + y^2 + z^2$. This region is said to be *timelike,* and all events in it can be causally related to event $(0, 0, 0, 0)$.

What about causality? That is, if event B follows event A in one system, does it follow it in all other systems? Again let event A be $(0, 0, 0, 0)$ and B be (x, y, z, t), as observed in S. The time interval between event A and event B, as observed in S, is $\Delta t = (t - 0) = t$. By Equations 2–15, the time interval between these two events, as observed in S', is

$$\Delta t' = t' - 0 = t' = \gamma\left(t - \frac{vx}{c^2}\right) \tag{2–22}$$

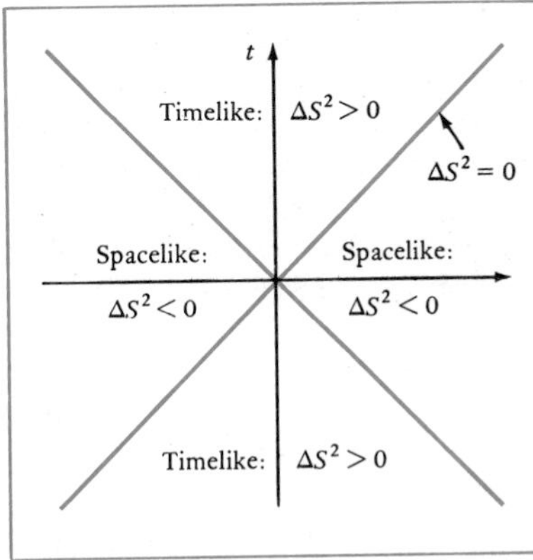

FIGURE 2–12. *Regions of space-time for the space-time interval ΔS^2 greater than, equal to, and less than zero.*

The quantities t, x, and v may have positive or negative values, but whatever their signs, we must also, by the invariance of the space-time interval, have

$$\Delta S^2 = c^2t'^2 - x'^2 = c^2t^2 - x^2$$

If in S, event B follows event A, then $t > 0$. Then by Equation 2–22 we shall have $t' > 0$ if $t > (v/c^2)x$, or rewritten, $c^2t^2 > (v/c)^2x^2$.

Now, $v/c < 1$; therefore for any timelike event $\Delta S^2 > 0$ (see Figure 2–12) and $t > 0$, we shall always have $t' > 0$; observers in all inertial systems always observe event B to follow event A. Similarly, in the timelike cone with $t < 0$, we shall always have $t' < 0$; then event B precedes event A in all systems. Causality *is* preserved for events within the timelike region.

On the other hand, there is no unique ordering of events in the spacelike region. Assume, for example, that in S event B follows event A. Then $t > 0$, and by Equation 2–22, t' will be negative if $t < (v/c^2)x$, or $ct < (v/c)x$. This can, of course, occur in the spacelike region; thus, whether or not there is a reversal of order depends on where the event is located in this region.

SUMMARY

Classical mechanics is covariant under the Galilean transformations and agrees with experiment for $v \ll c$.

The correct description of the physical universe is based on the invariance of the speed of light and the principle of relativity, which says that all the laws of physics are covariant in all inertial systems under the Lorentz transformations.

The Lorentz coordinate transformations are

$$\begin{aligned} x' &= \frac{x - vt}{\sqrt{1 - (v/c)^2}} \\ y' &= y \\ z' &= z \\ t' &= \frac{t - (v/c^2)x}{\sqrt{1 - (v/c)^2}} \end{aligned} \qquad (2\text{–}15)$$

No signal or particle can move at a speed exceeding the speed of light in vacuum.

Relativity physics is supported by the following experimental observations: the Michelson-Morley and related experiments concerning the invariance of the speed c, direct observation of time dilation in unstable particle decay, and the observation that the speed of particles accelerated in high-energy particle accelerators never exceeds c.

The transition from relativity physics to classical physics can be written symbolically:

$$\underset{v/c \to 0}{\text{Limit}} \text{ (relativity physics)} = \text{classical physics}$$

The transformation equations relating length and time intervals are the following:

	Relativistic form	*Classical form*
Length:	$L = L_0\sqrt{1-(v/c)^2}$	$L = L_0$
Time:	$T = \dfrac{T_0}{\sqrt{1-(v/c)^2}}$	$T = T_0$

Some invariants under the relativistic Lorentz transformations:

- The speed of light.
- That a single space-time event in S is a single event in S'.
- The square of the space-time interval, $\Delta S^2 = c^2\,\Delta t^2 - (\Delta x^2 + \Delta y^2 + \Delta z^2)$.

PROBLEMS

Problems are grouped by section title and number. Those with a black bullet before the problem number are short and uncomplicated; those without a symbol before the problem number are of average difficulty. Problems with a color square before the problem number are more difficult.

Galilean transformations, §2–2

● **2–1.** Describe two different methods of synchronizing two separated clocks in an inertial system in which both clocks are at rest.

● **2–2.** What changes would there be in relativistic space-time structure if the invariant speed—the speed that remains the same in all inertial systems—were $2c$ instead of c?

2–3. A strong wind moves the air at constant velocity along the x-axis. To measure (1) the speed of sound waves with respect to a system S' in which the air is at rest, and (2) the velocity $\mathbf{v}$ of the wind relative to the ground (system S), three people at rest on ground do the following. One person at the origin in S fires a gun at time $t = 0$; the other two people stand along the x-axis, one at -1000 ft and one at $+1000$ ft, and record the arrival times of the sound from the firing of the gun. The observer at -1000 ft hears the sound 0.952 s after the firing; the observer at $+1000$ ft 0.800 s after firing. (a) What is the speed of sound in S'? (b) What is the velocity of the wind relative to the ground? (c) In what single reference frame is the speed of sound the same, irrespective of the propagation direction of the sound wave?

Covariance of classical mechanics, §2–3

2–4. A particle of mass m moving initially with velocity $\mathbf{v}$ to the right collides head-on with an identical particle initially at rest. The collision is perfectly elastic. (a) What is the total momentum of the two particles before and after collision? (b) What is the total kinetic energy of the two particles before and after the collision? (c) Suppose that the same collision is viewed from a reference frame that is moving to the right at velocity $\mathbf{v}/2$. What would the total momentum of the two particles be before and after collision? (d) What would be the total kinetic energy of the two particles before and after collision?

Failure of Galilean transformations, §2–4

2–5. Assume that the Galilean transformations hold and that in the Michelson-Morley experiment (see Section 2–4) the length of the cylinders is 10.0 m and the wavelength of the light 5000 Å. (a) If the speed of the earth through the ether is 3.00×10^4 m/s, what will be the fraction of wavelength shift between the two beams when they recombine? (b) At what speed would the apparatus have to move through the ether to have the two beams recombine 180° out of phase with one another?

The second postulate and Lorentz transformations, §2–5

● **2–6.** In inertial system S, a baby is born at $x = 3.0 \times 10^6$ m at time $t = 0.01$ s. Where and when is the baby born according to the system S', which is moving along the positive x-axis at speed $v = 0.75c$?

2–7. Solve algebraically for each of the coordinates x, y, z, and t in terms of x', y', z', and t' in Equations 2–15. You will thereby confirm that the inverse Lorentz transformations result from changing $\mathbf{v}$ to $-\mathbf{v}$ and interchanging primes and unprimes.

2–8. Relative to an inertial system S in which our sun is at rest, the planet Mars moves in an elliptical orbit of semimajor axis $a = 2.28 \times 10^{11}$ m and semiminor axis $b = 2.24 \times 10^{11}$ m. Find the velocity (that is, the magnitude and direction) of the inertial system in which the orbit of Mars is circular. In terms of the semimajor and semiminor axes a and b, the equation of an ellipse centered at the origin can be expressed as $(x/a)^2 + (y/b)^2 = 1$.

2–9. Two inertial systems are oriented as in Figure 2–1, with system S' moving to the right at speed $(c/2)$ relative to system S. At the time $t = t' = 0$, a point source at the origin and at rest in S emits a light wave. Assume that the classical medium transmitting the light wave is at rest relative to system S. For both the Galilean representation and the Lorentz representation (*a*) show the wavefront and the $x'y'$ axes at time $t = 1.0 \times 10^{-8}$ s as observed in system S; (*b*) show the wavefront and the xy axes at time $t' = 1.0 \times 10^{-8}$ s as observed in system S'.

2–10. In inertial system S, a man at the origin holds a laser pointing along the $+y$ axis. With respect to system S, a fast car moves at constant speed $c/2$ in the negative x-direction. A passenger in the car measures the velocity of the light beam arriving from the laser when system S observers see the car pass the positive y-axis. For system S', in which the car is at rest, what is the speed and direction of the light beam?

The relativistic velocity relations, §2–6

● **2–11.** The nucleus of a particular atom, initially at rest in the laboratory system, is unstable and disintegrates into two particles, as shown in Figure 2–13. What is the velocity of particle 1 with respect to an observer at rest with particle 2?

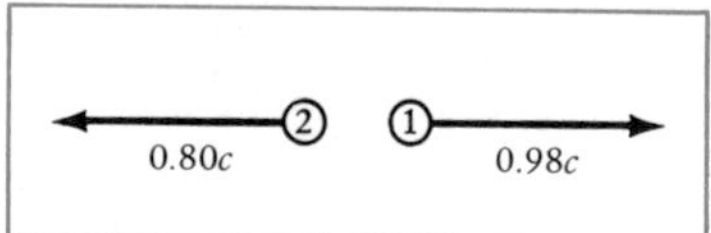

FIGURE 2–13.

2–12. Consider the three blocks labeled 1, 2, and 3 in Figure 2–14. Block 1 slides at speed $c/2$ relative to the ground; block 2 slides at speed $c/2$ relative to block 1; and block 3 slides at speed $c/2$ relative to block 2. Find the speed of (*a*) block 2 relative to the ground; (*b*) block 3 relative to the ground.

2–13. In a colliding-beam experiment at the Stanford Linear Accelerator, a high-energy electron beam collides head-on with a high-energy positron beam.† The positron, the electron's antiparticle, has the same mass as the electron but a positive instead of a negative electric charge (Section 4.5). In one experiment, the speed of the particles in each beam was $(1 - 10^{-8})c$, meaning less than c by only one part in 10^8. (*a*) Express the difference in meters per second between the speed of the particles and the speed of light. (*b*) From the point of view of an observer at rest with respect to the positrons, by what percentage did the speed of the electrons differ from c?

† See "Electron-Positron Annihilation and the New Particles," *Sci. Am.*, June 1975, p. 50.

2–14. In the laboratory two particles are observed to be emerging from a nuclear collision, each with speed $0.80c$ but one at an angle of 30° above the positive x-axis and the other 30° below the positive x-axis. (*a*) What is the velocity of the inertial system S' in which the two particles move outward along the $+y$- and $-y$-axes? (*b*) What is the speed of each particle relative to S'?

2–15. For the circumstances given in Problem 2–14, find (*a*) the velocity of the inertial system S'' and (*b*) the speeds of the two particles from the point of view of S'' observers if S'' observers are to see the two particles emerging at 30° above and below the *negative* x-axis.

Length and time intervals, §2–7

● **2–16.** An observer at rest in the laboratory system of Problem 2–11 (see Figure 2–13) finds that both outgoing particles are also unstable, particle 1 decaying after 6.6×10^{-6} s and particle 2 decaying after 6.0×10^{-6} s. What are the lifetimes of the two particles from a reference frame in which particle 2 is at rest?

● **2–17.** A meterstick is 100 cm long and 2 cm wide. At what velocity must the meterstick be moving so that its length measures the same as its width, 2 cm?

● **2–18.** A bomb, at rest in the earth frame, is set to explode in one hour. If the bomb is immediately put in motion at constant speed $0.98c$ relative to the earth frame, what distance does it travel before exploding?

● **2–19.** Because New York and Chicago are in different time zones, two o'clock in New York occurs an hour earlier than two o'clock in Chicago. The events are not simultaneous. Find the direction of motion (east or west) of a reference frame in which the events are more nearly simultaneous.

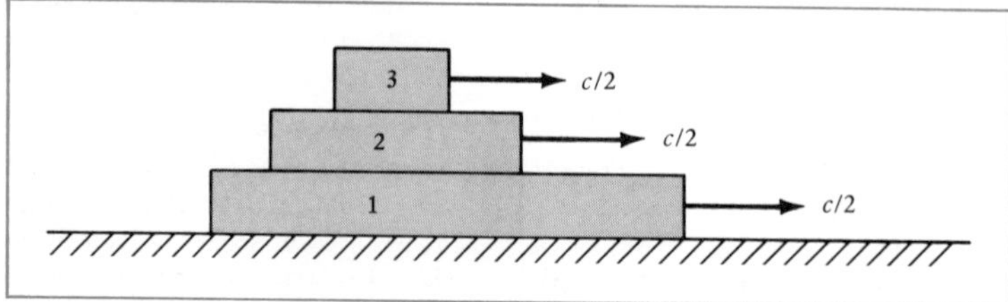

FIGURE 2–14.

● **2–20.** Observers at rest within a spacecraft find the time interval elapsing between the appearance of the spacecraft at point A fixed on Earth and later of point B fixed on Earth to be Δt. The spacecraft's speed relative to Earth is v. How long, relative to observers on Earth, does it take the spacecraft to travel from point A to point B?

● **2–21.** A car of rest length 5.0 m is traveling at 55 mi/hr = 25 m/s with respect to the ground. How much shorter is the car when ground observers measure it than when car observers measure it?

2–22. In the following, an alternative derivation of the time-dilation relation is given. Suppose that a pulse of light completes a round trip along the axis of a cylinder, as shown in Figure 2–5. (*a*) What is the time interval T_0, measured by an observer S' at rest with respect to the cylinder, taken by the pulse in completing the round trip. Express T_0 in terms of the cylinder's length l and the speed of light c? (*b*) Suppose that the cylinder is in motion at speed v, its axis at right angles to the direction of motion of an observer S, who clocks the round-trip time of the light pulse. The observer realizes that the speed of the light pulse must be c. Find the relation giving the round-trip time T in terms of l, v, and c. (*c*) Eliminate l in the two relations derived in parts (*a*) and (*b*) and find T in terms of T_0, v, and c.

2–23. Some unstable elementary particles decay with average lifetimes as short as 10^{-23} s (from the instant of the particle's creation to its decay into other particles). If a particle having a lifetime of 1.0×10^{-23} s in its own rest system is created in a bubble chamber and travels at 98 percent of the speed of light, how far (in angstroms) will it travel in the bubble chamber before decaying?

2–24. Two spacecraft pass one another moving in opposite directions at a relative speed of $0.98c$. An observer in spacecraft 1 measures the length of the other spacecraft as two-fifths the length of his own. (*a*) What is the relative rest length of craft 2 compared with the *rest* length of craft 1? (*b*) What will an observer in craft 2 determine as the length of moving craft 1; how does the measurement compare with his own craft's length?

2–25. What is wrong with the following argument? Meterstick B is in motion relative to meterstick A, so that the length of B is contracted relative to A. Meterstick A is also in motion relative to B, so that A is contracted relative to B. How can meterstick A be both longer and shorter than meterstick B? (*Hint:* What requirement must be met in properly measuring the length of a moving object? Are two spatially separated events simultaneous for all observers?)

2–26. In high-energy linear accelerators, electrons can be accelerated to speeds very near the speed of light. In a typical example, we consider a beam whose electrons have a final speed of $[1 - (2.45 \times 10^{-9})]c$. Assume that the electrons travel 100 m through an accelerator at this speed, relative to the laboratory system. (*a*) What is the length (in millimeters) of the accelerator tube to an observer at rest with respect to the electrons? (*b*) Observers at rest in the laboratory system watch a clock at rest with respect to the electrons. How long (in days) does it take for the minute hand of this clock to make one rotation, as observed in the laboratory system?

2–27. A car of rest length 20 ft moves at high speed through a garage of rest length 16 ft; the garage doors are open at both ends (see Figure 2–15). According to an observer S at rest with respect to the garage, there is one instant, say $t = 0$, in which the car is entirely inside the garage, the ends of the car coinciding simultaneously with the garage openings. (*a*) What is the speed of the car? (*b*) What is the length of the garage as viewed by the car's driver? (*c*) For the driver, will there ever be a time at which the car is entirely inside the garage?

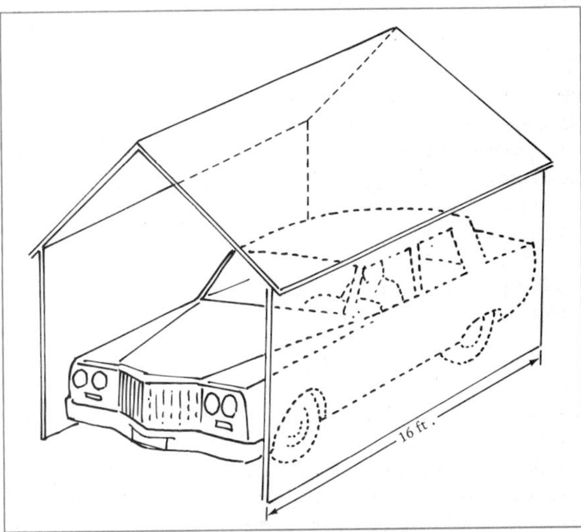

FIGURE 2–15. *Instantaneous position at time $t = 0$ of a 20-ft-long car moving through a 16-ft-long garage at speed v, as viewed by an observer S at rest with respect to the garage.*

2–28. Plans are made to send a spacecraft from Earth to a nearby star 10 light-years away and at rest with respect to Earth. The life-support systems within the spacecraft will last for a year. (*a*) What is the minimum speed of the spacecraft relative to the Earth-star system if the crew is to survive the trip? (*b*) If time were not dilated, what minimum speed would be necessary for the trip?

2–29. Although the speed of propagation of electromagnetic waves is c for all reference frames, the frequency and wavelength of monochromatic waves differ from one reference frame to another; that is, there is a *relativistic Doppler effect.* The disturbance y of a monochromatic wave traveling along the positive x-axis can be represented by the relation $y = y_0 \sin 2\pi(\nu t - x/\lambda)$, where ν and λ are the frequency and

wavelength measured by an observer S. The same wave, observed along the x'-axis in reference frame S', must have a similar form, namely, $y' = y_0 \sin 2\pi(\nu' t' - x'/\lambda')$. For electromagnetic waves $\nu\lambda = \nu'\lambda' = c$; and the disturbance y is perpendicular to the propagation direction.
(*a*) Use the Lorentz coordinate transformations in the equations for y and y' above to get the relativistic Doppler equation $\nu' = \nu\sqrt{[1 - (v/c)]/[1 + (v/c)]}$. (*b*) Show that the relativistic Doppler effect reduces to the classical Doppler effect, $\nu' = \nu[1 - (v/c)]$, when $v \ll c$.

2–30. Distant galaxies have been found that are moving away from Earth with speeds as high as $0.8000c$. If a star within a galaxy moving at this speed emits electromagnetic radiation of wavelength 6565 Å (red H_α-line of hydrogen) relative to an observer at rest on the emitting star, what would be the wavelength of this radiation relative to an Earth observer, and in what part of the electromagnetic spectrum would it fall? (See Problem 2–29.)

2–31. Monochromatic light passes through a single slit and produces a diffraction pattern on a distant screen, as shown in Figure 1–2. The light source, slit, and screen are all at rest with respect to a first observer, who marks the locations of intensity zeros on the screen. Suppose that a second observer is in motion at a very high speed along the line in which the light travels from the source to the slit. (*a*) Will this observer find the intensity zeros to come at the marks made by the first observer? (*b*) Is the distance between the slit and the observation screen changed? (*c*) A third observer travels upward along the sheet containing the slit at a very high speed. Does this observer find the intensity zeros at the locations marked by the first observer? (*d*) Is the slit width changed for the third observer?

■ **2–32.** See Problem 2–31. Suppose that the slit and screen are moved toward the source at a very high speed. Will an observer at rest with respect to the slit-screen system find the intensity zeros at the same marks they were at when the slit and screen were at rest relative to the source? (*Hint:* Observe the diffraction from the slit-screen system and use the Doppler equation of Problem 2–29.)

■ **2–33.** In system S, a straight rod is observed to be moving at a constant velocity $\mathbf{v}$ to the right, as shown in Figure 2–16. As measured in S, the rod's length is L and the rod makes an angle of 60° relative to the $+x$-axis. A camera located at point P in system S has its shutter opened for a time that is very short compared with the time the rod takes to pass the y-axis. (*a*) Because light from different points along the rod has to travel different distances to the camera, the light leaving the different points on the rod at the same instant of time does not arrive at the camera simultaneously. At what speed v must the rod be moving so that the camera would photograph the rod end-on, as though it were aligned along the y-axis? (*b*) Show that the exposure time of the camera must be shorter than $L \cos 60°/v$ for a sharp image to be produced on the film.

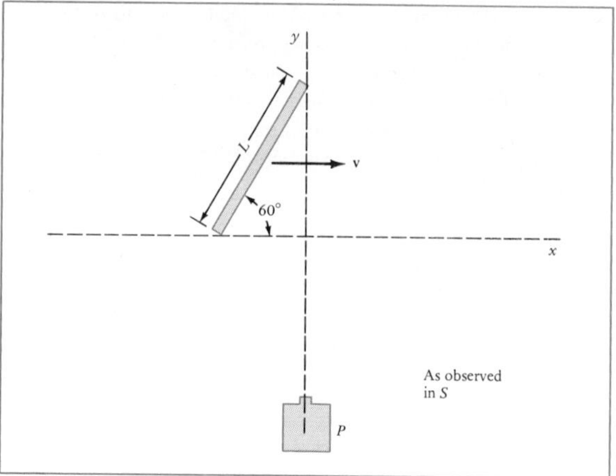

FIGURE 2–16.

■ **2–34.** A luminous cube (a die) has an edge length L_0 when at rest. A camera photographs the cube when the cube is traveling at a high speed v relative to it. The camera is far from the cube, which moves at right angles to the line joining cube and camera (see Figure 2–17*a*). We concentrate on the light that comes from cube edges A, B, and C and enters the camera during the very short time the shutter is open to record the photographic image. Light pulses that leave points A, B, and C simultaneously (in the reference frame of the camera) do *not* reach the camera simultaneously; light has a finite speed, and it travels farther from A than from B and C. Consequently, during the brief moment in which the camera shutter is open, light from the farther edge A enters the camera at the same time as light that has been emitted from edges B and C at a *later* time. The developed photograph appears as shown in Figure 2–17*b*; the rear face AB (with five dots) is visible, while the side face BC (with six dots) is shortened by space contraction. Show that a person examining such a photograph might infer that the cube was not moving with high speed at right angles to the camera but was at rest after being rotated through an angle θ, as shown in Figure 2–17*c*, where $\sin \theta = v/c$. This illustrates a general effect—a high-speed object moving transverse to the observation point is seen to be *rotated*.

■ **2–35.** According to observers in an inertial system S in which the earth is at rest, a galaxy has been moving away from the earth at constant speed $0.60c$. When the galaxy was 3.0×10^9 light-years from the earth, a star was born within the galaxy. The lifetime of the star according to an observer S' at rest relative to the star is 10×10^9 years. (*a*) What is the lifetime of the star relative to S? (*b*) According to S, how far from the earth is the star at the star's death? (*c*) Over what period would a single observer on the earth receive light from the star?

■ **2–36.** In an experimental test of the time-dilation phenomenon, the time interval registered by an atomic clock

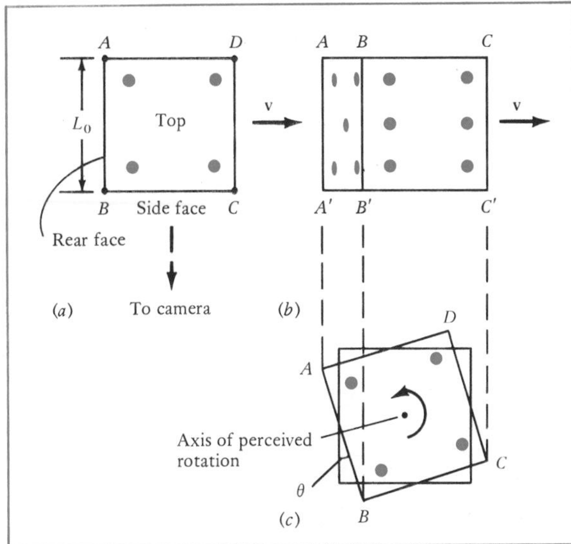

FIGURE 2–17. *A moving cube being photographed at right angles to its direction of motion: (a) view from above; (b) the photograph of the side of the cube; (c) an interpretation of (b) in which the cube has been rotated.*

carried on a jet plane flying on a coast-to-coast round trip was compared with the time interval registered on an atomic clock that remained at rest on the earth. Assume that the earth's surface is an inertial system and that the plane travels at 300 m/s relative to the surface. By how many seconds is the traveling clock behind the stationary clock on the earth after a round-trip flight of 8000 km?

The twin paradox, §2–8

2–37. It is estimated that the earth has been in orbit about the sun for 5×10^9 yr at the present orbital speed of 3×10^4 m/s. Over this period, how much time (in years) have clocks on the earth lagged behind clocks at rest relative to the inertial system of the sun?

■ **2–38.** Evidence for the twin paradox was obtained in 1972 when two scientists carried atomic clocks on jet flights around the world and compared them with an atomic clock at rest on the earth's surface. Assume that the flight was around the equator at a speed of 700 mi/hr = 312 m/s relative to the earth's surface and in the eastward direction. The radius of the earth is 6.37×10^6 m and the tangential speed of a point on the earth's surface at the equator is 463 m/s. It is assumed that the center of the earth is at rest in an inertial system S. (*a*) With respect to S, how long did it take for the plane to return to the stationary clock on the earth? (*b*) What would be the elapsed time on the clock that was stationary on the earth? (*c*) On the jet plane? (*d*) When the plane returns to the clock at rest on earth and the clocks are compared, by how much does the plane clock lag behind the surface clock? The experimental results agreed with relativity theory.

2–39. Triplets A, B, and C are born "simultaneously" on the earth. A stays at home, and B travels immediately after birth to a nearby star 100 light-years away at speed $0.8c$; C travels to another star 100 light-years away at speed $0.6c$. On reaching their respective destinations, B and C reverse directions and return to the earth at the same respective speeds. (*a*) What is A's age when B returns? (*b*) What is B's age when she meets A? (*c*) What is A's age when C returns? (*d*) What is C's age when he meets A? All space and time measurements must be made from a single inertial system, say that in which A is at rest.

Space-time events and the light cone, §2–9

2–40. In system S, the following three space-time events occur: Event 1 at (0, 0); Event 2 at (2 light-years, 1 year); and Event 3 at (2 light-years, 3 years). (*a*) Show these three events on a space-time graph for system S. (*b*) Find all inertial systems in which any two of the events are simultaneous and in these systems show the three events on a space-time diagram.

3
Relativistic Dynamics: Momentum and Energy

Hendrik A. Lorentz (1853, Netherlands—1928). Theory of electromagnetic radiation from atomic point of view. 1902, Nobel Prize (with Pieter Zeeman), theory of electromagnetic radiation. 1904, developed Lorentz coordinate transformations. (Photo, H. A. Lorentz, Collected Papers *(The Hague: Martinus Nihoff)—courtesy of the American Institute of Physics, Niels Bohr Library.)*

Relativistic kinematics, which deals with space intervals, time intervals, and velocities, has as its basis that all measurements of the speed of light give the same constant c. Special relativity requires, in addition, that the laws of physics be covariant for all inertial observers. Relativistic dynamics, which we study now, deals with the mechanics of particles moving at speeds up to c, and the appropriate relativistic forms for momentum and energy. We shall require that the basic laws of momentum and energy conservation be covariant under the Lorentz transformations.

The analogy between the invariant space-time four-vector and the momentum-energy four-vector is used to derive the transformation relations for energy and momentum components. One important application of special relativity to electromagnetism (apart from the constancy of the propagation speed c of electromagnetic radiation) involves the so-called magnetic force. This force, ordinarily taken as a distinctive type of electromagnetic interaction between charged particles in motion, originates from the strictly electric, or Coulomb, force.

3-1 Relativistic Momentum

By analogy with the classical relation for momentum, we take a particle's relativistic momentum $\mathbf{p}$ to be given by the relation

$$\mathbf{p} = \gamma m_0 \mathbf{v} \tag{3–1}$$

The subscript zero on the mass term m_0 in Equation 3–1 is there to remind us that the mass measurement is always made in a system in which the particle is at rest. The quantity γ is yet to be determined. The product γm_0 is sometimes referred to as the *relativistic mass* and written $m = \gamma m_0$. By its definition in Equation 3–1 it is that physical quantity by which the velocity $\mathbf{v}$ must be multiplied to yield the vector quantity $\mathbf{p}$ so that the total momentum $\Sigma\mathbf{p}$ of an isolated system is conserved.

We know that the *classical,* or Newtonian, law of momentum conservation is covariant under a *Galilean* transformation; the classical momentum of a particle is taken to be its velocity multiplied by an *invariant* mass only. For classical physics, $\gamma = 1$. Clearly, the classical relation for momentum *cannot* be covariant under a Lorentz transformation. Therefore, in imposing the requirements of relativity on the laws of dynamics for particles at high speed, we must be prepared to find that the quantity γ does not equal the invariant constant 1 but depends on the particle's speed.

In reexamining the law of momentum conservation for particles traveling at relativistic speeds, we insist that the two basic postulates of the special theory of relativity be satisfied. The physical laws must be covariant for all inertial frames; and the speed of light must be measured as the same constant by all observers. We know that whatever differences emerge between the relativistic and classical forms of momentum, the relativistic momentum $\gamma m_0 \mathbf{v}$ *must,* through the correspondence principle, reduce for low speeds to the familiar form $m_0\mathbf{v}$. Therefore, γ must approach 1 as v/c approaches zero.

Our strategy consists first in considering an elastic collision between two identical objects in a collision so completely symmetrical that we can be assured that the law of momentum conservation is valid. We then examine the same collision from the point of view of another inertial observer, with the requirement that the momentum-conservation law be covariant.

Suppose that we have two particles A and B with the same mass m_0. With respect to the laboratory system, the particles are projected at one another at equal speeds and collide elastically, as shown in Figure 3–1*a*. The velocities of the two particles are equal in magnitude but opposite in direction, both before and after the collision. Because of the symmetry of the collision, we can be sure that the collision is observed in the center-of-mass reference frame, which we designate S. In inertial frame S, the system's total momentum is zero at all times.

The collision appears even more symmetrical in this center-of-mass frame if we rotate the coordinate axes, as shown in Figure 3–1*b*, so that the velocities of both particles make the same angles with the x- and y-axes. In system S, the magnitude of the x-component of the velocity is the same for each particle, v_x. Similarly, the magnitude is the same for the y-components, v_y.

The collision looks different from other inertial frames. An observer in an inertial frame S', for example, who is moving to the right at speed v_x, views the collision as shown in Figure 3–1*c*. The spatial coordinates and velocity components of the two particles as viewed in S' follow directly from the Lorentz coordinate and velocity transformations (Equations 2–15 and 2–16). With $\dot{x}'_A = 0$, an observer in system S' sees particle A traveling back and forth along the y'-axis and B traveling obliquely before and after the collision. Likewise, an observer in a third system S'', which is moving to the *left* at speed v_x relative to system S, views the collision as shown in Figure 3–1*d*. Note that there is *no space contraction* at right angles to the relative motions of the inertial systems; therefore, the y-coordinates of each particle will be the same in all three systems.

We now use the postulates of relativity to find the functional form of γ in Equation 3–1 that relates the relativistic momentum $\mathbf{p}$ to the mass m_0 and velocity $\mathbf{v}$ of a particle. Simply on the basis of symmetry, we can assert for the center-of-mass frame S (Figure 3–1*b*) that both particles A and B were thrown *simultaneously,* and that they collided at the origin and were then caught *simultaneously* by the respective observers. In other words, the *time interval* T_{cm}*, as observed in the center-of-mass reference frame,* between the throwing and the catching of particle A is the *same* as the interval between the throwing and the catching of particle B.

Moreover, an observer in this center-of-mass reference frame can be certain that momentum is conserved in the collision; indeed, the total vector momentum of the system along any one direction must be zero. This implies that the change Δp_{yA} in the momentum component along the y direction for particle A before and after collision must equal in magnitude the change Δp_{yB} in the momentum component along y for particle B. Thus, the law of momentum conservation requires that

$$\Delta p_{yA} = \Delta p_{yB} \tag{3–2}$$

Particle A travels a distance $\frac{1}{2}y$ along y before collision, and another $\frac{1}{2}y$ after; all told, it travels a distance y in the time interval T, so that the y-component of A's velocity is y/T. The momentum change of A along y is twice the magnitude of the momentum component along y before or after collision. Therefore, using Equation 3–1, we have

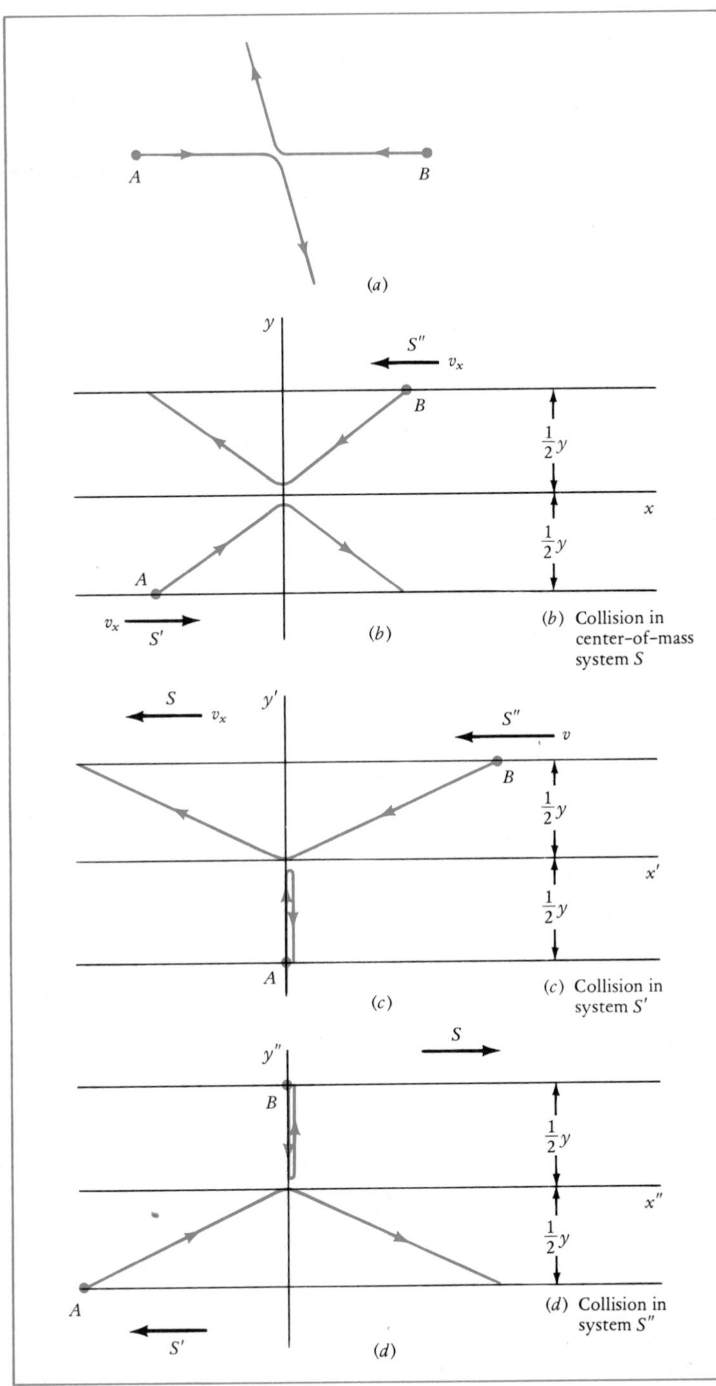

FIGURE 3–1. *Identical particles A and B colliding elastically at equal speeds: (a) viewed by an observer S in the center-of-mass reference frame; (b) same as (a), but with x- and y-axes so oriented that the collision is completely symmetrical for the observer in the center-of-mass reference frame; (c) viewed by an observer S′, who, relative to S, is moving along the x-axis at speed v_x; (d) viewed by an observer S″, who, relative to S, is moving along the −x-axis at speed v_x.*

$$\Delta p_{yA} = 2\gamma_A m_0 \left(\frac{y}{T}\right)$$

where $\gamma_A m_0$ is the relativistic mass of particle A. We also have

$$\Delta p_{yB} = 2\gamma_B m_0 \left(\frac{y}{T}\right)$$

where $\gamma_B m_0$ is the relativistic mass of particle B. Equation 3–2 then becomes, for an observer in the center-of-mass inertial frame,

$$\frac{\gamma_A m_0 y}{T} = \frac{\gamma_B m_0 y}{T}$$

Because of the symmetry of the collision, we know that the two particles have the same value for γ.

If relativistic momentum conservation is to hold for *all* inertial frames, then Equation 3–2 must also be satisfied for any other inertial frame moving at constant velocity relative to the system's center of mass. Two such frames are S' and S'' (see Figure 3–1). We will apply momentum conservation to the collision of the particles as viewed in system S'. For the y' components, Equation 3–2 gives

$$\Delta p'_{yA} = \Delta p'_{yB}$$

Recalling that in all three systems (S, S', and S'') the y-coordinate for an event is the same, and that the rest masses of the two particles are equal, we can write the magnitude of the momentum change as

$$\Delta p'_y = \gamma m_0 \left(\frac{\text{travel distance}}{\text{travel time}}\right)$$

as measured in S'. Therefore,

$$\Delta p'_{yA} = \frac{2(\gamma'_A m_0)y}{T'_A} \qquad \text{and} \qquad \Delta p'_{yB} = \frac{2(\gamma'_B m_0)y}{T'_B}$$

where γ'_A and γ'_B are the values of γ' (measured in S') for particles A and B, and T'_A and T'_B are the time intervals in S' between the releases and returns of particles A and B.

Equation 3–2 then becomes

$$\frac{\gamma'_A}{T'_A} = \frac{\gamma'_B}{T'_B} \tag{3–3}$$

Momentum conservation requires the ratio of γ'_A to γ'_B to equal the ratio of the time intervals, T'_A to T'_B.

It is easy to show by relativity theory that the time interval T'_A between the release and the return of particle A must differ from the interval T'_B between the release and return of particle B. The two events for particle A occur at the same location, $x'_A = 0$ and $y'_A = -y/2$. Therefore, both events, the release and the return of particle A, are measured by a clock at rest in S'; the time interval T'_A is then the proper time interval T_0.

The two events for particle B, however, do not occur at the same location in S'. Although they have the same y-coordinate, the release of B occurs to the right and the return of B to the left of the y-axis. To find the time interval T between these two events for particle B, recall that in system S'' (Figure 3–1d) the two events for particle B do occur at the same space location. Because of the symmetry of the collision, the proper time interval between the release and return of particle B in S'' is the same proper time interval T_0 as the interval between the release and the return of particle A in S'. Because S' is moving at speed $v_x = v$ (hereafter) relative to S, the interval T'_B between the two events for B will be longer when measured in S' and given by Equation 2–20:

$$T'_B = \frac{T_0}{\sqrt{1-(v/c)^2}}$$

Substituting for T'_A and T'_B in Equation 3–3 gives

$$\gamma'_B = \frac{\gamma'_A}{\sqrt{1-(v/c)^2}} \tag{3–4}$$

The time dilation effect can be regarded as a consequence of the Lorentz transformations. Therefore, Equation 3–4 must hold if momentum conservation is to be covariant under the Lorentz transformations. Clearly, this equation cannot be satisfied unless γ'_A and γ'_B are different. Recalling that the relativistic momentum $\mathbf{p} = \gamma m_0 \mathbf{v}$ must approach the classical $\mathbf{p} = m_0 \mathbf{v}$ for small v, we know that $\gamma \rightarrow 1$ as $v \rightarrow 0$.

In the collision shown in Figure 3–1c, *both* particles are moving. We want to find how γ depends on the particle's motion relative to any inertial system. To do this, consider a special collision for which particle A is at rest. The collision in Figure 3–1c approaches this situation if the y-components of the velocities of both A and B approach zero. Particle A is then at rest with $\gamma_A = 1$, and particle B approaches and then recedes along a straight line in a grazing collision, as shown in Figure 3–2. With $y_B = 0$, the speed of particle B becomes the same as the speed v of S' in Figure 3–1c. Equation 3–4 then reduces to the form

$$\gamma = \frac{1}{\sqrt{1-(v/c)^2}} \tag{3–5}$$

Figure 3–3 shows the functional dependence of γ on the speed of the particle. As we see from the figure, the value of γ does not deviate appreciably from its classical value 1 until the particle's speed becomes comparable to the speed of light. With $v = (c/10)$, for example, the value of γ exceeds 1 by only 0.5 percent.

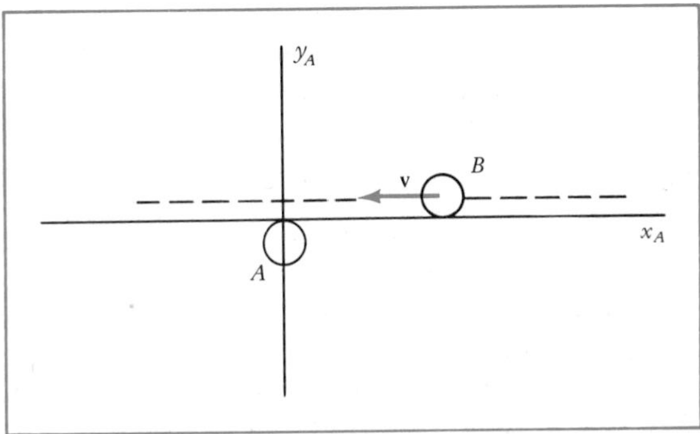

FIGURE 3–2. *Collision of moving particle B, with particle A at rest, as viewed by observer A; derived from Figure 3–1c in the limit of zero transverse velocity component.*

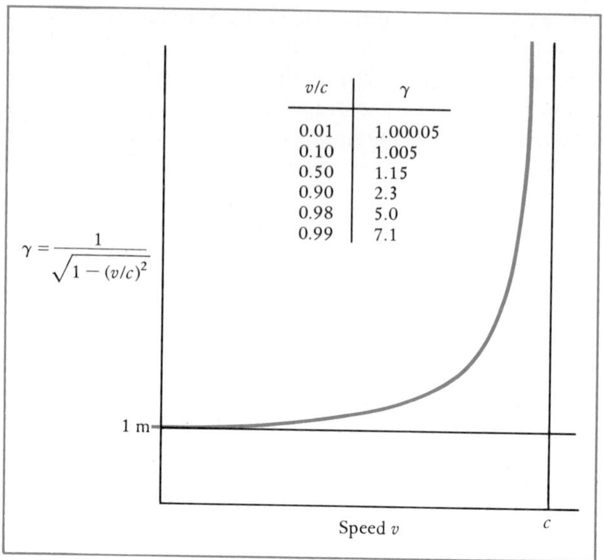

FIGURE 3–3. *Variation of γ with speed.*

Substituting Equation 3–5 into Equation 3–1 finally gives the relativistic form for a particle's momentum **p** in terms of its mass m_0 and velocity **v**:

$$\mathbf{p} = \gamma m_0 \mathbf{v} = \frac{m_0 \mathbf{v}}{\sqrt{1-(v/c)^2}} \tag{3–6}$$

Because γ increases with speed (see Figure 3–3), the relativistic momentum is always larger than the classical momentum $p_{\text{cl}} = m_0 v$; the difference between p and p_{cl} is negligibly small unless the particle speed approaches the speed of light. A graph of the magnitude of the relativistic momentum p versus speed v is shown in Figure 3–4. For comparison, the classical momentum is shown as a dotted line in Figure 3–4. Large differences between the two expressions occur when v approaches c. Classically, there is no upper limit to the speed of a particle and the momentum increases linearly with v for all speeds, no matter how large. Relativistic momentum, on the other hand, increases rapidly to infinity as the particle speed approaches the speed of light; beyond the speed of light the momentum would become imaginary, a physically impossible condition.

Experiments with particles traveling at speeds approaching that of light have shown conclusively that Equation 3–6 correctly gives a particle's momentum at all speeds up to c. For low-speed particles, with $v/c \ll 1$, the relativistic momentum, Equation 3–6, reduces to the classical form, $\mathbf{p} = m_0\mathbf{v}$.

Now that the expression for relativistic momentum has been found, it is natural to ask about the relativistic form of the laws of dynamics. Although relativistic dynamics is most easily treated in terms of relativistic energy and momentum, one important example can be solved by using Newton's second law written in relativistic form. Thus we should have

$$\mathbf{F} = \frac{d\mathbf{p}}{dt} = \frac{d}{dt}(\gamma m_0 \mathbf{v}) \tag{3–7}$$

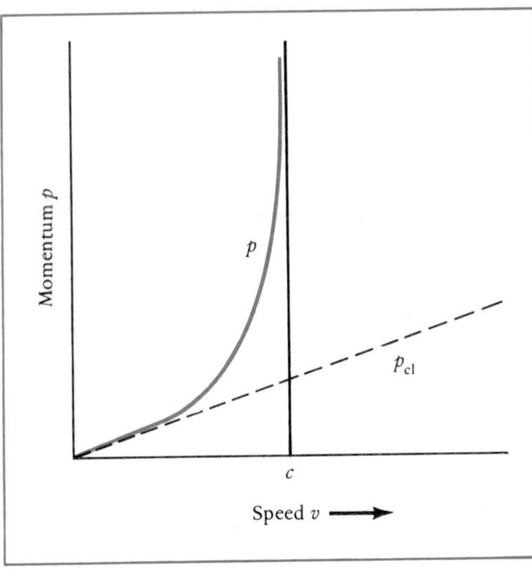

FIGURE 3–4. *Momentum versus speed for a particle. The solid curve is the relativistic momentum p; the dotted curve, the classical momentum* p_{cl}.

where force is the time rate of change of momentum, as it is in classical dynamics. With the mass m_0 a constant, but γ a function of the speed v according to Equation 3–5, Newton's law can also be written

$$\frac{\mathbf{F}}{m_0} = \gamma \frac{d\mathbf{v}}{dt} + \mathbf{v}\frac{d\gamma}{dt} \tag{3–8}$$

In classical mechanics, $\gamma = 1$ and $d\gamma/dt = 0$, so Equation 3–8 reduces to the familiar relation

$$\frac{\mathbf{F}}{m_0} = \frac{d\mathbf{v}}{dt} = \mathbf{a}$$

One very important situation, in which a particle's speed is constant while its velocity changes, concerns a particle traveling in a circular arc, under the influence of a force toward the center of the arc. Because the force is radial and at right angles to the particle's motion, the particle travels at constant speed but with an ever changing velocity direction.

Consider a particle of mass m_0 and electric charge Q moving with velocity $\mathbf{v}$ at right angles to a uniform magnetic field $\mathbf{B}$. The magnetic force, always perpendicular to the velocity, causes the particle to move in a circle at constant speed. The magnitude of the force is given by $F = QvB$. Because the particle moves with a constant speed, γ is constant (not necessarily equal to 1), and hence $d\gamma/dt = 0$. Equation 3–8 then becomes $\mathbf{F} = \gamma m_0 d\mathbf{v}/dt = \gamma m_0 \mathbf{a}$. The symbol $\mathbf{a}$ represents the radial acceleration of magnitude v^2/r, where r is the radius of the circular path. Combining the above relations, we have

$$F = \gamma m_0 \frac{dv}{dt}$$

$$QvB = \frac{\gamma m_0 v^2}{r}$$

Recalling that the relativistic momentum is given by $p = \gamma m_0 v$, we can rewrite this relation as

$$p = \gamma m_0 v = QBr \tag{3–9}$$

Except for the appearance of γ in the middle term, this equation has exactly the same form as the equation of classical mechanics.

Equation 3–9 is the basis of a simple method of determining the relativistic momentum of a charged particle. If Q is known, and B and r are measured, then p can be computed from the equation. Since a charged particle's speed v can be measured by using crossed electric and magnetic fields (Section 11–12), when the mass m_0 is known, γ can be computed from Equation 3–9. This was one of the first methods used to verify the variation of γ with speed (Figure 3–3) predicted by Einstein's relativity theory in 1905.

Example 3–1. An electron collides head on with a proton that is moving at $0.60c$. What must be the electron's initial speed if both particles are at rest after the collision? The proton's rest mass is 1836 times the mass of the electron.

The total momentum of the system remains constant. With both particles at rest after the collision, the momentum after collision is zero; therefore the total momentum before collision is zero. This requires that in magnitude

$$\text{Proton momentum} = \text{electron momentum}$$

Using Equation 3–6 for the relativistic momentum, we have

$$\frac{m_p v_p}{\sqrt{1 - (v_p/c)^2}} = \frac{m_e v_e}{\sqrt{1 - (v_e/c)^2}}$$

where m_p and v_p are the rest mass and speed of the proton before the collision; similarly, the subscript e refers to the electron. If we square both sides and use the given information, the above relation becomes

$$\frac{(1836)^2(0.60c)^2}{1 - 0.60^2} = \frac{v_e^2}{1 - (v_e/c)^2}$$

Solving for v_e, we have

$$v_e = c[1 + (5.3 \times 10^{-7})]^{-1/2} \approx [1 - (2.6 \times 10^{-7})]c$$

where the binomial expansion for $(1 + x)^{-1/2} \approx 1 - \frac{1}{2}x$, which is applicable for small x, has been used. Note that the incident speed of the electron is close to the speed of light but not equal to it. Contrast this with the prediction of classical dynamics; with the momentum magnitudes of the colliding particles related simply by $m_p v_p = m_e v_e$, it can be predicted that the electron speed will be $\frac{1836}{1} v_p$, or $1100c$! All experiments agree that a particle's momentum is given by Equation 3–6.

3–2 Relativistic Energy

We now have a relativistic expression for momentum, Equation 3–6, and one for Newton's second law of motion, Equation 3–7. We next ask, "What is the relativistic kinetic energy E_k?" To find this, we define kinetic energy, as in classical physics, as the total work done in bringing a particle from rest to the final speed v under a force F:

$$E_k = \int_0^s F\,ds = \int_0^s \frac{d}{dt}(\gamma m_0 v)\,ds$$

$$= m_0 \int_0^t \frac{d}{dt}(\gamma v) v\,dt = m_0 \int_0^{\gamma v} v\,d(\gamma v)$$

$$= m_0 \int_0^v (v^2\,d\gamma + \gamma v\,dv)$$

To integrate, we must recognize that both γ and v are variables, the dependence of one on the other being given by Equation 3–5.† We shall find it simpler here to express v in terms of γ and then integrate with respect to the variable γ. We can get the expressions for v and dv by rewriting Equation 3–5 in the form

$$\frac{v^2}{c^2} = 1 - \frac{1}{\gamma^2} \tag{3–10}$$

Differentiating, we have

$$2v\frac{dv}{c^2} = 2\frac{d\gamma}{\gamma^3}$$

or

$$\gamma v\,dv = \frac{c^2}{\gamma^2}\,d\gamma$$

By Equation 3–10, this relation becomes

$$\gamma v\,dv = (c^2 - v^2)\,d\gamma$$

Substituting for $\gamma v\,dv$ in the equation for the kinetic energy E_k above, we then have

$$E_k = m_0 \int_1^\gamma [v^2\,d\gamma + (c^2 - v^2)\,d\gamma] = m_0 c^2 \int_1^\gamma d\gamma$$

or

$$E_k = (\gamma - 1) m_0 c^2 = m_0 c^2 \left[\left(\frac{1}{\sqrt{1 - (v/c)^2}}\right) - 1\right] \tag{3–11}$$

It is obvious that the relativistic kinetic energy E_k is markedly different from the classical form $E_{k,cl} = \frac{1}{2} m_0 v^2$. Graphs of both E_k (solid curve) and $E_{k,cl}$ (dotted curve) are shown in Figure 3–5.

† For low speeds, $\gamma = 1$, and $d\gamma/dt = 0$, so that the integral reduces to

$$E_k = m_0 \int (v^2\,d\gamma + \gamma v\,dv) = m_0 \int_0^v v\,dv = \tfrac{1}{2} m_0 v^2$$

the classical relation for kinetic energy.

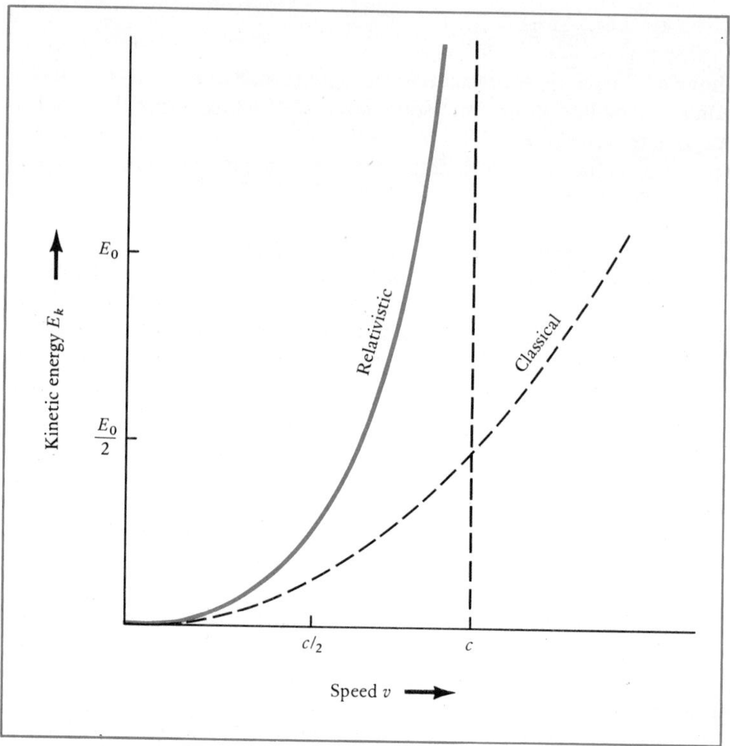

FIGURE 3–5. *Kinetic energy versus speed of a particle.*

Whereas the classical kinetic energy is proportional to v^2, the relativistic kinetic energy, since it is linear in γ, varies with speed as γ varies with v (see Figure 3–3).

For particle speeds comparable to c, the two curves in Figure 3–5 differ drastically, with E_k approaching infinity as $v \rightarrow c$. For speeds far less than c, there is little difference in the relativistic and classical values. This is as it must be; the correspondence principle requires that E_k reduce to $E_{k,cl}$ for $v \ll c$. This can be proved analytically by using Equation 3–5 in 3–11 to find E_k in terms of v^2. First, we solve for $(v/c)^2$ in Equation 3–5:

$$\gamma = \frac{1}{\sqrt{1 - (v/c)^2}} \tag{3–5}$$

Therefore,

$$\left(\frac{v}{c}\right)^2 = 1 - \frac{1}{\gamma^2} = \frac{\gamma^2 - 1}{\gamma^2} = \frac{(\gamma - 1)(\gamma + 1)}{\gamma^2}$$

Substituting for $(\gamma - 1)$ in Equation 3–11

$$E_k = (\gamma - 1)m_0c^2 \tag{3–11}$$

we have

$$E_k = \frac{\gamma^2}{(\gamma + 1)} m_0v^2$$

This expression for the relativistic kinetic energy is like the classical expression, $E_{k,cl} = \frac{1}{2}m_0v^2$. For low speeds, $v \ll c$ and $\gamma \rightarrow 1$, and Equation 3–11 becomes

$$\lim_{(v/c)\to 0} E_k = \lim_{\gamma\to 1} \frac{\gamma^2}{(\gamma + 1)} m_0 v^2 = \frac{1}{2} m v_0^2$$

In the extreme relativistic region, on the other hand, $v \approx c$, $\gamma \gg 1$, and Equation 3–11 reduces to

$$\lim_{v/c\to 1} E_k = \lim_{\gamma\to\infty} (\gamma - 1)m_0c^2 = \gamma m_0 c^2 = E$$

In this region, the kinetic energy E_k approaches the total energy E.

The relativistic relation for kinetic energy (Equation 3–11) suggests a new interpretation of mass and energy, physical properties that are classically considered separate and distinct. On the right side of Equation 3–11 appears the energy term m_0c^2, the product of the rest mass of the particle and the square of the fundamental constant c. Corresponding to every rest mass m_0 is an energy m_0c^2; it is natural to interpret this energy as the *rest energy* E_0 of the particle

$$E_0 = m_0c^2 \tag{3–12}$$

Equation 3–12 is an example of the famous Einstein relation implying the equivalence of mass and energy. As we shall see, rest mass can be regarded merely as another form of energy, and the separate laws of energy and mass conservation in classical physics become a single conservation law in relativity physics—the conservation of mass-energy. This combined conservation law holds in any inertial system, whereas the separate laws of energy and mass do not (Section 3–4).

Using Equation 3–12 in Equation 3–11, we can rewrite Equation 3–11 in energy relations:

$$E_0 + E_k = \gamma E_0 \tag{3–13}$$

On the left side of Equation 3–13 appear the particle's inherent energy E_0 (rest energy) and its energy arising from motion E_k (kinetic energy); this sum must yield the particle's total energy, symbolized E. Thus, the total energy E of a particle of rest mass m_0 moving at speed v relative to an inertial system can be expressed in the following equivalent forms:

$$E = E_0 + E_k = \gamma E_0 = \gamma m_0 c^2 = mc^2 \tag{3–14}$$

where $m = \gamma m_0$ is the so-called relativistic mass and $\gamma = 1/\sqrt{1 - (v/c)^2}$.

For a system of particles (say a ball composed of moving electrons and nuclei) the system's rest energy E_0 and rest mass m_0 are the total energy and total mass when the center of mass of the system is at rest. If the center of mass is moving, the total energy of the system is its rest energy E_0 plus the kinetic energy of its center of mass.

In physics, momentum is generally more fundamental than velocity; for example, we have momentum conservation but not velocity conservation. Therefore, it is convenient to express the total energy E in terms of p rather than v or γ. We can eliminate v as follows. Squaring Equation 3–1 and multiplying both sides by c^2, we have

$$p^2c^2 = \gamma^2 m_0^2 v^2 c^2 = \gamma^2 (m_0c^2)^2 \left(\frac{v}{c}\right)^2$$

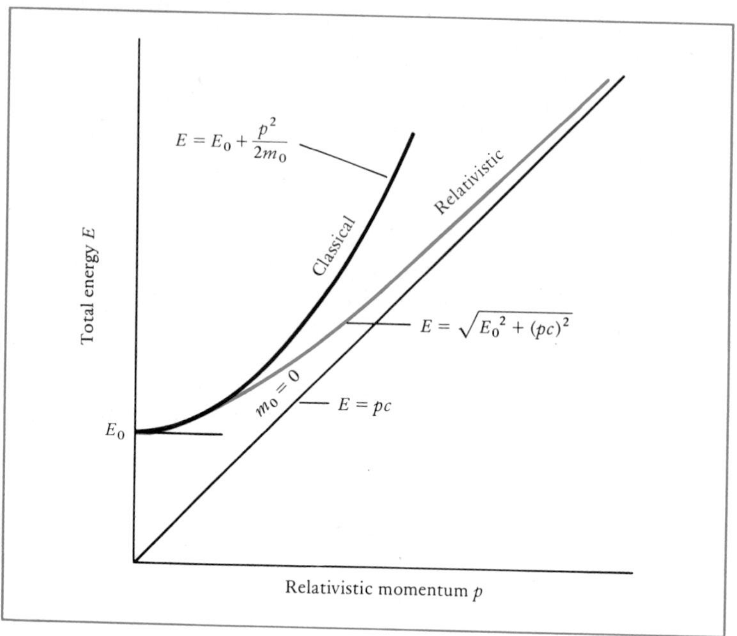

FIGURE 3–6. *Total relativistic energy versus relativistic momentum of a particle.*

Using Equation 3–12 for m_0c^2 and Equation 3–10 for $(v/c)^2$, we have

$$p^2c^2 = \gamma^2 E_0{}^2 - E_0{}^2$$

Finally, by Equation 3–14, $E = \gamma E_0$, the relation becomes

$$E^2 = E_0{}^2 + (pc)^2 \tag{3–15}$$

The particle's total energy $E = E_k + E_0$ is now related to its momentum p and rest energy E_0.

Figure 3–6 shows a particle's relativistic total energy E plotted as a function of its relativistic momentum p, following Equation 3–15. Also shown is the classical total energy, with $E_{k,cl} = \frac{1}{2}m_0v^2 = p^2/2m_0$. Note that for large magnitudes of momentum or energy the classical values and the relativistic values differ drastically.

It is instructive to examine the relativistic equations under two limiting conditions, very low speeds and very high speeds. First, here are the equations applicable at all speeds:

$$\text{Quantity:} \quad \gamma = \frac{1}{\sqrt{1-(v/c)^2}} \tag{3–5}$$

$$\text{Momentum magnitude:} \quad p = \gamma m_0 v \tag{3–1}$$

$$\text{Total energy:} \quad E = \gamma E_0 \tag{3–14}$$

$$= \sqrt{E_0{}^2 + (pc)^2} \tag{3–15}$$

$$\text{Kinetic energy:} \quad E_k = (\gamma - 1)E_0 \tag{3–11}$$

CLASSICAL REGION ($v \ll c$). For the classical region, Newtonian mechanics adequately describes the behavior of particles, and the relativistic quantities reduce to their familiar classical forms:

$$\gamma \approx 1$$

$$p \approx m_0 v$$

$$E_k \approx \frac{1}{2} m_0 v^2 = \frac{p^2}{2m_0}$$

$$E \approx E_0 + \frac{p^2}{2m_0}$$

At low speeds, the kinetic energy varies with the *square* of the momentum, and the kinetic energy is much less than the rest energy because

$$\frac{E_k}{E_0} = \frac{(1/2)m_0 v^2}{m_0 c^2} = \frac{1}{2}\left(\frac{v}{c}\right)^2 \ll 1$$

EXTREME RELATIVISTIC REGION ($v \approx c$). Newtonian mechanics is inapplicable, and the relativistic equations become

$$\gamma \gg 1$$

$$pc = \gamma m_0 v c \approx \gamma m_0 c^2 \gg E_0$$

$$E \approx pc$$

$$E_k \approx E$$

If a particle were to have nonzero energy E and momentum p but zero *rest* energy E_0 (and zero rest mass m_0)—a possibility that makes no sense from a classical point of view—then the last two equations above would be *exactly true.* To find the speed of such a particle, we should substitute Equation 3–5 in Equation 3–14 and rearrange terms to get $E\sqrt{1-(v/c)^2} = E_0$. From this relation, we see that for E to be nonzero even though $E_0 = 0$, the particle speed v must be c. Thus, for a particle of zero rest energy,

$$E_0 = 0 \qquad E = pc \qquad E_k = E \qquad v = c$$

As we shall see, real particles with just these properties do exist.

Suppose that a particle has some finite energy E and travels with the speed of light. Then its rest mass must be zero, since

$$E\sqrt{1-(v/c)^2} = E_0$$

$$E\sqrt{1-1} = 0 = E_0$$

A particle of zero rest mass, always traveling at speed c can have any energy E and any momentum $p = E/c$. A particle of nonzero rest mass, however, can approach speed c only as its total energy approaches infinity.

3–3 Equivalence of Mass and Energy, and Bound Systems

By examining *unbound systems* and *bound systems,* we can see how the equivalence of mass and energy and the conservation of mass-energy are significant.

UNBOUND SYSTEMS. Consider a collision between two particles, each having a rest mass m_0 and rest energy $E_0 = m_0c^2$. These particles are projected toward one another, each with speed v relative to an observer at rest in the center-of-mass reference frame. We assume the collision to be perfectly *inelastic;* thus, the two particles stick together to form a single particle, whose rest mass is designated M_0. We ask, How is the final rest mass M_0 related to the rest mass m_0 of each of the incident particles? Although classical physics implies merely that M_0 equals $2m_0$, we shall find that relativistically this is not so.

By linear momentum conservation, we know that the total momentum must always be zero. The total momentum is initially zero, since the particles before the collision have equal but opposite momenta; therefore after the collision, the amalgamated particles must be at rest.

The total energy of the two particles before the collision is $2E = 2(\gamma E_0) = 2\gamma m_0 c^2$, where $E = \gamma m_0 c^2$ is the rest energy *plus* the kinetic energy of each particle. By energy conservation, the total energy of the two particles before the collision must equal the total energy of the composite after the collision.

Since the composite object is at rest after the collision (center of mass at rest), the system's total energy after collision consists entirely of rest energy $M_0 c^2$. Total mass-energy conservation then requires that

$$E \text{ (before collision)} = E \text{ (after collision)}$$

$$2\gamma m_0 c^2 = M_0 c^2$$

or

$$M_0 = \frac{2m_0}{\sqrt{1 - v^2/c^2}}$$

The rest mass M_0 of the composite object after collision *exceeds* the rest mass $2m_0$ of the incident particles. The incoming kinetic energy of the two particles has been transformed into rest energy of the combined particles after the collision.

In this example, a collision was observed from a reference frame in which the center of mass was at rest. More generally, an observer in *any* inertial system finds that the total relativistic momentum is conserved and that the total mass-energy is conserved (see Section 3–4).

Example 3–2. Two satellites, each of rest mass 4000 kg and traveling at a speed of 8.0 km/s with respect to an earth observer, approach one another, collide, and stick together. Find the increase in the total rest mass of the system.

Because the satellites have equal but opposite momenta, their total momentum is zero; therefore after the collision, the composite object is at rest. Since total energy is conserved, all the kinetic energy of the incident satellites is converted into rest mass, and

$$\text{Increase in rest mass} = \Delta m = \frac{2E_k}{c^2}$$

where E_k is the initial kinetic energy of each satellite. The speed of each satellite, 8.0 km/s, is much lower than the speed of light, and we can use the classical expression for the kinetic energy, $E_k = \frac{1}{2}m_0 v^2$. Therefore,

$$\Delta m = \frac{2E_k}{c^2} = \frac{2(\frac{1}{2}m_0 v^2)}{c^2} = m_0\left(\frac{v}{c}\right)^2$$

$$= (4000 \text{ kg})\left(\frac{8.0 \times 10^3}{3.0 \times 10^8}\right) \approx 0.003 \text{ g}$$

Because of the low speeds of the incoming satellites (compared with c) the rest mass of the composite object after collision is only 3 mg larger than the rest masses of the satellites before collision (8000 kg)—hardly discernible.

Example 3–3. As in Example 3–2, two objects of equal rest masses m_0 move toward one another at equal and very high speeds. They collide head-on in a perfectly inelastic collision and come to rest. At what speed must each object be moving initially if the rest mass of the composite system after the collision is to be twice the sum of the rest masses of the two objects before collision (in other words, if there is to be a doubling of the system's rest mass)?

Designating the kinetic energy, total energy, and rest energy of each object before collision as E_k, E, and E_0, we have

$$E(\text{before collision}) = E(\text{after collision})$$

The total energy before collision is twice the rest energy plus kinetic energy for one object. After collision, there is rest energy only (the system is at rest); this is $2(2E_0) = 4E_0$, or four times the rest energy of one object before collision. Therefore,

$$2(E_0 + E_k) = 4E_0$$

or

$$E_k = E_0$$

The kinetic energy of each incoming object must equal its rest energy. Obviously we must use the relativistic relation for $E_k (E_k = E - E_0)$. With $E_k = E_0$ here, we then have

$$E - E_0 = E_0$$

$$E = 2E_0$$

and by Equation 3–14,

$$E = \frac{E_0}{\sqrt{1 - (v/c)^2}} = 2E_0$$

Solving for v in this equation, we find

$$v = 0.866c$$

BOUND SYSTEMS. One of the most important consequences of Einstein's relativity theory arises when particles are bound by some attractive force to form a single system. Examples are the sun and planets bound by gravitational forces to form the solar system, and electrons and nuclei bound by electrical forces to form a molecule. Here we consider the binding together of just *two* particles. Particles A and B are bound through an attractive force to form a single system. Once particles are bound, energy is needed to separate the system into its parts because of the attractive force. This breaking up of the bound system is shown symbolically in Figure 3–7, where E_b is the energy that must be added to the system to separate the particles

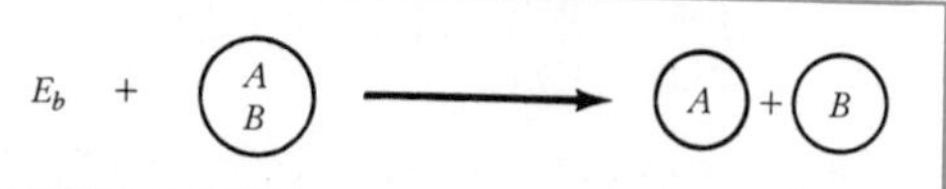

FIGURE 3–7. *Symbolic representation of the splitting of two bound particles.*

A and B completely with A and B both at rest. We apply the conservation of total relativistic energy to the situation of Figure 3–7. It is assumed that observations are made in the center-of-mass reference frame; the composite of A and B is at rest. Letting the rest mass of the composite system be M_0 and the rest masses of the separated individual particles be m_{0_A} and m_{0_B}, we have

$$M_0c^2 + E_b = m_{0_A}c^2 + m_{0_B}c^2 \tag{3–16}$$

The separated particles A and B have no kinetic energy because E_b is the minimum energy necessary for separating the particles from the attractive force. For a bound system, $E_b > 0$; and Equation 3–16 requires $M_0 < m_{0_A} + m_{0_B}$. That is, the rest mass of the bound system must be smaller than the sum of the rest masses of the individual particles when separated. This contrasts with unbound systems, in which the rest mass of the composite *exceeds* the rest masses of the separated particles (see Examples 3–2 and 3–3). In principle it is possible, using Equation 3–16, to calculate the binding energy E_b if we know merely the rest mass of the bound system and the rest masses of its constituents. We shall find, however, that only for nuclear forces is the binding energy between particles great enough to produce a measurable mass difference.

The total relativistic energies for the bound system and the separated system are shown on the left in Figure 3–8. It is often simpler to use what might be called the total mechanical energy E_m (the total kinetic energy and potential energy, with rest energy excluded) as the scale for energy instead of the total relativistic scale E. We are free to choose the zero of E_m at any value; for particles interacting with one another through a force that approaches zero as the separation distance approaches infinity, the most convenient zero value of E_m corresponds to the value for the particles infinitely separated and at rest. As shown on the right side of Figure 3–8, when particles are *bound* to one another, the mechanical energy E_m is negative (here, it is $-E_b$), since energy must be added to this bound system to separate the particles completely and bring the energy of the system up to zero.

Example 3–4. The binding energy of an electron and a proton together, as they form a stable hydrogen atom, is known experimentally to be 13.6 eV. This energy is also called the ionization energy, since it is the energy that must be added to a hydrogen atom to separate it into two oppositely charged particles. The total mechanical energy of a hydrogen atom is -13.6 eV. By Equation 3–17, it is then possible to compute the difference between the mass of the hydrogen atom, $M_{0_H} = 1.67 \times 10^{-27}$ kg, and the sum of the rest masses of the separated electron, m_{0_e}, and proton m_{0_p}:

$$m_{0_e} + m_{0_p} - M_{0_H} = \frac{E_b}{c^2} = \frac{13.6 \text{ eV}}{c^2}$$

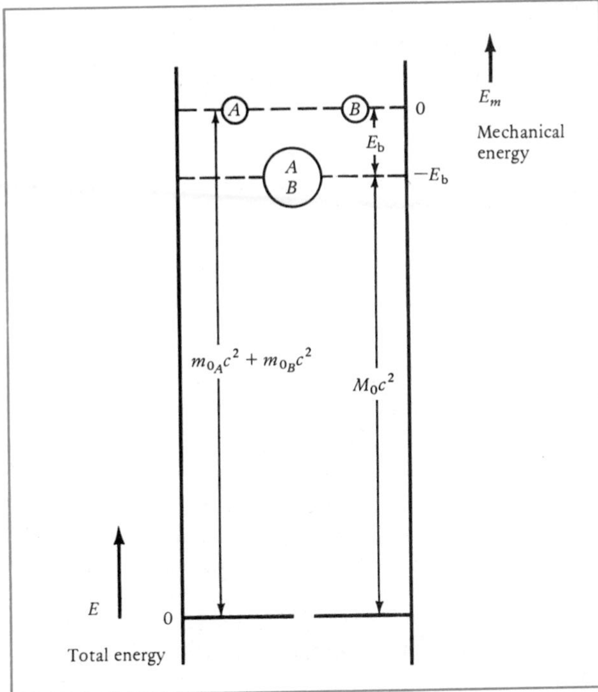

FIGURE 3–8. *Energy-level diagram showing the energy of two particles when bound together and when completely separated from one another and at rest. The total relativistic energy is plotted on the left, the mechanical energy on the right.*

The fractional change in the mass is

$$\frac{E_b/c^2}{M_0} = \frac{(13.6\ \text{eV})(1.60 \times 10^{-19}\ \text{J/eV})}{(1.67 \times 10^{-27}\ \text{kg})(3.00 \times 10^8\ \text{m/s})^2}$$

$$\frac{E_b}{M_0c^2} = 1.53 \times 10^{-8}$$

This fractional mass difference, slightly more than one part in 100 million, is much smaller than the experimental fractional error in the measurement of the masses of the hydrogen atom, proton, and electron. Therefore, in a reaction in which the binding energy is *several electron volts*—and all *chemical* reactions are of this order of magnitude—it is impossible to detect directly a change in the total mass of the system. A change in mass can, however, be detected in *nuclear* reactions, in which the binding energy typically is *several million electron volts.*

3-4 Momentum-Energy Four-Vector

Special relativity and the Lorentz coordinate transformations were constructed on this fundamental observation: The speed of light is invariant. Two other scalar quantities are invariants in relativity theory: the magnitude of the space-time four-vector and the magnitude of the momentum-energy four-vector.

SPACE-TIME FOUR-VECTOR. Consider the space and time intervals between any two events. According to observers in inertial system S, event 1 occurs at (x_1, y_1, z_1, t_1) and event 2 at (x_2, y_2, z_2, t_2). The space interval between the events can be represented by the displacement vector whose tail is located at (x_1, y_1, z_1) and whose head is at (x_2, y_2, z_2). If we designate the x-components of this displacement vector $\Delta x = x_2 - x_1$ (similarly for the y- and z-components), the length ΔL of the vector is

$$\Delta L = \sqrt{(\Delta x)^2 + (\Delta y)^2 + (\Delta z)^2} \tag{3–17}$$

The time interval between the events is

$$\Delta t = t_2 - t_1 \tag{3–18}$$

When the space and time intervals between these two events are viewed from any other inertial system, say S', they will be, similarly,

$$\Delta L' = \sqrt{(\Delta x')^2 + (\Delta y')^2 + (\Delta z')^2} \tag{3–19}$$

$$\Delta t' = t_2' - t_1' \tag{3–20}$$

What comparison can be made between the space intervals and the time intervals as they are viewed in the two systems S and S'? As usual, we have S' moving along the positive x-axis relative to S (see Figure 2–1) so that the transformation equations (Equations 2–1 or 2–15) apply.

Suppose first that the relative speed v between S and S' is much lower than c. The Galilean transformations suffice; and using Equation 2–1 in Equations 3–19 and 3–20, we immediately find that

$$\Delta L' = \Delta L \qquad \text{and} \qquad \Delta t' = \Delta t \tag{3–21}$$

For Euclidean space, the spatial length between two events is the same (invariant) for any inertial system; likewise, the time interval between the events is invariant. The speed of light is *not* invariant, however.

Now, consider the more general situation, in which v may be comparable to the speed of light. To guarantee the invariance of c, we must apply the Lorentz transformations, Equations 2–15. When these are used in Equations 3–19 and 3–20, one finds that no longer are space intervals and time intervals separately invariant. As proved in Section 2–8, there is instead a single space-time interval that is now the invariant quantity:

$$\begin{aligned}\Delta S &= \sqrt{c^2(\Delta t)^2 - [(\Delta x)^2 + (\Delta y)^2 + (\Delta z)^2]} \\ &= \sqrt{c^2(\Delta t')^2 - [(\Delta x')^2 + (\Delta y')^2 + (\Delta z')^2]}\end{aligned} \tag{2–21)(3–22}$$

The space-time interval ΔS between any two events is an invariant quantity; it has the same value in all inertial systems.†

Equations 3–19 and 3–22 are closely analogous, with an additional term c^2t^2, however, appearing in the relativistic space-time invariant. This suggests that we conceive of a relativistic space-time vector, a vector in four-

† It is arbitrary whether one chooses the space-time interval to be

$$\sqrt{c^2(\Delta t)^2 - [(\Delta x)^2 + (\Delta y)^2 + (\Delta z)^2]}$$

or the other way around,

$$\sqrt{(\Delta x)^2 + (\Delta y)^2 + (\Delta z)^2 - c^2(\Delta t)^2}$$

Both conventions are used; we shall always use the interval as defined in Equation 3–22.

dimensional space having components ix, iy, iz, and ct, where $i = \sqrt{-1}$. Squaring each term of this space-time *four-vector* and adding the squares gives $(\Delta S)^2$. The four-vector concept is a convenient mathematical model to use in relating space intervals and time intervals observed in different inertial systems. This does not imply, however, that we live in a four-dimensional world, with three imaginary spatial components and one time component. Of course, in any single system one measures separate space intervals (three-dimensional vector) and time intervals. But when observers in two different inertial systems want to compare space intervals and time intervals between any two events, then the four-vector model is useful.

MOMENTUM-ENERGY FOUR-VECTOR. Now consider the basic relation of relativistic dynamics, relating a particle's total energy E and momentum p to its rest energy E_0:

$$E^2 = E_0^2 + (pc)^2 \tag{3–15}$$

Although this equation was derived for motion along one direction only, it holds for any motion. In three-dimensional space, p represents the magnitude of the total momentum of the particle and has the rectangular components p_x, p_y, and p_z, where $p^2 = p_x^2 + p_y^2 + p_z^2$. Thus we can rewrite the momentum-energy relation (Equation 3–15) as

$$\left(\frac{E_0}{c}\right)^2 = \left(\frac{E}{c}\right)^2 - p^2 = \left(\frac{E}{c}\right)^2 - (p_x^2 + p_y^2 + p_z^2) \tag{3–23}$$

The left side of this equation depends only on the particle's rest mass, since $(E_0/c)^2 = (m_0c)^2$. The *rest* mass m_0 and *rest* energy E_0 of any one particle are always the same. Put differently, a particle's rest energy E_0 is invariant regardless of the inertial frame relative to which its momentum and energy are measured. Thus Equation 3–23, with an invariant quantity $(E_0/c)^2$ on its left side, is of the same form as Equation 3–22. A particle's relativistic momentum components and energy are related to an invariant in exactly the same way as an event's space and time coordinates are related to an invariant space-time interval.

Just as in relativistic kinematics we treat the three space coordinates and one time coordinate of an event as components of a four-dimensional space-time interval, so too in relativistic dynamics we regard the three components of momentum and the one value of energy in any one inertial frame as combining to yield a *momentum-energy four-vector.* The components of this four-vector in the inertial frame S are ip_x, ip_y, ip_z, and E/c, of which the first three give the particle's momentum components along the x-, y-, and z-axes, and E is the total energy of the particle in this inertial frame. Squaring the four components of the momentum-energy four-vector and adding yields Equation 3–23. Indeed, the fundamental result of relativistic dynamics is that the momentum-energy four-vector in *any* inertial frame yields the same invariant quantity $(E_0/c)^2$:

$$\text{Momentum-energy invariant} = \left(\frac{E_0}{c}\right)^2 = \left(\frac{E}{c}\right)^2 - (p_x^2 + p_y^2 + p_z^2)$$

$$= \left(\frac{E'}{c}\right)^2 - (p_x'^2 + p_y'^2 + p_z'^2) \tag{3–24}$$

Comparing the momentum-energy four-vector $(ip_x, ip_y, ip_z, E/c)$ with the space-time four-vector (ix, iy, iz, ct), we see that we can construct one from the other by using the replacements

$$\begin{aligned} &ip_x \quad \text{for } ix \\ &ip_y \quad \text{for } iy \\ &ip_z \quad \text{for } iz \\ &\frac{E}{c} \quad \text{for } ct \qquad \text{or} \qquad \frac{E}{c^2} \quad \text{for } t \end{aligned} \tag{3–25}$$

The Lorentz coordinate transformations let us compute the space and time coordinates of an event in system S' given the corresponding space and time coordinates of the event in S, and the converse. We can construct corresponding transformation relations that give the momentum components and the total energy of a particle in inertial frame S' in momentum components and energy of another inertial frame S. Using Equations 3–25 in the Lorentz coordinate transformations, Equations 2–15, we find

$$\begin{aligned} p'_x &= \frac{p_x - v(E/c^2)}{\sqrt{1-(v/c)^2}} \\ p'_y &= p_y \\ p'_z &= p_z \\ E' &= \frac{E - vp_x}{\sqrt{1-(v/c)^2}} \end{aligned} \tag{3–26}$$

We see from the first equation that the x-component of a particle's momentum as observed in S' depends not only on the corresponding momentum in S and the speed v of S' relative to S but also on the particle's total energy E. Similarly, the last equation shows that the total energy in S' depends on the total energy in S, on the relative speed v between the inertial systems, and also on the momentum along x in S.

The inverse transformations again result from interchanging primes and unprimes and replacing v with $-v$. Note that the velocity v appearing in Equations 3–26 is that of *inertial frame S' relative to that of inertial frame S*. The *velocity of a particle* that is being observed in S and S' is *not* the same as the v appearing in Equations 3–26. The particle's velocity is given by the particle's relativistic momentum divided by γm_0; in S, for example, the *particle's* velocity is given by

$$\mathbf{v} = \frac{\mathbf{p}}{\gamma m_0} = \frac{\mathbf{p}}{E/c^2} = \frac{\mathbf{p}c^2}{E} \tag{3–27}$$

To see how the momentum-energy transformation relations work, consider the following example.

Example 3–5. A high-energy particle, say A, of rest mass m_0 and moving at speed $v_A = 0.80c$ relative to the laboratory system S, collides with a target particle B, initially at rest and having rest mass $2m_0$.

(*a*) What is the total energy E of the particles in the lab system S?
(*b*) Find the velocity of the center-of-mass inertial system S' (in which the total momentum is defined to be zero) relative to the laboratory.
(*c*) What is the total energy E' in the center-of-mass system S'?

Choosing the positive x-axis to be the direction of motion of the incident particle A, we have

Particle A	*Particle B*
$m_{0_A} = m_0$	$m_{0_B} = 2m_0$
$\dot{x}_A = 0.80c$	$\dot{x}_B = 0$
$\dot{y}_A = \dot{z}_A = 0$	$\dot{y}_B = \dot{z}_B = 0$
$v_A = 0.80c$	$v_B = 0$
$\gamma_A = \left[1 - \left(\frac{v_A}{c}\right)^2\right]^{-1/2}$	$\gamma_B = \left[1 - \left(\frac{v_B}{c}\right)^2\right]^{-1/2}$
$\gamma_A = [1 - (0.80)^2]^{-1/2} = \frac{5}{3}$	$\gamma_B = [1 - (0.0)^2]^{-1/2} = 1$
$p_{x_A} = \gamma_A m_{0_A} \dot{x}_A = \frac{5}{3}(m_0)(0.80c)$	$p_{x_B} = \gamma_B (m_{0_B}) \dot{x}_B$
$p_{x_A} = \frac{4}{3} m_0 c$	$p_{x_B} = 0$
$p_{y_A} = p_{z_A} = 0$	$p_{y_B} = p_{z_B} = 0$
$E_A = \gamma_A m_0 c^2 = \frac{5}{3} m_0 c^2$	$E_B = \gamma_B m_{0_B} c^2 = 1(2m_0)c^2$

(*a*) The total energy of the two particles, as observed in S is

$$E = E_A + E_B = \tfrac{5}{3} m_0 c^2 + 2m_0 c^2$$

$$E = \tfrac{11}{3} m_0 c^2$$

(*b*) The relations between total energy and momentum components of each particle as viewed in the lab system S and as viewed in the center-of-mass system S' follow from Equations 3–26.

Using Equations 3–26, where **v** is the velocity of the center-of-mass system along the positive x-axis, we then have for the momentum components of particles A and B in system S':

Particle A	*Particle B*
$p'_{x_A} = \dfrac{p_{x_A} - vE_A/c^2}{\sqrt{1 - (v/c)^2}}$	$p'_{x_B} = \dfrac{p_{x_B} - vE_B/c^2}{\sqrt{1 - (v/c)^2}}$
$p'_{x_A} = \dfrac{(\frac{4}{3})m_0 c - v(\frac{5}{3})m_0}{\sqrt{1 - (v/c)^2}}$	$p'_{x_B} = \dfrac{0 - v(2m_0)}{\sqrt{1 - (v/c)^2}}$
$p'_{y_A} = p'_{z_A} = 0$	$p'_{y_B} = p'_{z_B} = 0$

By definition, the total momentum of the two particles in the center-of-mass system must be zero:

$$p'_{x_A} + p'_{x_B} = \frac{\frac{4}{3} m_0 c - \frac{5}{3} m_0 v - 2m_0 v}{\sqrt{1 - (v/c)^2}} = 0$$

This requires that the numerator be zero, which in turn requires $v = \frac{4}{11}c$.

(c) The total energy E' of the two particles in the center-of-mass system can be got by using the last equation of Equations 3–26 to find the total energy of each particle in S':

$$E'_A = \frac{E_A - vp_{xA}}{\sqrt{1-(v/c)^2}} = \frac{\frac{5}{3}m_0c^2 - (\frac{4}{11}c)(\frac{4}{3}m_0c)}{\sqrt{1-(\frac{4}{11})^2}}$$

$$E'_A = 1.27m_0c^2$$

$$E'_B = \frac{E_B - vp_{xB}}{\sqrt{1-(v/c)^2}} = \frac{2m_0c^2 - 0}{\sqrt{1-(\frac{4}{11})^2}} = 2.15m_0c^2$$

Thus the total energy of the two particles as viewed in the center-of-mass system is $E' = E'_A + E'_B = (1.27 + 2.15)m_0c^2 = 3.42m_0c^2$. This is less than the total energy $E = 3.67m_0c^2$ evaluated in part (a), as it must be. This difference represents the energy carried by the center of mass. As a check on the calculations in the above example, we should find that the momentum-energy invariant quantity, $(E/c)^2 - (p_x^2 + p_y^2 + p_z^2)$, for each particle is equal to the particle's rest mass divided by c^2, whether evaluated in system S or in S', an exercise you are urged to carry out.

3-5 Special Relativity and the Electromagnetic Interaction

The speed of propagation of electromagnetic waves is a relativistically invariant quantity, c. Electric charge is also a relativistic invariant; that is, the magnitude of a particle's electric charge does not depend on the speed of the particle. The simplest and most compelling experimental evidence for electric-charge invariance is that an electrically neutral system of particles of equal but opposite charge remains electrically neutral quite apart from the speeds of the particles within it. Thus, a hydrogen molecule with two protons and two electrons, and a helium atom also with two protons and two electrons, are both found to be electrically neutral to better than one part in 10^{20}, although the speeds of the component particles differ greatly for the two systems.

The electromagnetic interaction between a pair of electrically charged particles is customarily separated into two parts: the *electric* force, which always acts between any two charged particles whether each is at rest or in motion relative to the observer; and the *magnetic* force, which acts between them only when they are both in motion relative to the observer. It is easy to see that the magnetic force between any two charged particles Q_1 and Q_2 can be turned off, so to speak, merely by a proper choice of reference frame. For example, an observer who rides in a reference frame that is at rest with respect to charge Q_1 will measure no magnetic force between the two charges. First, the magnetic force on Q_2 due to Q_1 will be zero, because Q_1 creates no magnetic field at the site of Q_2 (or at any other point in space, since Q_1 is at rest). Second, even if Q_2 were moving with respect to the observer, there would be no magnetic force on Q_1 due to Q_2, since Q_1 is not moving. This example shows us that a magnetic force between two charged particles can exist for observers in some inertial systems but not others. The first postulate of relativity, the covariance of the form of a physical law in all inertial frames, requires that the laws of electromagnetism be of the same form in all inertial frames. It is possible to qualitatively relate electric and magnetic forces as they are viewed in different inertial systems.

Clearly, the magnetic interaction is intimately related to the choice of reference frame. Since special relativity deals with the transformation relations between reference frames, we expect that the appearance or non-appearance of the magnetic force between charged particles is basically a relativistic effect. It is easy to see, at least in a qualitative way, that

$$\text{Magnetism} = \text{electricity} + \text{relativity}$$

In this view, the magnetic interaction is not a separate and distinct type of fundamental force; rather, the so-called magnetic force originates from the strictly electric (or Coulomb) interaction and the requirements of special relativity.

To see the connection among magnetism, electricity, and relativity, let us look at what is usually described as a purely magnetic force. Two parallel, infinitely long, current-carrying conductors attract one another magnetically when the currents are in the same direction and repel one another when the currents are in opposite directions. It is thought that in this situation an electric force does not act between the two conductors, since each is taken to be electrically neutral. In the standard analysis, we say that the moving charges in one conductor generate a magnetic field at the site of the other conductor and the charges moving through the latter conductor are thereby acted on by this magnetic force. This force is transmitted ultimately to the crystalline lattice of the solid wire through which the charges move.

Let us reexamine this situation from the point of view of special relativity. For simplicity, we imagine that the current through each conductor exists because there are equal amounts of free *positive* charges moving in the direction of the current and free *negative* charges moving in the opposite direction. We know that in metals the current results from the motion of free negative charges (electrons) only, but the physical arguments concerning the more symmetrical situation chosen here are simpler, and the results are like what one would get by assuming the motion of particles with only one sign of charge.

Figure 3–9 shows positive and negative charges moving through the lattices of two conducting wires; each wire is assumed to be electrically neutral. The moving charges produce an electric current to the right in each wire, and classically this results in an attractive magnetic force between the two wires. Can we, by invoking special relativity, describe this effect by an electric force only? The answer is yes, but only if we are very careful. First, we must recognize that we have not considered how a force transforms between one inertial frame and another. A force on a particle does *not* have the same magnitude for all observers; that is because it is defined as the particle's momentum change per unit time interval, and both these quantities depend on the observer. We shall not derive the force-transformation relations in detail here but merely acknowledge as reasonable that although the magnitude of a force may change in the transformation between one inertial frame and another, its direction does not change.

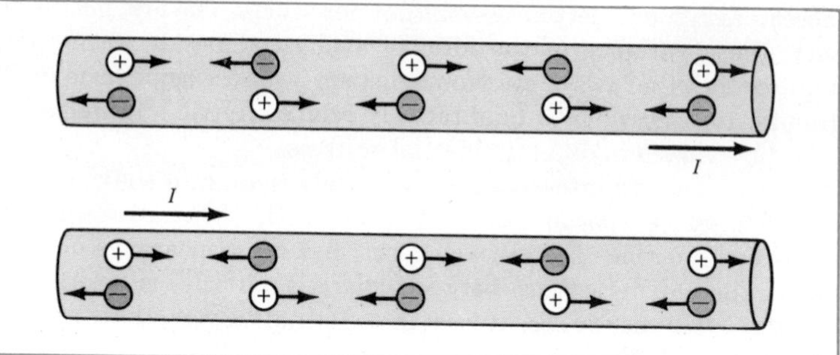

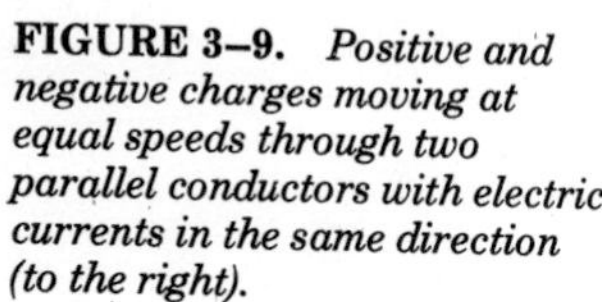
FIGURE 3–9. *Positive and negative charges moving at equal speeds through two parallel conductors with electric currents in the same direction (to the right).*

Let us focus on the net electric force on the positive and negative charges in one of the wires. In doing so, we intentionally *exclude* any magnetic interaction, concentrating on electric forces alone and invoking the requirements of relativity.

To find the force on any charge, we shall always transform the force to an inertial system in which the charge is at rest. First, we analyze the force on one of the positive charges in Figure 3–9. A view from the inertial frame in which the positive charges are at rest is shown in Figure 3–10*a*. In this frame the distance between adjacent positive charges is indicated by D_0. On the other hand, the negative charges now are moving to the left at a higher speed than in Figure 3–9; because of space contraction, the distance D between adjacent negative charges is smaller than D_0. The resultant electric force on a given positive charge is due to the other charges, both positive and negative, in the same wire and also to the positive and negative charges in the other wire. It is obvious that other positive charges in the same wire produce no net force, since there are equal numbers of uniformly spaced positive charges to the left and to the right; the same is true for negative charges in the same wire.

What about the resultant force due to the charged particles in the other wire? Because of the relativistic contraction of the distance between adjacent negative charges, there will be more negative charges per unit length than positive charges. Thus, to a positive charge in, say, the lower wire, the *net* charge of the upper wire is *negative*, and the chosen positive charge is attracted to the upper wire. Transforming back to the inertial system in which the wires are at rest (Figure 3–9) will still give an attractive force between the positive charge and the upper wire, although one of a different magnitude. (Likewise, a positive charge in the upper wire of Figure 3–10*a* finds the lower wire negatively charged, because of the contraction of the distance between the moving negative charges in it, and is therefore attracted to the lower wire.)

We next consider forces on the negative charges, following the same procedure as before. First we view the charges from an inertial frame in which the chosen negative charge is at rest; then we transform back to the system in which the wires are at rest. Figure 3–10*b* illustrates the motion of the charges from a reference frame in which the negative charges are at rest. Again the negative charge experiences no net force from the positive and negative charges in the same wire because of equal numbers in both directions along the wire; but because of space contraction for the moving positive charges in the upper wire, a negative charge in the lower wire finds the

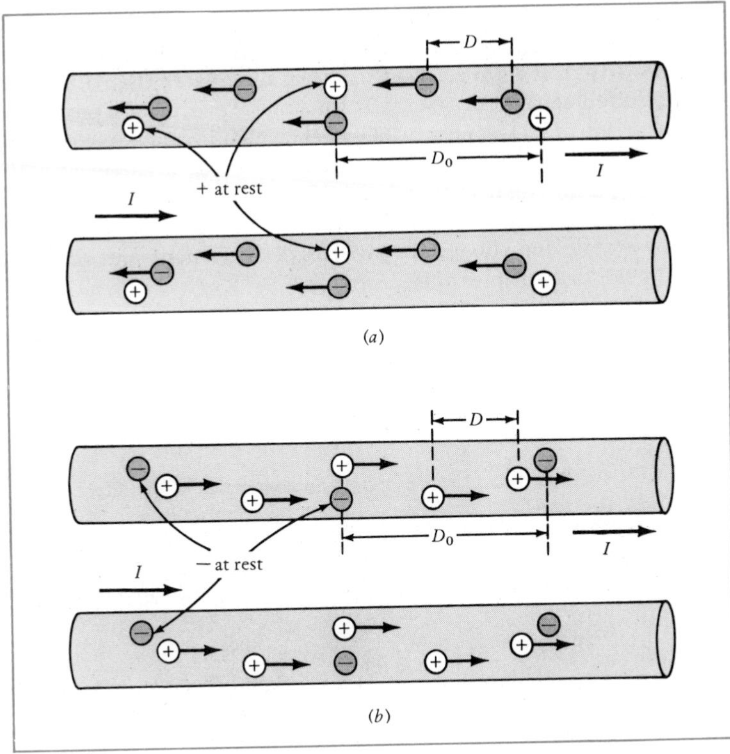

FIGURE 3–10. *The conductors and charge carriers of Figure 3–9, now as viewed by (a) an observer at rest with respect to the positive charges; (b) an observer at rest with respect to the negative charges.*

upper wire to have a net positive charge and is therefore attracted to it. (Similar arguments hold for negative charges in the upper wire.) Finally, transforming back to the reference frame in which the wires are at rest (Figure 3–9) still gives an attractive force between any negative charge and the other wire.

Both the positive and negative charges in one of the wires of Figure 3–9 are attracted to the other wire. These charges will then move across the width of their own wire until they arrive at the surface closest to the other wire, but because they are bound to the conducting wire, they cannot leave the surface; thus, the attractive force is transmitted to the entire wire. We have shown qualitatively that two wires carrying currents in the same direction attract one another *without* the action of a magnetic force, at least an explicit one.† Similar arguments would show that the resultant force between two parallel, infinitely long wires carrying currents in *opposite* directions would be *repulsive*.

3-6 Computations and Units in Relativistic Mechanics

The classical equations for the momentum and kinetic energy of a particle can be used only when the speed of the particle is much less than the speed of light; for high speeds the relativistic relations must be invoked. It is useful

† A detailed calculation of the attractive force based on relativistic considerations shows the force to be identical to what is given by the conventional analysis based on magnetic forces.

to have a rule of thumb for determining how a computation in a problem can safely be treated—whether relativistically or classically. Table 3–1 shows the conditions that, if fulfilled, lead to errors no greater than 1 percent in the computed momentum or energy. If the kinetic energy of a particle is a very small fraction of its rest energy, classical mechanics applies; contrarily, if the total energy or the kinetic energy greatly exceeds a particle's rest energy, then the extreme relativistic relation $E = pc$ (which holds strictly only for $m_0 = 0$) applies.

For atomic and subatomic particles, a convenient unit of energy is the electron volt or multiples of it:

$$\text{Kilo-electron-volt} = 1\ \text{keV} = 10^3\ \text{eV}$$

$$\text{Mega-electron-volt} = 1\ \text{MeV} = 10^6\ \text{eV}$$

$$\text{Giga-electron-volt}^\dagger = 1\ \text{GeV} = 10^9\ \text{eV}$$

TABLE 3–1. *Rule of thumb for computing momentum and kinetic energy*

	CONDITION	APPROXIMATE RELATION (ERROR $\leq$ 1 PERCENT)
CLASSICAL REGION	$E_k/E_0 < 0.01$ or $v/c < 0.1$	$E_k \approx \frac{1}{2}m_0v^2$ and $p \approx m_0v$
EXTREME RELATIVISTIC REGION	$E/E_0 > 7$ or $E_k/E_0 > 6$ or $v/c > 0.99$	$E \approx pc$

From Equation 3–15, the corresponding unit for momentum is the electron volt divided by the speed of light, eV/c. When momentum is expressed in these units, the unit for pc is just the electron volt, eV. The speed of an atomic particle is most conveniently given in units of the speed of light, that is, as v/c. In these units, the speed of any particle must lie somewhere between 0 and 1. These particular units (eV for energy, eV/c for momentum, and v/c for speed) simplify calculations in classical problems as well as relativistic ones. It is also preferable to use a particle's rest energy (in electron volts), rather than the rest mass (in kilograms).

The classical kinetic-energy and momentum relations can be written in terms of a particle's rest energy $E_0 = m_0c^2$, its speed v/c, and the constant c, as follows:

$$E_{k,cl} = \frac{1}{2}m_0v^2 = \frac{1}{2}(m_0c^2)\left(\frac{v}{c}\right)^2 = \frac{1}{2}E_0\left(\frac{v}{c}\right)^2 \tag{3–28}$$

$$p_{cl} = m_0v = \frac{(m_0c^2)(v/c)}{c} = \frac{E_0(v/c)}{c} \tag{3–29}$$

† The energy unit BeV (billion electron volts) is also used to denote 10^9 eV. The GeV is preferred, however, because in European usage a billion designates a million million (10^{12}), not a thousand million (10^9).

For example, the kinetic energy and momentum of an electron (rest energy, $E_0 = 0.51$ MeV) moving with a speed of $\frac{1}{100}c$, for which the classical relations apply, are easily found.

$$E_k = \frac{1}{2}E_0\left(\frac{v}{c}\right)^2 = \frac{1}{2}(0.51 \text{ MeV})(10^{-2})^2 = 0.26 \times 10^{-4} \text{ MeV} = 26 \text{ eV}$$

This is almost twice the kinetic energy of the electron in a hydrogen atom.

$$p = \frac{E_0(v/c)}{c} = \frac{(0.51 \text{ MeV})(10^{-2})}{c} = \frac{0.51 \times 10^{-2} \text{ MeV}}{c} = 5.1 \text{ keV}/c$$

The masses of particles in atomic physics are most often given in units of the *atomic mass unit* (unified), or "u." *One atomic mass unit is defined as one-twelfth the mass of a neutral carbon atom (isotope 12).* Avogadro's number, $6.022\,04 \times 10^{23}$, gives the number of atoms in 12 g of atomic carbon. Therefore,

$$1 \text{ u} = \frac{1}{12}\left(\frac{12 \text{ g}}{6.022\,04 \times 10^{23}}\right) = 1.660\,57 \times 10^{-27} \text{ kg}$$

The relation between the unified atomic mass unit, u, and the energy unit mega-electron-volt, MeV, is particularly useful. We find this from the general mass-energy relation, Equation 3–12:

$$\begin{aligned} E &= mc^2 = (1 \text{ u})c^2 \\ &= \frac{(1.660\,57 \times 10^{-27} \text{ kg})(2.997\,93 \times 10^8 \text{ m/s})^2}{(1.602\,19 \times 10^{-19} \text{ J/eV})(10^6 \text{ eV/MeV})} \\ &= 931.502 \text{ MeV} \end{aligned}$$

Therefore,

$$1 \text{ unified atomic mass unit} = 1 \text{ u} = 931.5 \text{ MeV}/c^2$$

This relation can be regarded as giving the basic conversion factor between mass and energy units. The rest energies of the electron and proton are:

$$\text{Electron rest energy} = 0.511\,003 \text{ MeV} = 5.486\,24 \times 10^{-4} \text{ u}\cdot c^2$$

$$\text{Proton rest energy} = 938.280 \text{ MeV} = 1.007\,28 \text{ u}\cdot c^2$$

It is worth memorizing that the rest energy of an electron is approximately one-half a mega-electron-volt, $\frac{1}{2}$ MeV, and that the rest energy of a proton is approximately one giga-electron-volt, 1 GeV. By convention, when a particle is described as, say, a 3.0-MeV particle, this means that the *kinetic* energy, not the total energy E, is 3.0 MeV.

Example 3–6. What is the speed of (*a*) a 2.0-MeV electron? (*b*) a 2.0-MeV proton?

(*a*) The rest energy of an electron is $E_0 = 0.51$ MeV. With a kinetic energy $E_k = 2.0 \text{ MeV} > E_0/100$, a relativistic calculation must be made. Combining Equations 3–5 and 3–14 we get the total energy E in terms of the speed v

$$E = \frac{E_0}{\sqrt{1 - (v/c)^2}} \tag{3–30}$$

or equivalently,

$$\frac{v}{c} = \sqrt{1 - \left(\frac{E_0}{E}\right)^2} = \sqrt{1 - \frac{E_0^{\ 2}}{(E_k + E_0)^2}}$$

These relations are often useful in relativistic calculations. For the 2.0-MeV electron we then have

$$\frac{v}{c} = \sqrt{1 - \left(\frac{0.51}{2.51}\right)^2} = 0.98$$

Thus, the electron has a speed that is 98 percent of the speed of light.

(*b*) For the proton, $E_k = 2.0$ MeV and $E_0 = 938$ MeV. Therefore, $(E_k/E_0) = 2.0\ \text{MeV}/938\ \text{MeV} = 0.002 < \frac{1}{100}$ and one can safely use the classical relation for kinetic energy, Equation 3–28, $E_k = \frac{1}{2}E_0(v/c)^2$, which yields

$$\frac{v}{c} = \sqrt{\frac{2E_k}{E_0}} = \sqrt{\frac{2(2.0\ \text{MeV})}{938\ \text{MeV}}} = 0.065$$

SUMMARY

The relativistic momentum of a particle is

$$p = \gamma m_0 v \tag{3–5}$$

where

$$\gamma = \frac{1}{\sqrt{1 - (v/c)^2}} \tag{3–6}$$

In relativistic dynamics the total energy of a particle is given by the Einstein equation,

$$E = \gamma m_0 c^2 = \frac{m_0 c^2}{\sqrt{1 - (v/c)^2}} = mc^2 \tag{3–14}$$

A particle's kinetic energy is

$$E_k = E - E_0 \tag{3–13}$$

where $E_0 = m_0 c^2$ is the rest energy of the particle.

The dynamical quantity

$$E_0^{\ 2} = E^2 - (pc)^2 \tag{3–15}$$

is invariant under the Lorentz transformations, having the same value in all inertial systems.

The rest mass of a bound system of particles is less than the total rest mass of the separated parts by E_b/c^2, where E_b is the total binding energy.

A particle's total energy and its momentum components transform from one inertial system to another in a way similar to that of the space-time Lorentz transformations:

$$\begin{aligned} p'_x &= \frac{p_x - v(E/c^2)}{\sqrt{1 - (v/c)^2}} \\ p'_y &= p_y \quad p'_z = p_z \\ E' &= \frac{E - vp_x}{\sqrt{1 - (v/c)^2}} \end{aligned} \tag{3–23}$$

Momentum conservation and total energy conservation laws are covariant under Lorentz transformations.

Maxwell's laws of electrodynamics are Lorentz-covariant. The magnetic interaction between moving charges arises, according to special relativity, when a strictly electric interaction is transformed to another inertial system.

PROBLEMS

Relativistic momentum, §3–1

● **3–1.** At what speed is a particle's momentum twice that given by the classical momentum relation?

● **3–2.** An unstable particle, initially at rest, decays into a proton (charge, $+e$) and a pion (charge, $-e$). Show that in the presence of a uniform magnetic field, the radius of curvature is the same for each particle.

● **3–3.** A particle's speed increases by a factor of 100, from $0.0098c$ to $0.98c$. By what factor does the particle's momentum increase?

3–4. What is the speed below which the error produced by using the classical relation rather than the relativistic momentum one will be less than 1.0 percent?

3–5. An incident particle having mass m_0 and going at a speed v much less than c collides with a second particle at rest and also of mass m_0. The collision is perfectly elastic. (*a*) Find the total momentum and total kinetic energy of the system. (*b*) Find the total momentum and total kinetic energy of the system before and after the collision, but now relative to a reference frame moving at speed $v/2$ in the same direction as the incident particle.

3–6. A particle of electric charge -1.6×10^{-19} C and of rest mass $m_0 = 1.0 \times 10^{-30}$ kg moves in a circular orbit about a particle of charge $+1.6 \times 10^{-19}$ C and mass $M_0 \gg m_0$. Assuming that the only force acting on the orbiting particle is the electrostatic attractive force, determine the radius of the orbit for a tangential speed of $0.8c$.

3–7. A particle of mass m_0, traveling at speed $0.98c$ collides head-on with a target particle of mass $3m_0$ initially at rest. If the target particle moves away from the collision at speed $0.89c$, what is the velocity of the incident particle after the collision?

Relativistic energy, §3–2

● **3–8.** A particle's speed increases by a factor of 100, from $0.0098c$ to $0.98c$. By what factor does the particle's kinetic energy increase?

● **3–9.** Show that $E = pc^2/v$.

● **3–10.** Protons at the Fermi National Accelerator Laboratory are accelerated to kinetic energies as high as 500 GeV $= 5.00 \times 10^{11}$ eV. What is the speed of 500-GeV protons? The rest energy of a proton is 0.938 GeV.

● **3–11.** What is the momentum, in units of keV/c, of a 0.40-keV electron (classical)?

● **3–12.** What is the momentum of a particle whose kinetic energy equals its rest energy E_0?

● **3–13.** (*a*) What is the radius of curvature of a beam of 0.51-MeV electrons injected perpendicularly into a uniform magnetic field of 0.01 T? (*b*) What would be the radius of the beam of 0.51-MeV protons injected into the same field?

● **3–14.** A 3.0-GeV electron (extremely relativistic) moves at right angles to the field lines of a uniform magnetic field in a path with a radius of curvature of 36 m. At what kinetic energy would an electron have an 18-m radius of curvature in the same magnetic field?

3–15. A proton (mass, $1840m_e$) is moving at speed $c/20$. At what speed will an electron (mass, m_e) have the same kinetic energy as the proton?

3–16. An alpha particle (the nucleus of a helium atom) has four times the mass of a proton. At what speed will a proton have the same momentum as an alpha particle moving at speed $0.80c$?

3–17. A beam of electrons is projected into a uniform magnetic field **B**, the beam perpendicular to **B**. (*a*) At what speed v will the radius of curvature of the electron beam in the magnetic field be twice that predicted by classical physics? (*b*) What is the kinetic energy (MeV) of the electrons in the beam?

3–18. An unstable particle, initially at rest, decays into a proton (charge, $+e$; rest energy, 938.2 MeV) and a pion (charge $-e$; rest energy, 139.6 MeV). A uniform magnetic field of 0.25 T exists, with the field perpendicular to the velocities of the created particles. The radius of curvature of each track is found to be 1.33 m. What is the rest mass of the original unstable particle?

3–19. Show that the slope of the kinetic-energy-versus-momentum curve is the speed of the particle both in relativistic mechanics and in classical mechanics.

3–20. The Soviet Union plans to build a 2000-GeV proton accelerator alongside a 20-GeV electron accelerator, arranged so that the beams will intercept head-on. In a head-on collision between a proton and an electron, what is the total momentum of the two-particle system?

3–21. For a highly relativistic particle ($E \gg E_0$), show that the particle's momentum can be approximated by

$$pc \approx E\left[1 - \frac{1}{2}\left(\frac{E_0}{E}\right)^2\right]$$

3–22. At the Stanford Linear Accelerator Center, electrons are accelerated to energies as high as 20 GeV. Find the percentage difference between the speed of a 20-GeV electron and the speed of light.

Mass = energy, §3–3

● **3–23.** Two identical particles, each of kinetic energy E_k and at room temperature T, collide head-on and are brought to rest. How much must the composite object's mass be reduced to bring its temperature back to T?

● **3–24.** The energy necessary to separate a water molecule, H_2O, into three isolated atoms is 9.6 eV. There are 6.0×10^{23} water molecules in 18 g of water. What is the fractional change in mass of a water molecule when it is broken into 2 hydrogen atoms and 1 oxygen atom?

3–25. A technologically advanced civilization on a nearby star 20 light-years from Earth wishes to send a spaceship having a final payload rest mass of 4000 kg on a fly-by trip past our planet. To have reasonable travel time, the ship is quickly accelerated to a speed of $c/2$. What minimum energy is required to accelerate the ship to this speed? Compare this with the annual 7.9×10^{19} J of energy used in the United States.

3–26. Take the rest energy of an electron as $\frac{1}{2}$ MeV and of a proton as 1 GeV. At what kinetic energy will an electron have the same momentum as (*a*) a 5-eV proton? (*b*) a 5-MeV proton? (*c*) a 5-GeV proton?

3–27. Water has a latent heat of fusion of 334 J/g and a latent heat of vaporization of 2260 J/g. Its specific heat is 4.18 J/g·°C. One mole (6.02×10^{23} molecules) of water has a mass of 18 g. (*a*) What is the difference in mass (g) between 1 mol of ice molecules (18 g) at 0°C and 1 mol of water vapor molecules at 100°C? (*b*) Compute the difference in the average binding energy (eV) per molecule between ice molecules at 0°C and water vapor molecules at 100°C.

3–28. A 12-V automobile battery is rated at 6.0×10^3 A·h. Indicate (*a*) whether the mass of the battery increases or decreases as it is completely discharged, and (*b*) what the amount of change is.

3–29. Two cubes of ice, each of mass m_0 and temperature 0°C, are projected toward one another, each moving at speed 25.9 m/s relative to a lab observer. They collide head-on in a perfectly inelastic collision. (*a*) Find the temperature and physical state of the composite object after collision. (*b*) What is the fractional increase in rest mass of the system arising from the collision? (The specific heat of ice is 2.30×10^3 J/kg·°C and of water is 4.19×10^3 J/kg·°C. The latent heat of fusion is 3.35×10^5 J/kg.

3–30. Antiprotons, particles like protons except oppositely charged, were first produced in the laboratory by projecting a beam of high-energy protons onto a target composed of protons at rest. The simplest reaction is

$$\mathrm{p} + \mathrm{p} = \mathrm{p} + \mathrm{p} + (\mathrm{p} + \bar{\mathrm{p}})$$

where both a proton (p) and an antiproton ($\bar{\mathrm{p}}$) are produced to conserve electric charge (and other quantities to be discussed later). What is the minimum kinetic energy of the incident proton necessary to produce this reaction? *Hint:* The minimum incident kinetic energy occurs when all outgoing particles are at rest relative to the center of mass.

Momentum-energy four-vector, §3–4

3–31. Particles with rest mass m_0 are accelerated to a kinetic energy E_k in a nuclear accelerator. These particles are then incident on a target composed of particles of the same rest mass m_0. (*a*) Show that in a collision between an incident particle and a target particle, the center of mass travels with speed $\gamma v/(\gamma + 1)$, where v is the speed of the incident particle and γ has the value $1/\sqrt{1 - (v/c)^2}$. Verify that this center-of-mass speed reduces to the classical result when $v \ll c$. (*b*) Assume that the collision in (*a*) is perfectly *inelastic*. Using momentum conservation and energy conservation, show that the mass M_0 after collision is given by $M_0 = m_0\sqrt{2(\gamma + 1)}$, and verify that this reduces to the classical value when $v \ll c$.

3–32. A particle of rest energy E_0 is at rest in the laboratory system S. Find (*a*) the momentum p', and (*b*) the total energy E' of this particle relative to a reference frame S' moving at speed $(c/2)$ with respect to the lab system. (*c*) Show that the quantity $[E^2 - (pc)^2]$ has the same value in both systems S and S'.

3–33. In a high-energy experiment, protons with incident momentum 100 GeV/c collide with neutrons at rest in a target. If there are only two outgoing particles after the collision, a proton and a particle X, what is the maximum rest energy of particle X? Take the rest energies of the proton and neutron to be 1.0 GeV each.

3–34. One member of the meson family of elementary particles is the neutral pion, which has zero electric charge and rest energy 135 MeV. This particle is unstable and quickly decays into two photons that (as mentioned at the end of Section 3–2) have zero rest mass but energy and momentum related by $E = pc$. When a neutral pion, initially at rest, decays, what is the momentum and energy of each of the two outgoing photons?

■ **3–35.** A particle of rest energy 140 MeV has a momentum of 100 MeV/c in the laboratory system. Find the value, relative to the laboratory, of the velocity of the inertial system in which the particle's momentum (*a*) is twice that in the lab system; (*b*) is twice that in the lab system but oppositely directed.

■ **3–36.** A particle of rest mass energy 140 MeV has a momentum of 100 MeV/c along the positive x-axis in the lab system. Find the values, relative to the lab system, of the velocities of the *two* inertial systems (moving along the x-axis) in which the particle's kinetic energy is twice that in the lab system.

3–37. Relative to the laboratory system, two electrons move perpendicular to one another at the same speed, $0.80c$, and collide at the origin in the laboratory system. Find the velocity of the center-of-mass system relative to the lab system.

■ **3–38.** If the neutral pion in Problem 3–34 were initially moving with kinetic energy 270 MeV in the laboratory system, what would be the momentum and energy of each of the two outgoing photons if the velocity of one of the outgoing photons is (*a*) parallel to the velocity of the neutral pion? (*b*) perpendicular to the velocity of the neutral pion?

The electromagnetic interaction, §3–5

3–39. Assume that the electric current through a long, straight conductor is due to the motion of equal amounts of free positive charges moving to the right at speed v and free negative charges moving to the left, as observed in system S (see Figure 3–11). (*a*) Show that the current is given by $I = 2ev/l_0\sqrt{1-(v/c)^2}$, where e is the magnitude of the charge of each charge carrier, and l_0 is the rest length between adjacent positive charges, or adjacent negative charges. (*b*) Find the current I', as measured in a reference frame S', in which the positive charge carriers are at rest, and show that an observer in S' measures a current $I' = I/\sqrt{1-(v/c)^2}$, a larger current than that measured in frame S.

Computations and units, §3–6

● **3–40.** What is the speed, within 1 percent, of a 5-GeV electron?

● **3–41.** (*a*) Show that the radius of curvature of an electron, with a kinetic energy $E_k = 6$ MeV, moving perpendicular to a uniform magnetic field can be approximated by $r = E_k/QBc$. (*b*) Show that the radius of curvature of a proton, with kinetic energy $E_k = 6$ MeV, moving perpendicular to a uniform magnetic field can be approximated by $r = \sqrt{2E_kE_0}/QBc$, where E_0 is the rest energy of the proton.

● **3–42.** (*a*) What is the minimum power necessary to accelerate protons (beam intensity $\sim 2 \times 10^{13}$ protons/s) to 500 GeV at the Fermi National Accelerator Laboratory? (*b*) What is the minimum power necessary to accelerate electrons (beam intensity $\sim 8 \times 10^{13}$ electrons/s) to 20 GeV at the Stanford Linear Accelerator Center?

■ **3–43.** A 20.0-MeV electrons makes a head-on elastic collision with a proton initially at rest. (*a*) Show that the center of mass moves with a speed $0.021c$. (*b*) Show that the electron, both before and after the collision, can be treated as extremely relativistic (see Table 3–1), and that the proton after collision can be treated classically.

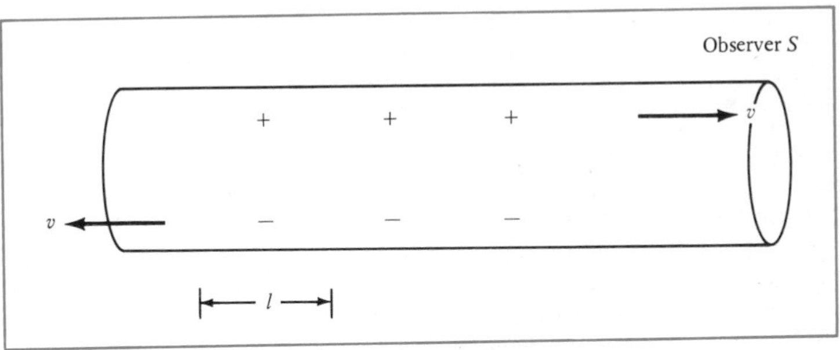

FIGURE 3–11.

4

Quantum Effects: The Particle Aspects of Electromagnetic Radiation

Max K. E. L. Planck (1858, Germany—1947). Ph.D., University of Munich, summa cum laude, *aged 21. 1900, quantum theory of blackbody radiation. 1918, Nobel Prize, discovery of elementary quanta. (Photo, Burndy Library—courtesy of the American Institute of Physics, Niels Bohr Library.)*

4-1 Quantization in Classical Physics

The theory of relativity and the quantum theory constitute the two great theoretical foundations of twentieth-century physics. Just as relativity theory leads to new insights into space and time and to profound consequences in mechanics and electromagnetism, so the quantum theory leads to drastically new ways of thinking about the behavior of atoms, nuclei, and solids. Some aspects of the quantum description of nature are not totally new, however, and are to be found in classical physics.

In the study of the physical world we find two general kinds of physical quantities: quantities that have a continuum of values and quantities that have only discrete, or *quantized,* values.

Figure 4–1 shows several examples of *classical* continuous, or non-quantized, physical quantities:

a. The speed of a free particle, which can range continuously from zero to the speed of light.
b. The magnitude of the angular momentum of a particle, which can have any value from zero to infinity.
c. The mechanical energy of a system of two particles, which can assume any negative value when the particles are bound together ($E_m < 0$) and any positive value when the particles are free ($E_m > 0$).
d. The angle between the direction of a magnet's dipole moment μ and an external magnetic field **B**, which may range continuously from 0 to 180°.

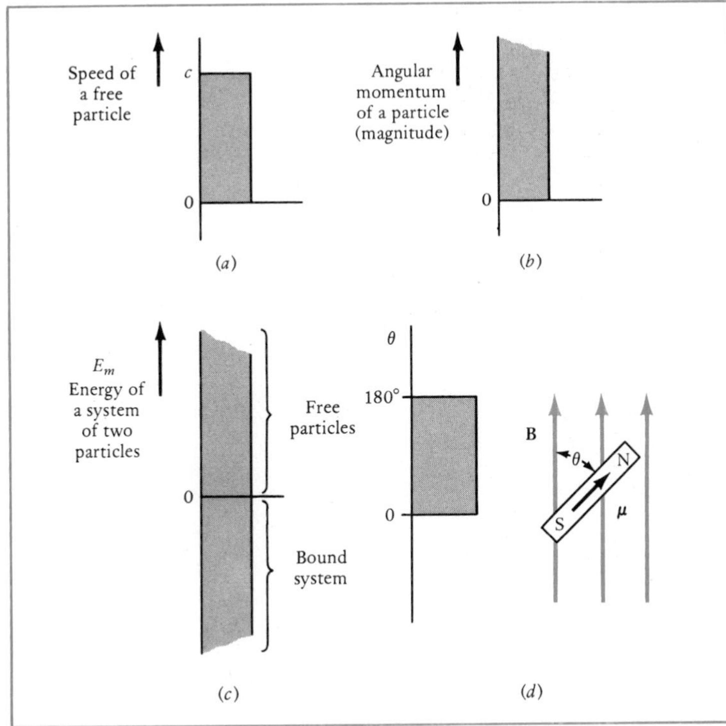

FIGURE 4–1. *Some examples of classical physical quantities having a continuum of allowed values.*

Figure 4–2 shows several examples of quantized physical quantities:

a. The speed of light in a vacuum, always c.

b. Electric charge, which is quantized in that the total charge of any body is precisely an integral multiple, either positive or negative, of the fundamental electron charge e. The quantization of charge, clearly revealed in the chemical idea of valence and in the laws of electrolysis, was most directly demonstrated in the oil-drop experiments of R. A. Milliken (1910), in which the charge of the electron was measured directly.

c. Standing waves and resonance, which are particularly striking manifestations of quantization in classical physics. The frequency of oscillation of a resonating vibrating string, fixed at both ends, can be only an integral multiple of the lowest, or fundamental, frequency of oscillation f_0. The fundamental frequency is determined in turn by the physical properties of the string and its length. The wave on the string is repeatedly reflected from the boundaries, or fixed ends, and constructively interferes with itself, so to speak, to produce standing waves. Resonance can be achieved only if the distance between the end points is precisely an integral multiple of half-wavelengths. It was argued in Section 1–4 that the frequency of a wave is precisely determined only when the wave has an infinite extension in space; this argument is valid even for a wave trapped between reflecting boundaries, because it can be imagined that the wave is folded on itself an infinite number of times.

d. The observed rest masses of atoms, which do not occur in a continuous range. This was first perceived in the fundamental studies of chemical combination that led to the atomic theory of Dalton. The masses of atoms occurring in nature are now known very precisely, and whereas they are nearly in the ratio of integers, they are not precisely so. One of the principal tasks of nuclear physics is to explain on some fundamental basis these departures from integral ratios.

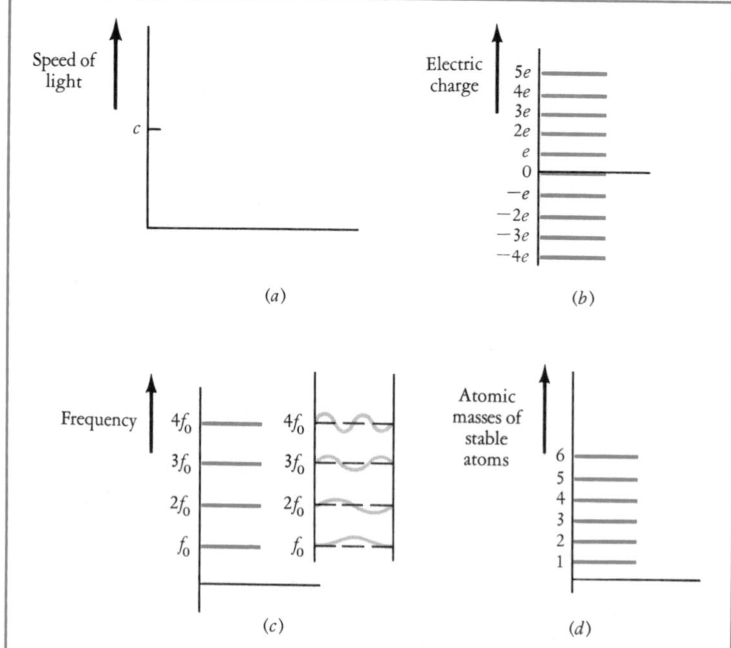

FIGURE 4–2. *Some examples of classical physical quantities having quantized values.*

The quantum theory is largely based on the discovery that certain quantities hitherto regarded in classical physics as continuous are, in fact, quantized. Historically the quantum theory originated in the theoretical interpretation of electromagnetic radiation from a blackbody (a perfect absorber and radiator). Near the end of the 1800s, it was found that the experimentally observed variation with wavelength of the intensity of electromagnetic radiation from a blackbody disagreed with the theoretical expectations of classical electromagnetism. Max Planck, formulator of the quantum theory, showed in 1900 that a revision of classical ideas, through the concept of energy quantization, led to satisfactory agreement between experiment and theory. Because a detailed analysis of blackbody radiation involves sophisticated arguments, we shall introduce the quantum concepts through the simpler and more compelling considerations that arise in the photoelectric effect.

4-2 The Photoelectric Effect

In 1887, the year Michelson and Morley performed their famous experiment on the constancy of the speed of light, Heinrich Hertz discovered the photoelectric effect. Eighteen years later, both of these experimental phenomena were explained theoretically by one man, Albert Einstein. He introduced relativity theory to describe the first phenomenon, and he extended Planck's quantum theory to explain the second.

The photoelectric effect is one of several different processes by which electrons can be removed from the surface of a substance. (We shall refer to metal surfaces in particular, since the effects were first observed in metals and still are most easily observed in them.) The valence electrons in a metal are free to move about through its interior but are bound to the metal as a whole; these relatively free electrons are the ones that are removed. Different ways of emitting electrons are:

1. *Thermionic emission.* Heating the metal gives thermal energy to the electrons and effectively boils them from the surface.
2. *Secondary emission.* Kinetic energy is transferred from particles that strike the metal surface to the electrons in the metal.
3. *Field emission.* Electrons are extracted from the metal by a strong external electric field.
4. *Photoelectric emission.*

The photoelectric effect occurs when electromagnetic radiation shines on a clean metal surface and electrons are released from the surface. A simple description is that a light beam supplies any electron with an amount of energy that equals or exceeds the energy that binds the electron to the surface; that electron is thus allowed to escape. A detailed description of the photoelectric effect requires a knowledge, based on experiment, of how the several variables involved in photoelectric emission are related. These variables are the frequency ν of the light, the intensity I of the light beam, the photoelectric current i, the kinetic energy $\frac{1}{2}m_0v^2$, and the chemical identity of the surface from which the electrons emerge. Kinetic energy refers to the energy the emergent electrons, called *photoelectrons,* have just after they are freed from the surface. (We shall see shortly that the use of the classical kinetic-energy formula is justified.)

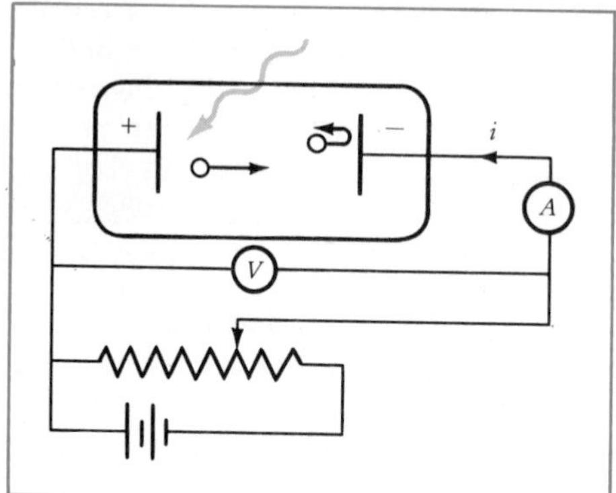

FIGURE 4–3. *Schematic experimental arrangement for studying the photoelectric effect. V, voltmeter; A, ammeter.*

Figure 4–3 shows a schematic diagram of an experimental arrangement for studying important aspects of the photoelectric effect. Monochromatic light shines on a positively charged metal surface, the anode, enclosed in a vacuum tube. (An evacuated tube is used so that collisions between photoelectrons and gas molecules are essentially eliminated.) When photoelectrons are emitted, some travel to the right toward the negatively charged cathode and on reaching it constitute the current flowing in the circuit (conventional current, as shown in the figure). It is here assumed that both anode and cathode surfaces are made of the same metal.† Photoelectrons leave the anode with a variety of kinetic energies, and the negatively charged cathode repels them. Photoelectrons liberated from the anode with a kinetic energy equal to the work done on each electron by the retarding electrostatic field of potential difference V will be brought to rest just in front of the cathode. For these, $eV = \frac{1}{2}m_0v^2$, where v is the speed of the photoelectron as it leaves the anode surface and V is the potential difference that stops the photoelectron of rest mass m_0 and charge e. When the most energetic photoelectrons, of speed v_{max}, are brought to rest in front of the cathode by a sufficiently large potential difference V_0, all the other photoelectrons are stopped too, and no photocurrent exists; $i = 0$. Then

$$eV_0 = \tfrac{1}{2}m_0v_{\text{max}}^2 \tag{4–1}$$

At still higher retarding potential differences, *all* photoelectrons are turned back before reaching the cathode.

After we examine the results of experiment, we shall give the results that might be expected on the basis of the classical theory of electromagnetism. The experimental results strongly disagree with the classical expectations. Finally, we shall see how the photoelectric effect can be understood from a quantum interpretation.

EXPERIMENTAL RESULTS OF THE PHOTOELECTRIC EFFECT. The results of experiments on the photoelectric effect are summarized in Figure 4–4. We shall take them up in the order in which they are given in the figure.

† When the anode has a surface of a different composition from the cathode surface, the description is more detailed, but straightforward. [See article by J. Rudnick and D. S. Tannhauser, *Am. J. P., 44,* 796 (1976).]

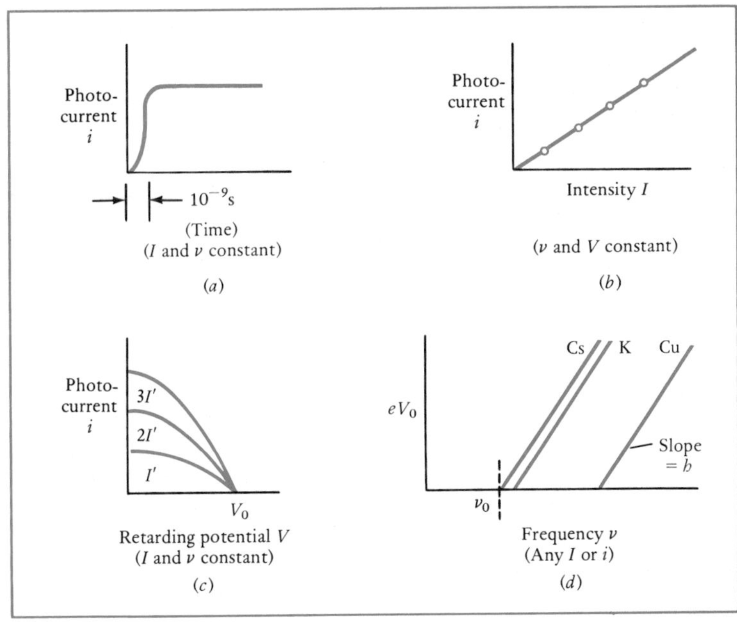

FIGURE 4–4. *Experimental results of photoelectric emission.*

a. When light shines on a metal surface and photoelectrons are emitted, the photocurrent begins *almost instantaneously,* even when the light beam has an intensity as low as 10^{-10} W/m^2 (the intensity at a distance of 200 mi from a 100-W light source). The delay in time, from the instant that the light beam first shines on the surface until photoelectrons are first emitted, is no larger than 10^{-9} s.

b. For any fixed frequency and retarding potential, the *photocurrent i* is directly *proportional to the intensity I* of the light beam. Since the photocurrent is a measure of the number of photoelectrons released per unit time at the anode and collected at the cathode, the relation signifies that the number of photoelectrons emerging per unit time is proportional to the light intensity. (The variation in photocurrent with intensity is used in practical photoelectric devices.)

c. For a constant frequency ν and light intensity I, *the photocurrent decreases with increasing retarding potential* V and finally *reaches zero when* V *is equal to* V_0 (see Equation 4–1). With a small retarding potential, the low-speed, low-energy photoelectrons are brought to rest and no longer contribute to the photocurrent. When the retarding potential equals V_0, even the most energetic photoelectrons are brought to rest, and $i = 0$.

d. For any particular surface, the value of the *stopping potential* V_0 *depends on the frequency of the light but is independent of the light intensity* and therefore, from (*b*), independent also of the photocurrent. Figure 4–4*d* shows experimental results for the three metals cesium, potassium, and copper. For each metal there is a well-defined frequency ν_0, the *threshold frequency,* which must be exceeded for photoemission to occur at all; that is, no photoelectrons are produced, however great the light intensity, unless $\nu > \nu_0$. For most metals the threshold frequency lies in the region of ultraviolet light, and a typical stopping potential is several volts. The emitted photoelectrons have energies of several electron volts; therefore, we are justified in using the classical kinetic-energy formula for the photoelectrons.

For any one type of metal the experimental results of Figure 4–4*d* can be represented by the straight-line equation

$$eV_0 = h\nu - h\nu_0$$

where h, representing the slope of the straight line, is found to be the *same* for *all* metals, and ν_0 is the threshold frequency for the particular metal. Rearranging terms and using Equation 4–1 gives

$$h\nu = \tfrac{1}{2}m_0v_{\max}^2 + h\nu_0 \tag{4–2}$$

Since $\tfrac{1}{2}m_0v_{\max}^2$ has the dimensions of energy, the terms $h\nu$ and $h\nu_0$ must also have the dimensions of energy.

CLASSICAL INTERPRETATION OF THE PHOTOELECTRIC EFFECT. Certain effects can be expected on the basis of the classical properties of electromagnetic waves for each of the four experimental results on the photoelectric effect just described. We consider them with reference to Figure 4–4 and in the same order.

a. Because of the apparently continuous nature of light waves, the energy absorbed on the photoelectric surface should be proportional to the intensity of the light beam (the power per unit area), the area illuminated, and the time of illumination. All electrons that are bound to the surface of the metal with the same energy must be regarded as equivalent, and any one electron will be free to leave the surface only after the light beam has been on long enough to supply the electron's binding energy. Moreover, since any one electron is equivalent to any other electron bound with the same energy, when one electron has accumulated enough energy to be freed, numerous other electrons should have, too. Independent experiments show that in a typical metal, the least energy with which an electron is bound to the surface is a few electron volts. A conservative calculation (see Problem 4–7) shows that in the case of an intensity as low as 10^{-10} W/m^2, for which delay times no larger than 10^{-9} s have been observed, no photoemission can be expected until at least several hundred hours have elapsed! Clearly, the classical theory can't account for the essentially instantaneous photoelectric emission.

b. Classical theory predicts that as the light intensity is increased, so is the energy absorbed by electrons at the surface. Hence, the number of photoelectrons emitted, or the photocurrent, is expected to increase proportionately with the light intensity. Here classical theory agrees with the experimental result.

c. The results of these observations show that there is a distribution in the speeds, or energies, of the emitted photoelectrons; the distribution is in itself not incompatible with classical theory, because it can be attributed to the varying degrees of binding of electrons at the surface or to the varying amounts of energy extracted by electrons from the incident light beam. That there is, however, a very well-defined stopping potential V_0 for a given frequency, independent of the intensity, indicates that the maximum energy of released electrons in no way depends on the total amount of energy reaching the surface per unit time. Classical theory predicts no such effect.

d. The existence of a threshold frequency for a given metal, a frequency below which no photoemission occurs however great the light intensity, is completely inexplicable in classical terms. In the classical view, the primary circumstance that determines whether or not photoemission will occur is the energy reaching the surface per unit time (or the intensity), and *not* the frequency. Further, the appearance of a single constant h that relates, through Equation 4–2, the maximum energy of photoelectrons to the frequency for any material cannot be understood in terms of any constants of classical electromagnetism.

In short, *classical electromagnetism cannot give a reasonable basis for understanding the experimental results illustrated in Figure 4–4a, c, and d.*

QUANTUM INTERPRETATION OF THE PHOTOELECTRIC EFFECT. An understanding of the photoelectric effect is to be found through the quantum theory. Albert Einstein first applied the quantum theory to electromagnetic radiation in 1905; this led to a satisfactory explanation of photoelectric effect experiments that R. A. Millikan carried out in 1916.

According to the quantum theory, the apparently continuous electromagnetic waves are quantized and consist of discrete *quanta*, called *photons*. Each photon has an energy E that depends only on the frequency (or on the wavelength) and is given by

$$E = h\nu = h\frac{c}{\lambda} \tag{4–3}$$

The constant h is the very same h that appears in Equation 4–2, which summarizes the results of experiments on the photoelectric effect. This fundamental constant of the quantum theory is called *Planck's constant* because its value was first determined and its significance first appreciated by Planck in 1900 in interpreting blackbody radiation. The present value of Planck's constant is found by experiment to be

$$h = 6.6262 \times 10^{-34}\ \text{J}\cdot\text{s}$$

From the quantum point of view, a beam of light of frequency ν consists of particlelike photons, each with energy $h\nu$. A single photon can interact only with a single electron at the metal surface of a photoemitter; it cannot share its energy among several electrons. Since a photon travels with the speed of light, it must, by relativity theory, have zero rest mass and an energy that is then entirely kinetic. When a particle with a zero rest mass ceases to move with a speed c, it ceases to exist; so long as it exists, it moves at the speed of light. Thus, when a photon strikes an electron bound in a metal and no longer moves at the unique speed c, it relinquishes its entire energy $h\nu$ to the single electron it strikes. If the energy the bound electron gains from the photon exceeds the energy binding it to the metal surface, the excess energy becomes the kinetic energy of the photoelectron.

By the quantum theory, we are prepared to interpret the experimental results of the photoelectric effect. We examine them in reverse order for convenience, referring to Figure 4–4.

d. The terms in Equation 4–2 now give a simple meaning to the energies of the photon and photoelectron:

$$h\nu = \tfrac{1}{2}m_0v_{\text{max}}^2 + h\nu_0 \tag{4–2}$$

The left side of this equation gives the energy carried by a photon and supplied to a bound electron. The electrons that are least tightly bound leave the surface with maximum kinetic energy. The right side of Equation 4–2 gives the energy the electron gains from the photon, namely, the kinetic energy and the binding energy. The binding energy of the electrons least tightly bound to the metal surface, often represented by φ, is called the *work function*; it represents the work needed to remove the least tightly bound electron. Therefore,

$$\varphi = h\nu_0 \tag{4–4}$$

and Equation 4–2 can be written in the form

$$h\nu = \tfrac{1}{2}m_0 v_{\max}^2 + \varphi \tag{4–5}$$

The value of φ for a particular material, determined from the photoelectric effect, agrees with the value of the work function obtained through independent experiments based on different physical principles (Section 8–5). An electron bound with an energy φ can be released only if a single photon supplies at least this much energy, that is, only if $h\nu > \varphi = h\nu_0$ or $\nu > \nu_0$. Figure 4–4*d* then takes on new meaning. The ordinate can now be identified with the photon energy, as shown in Figure 4–5. (Figure 4–5 should also be compared with the right-hand side of Figure 3–8, where for the photoelectric effect, $E_b = \varphi$.)

c. A well-defined maximum kinetic energy of photoelectrons exists for any given frequency, because the frequency of the electromagnetic radiation determines precisely the photon energy ($E = h\nu$).

b. The intensity of a monochromatic electromagnetic wave takes on a new meaning. It is, in the quantum view, the energy of each photon multiplied by the number of photons crossing a transverse unit area per unit time. An increase in the intensity of a light beam means, therefore, a proportionate increase in the number of photons striking the metal surface. It is then expected that the number of photoelectrons or the photocurrent i will be proportional to I.

a. Photoemission occurs with no appreciable delay, because whether an electron is released depends, even at the smallest intensity, not on its accumulating energy but simply on its being hit by a photon, which on stopping relinquishes all of its energy to the electron.

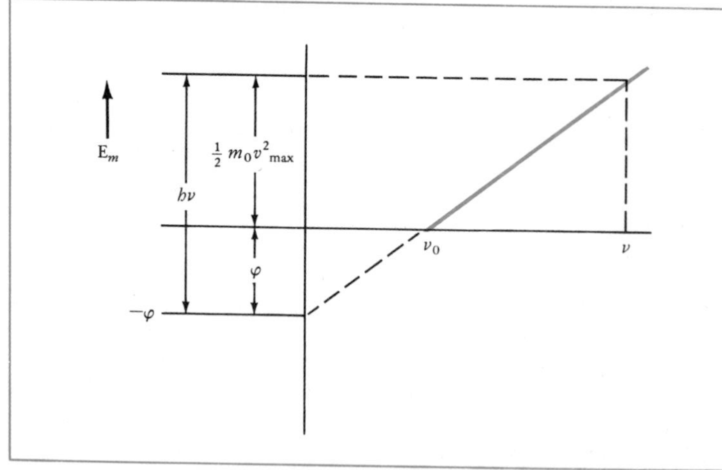

FIGURE 4–5. *Maximum kinetic energy of photoelectrons as a function of the frequency of the incident photons for a particular material.*

TABLE 4–1

EFFECT (FIGURE 4–4)	EXPERIMENT	CLASSICAL ELECTROMAGNETISM	QUANTUM THEORY
a	Essentially instantaneous photoemission (10^{-9} s), even for very low intensities	Emission only after several hundred *hours* (10^6 s) for low intensities	A single photon gives its energy to a single electron essentially instantaneously
b	$I \propto i$	Energy/area-time $\propto i$	$I \propto$ number of photons $\propto i$/area-time.
c	A well-defined $\frac{1}{2}m_0v_{max}^2$, dependent only on ν	Inexplicable	A photon gives all its energy to a single electron
d	A threshold for photoemission, independent of I and i: $h\nu = \frac{1}{2}m_0v_{max}^2 + h\nu_0$	Inexplicable	Photon energy $= h\nu$; work function $= \varphi = h\nu_0$

Table 4–1 summarizes the results of experiment, the classical interpretation and the quantum interpretation, for each of the four effects shown in Figure 4–4.

Our discussion of the photoelectric effect has thus far concerned the effects found when visible or ultraviolet light shines on a *metal* surface. The first detailed experiments that led historically to Einstein's quantum interpretation were performed with metal surfaces. The effect occurs on materials other than metals and with photons of different frequencies and energies, however. The photoelectric effect can occur whenever a photon strikes a *bound* electron with enough energy to exceed the binding energy of the electron. An example is a photon freeing a bound electron from a single atom. The phenomenon is one of the most important interactions between short-wavelength electromagnetic radiation and atoms. When a high-frequency (high-energy) photon, such as an x-ray or a γ-ray, strikes an atom, an electron bound with an energy E_b can be released, provided that $h\nu > E_b$; then the photoemission results in atomic ionization. The kinetic energy of the released photoelectron generally must be written in the relativistic form $E - E_0$, and so the general form of Equation 4–2, the energy equation of the photoelectric effect, becomes

$$h\nu = (E - E_0) + E_b \tag{4–6}$$

The photoelectric effect thus provides an indirect method of measuring the energy of a photon. Suppose that the photoelectron's kinetic energy $E - E_0$ is measured and the binding energy E_b is known on some other basis; then $h\nu$ can be computed from Equation 4–6. Conversely, if $h\nu$ and $E - E_0$ are measured, then E_b can be determined.

The photoelectric effect is only one of several ways in which photons can be removed from a beam of electromagnetic radiation. The photoelectric effect can occur simultaneously with the competitive processes of the Compton effect and pair production (for details see Sections 4–4 and 4–5).

The new and fundamental insight into electromagnetic radiation that the photoelectric effect provides is the quantization of electromagnetic waves, or the existence of photons. We can properly speak of the quantiza-

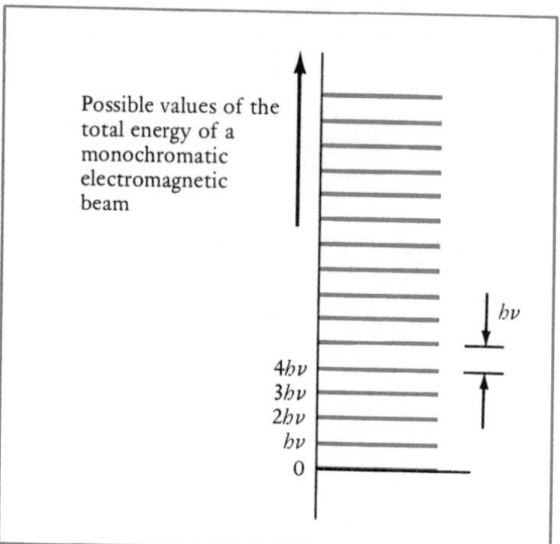

FIGURE 4–6. *Allowed energies of a beam of monochromatic electromagnetic radiation.*

tion of electromagnetic waves since the radiation can be regarded as a collection of particlelike photons, each of energy $h\nu$. When the frequency of the radiation is specified as ν, the photon can have but one energy, $h\nu$. The total energy of a beam of monochromatic electromagnetic radiation is always precisely an integral multiple of the energy $h\nu$ of a single photon (Figure 4–6).

The granularity of electromagnetic radiation is not conspicuous in ordinary observations since the energy of any one photon is very small and the number of photons in a light beam of moderate intensity is enormous. The situation is like that found in the molecular theory—the molecules are so small and their numbers so great that the molecular structure of all matter is disclosed only in very subtle observations.

The electromagnetic spectrum, often scaled in units of frequency, can be scaled by the quantum theory in units of energy per photon (see Figure 4–7).

Electromagnetic waves are commonly characterized by their wavelength. It is useful to have a relation between the photon energy in electron volts and the corresponding wavelength in angstroms. For example, x-radiation of wavelength 1.0 angstrom (abbreviated Å) consists of photons whose energy E is

$$E = h\nu = \frac{hc}{\lambda} = \frac{(6.626 \times 10^{-34}\ \mathrm{J \cdot s})(2.998 \times 10^{8}\ \mathrm{m/s})}{(1.0\ \text{Å})(10^{-10}\ \mathrm{m/\text{Å}})(1.602 \times 10^{-19}\ \mathrm{J/eV})}$$
$$= 1.240 \times 10^{4}\ \mathrm{eV} = 12.40\ \mathrm{keV}$$

More generally, the photon energy E, in electron volts, can be written in terms of the wavelength λ in angstroms:

$$E = \frac{1.240 \times 10^{4}\ \mathrm{eV \cdot \text{Å}}}{\lambda} = \frac{0.01240\ \mathrm{MeV \cdot \text{Å}}}{\lambda} \qquad (4\text{–}7)$$

The energy per photon is smallest for radio-wave photons ($\sim 10^{-12}$ eV) and largest for γ ray photons (~ 1 GeV). The electromagnetic frequency spectrum corresponds exactly to the energy spectrum of a zero-rest-mass particle, or photon, whose energy can extend from zero to infinity, as shown in Figure 4–7. The rest energies of the electron and proton are also shown for comparison.

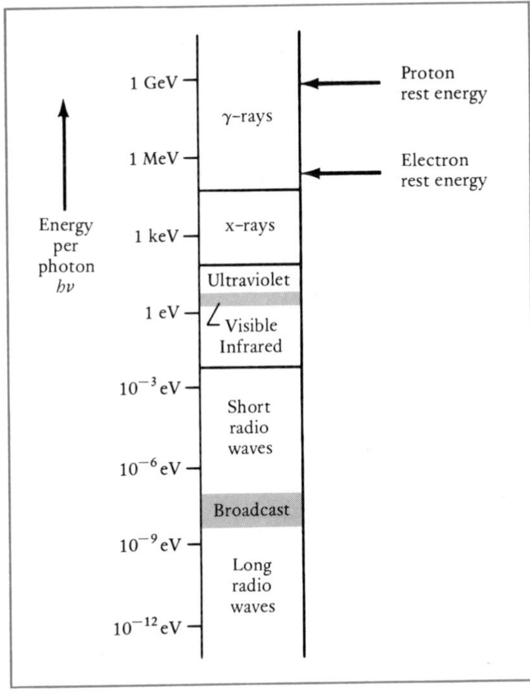

FIGURE 4–7. *Spectrum of electromagnetic radiation on scale of photon energy.*

As we saw in Section 1–4, the ideas of wave and particle seem mutually incompatible, even contradictory. That the photoelectric effect shows light behaving as if it consisted of particles or photons does not mean that we can dismiss the incontrovertible experimental evidence of the wave properties of light. Both descriptions must be accepted. An account of the way in which this dilemma is resolved is postponed (Section 5–4) until after we have explored more fully the quantum attributes of light.

Example 4–1. A 2.00-Å photon interacts with a bound electron in the hydrogen atom (binding energy, 13.6 eV). If the photoelectric effect occurs and the electron moves forward in the same direction as the incident photon, what are the energy (eV) and momentum (eV/c) of (a) the electron? (b) the proton?

(a) In the photoelectric effect, the photon (γ) gives all its energy and momentum to the bound system:

$$E_\gamma\ (\text{eV}) = \frac{12\,400\ \text{eV}\cdot\text{Å}}{\lambda\ (\text{Å})}$$

$$= \frac{12\,400\ \text{eV}\cdot\text{Å}}{2.00\ \text{Å}} = 6200\ \text{eV}$$

$$p_\gamma = \frac{E_\gamma}{c} = 6200\ \text{eV}/c$$

Some of the energy, 14 eV, is used in unbinding the electron and proton. Most of the remaining energy, (6200 − 14) eV = 6186 eV, will appear as kinetic energy of the released electron because it is much lighter than the proton. Thus, E_k (electron) $\approx 6.19 \times 10^3$ eV. Since this is much less than the electrons rest energy, $E_0 = 5.11 \times 10^5$ eV, we can treat the electron classically to get its momentum:

$$E_k = \frac{p^2}{2m_0} = \frac{(pc)^2}{2E_0}$$

or

$$p_e c = \sqrt{2E_k E_0} = \sqrt{2(6.19 \times 10^3 \text{ eV})(5.11 \times 10^5 \text{ eV})}$$

$$p_e = 7.95 \times 10^4 \text{ eV}/c$$

(*b*) By conservation of momentum, $p_\gamma = p_e + p_p$, or

$$p_p = (6.20 \times 10^3 - 7.95 \times 10^4) \text{ eV}/c$$

$$= -7.33 \times 10^4 \text{ eV}/c$$

The proton recoils in a direction opposite to that of the electron and with a momentum of great magnitude. Its kinetic energy is

$$E_k \text{ (proton)} = \frac{p^2}{2m_0} = \frac{(pc)^2}{2E_0}$$

$$= \frac{(7.33 \times 10^4 \text{ eV})^2}{2(9.38 \times 10^8 \text{ eV})}$$

$$= 2.9 \text{ eV}$$

As we assumed initially, the proton's kinetic energy is much smaller than the kinetic energy of the electron.

Example 4–2. Is it possible for the photoelectric effect to occur when the photon is incident on a single *unbound* charged particle?

We assume that an incident photon with energy $h\nu$ and momentum $h\nu/c$ collides with a free particle of rest mass m_0 initially at rest. Suppose that the photon is annihilated and gives all its energy and momentum to the free particle. Designating the energy and momentum of the *particle* after collision by E and p, we have

Conservation of energy: $$h\nu = E - E_0$$

Conservation of momentum: $$\frac{h\nu}{c} = p = \frac{\sqrt{E^2 - E_0^2}}{c}$$

Substituting $h\nu$ from the first equation into the second, and rewriting, we get

$$E - E_0 = \sqrt{E - E_0}\sqrt{E + E_0}$$

or

$$\sqrt{E - E_0} = \sqrt{E + E_0}$$

which is true only if $E_0 = 0$. Since any material particle has a nonzero rest mass, $E_0 \neq 0$, the photoelectric effect cannot take place with a free particle. In an actual photoelectric effect, one with bound electrons, the electron's mass is much smaller than the mass of the object to which it is bound. Then nearly all the photon's energy is transferred to the electron, and only a small fraction to the object to which the electron is bound. On the other hand, only a fraction of the photon's momentum is transferred to the bound electron.

4-3 X-Ray Production and Bremsstrahlung

In the photoelectric effect a photon transfers all its electromagnetic energy to a bound electron; the photon's energy appears as the binding energy and kinetic energy of the photoelectron. The inverse effect is that in which an electron loses kinetic energy and in so doing creates one or more photons. The process is most clearly illustrated in the production of x-rays.

First consider the fundamental process occurring when a fast-moving electron comes close to the positively charged nucleus of an atom and is deflected by it. The electron experiences a large attractive force in consequence of its near collision with the heavy positively charged nucleus and is diverted from its straight-line path; that is, it is accelerated. Classical electromagnetic theory predicts that any accelerated electric charge will radiate electromagnetic energy. Quantum theory also predicts this but requires that any radiated electromagnetic energy should consist of discrete quanta, of photons. It is expected, then, that a deflected and therefore accelerated electron will radiate one or more photons and that the electron will leave the site of the collision with less kinetic energy than it had.

The radiation produced in such a collision is often referred to as *bremsstrahlung* (German for "braking radiation"). A bremsstrahlung collision is shown schematically in Figure 4–8. In the figure, an electron approaches the deflecting atom with a kinetic energy E_{k1} and recedes with a kinetic energy E_{k2} after having produced a single photon of energy $h\nu$. The conservation of energy law requires that

$$E_{k1} - E_{k2} = h\nu \tag{4–8}$$

Because the atom's mass is at least 2000 times the electron's mass, we can ignore the very small energy of the recoiling atom. Whereas classical electromagnetic theory predicts continuous radiation throughout the time in which the electron is accelerated, quantum theory requires the radiation of a discrete number of photons. That this occurs in the bremsstrahlung process is clearly illustrated in the production of x-ray photons.

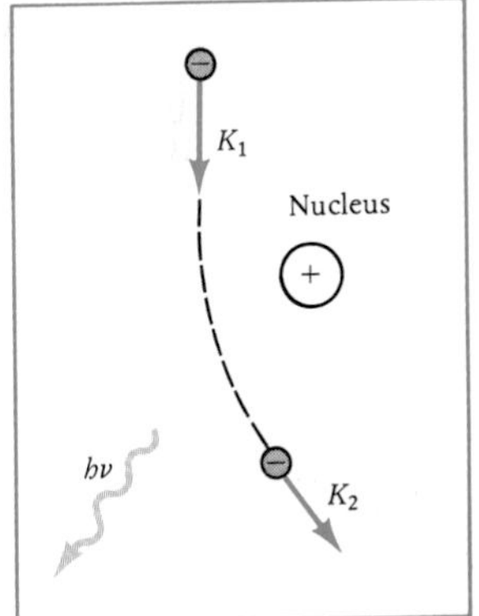

FIGURE 4–8. Bremsstrahlung *collision between an electron and a positively charged nucleus, with the emission of a single photon.*

X-rays were discovered and first investigated in 1895 by Wilhelm Roentgen, who assigned this name because the true nature of the radiation was at first unknown. X-rays are now known to consist of electromagnetic waves, or photons, having wavelengths about 10^{-10} m = 1 Å. It has been experimentally confirmed that they exhibit the wave phenomena of interference, diffraction, and polarization. Because they pass readily through many materials that are opaque to visible light, and because a typical x-ray wavelength is far shorter than the wavelengths of visible light, the experiments require considerable ingenuity. Our chief concern here is the energy characteristics of x-ray production. (We shall postpone discussion of x-ray absorption and intensity to Section 4–7 and of x-ray wavelengths to Section 5–2.)

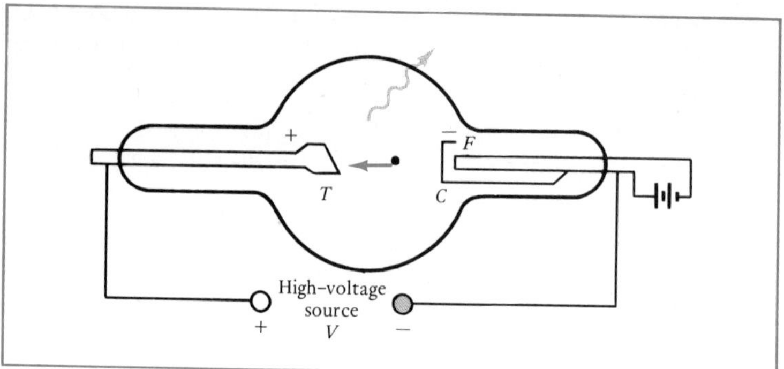

FIGURE 4–9. *Essential parts of an x-ray tube.*

The essential parts of a simple x-ray tube are shown in Figure 4–9. Electric current through the filament F heats the cathode to a high temperature so that the electrons have enough kinetic energy to overcome their binding to the cathode surface and thus be released in thermionic emission. These electrons are then accelerated through a vacuum by a large electrostatic potential difference V, typically several thousand volts, and strike the target T, which is the anode. While going from the cathode to the anode and before striking the target, each electron attains a kinetic energy E_k, given by

$$E_k = eV$$

where e is the electron charge.†

On striking the target, the electrons are decelerated and brought essentially to rest in collisions. Each electron loses its kinetic energy $E_k = eV$ because of its impact with the target. Although most of this energy appears as thermal energy in the target, a small fraction goes into the production of electromagnetic radiation through the bremsstrahlung process. Any electron striking the target can make numerous bremsstrahlung collisions with atoms in the target, thereby producing many photons. Occasionally, an electron is brought to rest in a single collision, and its *entire* kinetic energy is converted into the electromagnetic energy of a *single* photon. For this, the *most* energetic photon is produced and with $E_{k1} = eV$ and $E_{k2} = 0$, Equation 4–8 becomes

$$eV = E_k = h\nu_{max}$$

where ν_{max} is the maximum frequency of the x-ray photons produced. More often, electrons lose their energy at the target by heating it or by producing two or more photons, with frequencies lower than ν_{max}. We expect, then, a distribution in photon energies with a well-defined maximum frequency ν_{max} or a minimum wavelength $\lambda_{min} = c/\nu_{max}$ given by

$$E_k = h\nu_{max} = \frac{hc}{\lambda_{min}} = eV \qquad (4\text{–}9)$$

Note that this equation is equivalent to Equation 4–5 for the photoelectric effect when the binding-energy term is ignored.

† We have ignored the electron's kinetic energy as it left the cathode, because it is typically much less than Ve. When the electron strikes the target, it acquires an additional energy, the energy that binds it to the target surface; because the binding energy is always only a *few* electron volts, whereas E_k is at least several *thousand*, we may properly ignore it, too.

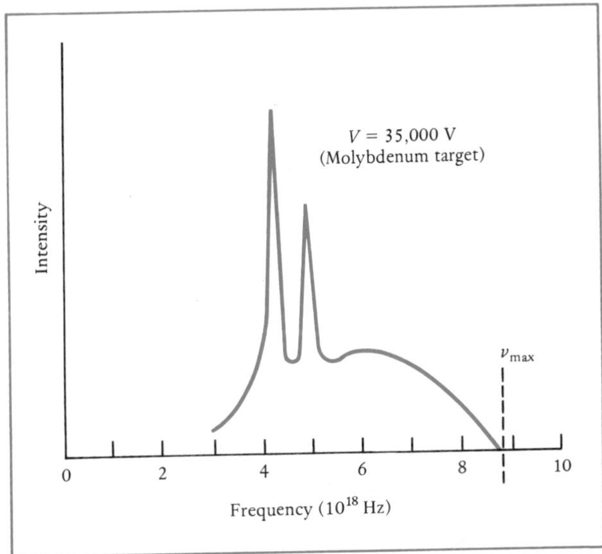

FIGURE 4–10. *Intensity of x-rays as a function of frequency.*

Figure 4–10 shows the variation in the intensity of emitted x-rays as a function of frequency under typical operating conditions. An abrupt cutoff appears at the limit ν_{max} of the *continuous x-ray spectrum* (dotted line); this limit is determined only by the accelerating potential V of the x-ray tube. The value of hc/e can be determined precisely by using Equation 4–9 and simultaneous measurements of λ_{min} and V. The value obtained for Planck's constant h agrees completely with values deduced from experiments on the photoelectric effect and other experiments.

Superimposed on the continuous spectrum are sharp increases in the intensity, or peaks, whose wavelengths characterize the target material; the explanation of these characteristic x-ray lines is found in the quantum description of the atomic structure of the target atoms. When the accelerating voltage V is changed but the target material is not, the high-frequency limit of the continuous x-ray spectrum changes but the characteristic x-ray frequencies do not. Conversely, when the target material is changed but the accelerating voltage is not, the characteristic x-ray spectrum changes but the limit of the continuous x-ray spectrum does not.

It is found that appreciable x-ray production occurs only if the accelerating potential V is of the order of 10,000 V or larger. Even at 10 kV (λ_{min} = 1.24 Å by Equation 4–9) somewhat less than 1 percent of the total energy appears as electromagnetic radiation; the remainder appears as thermal energy in the target.

4-4 The Compton Effect

In the photoelectric effect, a photon gives all its energy to a bound electron-metal system; it is also possible for only a part of a photon's energy to be given to a charged particle. This kind of interaction between electromagnetic waves and a material substance is a *scattering* of the waves by the charged particles of the substance. The quantum theory of the scattering of electro-

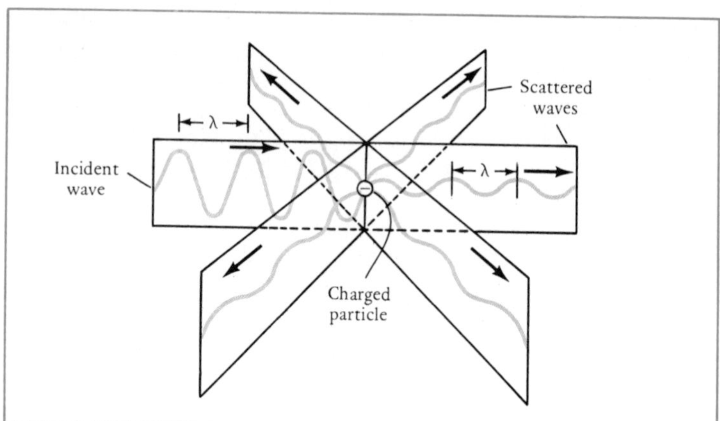

FIGURE 4–11. *Classical scattering of electromagnetic radiation by a charged particle.*

magnetic waves is known as the *Compton effect.* We shall first review briefly the classical theory of the scattering of electromagnetic waves by charged particles.

When a monochromatic electromagnetic wave impinges upon a charged particle whose size is much smaller than the wavelength of the radiation, the charged particle will be acted on principally by the sinusoidally varying electric field of the wave. Under the influence of this changing electric force, the particle will oscillate in simple harmonic motion at the frequency of the incident radiation (see Figure 4–11). Since the charge is accelerated continuously, it will produce electromagnetic radiation of the *same* frequency in all directions; the intensity is greatest in the plane perpendicular to the direction of motion of the oscillating charge and zero along the line of oscillation.† Classical theory predicts, then, that the scattered radiation will have the *same* frequency as the incident radiation. The charged particle becomes the transfer agent, absorbing some of the energy from the incident beam and reradiating this energy at the same frequency (or wavelength), although scattering it in all directions. The scattering particle neither gains nor loses energy, since it reradiates at the same rate as it absorbs. The classical scattering theory agrees with experiment for wavelengths of visible light and all other longer wavelengths of radiation. A simple example of the unchanged frequency of scattered radiation is light reflected from a mirror (a collection of scatterers); it undergoes *no* apparent change in frequency.

The magnetic field of an incident electromagnetic wave also affects a charged particle. A charge moving in the transverse magnetic field of the electromagnetic wave is acted on by a magnetic force *along* the direction of wave propagation.‡ When absorption is complete, this results in a radiation force F_r on the charged particle, given by $F_r = P/c$, where P is the power of the incident wave. Moreover, since a classical electromagnetic wave can exert a force on a scattering center, we attribute momentum p to the wave,

$$p = \frac{E}{c}$$

where E represents the electromagnetic energy of the incident wave.

† See Weidner and Sells, *Elementary Classical Physics,* 2nd ed., Section 35–7.
‡ See Weidner and Sells, *Elementary Classical Physics,* 2nd ed., Section 35–5.

Now we consider scattering from the point of view of the quantum theory. Using Einstein's successful photon interpretation of the photoelectric effect, Arthur H. Compton in 1922 used the particlelike, quantum nature of electromagnetic radiation to explain the scattering of x-rays. In the quantum theory, electromagnetic radiation consists of photons, each with an energy given by $E = h\nu$. Because a photon may be regarded as a zero-rest-mass particle moving at speed c, Equation 3–15 shows that the magnitude of the corresponding linear momentum $\mathbf{p}$ is given by E/c, in agreement with the classical result. For photons, we then have

$$p = \frac{E}{c} = \frac{h\nu}{c} = \frac{h}{\lambda} \tag{4–10}$$

Each photon in a beam of monochromatic electromagnetic radiation of wavelength λ has a momentum equal to h/λ. Equation 4–10 shows that the momentum of a photon is precisely specified when the wavelength, the frequency, or the energy of the photon is known. The direction of $\mathbf{p}$ is along the direction of propagation of the wave.

The distinctive feature the quantum theory introduces is that for monochromatic waves the electromagnetic momentum occurs not in arbitrary amounts but only in integral multiples of the momentum h/λ carried by a single photon. From Equation 4–10 we see that a photon's momentum increases with frequency, just as its energy increases with frequency. Therefore, the momentum of a high-frequency, high-energy photon, such as a γ ray, will far exceed the momentum of a low-frequency, low-energy photon, such as a radio photon.

When we regard a monochromatic electromagnetic beam as consisting of a collection of particlelike photons, each with a precisely defined energy and momentum, the scattering of electromagnetic radiation becomes a problem involving the collision of a photon with a charged particle. Then the problem is solved merely by applying the laws of energy and momentum conservation. Figures 4–12*a* and *b* show the incident photon and particle (at rest) before the collision and the moving particle and scattered photon after collision. In applying the conservation laws, we need to be concerned not with the details of the interaction between the photon and particle during the collision, but with only the total energy and momentum going into and coming out of the collision.

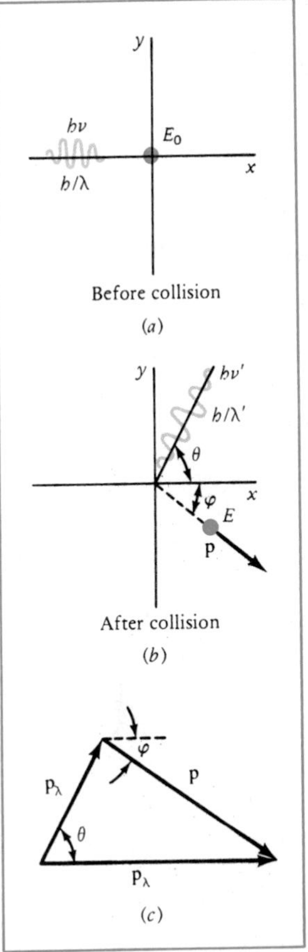

FIGURE 4–12. *Collision of a photon and a particle initially at rest.*

Unlike the classical scattering of electromagnetic waves, in which the particle after collision is assumed to gain essentially no energy, the quantum treatment calls for it to gain some; because its kinetic energy may be large, we shall treat it relativistically.

We take the particle, of rest mass m_0 and rest energy $E_0 = m_0c^2$, to be *free* and initially at rest. Then, applying energy conservation to the collision of Figure 4–12 gives

$$h\nu + E_0 = h\nu' + E \tag{4–11}$$

Here E is the energy of the recoiling *particle* after collision, $h\nu$ is the energy of the incident photons, and $h\nu'$ is the energy of the scattered photons. Since the final energy (rest energy plus kinetic energy) of the recoil particle, $E = \gamma E_0$, must exceed its initial energy E_0, we immediately see from Equation 4–11 that $h\nu' < h\nu$. Consequently, the scattered photon has *less* energy, a lower frequency, and a longer wavelength than the incident photon. This disagrees with the classical prediction of no frequency change on scattering. Do not think that the scattered photon is merely the incident photon moving in a different direction. Instead, the incident photon is annihilated, and the scattered photon is created.

Momentum conservation is implied by the vector triangle of Figure 4–12*c*, where $\mathbf{p} = \gamma m_0 \mathbf{v}$ is the relativistic momentum of the recoiling particle. The magnitude of the momentum of the incident photon is $p_\lambda = h\nu/c = h/\lambda$ and of the scattered photon is $p_{\lambda'} = h\nu'/c = h/\lambda'$. The scattering angle θ is the angle between the directions of $\mathbf{p}_\lambda$ and $\mathbf{p}_{\lambda'}$, the directions of the incident and scattered photons.

We want to solve for the change in wavelength $\lambda' - \lambda = \Delta\lambda$ in terms of θ. Applying the law of cosines to the triangle in Figure 4–12*c*, we have

$$p_\lambda{}^2 + p_{\lambda'}{}^2 - 2p_\lambda p_{\lambda'} \cos\theta = p^2 \tag{4–12}$$

Multiplying both sides of this equation by c^2 and recalling that for a photon $pc = h\nu$, we have

$$h^2\nu^2 + h^2\nu'^2 - 2h^2\nu\nu' \cos\theta = p^2c^2 \tag{4–13}$$

We can arrive at a similar expression from the energy-conservation relation, Equation 4–11. We place $h\nu$ and $h\nu'$ on one side of Equation 4–11 and E and E_0 on the other; then squaring the equation, we get

$$\begin{aligned} h^2\nu^2 + h^2\nu'^2 - 2h^2\nu\nu' &= E^2 + E_0{}^2 - 2EE_0 \\ &= 2E_0{}^2 + p^2c^2 - 2EE_0 \end{aligned} \tag{4–14}$$

where we have replaced E^2 with $E_0{}^2 + p^2c^2$, using Equation 3–15. Subtracting Equation 4–13 from Equation 4–14, we have

$$-2h^2\nu\nu'(1 - \cos\theta) = 2E_0{}^2 - 2EE_0$$

$$h^2\nu\nu'(1 - \cos\theta) = E_0(E - E_0) = m_0c^2(h\nu - h\nu')$$

$$\frac{h}{m_0c}(1 - \cos\theta) = c\frac{(\nu - \nu')}{\nu\nu'} = \frac{c}{\nu'} - \frac{c}{\nu} = \lambda' - \lambda$$

The increase in wavelength $\Delta\lambda$ then, is

$$\Delta\lambda = \lambda' - \lambda = \frac{h}{m_0c}(1 - \cos\theta) \tag{4–15}$$

This is the basic equation of the Compton effect. It gives the increase $\Delta\lambda$ in the wavelength of the scattered photon over the value of the wavelength of the incident photon. We see that $\Delta\lambda$ depends only on the rest mass m_0 of the recoiling particle, and on Planck's constant h, the speed c of light, and the angle θ of scattering. It is perhaps surprising to find that $\Delta\lambda$ is *independent* of the incident photon's wavelength λ. The quantity h/m_0c, appearing on the right-hand side of Equation 4–15 and having the dimension of length, is known as the *Compton wavelength*. Although the scattering angle θ determines the wavelength increase $\Delta\lambda$ unambiguously, we cannot predict in advance the angle at which any one photon will emerge.

If the recoiling particle is a free electron within the scattering material, then $m_0 = 9.11 \times 10^{-31}$ kg and $h/m_0c = 0.024\,26$ Å. When a photon emerges at, for example, $\theta = 90°$ with respect to the incident-photon direction, the wavelength change, by Equation 4–15, is 0.024 Å. When a scattered photon emerges at $\theta = 180°$, traveling, in other words, in the backward direction, with the recoil electron traveling in the forward direction, the collision can be characterized as a head-on collision. Then the wavelength change is a maximum and equal to 0.049 Å. In such a collision, the electron's kinetic energy is also a maximum.

For the 90° scattering, by a free electron, of incident radiation in the visible region, say 4000 Å, the *fractional* increase in wavelength $\Delta\lambda/\lambda$ is only 0.006 percent. Such a shift in wavelength is completely masked in visible light since electrons in an ordinary scattering material are not at rest but in thermal motion. At room temperature this results in a spread $\Delta\lambda/\lambda \approx 0.3$ percent arising from thermal motion of electrons. An observable shift of, say, $\Delta\lambda/\lambda = 2$ percent can be obtained by using incident radiation of wavelength $\lambda = 1$ Å; then $\Delta\lambda = 0.024$ Å. Thus, there is an easily observed shift for x-ray photons and photons of shorter wavelength. For photons of longer wavelength, the fractional change in wavelength is very small, and the scattered radiation has nearly the same wavelength and frequency as the incident radiation. Classically, the wavelengths of the incident and scattered radiations are equal; hence Compton scattering agrees with classical scattering in the region of $\Delta\lambda/\lambda \ll 1$. This is another example of the correspondence principle; by Equation 4–15,

$$\lim_{\substack{m_0 \to \infty \\ \text{or } h \to 0}} \frac{\Delta\lambda}{\lambda} = \lim_{\substack{m_0 \to \infty \\ \text{or } h \to 0}} \frac{h}{m_0 c\lambda} = 0$$

That the scattering of x-rays agrees with the photon model instead of the classical model (which predicts no change in wavelength) was shown first by A. H. Compton in 1922. Figure 4–13 gives schematically the experimental arrangement for monochromatic x-rays incident on a target of carbon, a substance having many free electrons. For any fixed angle θ, the detector (see Sections 5–2 and 8–3) can measure the scattered radiation's intensity as a function of wavelength. (Compare Figure 4–13 with Figure 4–11, where $\lambda = \lambda'$ and $\Delta\lambda = 0$.) Figure 4–14 shows the intensity of the scattered radiation versus the scattered wavelength for several fixed angles θ.

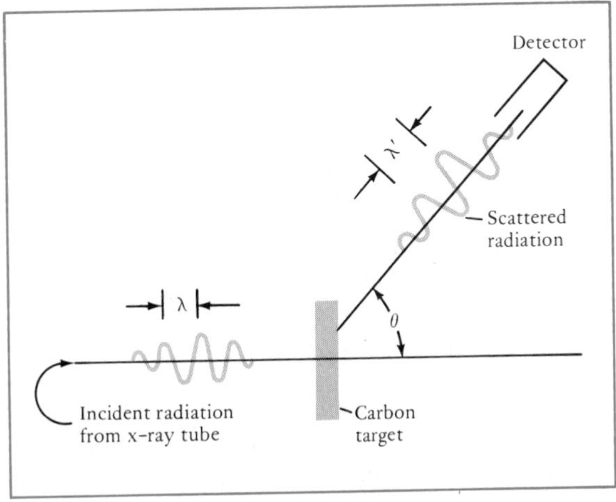

FIGURE 4–13. *Schematic of experimental arrangement for the Compton effect.*

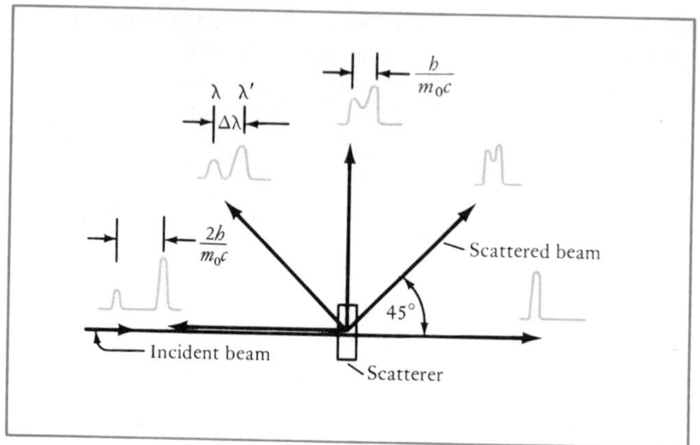

FIGURE 4–14. *Intensity of scattered radiation versus wavelength of scattered radiation for different angles θ.*

For any particular scattering angle θ, *two* predominant wavelengths are present in the scattered radiation: one of the same wavelength λ as the incident beam, the *unmodified wave,* and a second, of longer wavelength λ', the *modified wave,* given by the Compton equation (4–15). The unmodified wavelength results from the coherent scattering of the incident radiation by the inner electrons of atoms; these electrons are so tightly bound to the atoms that a photon cannot strike one of them without simultaneously moving the entire atom. The mass m_0 of one of these tightly bound electrons is, then, *effectively* the mass M_0 of the atom. Therefore, in a Compton collision between a photon and a tightly bound electron, the wavelength change $\Delta\lambda$ is $(h/M_0c)(1 - \cos\theta) \approx 0$, because M_0 is always thousands of times greater than m_0.

The Compton effect provides a simple way of determining the energy of a photon. From Equation 4–11, we have

$$E_k = E - E_0 = h\nu - h\nu'$$

Because $\nu = c/\lambda$ and $\nu' = c/\lambda' = c/(\lambda + \Delta\lambda)$, we can write this as

$$E_k = h\nu \frac{\Delta\lambda}{\lambda + \Delta\lambda} = h\nu \frac{\Delta\lambda/\lambda}{1 + (\Delta\lambda/\lambda)} \tag{4–16}$$

where $\Delta\lambda$ depends on the scattering angle θ (Equation 4–15). The kinetic energy of the recoil electron is a maximum, $E_{k,max}$, when a "head-on" collision occurs, the electron recoiling in the forward direction and the scattered photon traveling in the backward direction. In such a collision, $\theta = 180°$, and $\Delta\lambda = 2h/m_0c$; and Equation 4–16 becomes

$$E_{k,max} = h\nu \frac{2h\nu/m_0c^2}{1 + 2h\nu/m_0c^2} \tag{4–17}$$

Therefore, if we measure the energy of the most energetic recoil electrons, we can compute the energy $h\nu$ of the incident photon from Equation 4–17; the converse is also true.

The Compton effect shows clearly the particlelike aspects of electromagnetic radiation—not only a precise energy $h\nu$ but also a precise momentum h/λ can be assigned to a photon. Along any direction, the total momentum of a monochromatic electromagnetic beam can then assume *not* any value but only an exactly integral multiple of h. The linear momentum as well as the energy of electromagnetic radiation is quantized.

Example 4–3. A 2.00-Å photon interacts with the bound electron in a hydrogen atom (binding energy, 13.6 eV). A Compton collision occurs and the electron moves forward in the same direction as the incident photon. (*a*) What is the energy of the electron? (*b*) What is the energy of the scattered photon?

(*a*) Since the energy of the incident photon,

$$E = \frac{12\,400\ \text{eV}\cdot\text{Å}}{\lambda} = \frac{12\,400\ \text{eV}\cdot\text{Å}}{2.00\ \text{Å}} = 6200\ \text{eV}$$

is much larger than the binding energy of the electron, we can ignore the binding energy. Then, by Equation 4–17,

$$E_k = h\nu \frac{(2h\nu)/m_0c^2}{1 + 2h\nu/m_0c^2}$$

$$= (6.2 \times 10^3)\frac{(2 \times 6.2 \times 10^3)/(5.1 \times 10^5)}{1 + [(2 \times 6.2 \times 10^3)/(5.1 \times 10^5)]}\ \text{eV}$$

$$= 147\ \text{eV}$$

(*b*) Since the electron moves forward, the scattered photon must move backward. Then $\theta = 180°$, and $\Delta\lambda = 2h/m_0c = 0.049$ Å. Therefore,

$$\lambda' = \lambda + \Delta\lambda = 2.0 + 0.049 \approx 2.049\ \text{Å}$$

$$E'_\gamma = \frac{12\,400}{2.049\ \text{Å}} = 6052\ \text{eV}$$

These results are consistent with energy conservation; the total kinetic energy after collision is 147 eV (electron) plus 6052 eV (scattered photon) = 6200 eV, which was the energy of the incident photon.

Compare this Compton scattering of a 2.00-Å photon with the photoelectric effect of a 2.00-Å photon with a hydrogen atom in Example 4–1.

Example 4–4. Classical electromagnetic theory predicts that the radiation pressure P_r of a completely absorbed electromagnetic beam of intensity I will be $P_r = I/c$ for incidence along the normal to the absorbing surface.† Derive this relation from the quantum theory by assuming that the radiation consists of photons, each with momentum $h\nu/c$.

We designate the flux of the photon beam to be N photons per unit time per unit area perpendicular to the beam direction. Then, with an energy $h\nu$ carried by each photon in the beam, the total energy per unit time per unit transverse area, or the intensity I, is given by

$$I = N(h\nu)$$

If the beam strikes a surface that absorbs it completely, each photon transmits a momentum $h\nu/c$ to the surface. The total momentum trans-

† See Weidner and Sells, *Elementary Classical Physics*, 2nd ed., Equation (35–31).

mitted per unit time per unit area is then $N(h\nu/c)$; but the total momentum transferred per unit time is just the radiation force on the absorbing surface, and this force per unit area is the radiation pressure P_r:

$$P_r = \frac{Nh\nu}{c}$$

Eliminating $Nh\nu$ from the two equations, we have

$$P_r = \frac{I}{c}$$

for complete absorption. If the beam strikes a totally reflecting surface along the normal to this surface, the momentum is transferred *twice* to the surface, once when an incident photon is annihilated on striking the surface and again when a second photon of the same frequency is created and travels outward from the surface. Consequently, the radiation pressure is given by

$$P_r = \frac{2I}{c}$$

for complete reflection.

Emission of photons is the reverse of photon absorption. When a source emits photons, momentum conservation requires that the emitter recoil in the opposite direction. An example of this is the "photon rocket," which consists simply of a unidirectional source of electromagnetic radiation, like the laser. The rocket gains momentum in the forward direction because photons are emitted to the rear. Notice that a photon carries more momentum per total energy than a particle of nonzero rest energy does; that is, from $p = (E^2 - E_0^2)^{1/2}/c$, we see that for a given energy E the momentum p is a maximum for $E_0 = 0$.

As a zero-rest-mass particle, a photon always is observed to travel at the single speed c in any reference frame. Its energy $h\nu$ and momentum $h\nu/c$, on the other hand, depend on the observer's reference frame. Both quantities are proportional to the photon's frequency, and the frequency depends in turn on the reference frame by the (relativistic) Doppler effect. See Problem 2–29. Thus, an observer traveling in the direction (relative to the source) in which a photon is moving finds the photon's frequency, energy, and momentum all to have decreased relative to their values measured in a reference frame at rest with respect to the source of photons.

4-5 Pair Production and Annihilation

The photoelectric effect, the bremsstrahlung process, and the Compton effect all exemplify the conversion of the electromagnetic energy of photons into the kinetic energy and potential energy of material particles and vice versa. It is natural to ask whether it is possible to convert a photon's energy into *rest* mass—that is, to create pure matter from pure energy—or on the other hand, to convert rest energy into electromagnetic energy. The answer is yes, provided such conversions do not violate the conservation laws, including those of energy, momentum, and electric charge.

PAIR PRODUCTION. Consider first the minimum energy required to create a single material particle. Since the electron has the smallest nonzero rest mass of all known particles, it requires the least energy for its creation. Because a photon has zero electric charge, the law of charge conservation precludes the creation of a *single* electron from a photon. The creation of an electron pair, however, consisting of two particles with opposite electric charges, is possible and has been observed. The positively charged particle is called a *positron* and is said to be the *antiparticle* of the electron. The electron and the positron are similar in all ways except in the signs of their charges, $-e$ and $+e$ (and the effects of this difference). The minimum energy $h\nu_{\text{min}}$ needed to create an electron-positron pair is, by the conservation of energy,

$$h\nu_{\text{min}} = 2m_0c^2 \tag{4–18}$$

Since the rest energy m_0c^2 of an electron or of a positron is 0.51 MeV, the threshold energy $2m_0c^2$ for pair production is 1.02 MeV. The photon's wavelength corresponding to this threshold energy is 0.012 Å; hence electron pairs can be produced only by photons of very short wavelength (see Figure 4–7). The process in which matter is created from electromagnetic radiation is called *pair production,* because a particle and its antiparticle must always be created together if the conservation laws are to be satisfied. This phenomenon is a very emphatic demonstration of the interconvertibility of mass and energy.

If a photon's energy exceeds the threshold energy $2m_0c^2$, the excess can appear as kinetic energy of the created pair. Applying the conservation of energy to pair production gives

$$h\nu = m^+c^2 + m^-c^2 = (m_0c^2 + E_{\text{k}}{}^+) + (m_0c^2 + E_{\text{k}}{}^-)$$

$$h\nu = 2m_0c^2 + (E_{\text{k}} + E_{\text{k}}{}^-)$$

where ν is the frequency of the incident photon, and $E_{\text{k}}{}^+$ and $E_{\text{k}}{}^-$ are the kinetic energies of the created particles. The minimum energy $h\nu_{\text{min}}$, just enough to produce the pair, is obtained by setting the kinetic energies of the created particles equal to zero: $E_{\text{k}}{}^+ + E_{\text{k}}{}^- = 0$.

For pair production, both momentum conservation and energy conservation must hold. It is easy to prove that total momentum cannot be conserved in the disintegration of a high-energy photon into a particle-antiparticle pair. Suppose that in inertial system S, a photon of energy $E_\gamma = p_\gamma c$ is annihilated and a pair is produced, as shown in Figure 4–15. Since the conservation laws must hold in any inertial system, we can view the process in the center-of-mass system S' of the created pair (Figure 4–15*b*). Then the total momentum after collision is zero; but the momentum of the photon before the collision is not zero, since it must be moving at speed c in any reference frame. Therefore, a photon cannot decay spontaneously into a particle-antiparticle pair in free space.

Pair production can occur if a photon (of sufficient energy) passes near some particle, such as a massive atomic nucleus. This particle can then carry away some of the photon's momentum to assure momentum conservation.

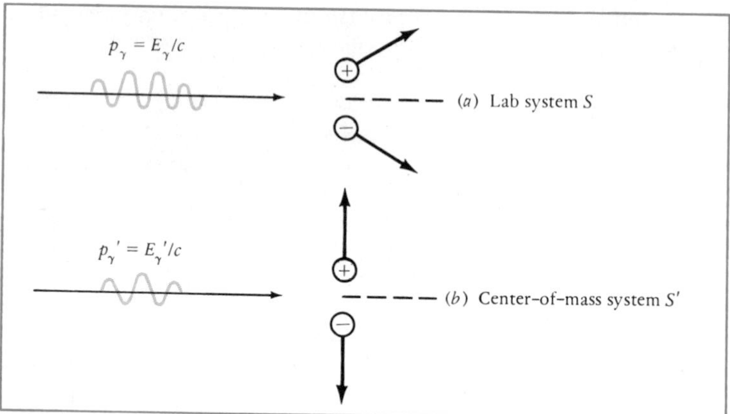

FIGURE 4–15.

If the particle's mass is very large compared with the mass of the created particles, the energy of the recoiling heavy particle would be so small as to be excludable from Equation 4–18.

Figure 4–16 is a schematic drawing of pair production, and Figure 4–17 is a bubble-chamber photograph showing the creation of electron-positron pairs. Figure 4–17 shows that high-energy γ-ray photons entered the area (top of the photograph), were annihilated, and electron-positron pairs were created. Although the incoming photons leave no tracks in the bubble chamber, the paths of the charged particles are visible because of the ionization effects they produce as they travel through the liquid hydrogen. The trajectories of the oppositely charged particles (with approximately equal kinetic energies) show opposite curvatures;† the particles were deflected into oppositely directed circular arcs by a uniform magnetic field.

The energy of a photon producing an electron-positron pair can be computed by means of Equation 4–19 if the kinetic energies of the electron and the positron are known. These energies can be determined from a photograph such as Figure 4–17 if the magnetic field B and the radius r of curvature of the trajectories are measured. The relativistic momentum p of each particle is given by

$$p = \gamma m_0 v = QBr \tag{3–9}$$

and the total energy E or the kinetic energy $E - E_0$ of the particle can then be computed by using Equation 3–15, $E^2 = E_0^2 + (pc)^2$.

The existence of positrons was predicted on theoretical grounds by P. A. M. Dirac in 1928. Four years later, C. D. Anderson observed and identified a positron during his studies of cosmic radiation. Shortly thereafter, electron-positron pairs were produced in the laboratory by means of particle accelerators operating at a few MeV of energy; they are now a commonly observed phenomenon in the interaction of high-energy photons and matter. Proton-antiproton and neutron-antineutron pairs were first created in the laboratory in 1955. Their threshold energies are several GeV. (The proton and neutron masses are approximately equal to 1 GeV and therefore require accelerating machines of very high energy.)

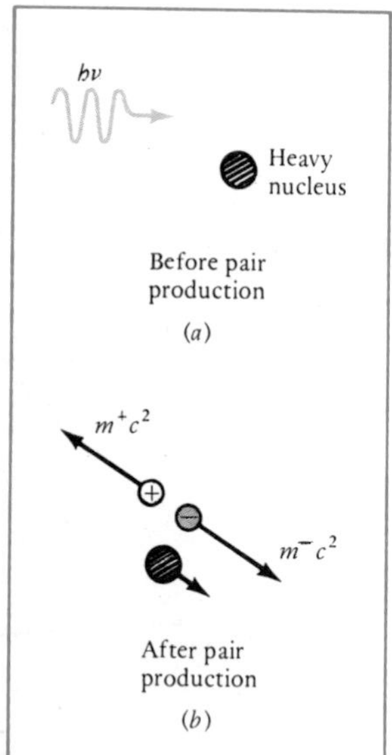

FIGURE 4–16. *Schematic diagram of pair production.*

† To be precise, the positron has, on the average, a greater kinetic energy than the electron, because the positively charged nucleus repels the positron and attracts the electron.

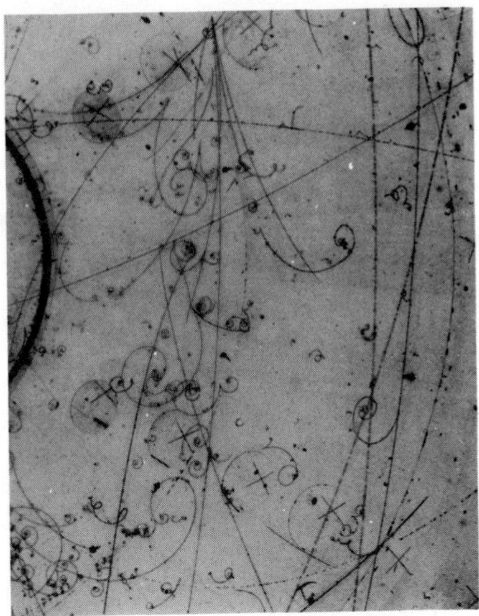

FIGURE 4–17. *Bubble-chamber photograph showing the creation of electron-positron pairs by the annihilation of photons. An electron or positron leaves a wake of bubbles along its path through the liquid hydrogen; a photon produces no track. Each electron-positron pair appears as a* V *with curled ends. An external magnetic field deflects an electron and a positron in opposite directions; as each electron and positron loses kinetic energy, its radius of curvature decreases. (Courtesy, Fermi National Accelerator Laboratory)*

PAIR ANNIHILATION. The annihilation of particle-antiparticle pairs and the concomitant creation of photons is the inverse of pair production. Consider the annihilation of matter and the creation of electromagnetic energy that may occur when an electron and positron are close and essentially at rest. The total linear momentum of the two particles is initially zero; therefore a *single* photon cannot be created when the two particles unite and are annihilated since that would violate momentum conservation. Momentum can, however, be conserved when *two* photons are created and move in opposite directions with equal momenta. Such a pair of photons would have equal frequencies and energies; see Figure 4–18. (Actually, three or more photons can be created, but the probability is *much* smaller than that for two photons. Similarly, when many electron-positron pairs are annihilated near a heavy nucleus, a single photon may be produced in a fraction of the annihilations.)

When two photons are created by pair annihilation of an electron (mass m_0) and positron (mass m_0) at rest, energy conservation requires

$$2m_0c^2 = h\nu_1 + h\nu_2$$

and momentum conservation,

$$0 = \frac{h\nu_1}{c} - \frac{h\nu_2}{c}$$

where ν_1 and ν_2 are the respective frequencies of the two created photons. It follows that

$$h\nu_1 = h\nu_2 = h\nu_{\text{min}} = m_0c^2$$

where $h\nu_{\text{min}}$ is the minimum energy of the outgoing photons and equals the rest mass of the electron (or positron), 0.51 MeV. If the particle and the antiparticle are moving together and annihilate each other in flight, the total energy of the created photons will be greater than $2m_0c^2$.

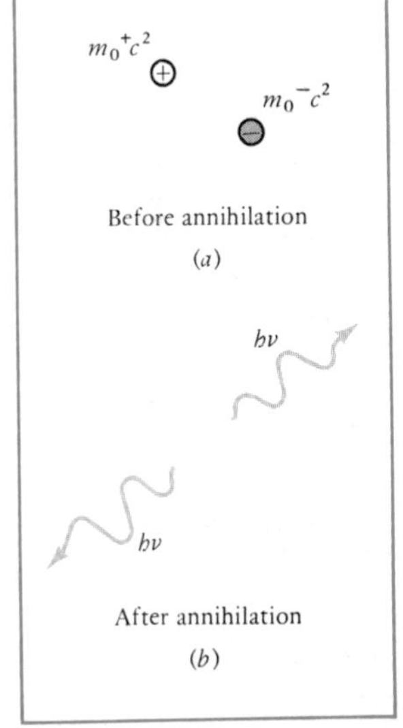

FIGURE 4–18. *Pair annihilation and the creation of two photons.*

Annihilation is the ultimate fate of positrons. When a high-energy positron appears, as in pair production, it loses its kinetic energy in collisions as it passes through matter, finally moving at low speed. It then combines with an electron, forming a bound system called a positronium atom; this decays very quickly (10^{-10} s) into two photons of equal energy. Thus, the death of a positron is signaled by the appearance of two annihilation quanta, or photons, moving in opposite directions and each having $\frac{1}{2}$ MeV of energy. The transitoriness of positrons is due not to an intrinsic instability but to the high risk of their collision with electrons and subsequent annihilation.

In our part of the universe there is a preponderance of electrons, protons, and neutrons; their antiparticles, when created, quickly combine with them in annihilation processes. It is conceivable, although at present purely conjectural, that there exists a part of the universe in which positrons, antiprotons, and antineutrons predominate.

Pair production and annihilation are particularly striking examples of mass-energy equivalence and provide still another confirmation of the theory of relativity.

4-6 Photon-Electron Interactions

Figure 4–19 summarizes the important photon-electron interactions, or collisions. In each instance a photon, an electron, or a positron approaches a slab of material and interacts with a particle in the slab, and one or more particles emerge. The salient features of each of these interactions, in the order in which they appear in the figure are:

a. The photoelectric effect: A photon strikes a bound electron and disappears, and the dislodged electron emerges.
b. The Compton effect: A photon collides with a free electron, thereby effecting the creation of a second photon of lower energy. The newly created scattered photon and recoil electron emerge.
c. Pair production: A photon is annihilated near a heavy particle, and an electron-positron pair is created and emerges.
d. Bremsstrahlung: An electron is deflected near a heavy particle, a photon is created, and the photon and deflected electron emerge.
e. Pair annihilation: A positron combines with an electron and they annihilate each other. Photons are produced and emerge.

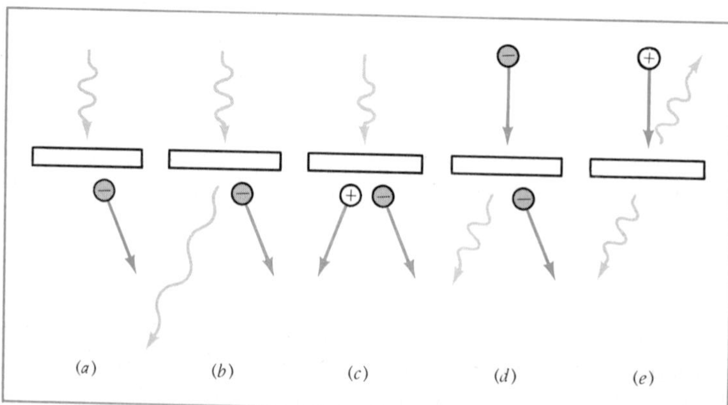

FIGURE 4–19. *Photon-electron interactions: (a) photoelectric effect; (b) Compton effect; (c) pair production; (d) bremsstrahlung; and (e) pair annihilation.*

All these photon-electron interactions exemplify just *one* basic interaction between the particle of the electromagnetic field (the photon) and a particle that can create an electromagnetic field (an electron or any other electrically charged particle). Even the ordinary electric, or Coulomb, force between electrically charged particles, and indeed all other electromagnetic effects, can be shown to result from an interchange of photons between charged particles (Section 12–1).

Note that the main features of the photon-electron collisions were derived simply by applying the laws of the conservation of energy, momentum, and electric charge and by assuming the existence of photons of energy $h\nu$ and momentum h/λ. We never considered the details of the interaction. Further, we did not calculate the probability of the occurrence of any of these processes. For the Compton effect, for example, we can predict the wavelength of a scattered photon for a particular direction, but we can't predict the direction of a particular photon. The probabilities of the occurrence of a specific photon-electron interaction, however, can be calculated very precisely by the methods of quantum electrodynamics.

Consider one final example of electron-photon interactions. Highly energetic (as high as 10^{19} eV) charged particles from the cosmic radiation enter the earth's atmosphere. They may produce a whole succession of electron-photon interactions, as follows. A collision between a cosmic ray particle and a nucleus may produce a high-energy γ ray through bremsstrahlung. This created γ ray may be annihilated on passing near a nucleus and so produce an electron-positron pair. The created charged particles, having large kinetic energies, may collide with nuclei they encounter on their way to the earth's surface and be deflected by them; by virtue of their acceleration during the collisions, they radiate high-energy photons by the bremsstrahlung process. Another possibility is that the positron may combine with the electron, both be annihilated, and two photons be created. The secondary photons may have energies exceeding 1.02 MeV and so produce more electron pairs. Thus, by the repeated occurrence of pair production, pair annihilation, bremsstrahlung, and to a lesser extent Compton and photoelectric collisions, a *cascade shower* of electrons, positrons, and photons is produced; the energy of the original photon becomes degraded and spread among many particles. The shower is effectively extinguished when pair production is energetically impossible. A diagram of photon-electron interactions is shown in Figure 4–20.

4–7 Absorption of Photons

The important processes that remove photons from a beam of electromagnetic radiation are the photoelectric effect, the Compton effect, and pair production (see Figures 4–19*a, b, c*). In each of the processes, a photon is removed from a forward-moving beam and an electron appears. Furthermore, each process occurs only when there are atoms with which the oncoming photons can collide and interact. The atoms provide bound electrons for the photoelectric effect, nearly free electrons for the Compton effect, and atomic nuclei for pair production. The intensity of the photon beam is reduced, then, only so far as the photons encounter and interact with

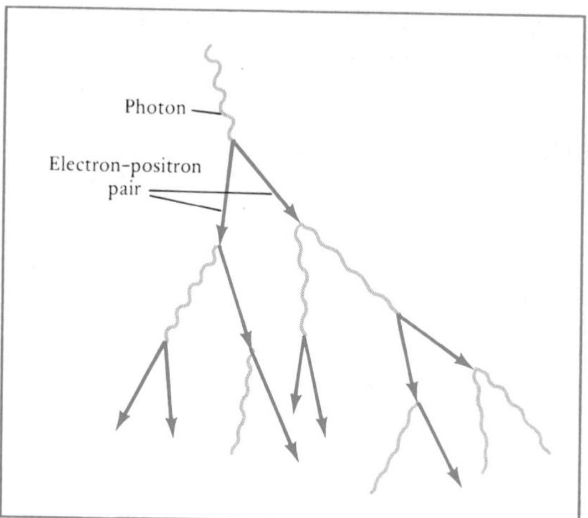

FIGURE 4–20. *Schematic representation of a cascade shower.*

atoms. (There is a process, which for the moment we ignore, in which a photon is selectively absorbed by a material according to its energy, and the total internal energy of the absorbing atom is thereby increased.)

The intensity I of electromagnetic radiation is defined as the energy per unit time passing through a unit area at right angles to the direction of propagation. Put in terms of a beam of monochromatic photons, it is the energy $h\nu$ of a single photon times the number of photons per unit time crossing a unit area perpendicular to the direction of the beam:

$$\text{Intensity of a photon beam} = \frac{\text{energy}}{\text{photon}} \times \frac{\text{number of photons}}{\text{area} \times \text{time}}$$

The last term on the right is, by definition, the *photon flux N*. Then,

$$I = (h\nu)N \tag{4–20}$$

When a photon beam traverses a material, the photon flux N is reduced, because photons are removed or are deflected from the forward direction. The absorption of photons by a material is shown schematically in Figure 4–21. Clearly, the probability that a photon will be removed from the beam increases with the number of atoms the beam encounters; therefore, the probability increases with the thickness of the absorber. In the figure, photons with a flux N are incident on a very thin absorber of thickness dx, and photons with a flux $N - dN$ emerge from the absorber in the forward direction. The number of photons removed per unit time by a unit area of the absorber is then dN. The number of photons removed in encounters with atoms in the absorber dN is proportional to the photon flux N. Further, since the number of atoms the beam encounters is directly proportional to the thickness of the absorber, dN is proportional also to dx. Therefore,

$$dN = -\mu N\,dx$$

where the proportionality constant is μ, the *absorption coefficient.* The minus sign appears because N decreases as x increases.

Let us rearrange terms and integrate x from zero thickness to a finite

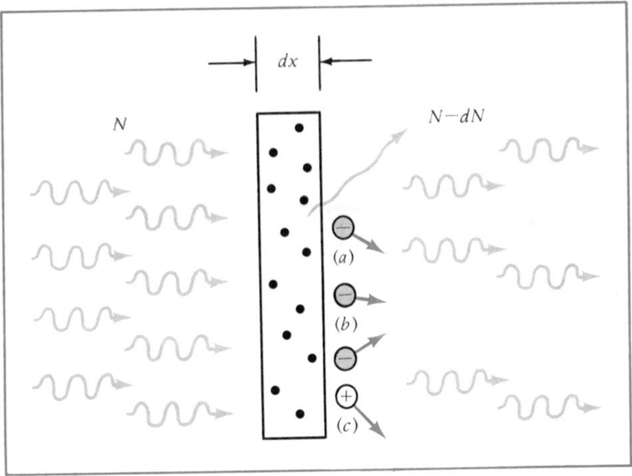

FIGURE 4–21. *Schematic representation of the absorption of photons by a material substance. (a) Compton effect; (b) photoelectric effect; (c) pair production.*

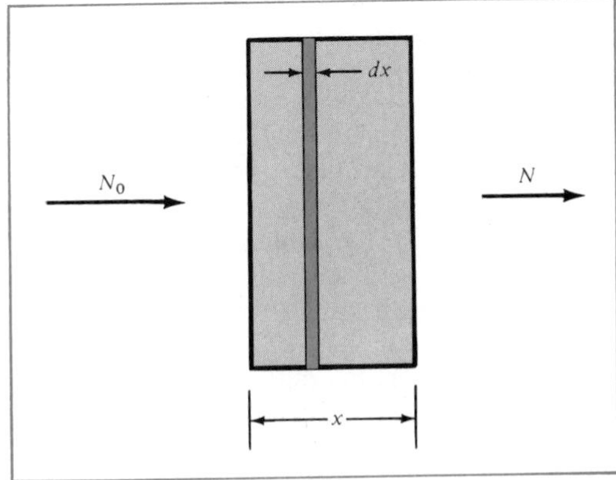

FIGURE 4–22. *Change in photon flux through an absorber.*

thickness x (Figure 4–22). Let us also integrate the flux from N_0, the flux incident on the absorber, to N, the flux emerging from the absorber of thickness x. Then we have

$$\int_{N_0}^{N} \frac{dN}{N} = -\mu \int_0^x dx$$

$$\ln \frac{N}{N_0} = -\mu x$$

$$N = N_0 e^{-\mu x} \tag{4–21}$$

Using Equation 4–20, we can write Equation 4–21 as

$$I = I_0 e^{-\mu x} \tag{4–22}$$

where $I_0 = (h\nu)N_0$ is the intensity incident on the absorber and $I = (h\nu)N$ is the intensity at a distance x from the front surface. This equation shows that the intensity of monochromatic electromagnetic radiation falls off exponentially through an absorber. The absorption increases with the thickness

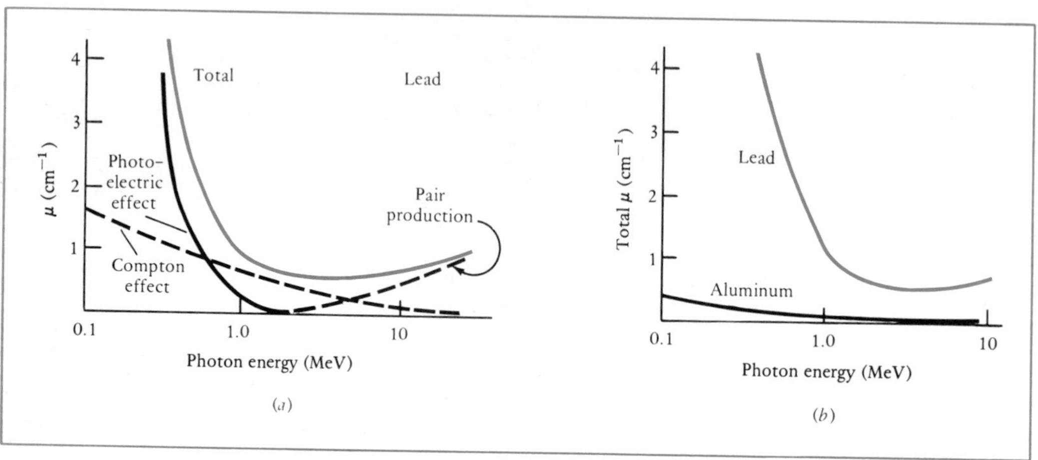

FIGURE 4–23. *Absorption coefficients as a function of photon energy; (a) for lead with several contributions to the total absorption coefficients; (b) for lead and aluminum.*

of the absorber (as x increases), or as the absorption coefficient μ increases. We also see from Equation 4–22 that when $\mu x = 1$, or $x = 1/\mu$, then $I = I_0/e$. Therefore, the quantity $1/\mu$ represents the absorber thickness at which the intensity I is $1/e$, or 37 percent, of the incident intensity I_0.

For a particular photon energy and for a particular absorbing material, the absorption coefficient μ is a constant having the unit reciprocal length. Its value does, however, change from one material to another, and it also depends, for a given absorbing material, on the energy (or frequency) of the radiation. Figure 4–23a shows the various contributions to the total absorption coefficients for lead as a function of the photon energy (plotted on a logarithmic scale). For low-energy photons, whose removal occurs principally through the process of the photoelectric effect, the absorption coefficient μ is large. It is smaller at intermediate energies, at which Compton collisions are the most effective in absorbing photons. It reaches a minimum in the vicinity of a few MeV and then rises again with photon energy. A little before its minimum, at 1.02 MeV (where the photon's wavelength is 0.012 Å), the threshold for pair production occurs, and at very high energies, where μ has increased somewhat, pair production predominates. The total absorption coefficients for two different absorbing materials, lead and aluminum, are shown in Figure 4–23*b*.

Figures 4–24 and 4–25 show how the absorption of electromagnetic radiation varies with the energy of the photons and with the identity of the absorbing material. They show the intensity of the incident beam, of the beam at various depths of penetration in the absorber, and of the emergent beam. Figure 4–24 shows the absorption of long-wavelength x-rays, which are appreciably absorbed in moderate thicknesses of lead, and so are called *soft*, and the absorption of short-wavelength, or *hard*, x-rays, which are only slightly attenuated through the same thickness of lead. This illustrates that for a particular absorbing material and for photons of relatively low energy (less than 4 MeV), the coefficient decreases as the photon energy increases. Figure 4–25 shows the absorption of x-rays of the same energy in lead and

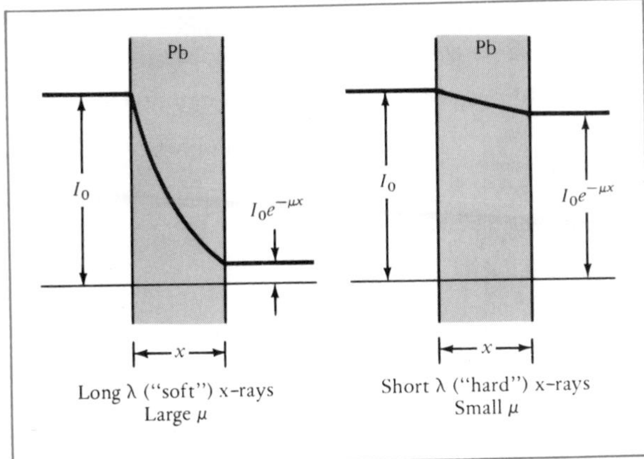

FIGURE 4–24. *Exponential absorption of soft and hard x-rays by a lead absorber.*

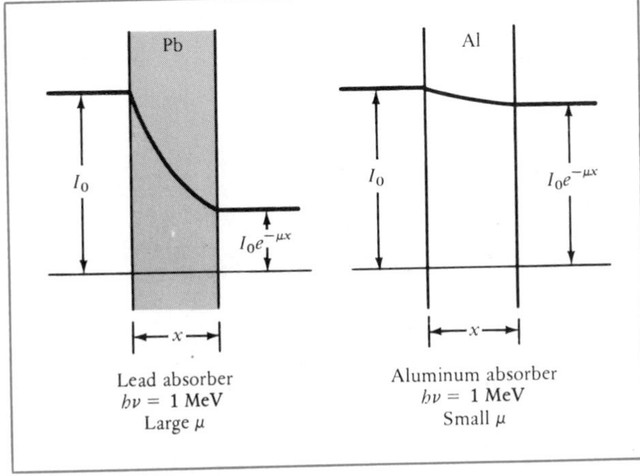

FIGURE 4–25. *Exponential absorption of 1-MeV photons by lead and by aluminum.*

in aluminum of the same thickness; clearly, lead is much more effective as an x-ray absorber and therefore has a much larger absorption coefficient for a given photon energy. It is generally true that dense materials, such as lead, are more effective absorbers than light materials, the basis of x-ray photography; the relative darkening of the photographic emulsion is a measure of the x-ray intensity, and the relative absorption properties of light and dense materials are the reason for the differential darkening of the image.

SUMMARY

Two kinds of physical quantities can be distinguished; those with a continuous range of values and those with a discrete, or quantized, set of values. The quantum theory, formulated by Max Planck in 1900, has shown that many quantities that appear superficially to have continuous values actually have discrete values only.

Monochromatic electromagnetic radiation, when interacting with matter, must be considered to consist of photons, each photon having a discrete energy and momentum:

$$E = h\nu \quad \text{and} \quad p = \frac{h}{\lambda} \qquad (4\text{–}3), (4\text{–}10)$$

A useful relation between the energy and the wavelength of a photon is

$$E = \frac{0.0124 \text{ MeV} \cdot \text{Å}}{\lambda} \qquad (4\text{–}7)$$

The following are the basic photon-particle interactions.

Photoelectric effect: The *complete* transfer of *electromagnetic energy* to a *bound electron:*

$$h\nu = E_b + (E - E_0) \qquad (4\text{–}6)$$

where E_b is the binding energy (work function) of an electron.

Bremsstrahlung: The *partial* or *complete* transfer of a particle's *kinetic energy* to *electromagnetic energy*. Photons of maximum frequency (minimum wavelength) are produced when an electron is brought to rest in a single collision:

$$eV = E_k = h\nu_{\text{max}} = \frac{hc}{\lambda_{\text{min}}} \qquad (4\text{–}9)$$

The Compton effect: The *partial* transfer of *electromagnetic energy* to the *kinetic energy* of a particle. When a photon of wavelength λ interacts with a free particle essentially at rest, a photon emerges at an angle θ (scattering), and the particle recoils with kinetic energy E_k:

$$\Delta\lambda = \lambda' - \lambda = \frac{h}{m_0 c}(1 - \cos\theta) \qquad (4\text{–}15)$$

$$E_k = h\nu \frac{\Delta\lambda}{\lambda + \Delta\lambda} \qquad (4\text{–}16)$$

where m_0 is the rest mass of the recoil particle.

Pair production and annihilation: The *complete* conversion of *electromagnetic energy* into the *rest energy* and *kinetic energy* of the created particles, and the reverse:

Pair production: $h\nu = 2m_0c^2 + (E_k^+ + E_k^-)$

Pair annihilation: $(m^+ + m^-)c^2 = 2h\nu$

The intensity of a monochromatic photon beam is the product of the photon energy and the photon flux N:

$$I = (h\nu)N \qquad (4\text{–}20)$$

In materials of thickness x, the absorption of monochromatic electromagnetic radiation (by the processes of the photoelectric effect, the Compton effect, and pair production) follows the relation

$$I = I_0 e^{-\mu x} \qquad (4\text{–}22)$$

where μ is the absorption coefficient, whose value depends on the nature of the absorber and on the photon energy.

PROBLEMS

Photoelectric effect, §4–2

● **4–1.** A laser produces a pulse of duration 1.0×10^{-9} s. The average power output during the pulse is 6.0×10^{6} W. What are the momentum and the energy of the laser pulse?

● **4–2.** A monochromatic point source of total power output P radiates at wavelength λ uniformly in all directions. Find the number of photons arriving per unit time upon a small surface of area A held at right angles to the incident photons and at a distance R from the point source (in terms of P, λ, R, A, and constants).

● **4–3.** When a 5.0-eV photon strikes a surface, the maximum kinetic energy of a photoelectron is 2.0 eV. What is the maximum kinetic energy of an electron emitted from this surface when it is illuminated by photons with an energy of 8.0 eV?

● **4–4.** No photoelectrons are released from a certain metallic surface when the wavelength of the monochromatic electromagnetic radiation striking the surface is larger than λ_0. If this photoemitting surface is then irradiated with light having a wavelength $\frac{1}{3}\lambda_0$, what is the maximum kinetic energy of the emitted photoelectrons (in terms of λ_0 and constants)?

● **4–5.** The work function of a cesium surface is 1.9 eV. Light from what portion of the visible spectrum (4000 Å to 7000 Å) will produce photoelectrons from a cesium surface?

4–6. The He-Ne gas laser commonly used in the physics laboratory produces a narrow beam of monochromatic light (6328 Å). For a laser beam power of 1.0 mW, find (*a*) the number of wave crests per second passing any point along the beam; (*b*) the number of photons per second.

4–7. Light of intensity 1.0×10^{-10} W/m^2 falls normally upon a silver surface that has one free electron per atom. The atoms are approximately 2.6 Å apart. Treat the incident radiation classically (as waves); assume the energy to be uniformly distributed over the surface; and assume all the light to be absorbed by the surface electrons. (*a*) How much energy does each free electron gain per second? (*b*) The binding energy of an electron at a surface is 4.7 eV. How long must one wait after the beam is switched on before any one electron gains enough energy to overcome its binding energy and be released as a photoelectron? Compare this with the experimental results.

4–8. A monochromatic (4000-Å) beam of very low intensity radiation (1.0×10^{-10} W/m^2) is incident on the positive anode of Figure 4–3. Assume that the 1.0-cm^2 anode is uniformly illuminated. (*a*) How many photons per second strike the anode? (*b*) If 0.10 percent of the incident photons produce photoelectrons reaching the negative cathode, what is the photocurrent (A) in the circuit?

4–9. A human eye with good accommodation can detect 100 photons of visible light. At what distance from the eye, which has a pupil diameter of 4 mm, would a light bulb radiating 100 W of 5000-Å light in all directions have to be placed so that the average number of photons entering the eye is 100 per second?

4–10. When a certain metallic surface is illuminated with monochromatic radiation of wavelength λ, the maximum kinetic energy of photoelectrons released from the surface is 30 eV. When the same surface is illuminated with radiation of wavelength 2λ, the maximum kinetic energy of photoelectrons is 10 eV. What is the maximum wavelength of incident radiation on this metallic surface that can cause the release of electrons?

4–11. A beam of 1240-Å light is incident on a thin tungsten target ($\varphi = 4.52$ eV), as shown in Figure 4–26. Some of the photoelectrons are emitted in the forward direction. What must the magnitude be for a uniform magnetic field **B** applied in the direction into the paper for the emitted photoelectrons to have a maximum radius of curvature of 10.0 cm?

4–12. The threshold wavelength for the photoemission of electrons from a gold surface is 2573 Å. (*a*) Calculate the work function φ (in eV) of an electron at the gold surface. (*b*) Determine the maximum kinetic energy and the speed of photoelectrons emitted when ultraviolet light of 1000 Å is incident on the gold surface.

4–13. When monochromatic light of 5690 Å shines on a potassium surface, the photoelectric current is stopped by a retarding voltage of 0.10 V. When 4050-Å light is used, the stopping potential is 0.99 V. (*a*) Assuming that h and e are unknown, compute the value of h/e from these data and compare this with the accepted value. (*b*) What is the computed work function of potassium (in eV)?

4–14. At what speed will an electron have the same momentum as (*a*) a 2.0-eV photon? (*b*) a 2.0-MeV photon?

4–15. The eye provides information about the external surroundings by focusing light waves onto the eye's retina.

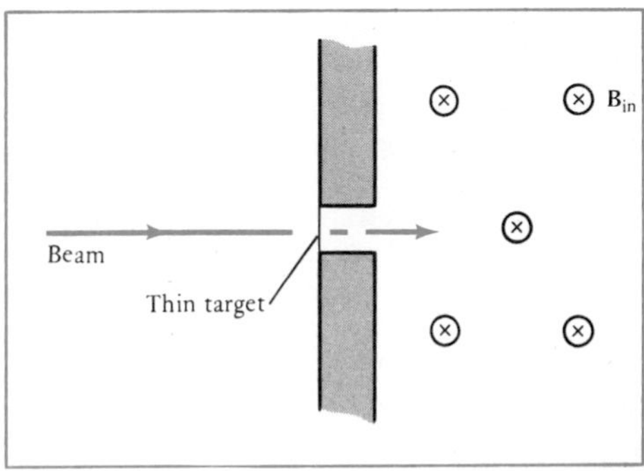

FIGURE 4–26.

There, a light photon is absorbed by a rhodopsin molecule (molecular mass ~36,000 μ) and the resultant chemical change in the molecule is responsible for generating the neural response. Assume that a photon of wavelength 6200 Å is absorbed by a rhodopsin molecule. (*a*) What is the recoil speed of the molecule? (*b*) What fraction of the photon energy is available for producing a chemical change in the molecule?

4–16. In an experimental arrangement of the photoelectric effect (Figure 4–3), the anode and cathode are 5.0 cm apart. An applied retarding potential of 5.0 V produces a uniform electric field between the electrodes. Monochromatic photons incident on the anode release photoelectrons with a maximum kinetic energy of 10 eV at the anode surface. (*a*) How long does it take the fastest photoelectrons to travel from the anode to the cathode? (*b*) How can the observed photocurrent begin in a time less than 10^{-9} s, shorter than that found in part (*a*)?

■ **4–17.** A 2000-kg spaceship, initially at rest in an inertial system, is accelerated along a straight line for 1.0 yr (3.1×10^7 s) by means of a photon rocket of 300-MW power. (*a*) What is the loss in mass of the spaceship during this time? (*b*) What is the speed of the spaceship after one year, and (*c*) what distance does it travel? (*d*) What fraction of the energy released is in kinetic energy of the spaceship?

X-ray production, §4–3

● **4–18.** In Figure 4–10 the intensity of emitted x-rays at frequency 4×10^{18} Hz is equal to the intensity at frequency 6×10^{18} Hz. Find the ratio of the number of photons per second with frequency 4×10^{18} Hz to the number per second with frequency 6×10^{18} Hz.

● **4–19.** A proton is accelerated from rest by a 1.0-kV electric potential difference. The proton then strikes a target and comes to rest abruptly, while creating a single photon. What is the momentum of this photon?

4–20. The separate conservation laws of total energy and total momentum must hold in all inertial systems. Use this fact to show that it is impossible for a moving, unbound point charge to slow down and emit a photon. (*Hint:* View in a system where the charge is initially at rest.)

■ **4–21.** A 5.0-MeV electron traveling east passes near a copper nucleus (rest energy, 6.0×10^4 MeV) and is deflected by the large electrostatic force. The electron is observed to emerge from the collision traveling south with an energy of 4.9 MeV. A single created photon travels away in the northeast direction. (*a*) Find the energy and momentum of the created photon. (*b*) Show that the recoil copper nucleus has a momentum 76 times that of the photon but a kinetic energy only $\frac{1}{200}$ that of the photon.

Compton effect, §4–4

4–22. A photon with wavelength 1.000 Å is incident on an isolated electron initially at rest. The electron, after its "collision" with the incident photon, is set in motion in the same direction as the incident photon. What is the wavelength of the scattered photon?

● **4–23.** A photon is incident on a free *proton* initially at rest, and the proton is set in motion along the same direction as the incident photon. By how much does the wavelength of the scattered photon exceed the wavelength of the incident photon? (The proton mass is 1836 times the electron mass.)

4–24. An x-ray photon of wavelength 1.24 Å collides head-on with an inner electron in an aluminum atom. The binding energy of this inner electron is 2300 eV. Show that when Compton scattering occurs, the backward-scattered photon will always have a wavelength of 1.24 Å (unmodified wave)—that it will never be 1.29 Å (modified wave).

■ **4–25.** One reason for the broadening of the lines in the Compton scattering of photons by free electrons (modified line) and bound electrons (unmodified line) shown in Figure 4–14 is the thermal motion of the scattering particles. Consider, for illustration, the 90° scattering of 1.0-Å photons by an entire atom in a copper target. The mass of a copper atom is 1.2×10^5 times the mass of the electron. (*a*) What is the shift in wavelength $\Delta\lambda$ for 90° scattering of photons by a copper atom at rest? (*b*) Because of thermal motion the copper atoms move randomly with average speeds approximately $\frac{3}{2}kT = \frac{1}{2}mv^2$, where k (Boltzmann's constant) $= 1.4 \times 10^{-23}$ J/K. Find the average speed of copper atoms at room temperature (300 K) and show that to copper atoms moving toward the x-ray source at speed v, the wavelength will appear shorter (Doppler shift, $\Delta\lambda/\lambda = v/c$), whereas to those moving away from the source the wavelength will appear longer; thus the scattered wavelength is $\lambda' = (1.0 \pm 1.1 \times 10^{-6})$ Å. Compare this with the Compton shift found in part (*a*).

4–26. Show that for Compton scattering from a free particle of mass m_0, the scattered photon always has energy less than $2m_0c^2$ if the angle between the scattered photon and the incident photon is larger than 60°, irrespective of how large the energy of the incident photon is. Thus, only high-energy, Compton-scattered photons that are moving forward in a 60° cone can produce particle-antiparticle pairs.

4–27. One method of determining the wavelength of a monochromatic beam of x-rays is by measuring the maximum kinetic energy of the recoil electrons. If a monochromatic x-ray beam strikes a metal target and the maximum kinetic energy of the recoil electrons is 425 keV, what is the wavelength of the x-rays?

4–28. A photon of momentum p_γ and energy E_γ approaches a charged particle of rest energy E_0, which is initially at rest. Show that the center of mass moves at speed $[E_\gamma/(E_\gamma + E_0)]c$.

■ **4–29.** A 0.511-MeV photon collides head-on with a free electron at rest. Assume Compton scattering. (*a*) Show that the center of mass moves at speed $c/2$. (*b*) Show that the total energy of the recoil electron in the center-of-mass reference frame is $2E_0/\sqrt{3}$. (*c*) Use the Lorentz transformation equations (3–26) to find the energy of the recoil electron in the laboratory system, and show that this agrees with Equation 4–17.

Pair production, §4–5

● **4–30.** A 2.0-MeV electron collides head-on with a 2.0-MeV positron and two photons are created. What is the energy of either photon?

● **4–31.** A π^0 particle is unstable and may decay into two or more photons. The rest energy of a π^0 particle is 135 MeV. If this unstable particle is initially at rest, what is the maximum energy of a created photon?

■ **4–32.** A photon collides with a free electron at rest. Show that the minimum photon energy necessary for this interaction to create an electron-positron pair is four times the rest energy of an electron.

4–33. An electron-positron pair moving at speed $0.98c$ along the $+x$-axis becomes annihilated, and two photons of equal energy are created. For each created photon, find (*a*) the momentum, (*b*) the energy, and (*c*) the direction relative to the initial velocity of electron-positron pair.

4–34. The binding energy of the least tightly bound electron in a neon atom is approximately 20 eV. Find the minimum wavelength of incident electromagnetic radiation shining on a neon gas that will produce the following: (*a*) photoelectric effect; (*b*) Compton scattering (assuming the electron to be free); (*c*) pair production.

■ **4–35.** A beam of 2.04-MeV photons is incident on a thin copper target. One of the photons passing near the nucleus of a copper atom becomes annihilated and produces an electron-positron pair at a distance of 1.0×10^{-3} Å from the nucleus. The created particles have the same kinetic energy E_k at the point of creation and move away in the direction in which the incident photon is moving. Because of the strong Coulomb force between each particle and the copper nucleus (charge $+29e$) the two oppositely charged particles will have different kinetic energies $E_k{}^+$ and $E_k{}^-$ when they have moved far from the nucleus. (*a*) Show that after the created particles have moved away a distance much larger than 1.0×10^{-3} Å, the difference in kinetic energies will be $E_k{}^+ - E_k{}^- = 0.84$ MeV. (*b*) If the two particles enter perpendicularly into a uniform magnetic field of 0.15 T, what will the radius of curvature for each be?

Photon-electron interactions, §4–6

● **4–36.** Identify the particular photon-electron interaction at each point in the cascade shower in Figure 4–20.

4–37. Describe three experimental methods for determining Planck's constant.

4–38. In each of the five photon-electron interactions in Figure 4–19, what changes (net charge, momentum, kinetic energy) occur in the material slab?

Photon absorption, §4–7

● **4–39.** Show that the thickness of absorbing material necessary to reduce the intensity of a beam of radiation to one-half its original intensity is $(\ln_e 2)/\mu = 0.693/\mu$.

● **4–40.** Two parallel slabs (absorption coefficients μ_1 and μ_2), placed back to back, are used to reduce the intensity of an x-ray beam. If the slabs are interchanged, (*a*) will the ratio of the intensity of the emerging beam to the intensity of the incident beam change? (*b*) will the ratio of the number of photons absorbed by slab 1 to the number absorbed by slab 2 change?

4–41. Monochromatic x-rays of wavelength 0.50 Å are incident on three different absorbers: water, aluminum, and lead. At this wavelength, the absorption coefficient of water is 0.491 cm^{-1}; of aluminum, 5.02 cm^{-1}; and of lead, 667 cm^{-1}. Calculate the thickness of each absorber necessary to reduce the intensity of the x-ray beam to $\frac{1}{100}$ its original intensity.

4–42. Polychromatic x-rays with photon energies ranging from 24.8 keV to 62 keV are incident on a lead absorber. In this energy range, the absorption coefficient of lead decreases from 667 cm^{-1} at 24.8 keV to 50.6 cm^{-1} at 62 keV. (*a*) Calculate the thickness of lead necessary to reduce the 62-keV intensity component to one-tenth its incident value. (*b*) For the lead thickness in part (*a*), what is the reduction in the 24.8-keV x-ray component?

4–43. A lead plate and an aluminum plate are placed flush against one another, and a beam of 62-keV photons is incident on the parallel plates, hitting the lead plate first. Determine the thickness of each plate necessary for both plates to absorb the same number of photons per second and for the intensity of the emerging beam to be one-half the incident intensity. For 62-keV photons, the coefficient of absorption of lead is 50.6 cm^{-1} and of aluminum is 0.737 cm^{-1}.

5
Quantum Effects: The Wave Aspects of Material Particles

Werner K. Heisenberg (1901, Germany—1976). 1925, matrix mechanics, an alternative form of wave quantum mechanics. 1927, uncertainty principle. 1932, Nobel Prize, creation of quantum mechanics. Significant contributions to nuclear physics, ferromagnetism, elementary particles, and the philosophical foundation of quantum theory. (Photo courtesy of the American Institute of Physics, Meggers Gallery of Nobel Laureates.)

5–1 De Broglie Waves

Electromagnetic radiation has two aspects, a wave aspect and a particle aspect. Experiments that show the interference and diffraction of electromagnetic radiation can be explained only if the radiation is assumed to consist of waves having frequency ν and wavelength λ. The distinctively quantum effects of electromagnetic radiation, such as the photoelectric and Compton effects, can be explained only if light is assumed to consist of particlelike photons, each with energy E and momentum p. The wave and particle properties of the radiation are related by:

$$E = h\nu \qquad (4\text{–}3), (5\text{–}1)$$

$$p = \frac{h}{\lambda} \qquad (4\text{–}10), (5\text{–}2)$$

In these equations, the two particlelike quantities appear on the left, and the two wavelike quantities appear on the right. Thus, the wave-particle duality of electromagnetic radiation is implied in these relations. The fundamental constant of the quantum theory, Planck's constant h, relates the wave to the particle characteristics. We can say that electromagnetic radiation under some circumstances will behave like particles (photons of zero rest mass) and under other circumstances like waves.

Does this dual wave-particle description, found necessary for radiation, have an even greater generality, holding for *all* particles—particles with a *finite* rest mass as well as those with zero rest mass? Louis de Broglie first posed this question in 1924. He conjectured, acknowledging the symmetry of nature, that a material particle might also show wave properties. De Broglie assumed further that the equations that give the particle characteristics of electromagnetic waves give also the wave characteristics of material particles, such as electrons. Although he did not develop quantum-mechanical equations describing general wave-particle behavior of a material particle, experiments have emphatically confirmed his hypothesis; the wave character of material particles is now well established. Because the wavelength can be measured from interference or diffraction effects, we shall concentrate our attention on the second of the relations, Equation 5–2.

For nonrelativistic, free particles, the wavelength λ of a material particle having a momentum $\mathbf{p} = m\mathbf{v}$ (where m is the particle's rest mass and $\mathbf{v}$ its velocity) is, by Equation 5–2,

$$\lambda = \frac{h}{p} = \frac{h}{mv} \qquad (5\text{–}3)$$

This is known as the *de Broglie relation.*

One might well ask, If an electron is to be regarded as a wave, at least under some circumstances, what is it that is waving? That same kind of question was asked about the fundamental nature of light. Not until after the electromagnetic theory of Maxwell (1865) and the subsequent experiments of Hertz could physicists assert that the wave properties of light corresponded to oscillations of the electric and magnetic fields. Ignorance of light's electromagnetic nature, however, did not prevent physicists long before

Maxwell and Hertz, for instance Young in 1801, from discovering the wavelike properties of light and interpreting interference and diffraction on this basis. Therefore, to establish whether a material particle has a wave nature, it is *not* necessary to know first what the wave phenomenon is. To test the de Broglie hypothesis is to determine by experiment whether material particles show interference and diffraction effects. Although the physical nature of a material particle's wave aspect is a critical consideration, we shall postpone it until after we have discussed the experiments that confirm that a material particle has a wavelength $\lambda = h/mv$.

The first particle with finite rest mass experimentally shown to have wave properties was the electron. J. J. Thomson discovered the electron as a particle in 1897. It was found to follow well-defined paths and to have a well-defined charge-to-mass ratio (for $v \ll c$), as well as mass, momentum, and energy that could be localized in space. In short, electrons exhibit the attributes of a particle.

The wave nature of electrons was not discovered until 1927, when the electron-diffraction experiments of C. Davisson and L. H. Germer confirmed the de Broglie relation. Why were the wave characteristics of electrons discovered only many years after their particle nature had been established? The difficulties in experimentation came from the very small wavelength for ordinary particles. As Equation 5–3 shows, a 1.0-kg particle moving at 1.0 m/s has a wavelength of only 6.6×10^{-34} m $= 6.6 \times 10^{-24}$ Å. Clearly, we should not expect, in throwing baseballs through an open window, to find a discernible diffraction pattern of hits on a distant wall any more than we should expect to see one when visible light passes through such a wide "slit." The wavelength of an ordinary material particle is very small compared with the dimensions of ordinary apertures, so that interference and diffraction effects are subtle. Therefore, if an object's wavelength is to be large enough to produce observable wave effects, its mass and velocity, as we see from Equation 5–3, must be small.

The diffraction grating having the smallest distance between "lines" is a crystal, a solid in which the atoms are located in a three-dimensional geometrical array. A typical distance between adjacent atoms is of the order of 10^{-10} m, or 1 Å. The most favorable conditions for observing the diffraction of particles are those in which the particle has a wavelength of comparable size. Since $\lambda = h/mv$, to have λ large we must choose a particle with small mass; the electron is such a particle.

Let us compute the kinetic energy of an electron with a wavelength λ of 1.00 Å. An electron with electric charge e, accelerated from rest by an electrostatic potential difference V, acquires a final kinetic energy $\frac{1}{2}mv^2$ when

$$eV = \tfrac{1}{2}mv^2 = \frac{1}{2m}p^2 = \frac{1}{2m}\left(\frac{h}{\lambda}\right)^2$$

$$V = \frac{h^2}{2me\lambda^2} = \frac{(6.62 \times 10^{-34}\ \mathrm{J \cdot s})^2}{2(9.11 \times 10^{-31}\ \mathrm{kg})(1.60 \times 10^{-19}\ \mathrm{C})(1.00 \times 10^{-10}\ \mathrm{m})^2}$$

$$= 150\ \mathrm{V}$$

An electron of 150 eV has a wavelength of 1 Å. Since this wavelength is comparable to that of a typical x-ray photon, we can expect both electrons and x-rays to show similar diffraction effects when passing through a crystal.

5-2 The Bragg Law

Max von Laue was the first to suggest in 1912 that crystalline solids, in which the arrangement of atoms follows a regular pattern, and in which the distance between atoms is approximately 1 Å, might be used as diffraction gratings for measuring x-ray wavelengths.

A crystal of sodium chloride, which has a particularly simple structure, is a standard material in x-ray diffraction. The external geometrical features of a rock-salt crystal suggest that the sodium and chlorine atoms (strictly, the ions Na^+ and Cl^-) are arranged in a simple cubic lattice (Figure 5–1). The sodium and chlorine atoms are located at alternate corners of identical elementary cubes, each with a distance d along an edge.

It is easy to compute the *lattice spacing* d from the density of the sodium chloride crystal and the atomic weights of sodium and chlorine. If d is the distance in centimeters from a sodium atom to the nearest chlorine atom, then there are $1/d$ atoms (half Na, half Cl) along an edge of a cube 1 cm long. Therefore in a cube of sodium chloride crystal 1 cm along an edge there are altogether $1/d^3$ atoms, and the total number of atoms per unit volume is $1/d^3$. The measured atomic weight of Na is 23.00 and of Cl is 35.45, so the molecular weight of NaCl is 58.45. Because Avogadro's number, 6.022×10^{-23}, gives the number of atoms in 1 g·mol, it follows that there are 6.022×10^{23} Na atoms in 23.00 g of Na, 6.022×10^{23} Cl atoms in 35.45 g of Cl, and therefore $2 \times 6.022 \times 10^{23}$ atoms (half Na, half Cl) in 58.45 g of NaCl. The measured density of sodium chloride in the crystalline form of rock salt is 2.163 g/cm³. Therefore we can write

$$\frac{\text{Atoms}}{\text{Volume}} = \frac{2 \times 6.022 \times 10^{23}\ \text{atoms/g·mol} \times 2.163\ \text{g/cm}^3}{58.45\ \text{g/g·mol}} = \frac{1}{d^3}$$

$$d = 2.820\ \text{Å}$$

This measure is the lattice spacing of sodium and chlorine atoms in a crystal of rock salt. It is typical of the interatomic spacing of atoms in any solid and is comparable to the wavelength of x-rays or of 150-eV electrons. We

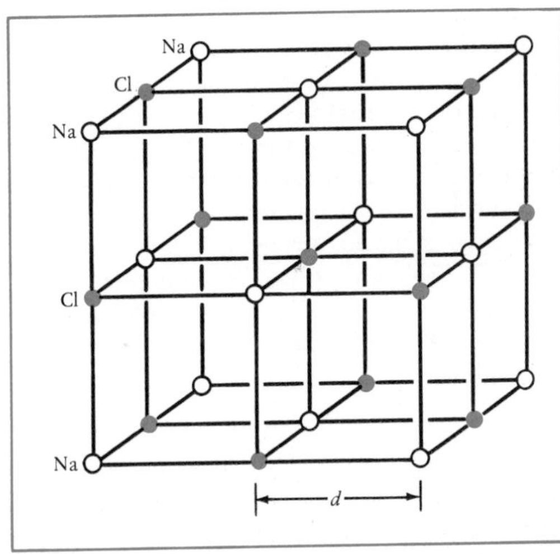

FIGURE 5–1. *Crystal structure of rock salt (NaCl).*

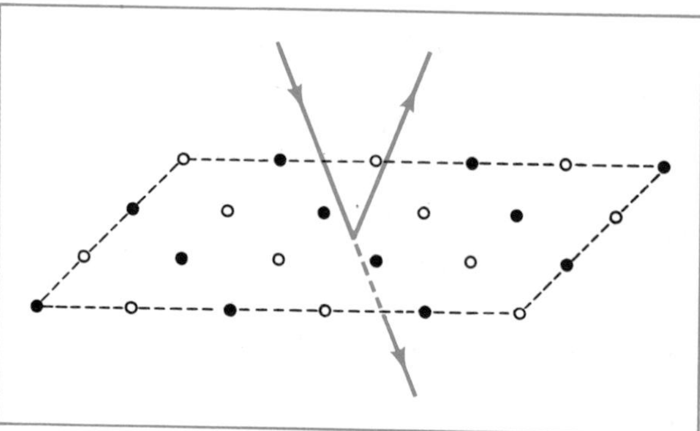

FIGURE 5–2. *Atoms in a Bragg plane.*

shall see how it is possible to determine x-ray and electron wavelengths from the lattice spacing and the geometrical character of the atomic arrangement.

When a wave impinges upon a collection of scattering centers, such as the atoms in a crystalline solid, each scattering center generates waves, radiating outward in all directions. The resultant wave from all scattering centers, measured in any one direction, depends on the interference between all the separate centers. It is remarkable that the atoms lying on any one plane within the crystal act toward the incident wave as a partially silvered mirror acts toward visible light; that is, they reflect a portion of the wave while allowing the remainder to pass through (see Figure 5–2).† These *Bragg planes* and *Bragg reflections* are named after W. H. Bragg, who with his son, W. L. Bragg, developed the fundamental theory of x-ray diffraction by crystals in 1913. Since the atoms do act like this, we can deal not with the interference between the waves generated by all the scattering centers individually but more simply with the interference between the waves reflected from parallel Bragg planes.

The reflection of waves from two adjacent and parallel Bragg planes is shown in Figure 5–3. The directions of the incident and reflected waves, both denoted by θ, are specified by the angle between the direction of propagation of the waves and the Bragg plane (not the normal to the reflecting planes). At each plane we regard the incident wave as transmitted partly undeviated and partly reflected.

The incident ray is partially reflected at the first Bragg plane; the reflected ray AB makes an angle θ with plane 1. That part AC of the incident ray that is transmitted through the first plane is partially reflected from the second plane, also in the direction θ. We concentrate on the wavefront BD perpendicular to the two reflected rays. The reflected rays will constructively interfere at some distant point only if they have the same phase at the points B and D. The points B and D are in phase when the path difference $ACD - AB = 2d \sin \theta$ is an integral multiple n of the wavelength λ. The condition, then, for constructive interference of waves reflected from adjacent parallel Bragg planes is

$$n\lambda = 2d \sin \theta \tag{5–4}$$

† For a simple proof, see the article by L. Elton and D. Jackson, *Am. J. Phys. 34,* 1036 (November 1966).

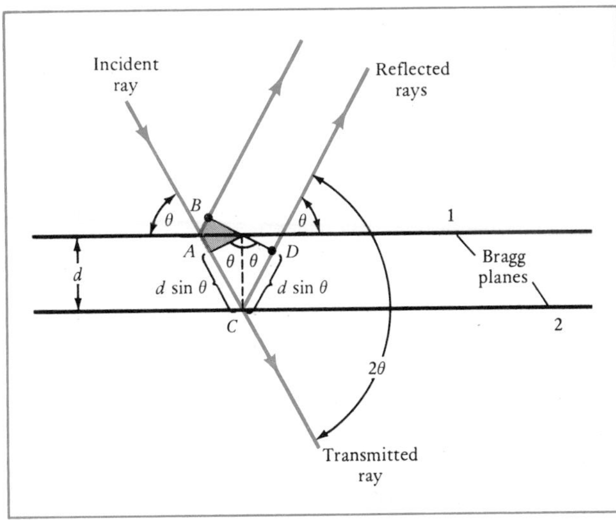

FIGURE 5–3. *Reflection of waves from two adjacent, parallel Bragg planes.*

where n, the *order of the reflection,* can have the values 1, 2, 3, ... This equation known as *Bragg's law,* is the basis of all coherent x-ray and electron diffraction effects in crystals.† Rays reflected at any angles except those satisfying Equation 5–4 interfere destructively, and the incident beam is completely transmitted. The Bragg law is a way of measuring wavelengths comparable to interatomic distances, for clearly, if n, d, and θ are known, λ can be computed. Note in Figure 5–3 that the angle between the transmitted and the reflected rays is 2θ, with the Bragg planes bisecting this angle.

5–3 X-Ray and Electron Diffraction

The essential elements of an *x-ray spectrometer,* a device for measuring x-ray wavelengths, are shown in Figure 5–4. A source of monochromatic x-rays shines on a crystal whose structure and interatomic dimensions are known. A detector, such as a chamber that is sensitive to x-ray ionization effects, measures the intensity of x-rays entering it. Both the crystal and the detector are rotatable, but the detector is always set at an angle 2θ from the forward beam; θ is the angle between the incident rays and the Bragg planes from which reflection is to be observed. The x-ray intensity in the detector will indicate a strong maximum only when the conditions of Equation 5–4 for Bragg reflection are satisfied. Since d is known from the crystalline structure, and θ is measured, the wavelength λ can be computed. For a given set of Bragg planes, and grating space d, wavelength λ, and order n, there is a *single direction* 2θ away from the direction of the incident beam in which the diffracted beam is strong.

Now consider a thin metallic foil through which a monochromatic x-ray beam is sent. The foil is made so that it consists of very many simple, perfect crystals, randomly oriented to one another. Only those particular micro-

† Equation 5–4, Bragg's law, resembles the equation that applies to an ordinary ruled diffraction grating. The two equations, however, are *not* the same.

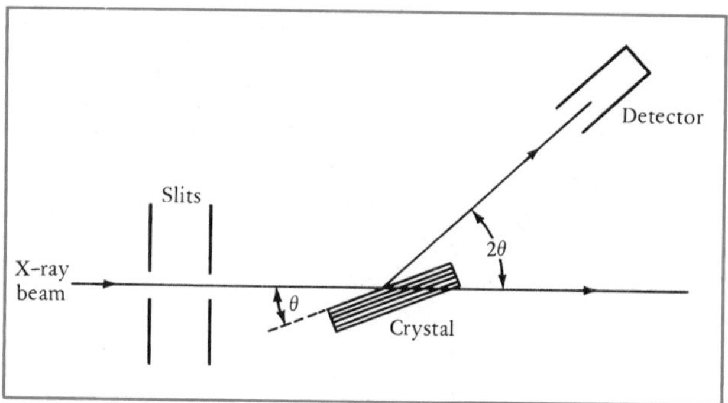

FIGURE 5–4. *Schematic diagram of an x-ray crystal spectrometer.*

crystals so oriented that the Bragg condition is fulfilled will produce a strongly diffracted beam; the other microcrystals will not diffract the incident beam coherently. Therefore the emerging beam will consist of two parts: an intense, central, undeviated beam and a beam scattered in a conical shape that makes an angle 2θ with the incident beam (see Figure 5–5). For each order, the angle θ is uniquely determined by the Bragg relation. When the scattered beam strikes a flat photographic plate, a pattern of intensities appears; a strong central spot is surrounded by a series of concentric circles (only one of which is shown in Figure 5–5). Since the radius of any circle and the distance from the scattering foil to the photographic plate are easily measurable, the angle 2θ can be computed directly. Finally, if the order n of the Bragg reflection is known, the wavelength of the x-rays can be computed from the Bragg relation.

To see how the many sets of Bragg planes in any single crystal affect x-ray and electron diffraction, consider again the arrangement of atoms in the sodium chloride crystal. A Bragg plane is any plane that contains atoms; there are many such planes in a cubic crystal, as shown by Figure 5–6. It is clear that the various planes, of which only a very few are shown in the figure, will differ in the value of the grating space d; consequently, there will be numerous Bragg angles θ, each satisfying the Bragg relation for a particular set of Bragg planes. The x-ray diffraction pattern will therefore be more complicated than that indicated in Figure 5–5; it will commonly show not a single circle but numerous concentric circles, each corresponding to diffraction from a particular set of Bragg planes. The intensities of the reflected beams from the different planes will, however, not be the same; the

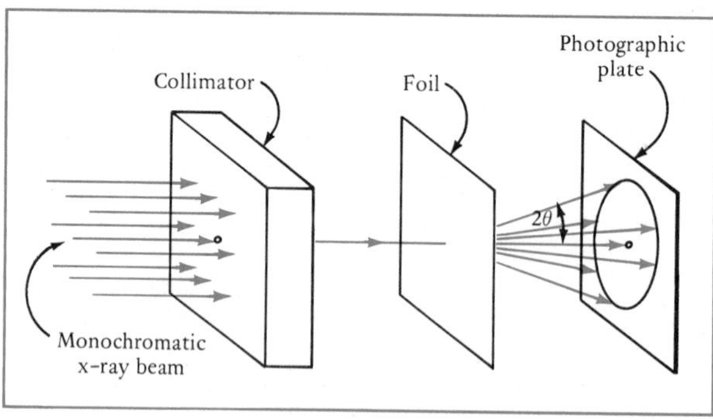

FIGURE 5–5. *Scattering of monochromatic x-rays by a thin metallic foil from one set of Bragg planes.*

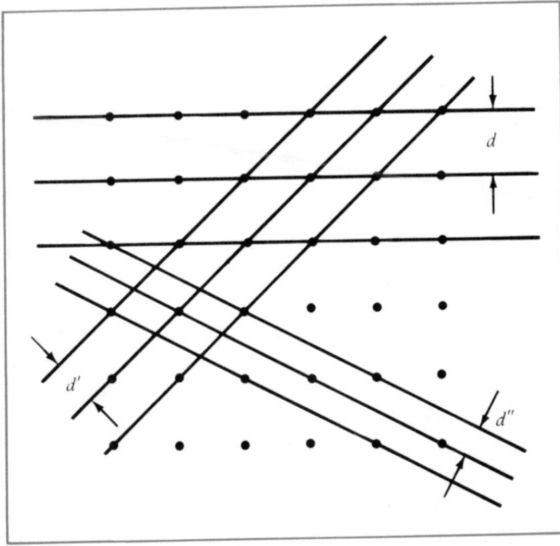

FIGURE 5–6. *Three sets of parallel Bragg planes with different grating spacings.*

intensity of a reflected beam depends on the number of atoms per unit area, and this number differs among planes. Figure 5–7 is an actual x-ray diffraction pattern of a polycrystalline sample.

Our discussion of diffraction from crystals has concerned the use of x-rays. As we have seen, electrons with a kinetic energy of 150 eV have the same wavelength as 1-Å x-rays, and a monoenergetic electron beam should and does show essentially the same diffraction effects as x-rays do†. Figure 5–8 is an electron diffraction pattern from a metallic foil; the pattern, like that in Figure 5–7 for x-ray diffraction, is in complete accord with the Bragg and de Broglie relations. In short, *electron diffraction experiments confirm the relation* $\lambda = h/mv$.

Electron and x-ray diffraction patterns can be used for measuring the wavelengths of electrons and x-rays when the crystalline structure is known; conversely, when the wavelengths of the x-rays and electrons are known, the diffraction patterns can be used to deduce the geometry of the crystalline structure and the interatomic spacings of the solids. Two applications of the principles of x-ray diffraction are, briefly:

1. The Compton effect, in which scattered photons appear with an unmodified and a modified wavelength, can be verified by using an x-ray spectrometer to measure the wavelengths of the radiation scattered by a target.
2. When a beam of x-rays having a continuous range of wavelengths is incident on a crystal, there is constructive interference leading to a beam deviated by an angle 2θ only if the Bragg law is satisfied. For a given angle θ, order n, and interatomic spacing d, the value of the wavelength is uniquely specified by the Bragg law, Equation 5–4, and only a single wavelength will be strongly reflected at the angle 2θ. A crystal in this arrangement acts as a *monochromator;* that means that it selects from the continuous range of wavelengths incident on the crystal a single monochromatic beam emerging at the angle 2θ.

† A monoenergetic beam of particles has not only a single energy but also a single wavelength and is therefore usually spoken of, with respect to its wave aspect, as "monochromatic."

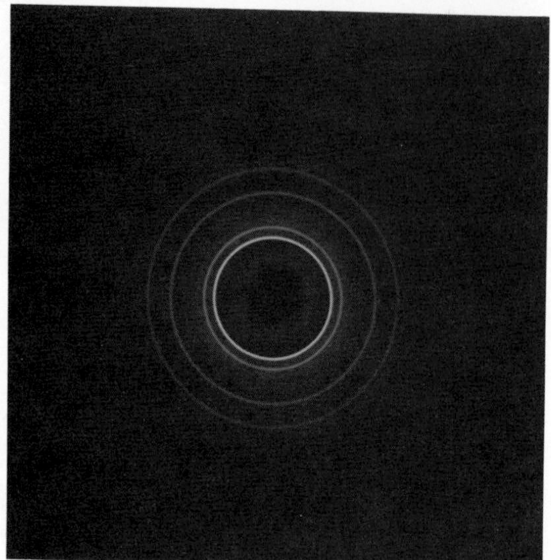

FIGURE 5–7. *X-ray diffraction pattern of a polycrystalline sample. The center is dark because a hole was cut in the photographic plate to allow the strong central beam to pass through it. (Courtesy Bell Laboratories [M. H. Read].)*

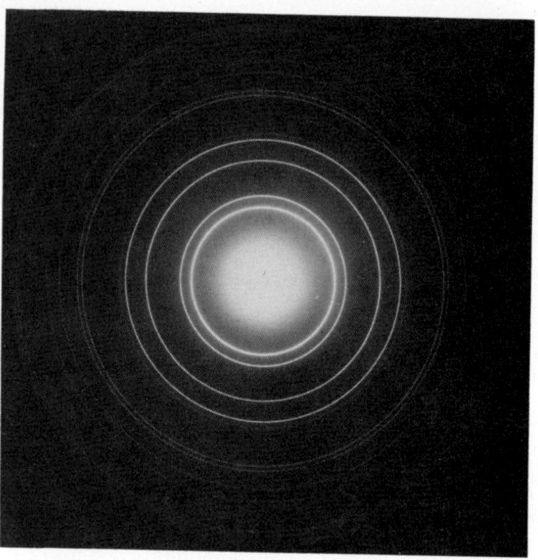

FIGURE 5–8. *Electron diffraction pattern of a polycrystalline sample. (Courtesy RCA Laboratories.)*

The experiment of Davisson and Germer first confirmed the wave properties of electrons (see Section 5–1). A beam of electrons of energy 54 eV is incident on a single crystal of nickel (see Figure 5–9). Electrons leave the nickel surface for two reasons. The first is by secondary emission, in which the incident electrons impart their kinetic energy to loosely bound electrons in the metal. Some of these electrons may then be released and leave the nickel surface in all directions with a continuous range of energies. The second is by electron diffraction, in which the incident electrons are diffracted by reflection from the Bragg planes within the nickel crystal. Davisson and Germer found that there was, besides the smooth variation in electron intensity arising from secondary emission, a pronounced peak at $\varphi = 50°$, which could be attributed to electron diffraction. Using the known geometry of crystalline nickel (from x-ray diffraction experiments), they computed the direction for strong reflection of electrons of 54 eV in nickel and got 50°. Thus the de Broglie relation was confirmed. Analysis of the experiment was complicated by the difference of the wavelength of electrons within a crystal from the wavelength in free space. The difference arises from a change in the speed of electrons as they enter or leave the nickel surface; they move faster as they pass into the interior of the material because of work done on them by an electric field at the surface (the work function of the particular material). Because of the change in speed, the electrons are refracted at the surface.

Shortly after Davisson and Germer's confirmation of the de Broglie relation for electrons diffracted by a large, single crystal, G. P. Thomson observed diffraction rings from the transmission of electrons through a thin

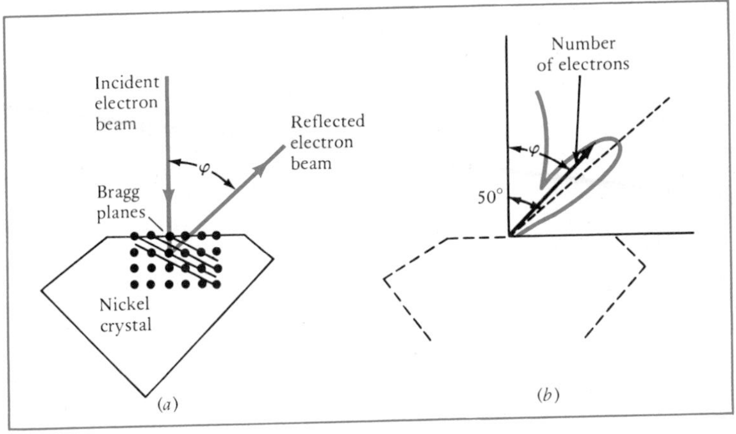

FIGURE 5–9. *(a) Reflection of electron waves by one set of Bragg planes in a crystal of nickel. (b) Number of electrons reflected from a nickel crystal as a function of angle φ, as shown by a polar plot.*

metallic foil composed of randomly oriented microcrystals.† Figure 5–8 is an example of *electron* diffraction through a thin film. This is similar to Figure 5–7, an example of *x-ray* diffraction. Both experiments are describable in terms of the wavelike behavior of the incident beams.

That an object of mass m and velocity v has a wavelength h/mv has been established by experiments not only with electrons but also with atoms, molecules, and even the uncharged nuclear particle, the neutron. Take as an example a neutron beam emerging from a nuclear reactor. Neutrons produced within the reactor (see Section 11–7) collide with the atoms of the reactor material and quickly reach thermal equilibrium with the material. A neutron behaves like a molecule in a gas, and a temperature can be attributed to a collection of neutrons. The relation between average translational kinetic energy per particle and the absolute temperature is

$$E_k = \tfrac{1}{2}mv^2 = \tfrac{3}{2}kT$$

When the temperature T of the material is room temperature, 300 K, a neutron with this average kinetic energy is said to be a *thermal neutron*. The wavelength of a thermal neutron whose mass is 1.67×10^{-27} kg is, then,

$$\frac{1}{2}mv^2 = \frac{p^2}{2m} = \frac{1}{2m}\left(\frac{h}{\lambda}\right)^2 = \frac{3}{2}kT$$

$$\lambda = \frac{h}{\sqrt{3mkT}} = 1.4\ \text{Å}$$

This wavelength is comparable to the distance between atoms in a crystalline solid, and *neutron diffraction* of monoenergetic thermal neutrons is observed.

The de Broglie relation attributes a wavelength to *any* particle that has momentum. Since neutrons, which are uncharged particles, can be diffracted, their wave properties do not depend on their having an electric charge. Furthermore, atoms and molecules, which have internal structure, show diffraction effects; therefore, the de Broglie relation applies even to systems of particles. All physical entities manifest a wavelike behavior.

† Thomson, whose experiments showed the wave properties of electrons (1927), was the son of J. J. Thomson, whose cathode ray experiments showed the particle properties of electrons (1897).

5-4 The Principle of Complementarity

We attribute both wave and particle characteristics to electromagnetic radiation and to material particles. This wave-particle duality makes us uneasy at first sight. Why are we uneasy? And how can we resolve the dilemma?

The concepts of particle and wave are basic in physics—they represent the only two possible modes of energy transport. In describing any ordinary large-scale phenomenon of energy transport in classical physics, we are always successful in applying one of the descriptions. For example, a disturbance that travels on the surface of a pond is certainly a wave, and a thrown baseball illustrates energy transport by a "particle." There is never any doubt about which description we should apply in such instances.

In a less direct illustration of wave and particle behavior, we can understand the propagation of sound through an elastic medium as a wave disturbance. We do not "see" the waves, as we did the water waves; nevertheless, we confidently apply the wave description to sound propagation because it alone describes diffraction and interference. A *wave model* for the propagation of sound agrees with *all* experimental observations of sound. Next let us consider particle behavior as it appears in the kinetic theory of gases. Although we never see the molecules, we are quite sure that their behavior is like that of very small, hard spheres since various experiments show it to be so. A *particle model* is the only appropriate way of describing gas behavior. When we describe phenomena remote from ordinary experience, we still apply one or the other of the two modes of description, because one of them is always successful in accounting for the experimental facts.

The wave and particle descriptions are mutually incompatible and contradictory. If a wave is to have its frequency or its wavelength given with infinite precision, then it must have an infinite extension in space. On the contrary, if it is confined to some limited region of space, it *resembles* a particle by its localizability, but it cannot be characterized by a single frequency and a single wavelength. An ideal wave, one whose frequency and wavelength are known with certainty, is altogether incompatible with an ideal particle, which has a zero extension in space.

Any energy-transport phenomenon must be described in terms of waves or of particles, but we cannot *simultaneously* apply a particle description and a wave description. We can and must use one or the other, never both at the same time.

What is disturbing about the descriptions of electromagnetic radiation and of material particles is that we apply *both* models. If we review our interpretation of the experiments discussed thus far, however, we find that we have *never* applied the descriptions *simultaneously*. Such application is logically impossible.

Consider electromagnetic radiation. The wave model describes experiments in interference and diffraction, in which we are confronted with alternate light and dark bands that are predicted and accounted for by wave theory. We never apply a particle description to interference and diffraction. Our confidence in the wave model for *propagation* of light is strengthened by Maxwell's classical electromagnetic theory, which predicts all the *wave* phenomena. But it would be rash, in view of the open-ended, tentative, and incomplete nature of all physical theory, to conclude that Maxwell's equations are the last word on electromagnetic theory. In fact, classical electromagnetism is incomplete to the degree that it cannot account for quantum effects. In summary, we use the wave model to describe the *propagation* of light; we do not, need not, and cannot apply the particle model to interference and diffraction effects.

Now consider phenomena that call for a particle model of electromagnetic radiation. They are the quantum effects, showing electromagnetic radiation in *interaction* with particles (Chapter 4). The radiation consists of photons, each with a specific energy and momentum, with the electromagnetic *particle* localized at a particular point in space, namely the site of the interaction. Interactions between radiation and matter require the particle description; such interactions are best described as collisions. We can make sense of the photon-interaction experiments only when electromagnetic radiation is assumed to consist of particles in collision. In short, when we visualize photon-electron interaction, we use a *particle* model; we cannot apply the wave model, and we don't need to.

Electromagnetic radiation shows both wave and particle aspects but not in the same experiments. Interference or diffraction experiments require a wave interpretation, and it is impossible to apply simultaneously a particle interpretation; a photon-interaction experiment requires a particle interpretation, and it is impossible to apply simultaneously a wave interpretation. Both the wave and the particle aspects are essential features of electromagnetic radiation, and we must accept both. According to the *principle of complementarity,* which Niels Bohr enunciated in 1927, *the wave and particle aspects* of electromagnetic radiation *are complementary.* To interpret the behavior of electromagnetic radiation in any one experiment, we must choose either the particle or the wave description. The wave and the particle aspects are *complementary* in that our knowledge of electromagnetic radiation is partial unless both wave and particle aspects are known; but choosing one description precludes simultaneously choosing the other. We choose the description by the experimental arrangement. Electromagnetic radiation is a more complicated entity than what can be comprehended in the simple and extreme notions of wave and particle—notions borrowed from our direct, ordinary experience with large-scale phenomena.

The complementarity principle applies also to the wave-particle duality of particles, such as electrons. For example, electrons in a cathode ray tube follow well-defined paths and indicate their collisions with a fluorescent screen by very small, bright flashes. Electrons are particles (or more properly, a particle model can be used to describe their behavior) in cathode ray experiments because all the electron energy, momentum, and electric charge are assigned at any one time to a small region of space. When electrons interact with other objects, they behave like particles. The particle nature of electrons is revealed in the cathode ray experiments, and therefore, by complementarity, the wave nature of electrons *must* be suppressed.

The wave nature of electrons appears in the experiments showing electron diffraction. Electrons are propagated as waves with an indefinite extension in space, and it is impossible to specify the location of any one electron. In short, the electron diffraction experiments show the wave nature of electrons, and by complementarity, the particle nature is necessarily suppressed in these experiments.

5-5 Probability Interpretation of De Broglie Waves

The wave nature of electromagnetic radiation can be described by oscillating electric and magnetic fields in space; the wave associated with a photon is this electromagnetic field. What about the waves associated with a material particle? What is it that is waving when we say that an electron or any other material particle shows wave properties? To answer this, we first relate the wave and particle descriptions of electromagnetic radiation.

When the intensity of a monochromatic beam of electromagnetic radiation directed at a screen is great, the screen appears to the eye to be uniformly illuminated; equivalently, a photographic plate placed at the screen will show a uniform darkening after it has been exposed and developed. The high intensity of the light beam means that so many photons arrive at the screen that they obscure the esentially granular and discrete nature of the electromagnetic radiation; the distinct and randomly arranged bright flashes merge into a seemingly continuous and constant illumination.

The intensity I of the illumination, the energy per unit area per unit time, is given by

$$I = \epsilon_0 E^2 c$$

where E is the magnitude of the instantaneous electric field at any point on the screen and ϵ_0 is the electric permittivity of free space.† Suppose that the intensity of the beam is made extremely weak. Instead of the uniformly illuminated area observed with the high-intensity beam, distinct bright flashes appear randomly on the plate; each bright flash corresponds to the arrival of a single photon.‡ Neither the position nor the time at which a single photon will strike the screen can be determined. Although the distribution of photons is completely random, the *average* number of photons arriving per unit area per unit time *can* be predicted; this number is the photon flux N. Since $h\nu$ is the energy per photon, the intensity of a monochromatic beam in terms of the photon flux is

$$I = (h\nu)N \tag{4–20}$$

If the positions of the individual flashes are recorded over a long period, the screen is again found to be uniformly covered.

This description of photon activity is like the kinetic theory of gases, which attributes the apparently continuous pressure of a gas to the effect of individual molecules. The molecules strike the walls of a container randomly and discretely, creating an effect of continuous pressure by their large numbers.

The intensity of monochromatic electromagnetic radiation can be expressed by either the wave description $I = \epsilon_0 E^2 c$ or the particle description $I = h\nu N$. We have in the intensity a quantity that has a precise meaning in both descriptions. The intensity bridges the gap between the two disparate models.

† We could express the intensity equally well by the magnetic field instead of the electric field: $I = B^2 c/\mu_0$.

‡ Actually, sophisticated instruments, rather than the eye or a photographic plate, can record the arrival of photons one by one.

Let us see what new meaning can be assigned to the square of the electric field strength E^2 in the photon description of light. Equating the two expressions for the intensity gives

$$I = h\nu N = \epsilon_0 E^2 c$$

Since the probability of observing a photon on a unit area per unit time is proportional to N, the probability of observing a photon $\propto N \propto E^2$.

The probability of observing a photon at any point in space is proportional to the square of the electric field strength at that point.

By the quantum theory, the electric field is not only the quantity that gives the electric force per unit electric charge but also the quantity, or function, whose square gives the probability of observing a photon at any given place. Classical electromagnetic theory can yield, through computed values of E^2, the probability of observing photons, although it is incapable of yielding the strictly quantum features of electromagnetic radiation.

We can now give meaning to the wave nature of a material particle, like an electron, in the following way. We assume that the relation between the probability of observing a particle and the square of the amplitude of its wave is exactly analogous to the relation between the probability of observing a zero-rest-mass photon and the square of the amplitude E^2 of its wave (the electric field). The amplitude of the wave associated with a particle is represented by ψ, called simply the *wave function.*

The wave function ψ is that quantity whose square ψ^2 is proportional to the probability of observing a material particle.

Thus, if ψ represents the wave function at the location x, the probability that the particle could be observed between x and $x + dx$ is given by $\psi(x)^2\,dx$:

$$\text{Probability of observing a particle in the interval } dx \propto \psi^2\,dx$$

The wave function of a particle, then, is analogous to the electric field of a photon. Just as $\mathbf{E}$ will generally be a function of both position and time, so will the wave function ψ.

It is not possible to specify the exact position of a *photon* at a particular time with complete certainty, but it is possible to specify the probability E^2 of observing it. Similarly it is not possible to specify the exact position of a *particle* at a particular time with complete certainty, but by ψ^2 one can specify the probability of observing it. Thus, a particle's wave function essentially gives rise to a *probability interpretation* of the position of a particle.

The interpretation of waves associated with material particles through probabilities was first given in 1926 by Max Born. The branch of quantum physics that deals with finding the values of ψ is known as *wave mechanics,* or *quantum mechanics.* The two principal originators of the wave mechanics of particles were Erwin Schrödinger (1926) and Werner Heisenberg (1925), who independently formulated quantum mechanics in different but equivalent mathematical forms.

Just as the electromagnetic theory of Maxwell is summarized in the Maxwell equations, which are the basis for computing values of $\mathbf{E}$, the wave mechanics of matter is governed by the *Schrödinger equation,* which is the basis for computing values of ψ in any problem in quantum physics. Here the parallel stops, however. Whereas the electric field, which has its origin in electric charges, gives not only the probability of observing a photon but also the electric force on a unit positive electric charge, the wave function of the Schrödinger equation has a physical meaning *only* for probability interpretation—it does *not* indicate any sort of force. The wave function is not directly measurable or observable; it does, however, give the most information one can extract concerning any system of objects; and *all* measurable quantities, such as the energy and momentum, as well as the probability of location, can be found from it. (The Schrödinger equation is discussed and applied to several simple situations in Section 5–9.)

Interference experiments have been performed with a Michelson interferometer and very weak sources of light, so weak that generally only a *single photon* was to be found between the light source and the observation screen at any one time. In the Michelson instrument, the interference pattern arises from the interference between *two* light beams. The light follows two paths going at right angles to one another from a partially silvered mirror.† We might imagine that when a single photon passes through the instrument, it travels along only one of the two possible paths; yet experimental results, from data collected over a long time, show the customary interference pattern of alternating light and dark bands, implying that the single photon has traveled both routes simultaneously and has interfered with itself. The complementarity principle resolves the apparent paradox. If we speak of the photon as traveling along one of two possible routes, we are localizing it, regarding it as a particle, and precluding consideration of its wave aspects; on the other hand, if we speak of its interference pattern, we are regarding it as a wave. And to regard it both ways at once is meaningless.

A similar situation arises when waves pass through two parallel slits. When either one of the two slits is closed, the pattern is the typical single-slit diffraction pattern: a broad, central maximum flanked by weaker, secondary maxima, as in Figure 1–2*a*. When both slits are open, the pattern is as shown in Figure 5–10: interference fine structure within a diffraction envelope. The pattern is not merely two single-slit diffraction patterns superposed; the interference between waves traveling through *both* of the slits is responsible for the rapid variations in intensity. In short, in a case in which waves can take two or more routes from a source to an observation point, we solve the problem by first superposing the wave functions (or electric fields for photons) from the two separate routes to find the resultant wave function (or electric field) and then squaring to find the probability (or intensity). That is to say, if ψ_1 and ψ_2 represent the wave functions for passage through slits 1 and 2 separately, then $(\psi_1 + \psi_2)^2$, not $\psi_1^2 + \psi_2^2$, gives the probability of observing a particle on the screen. Thus if a single electron or photon is directed toward a pair of slits, we cannot say which of the two slits it will pass through; we must speak in the language of waves and say that in effect it passes through *both* slits.

Example 5–1. A uniform beam of monochromatic light of wavelength 4130 Å and intensity 3.00×10^{-14} W/m² is incident normally on a 10×10 cm screen. Assume that the light is completely absorbed by the screen, each

† See Weidner and Sells, *Elementary Classical Physics,* 2nd ed., Section 38–8.

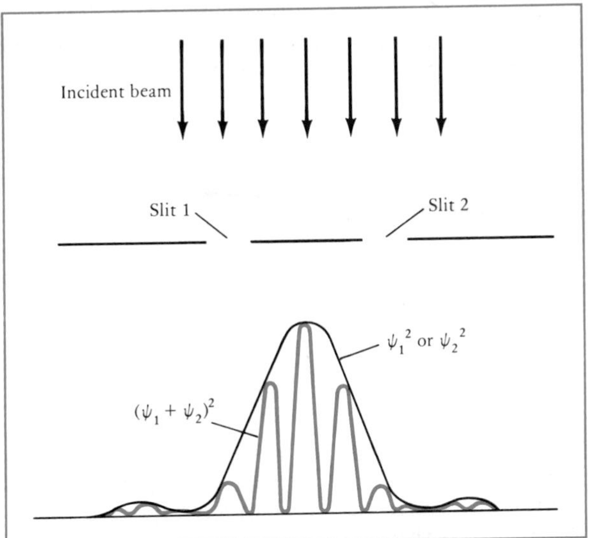

FIGURE 5–10. *Double-slit diffraction of particles. The wave functions ψ_1 and ψ_2 give the diffraction pattern when either slit 1 or slit 2 is open; the superposed wave function $\psi_1 + \psi_2$ gives the pattern when both slits are open. (The distance between the slits is grossly exaggerated in the figure.)*

photon producing a measurable flash at the point of absorption. Find the average photon flux on any 1.0 cm² of surface.

With the energy of each photon given by Equation 4–7,

$$h\nu = \frac{12\,400\ \text{Å}\cdot\text{eV}}{\lambda} = \frac{12\,400\ \text{Å}\cdot\text{eV}}{4130\ \text{Å}} = 3.00\ \text{eV}$$

the average photon flux follows from Equation 4–20:

$$N = \frac{I}{h\nu} = \frac{3.00 \times 10^{-14}\ \text{W/m}^2}{(3.00\ \text{eV})\,(1.60 \times 10^{-19}\ \text{J/eV})} = 6.25 \times 10^4\ \text{photons/m}^2\cdot\text{s}$$

or

$$N = 6.25\ \text{photons/cm}^2\cdot\text{s}$$

During a 1.00-s time interval, the average number of flashes on any 1.00-cm² area is 6.25. Since photons are quantized, here always with an energy of 3.00 eV, it is impossible to observe a fraction of a photon. In any 1-s period, one always measures an integral number of flashes—in one interval five might be seen, in another eight, and so on; over a long time interval the average will be 6.25 flashes/s. Furthermore, the spatial distribution of photons over the 1-cm² area will not be the same for all 1-s intervals but distributed randomly. Over a long period the flashes will be distributed uniformly over the area. Thus, the photon flux does not give precisely the time and location of the arrival of any photon but does give the probability of the space-time arrival.

5-6 The Uncertainty Principle

The principle of complementarity shows that it is impossible to apply simultaneously the wave and particle descriptions to a material particle or to a photon. If we choose one description, we preclude the other. If we describe, for example, electromagnetic radiation in the language of particles and locate a photon at any instant with complete precision, then the *uncer-*

tainties in position and time are both zero, $\Delta x = \Delta t = 0$; on the other hand, the uncertainties in the photon's wave attributes, wavelength λ and frequency ν, are then infinitely great, $\Delta\lambda = \infty$ and $\Delta\nu = \infty$.

Now consider a less extreme situation, one in which we are content to locate a photon in position and time, not with complete certainty, but with finite uncertainties Δx and Δt.

Earlier analysis (Section 1–4) showed that if the frequency ν of a wave is measured for only a *finite* time interval Δt, then the frequency is uncertain by an amount $\Delta\nu$:

$$\Delta\nu\,\Delta t \geq 1 \qquad (1\text{–}2),\ (5\text{–}5)$$

Similarly, if the wavelength λ is measured over a *finite* distance Δx along the direction of wave propagation, the wavelength is uncertain by an amount $\Delta\lambda$:

$$\Delta x\,\Delta\lambda \geq \lambda^2 \qquad (1\text{–}5),\ (5\text{–}6)$$

Note that these relations were derived on a strictly classical basis. The important implication of these relations for quantum physics is that we must associate two particlelike properties (energy E and momentum p) with two wavelike properties (frequency ν and wavelength λ) according to Equations 5–1 and 5–2:

$$E = h\nu \qquad \text{and} \qquad p = \frac{h}{\lambda}$$

Taking the differential of the first equation gives the uncertainty ΔE in energy in terms of the uncertainty $\Delta\nu$ in frequency,

$$\Delta E = h\,\Delta\nu \qquad (5\text{–}7)$$

Eliminating $\Delta\nu$ from Equations 5–5 and 5–7, we have

$$\Delta E\,\Delta t \geq h \qquad (5\text{–}8)$$

Thus, the product of the uncertainties in energy and in time is at least as large as Planck's constant h.† The meaning, in words, of Equation 5–8 is as follows. If an object—a photon, an electron, or even a system of particles—is known to exist in a state of energy E over a limited period Δt, then this energy is uncertain by at least an amount $h/\Delta t$; therefore, the energy of an object or system can be given with infinite precision ($\Delta E = 0$) only if the object or system exists for an infinite time ($\Delta t = \infty$). Equation 5–8 is one form of the celebrated *uncertainty principle*, or *principle of indeterminacy*, which Werner Heisenberg first introduced in 1927. We shall explore its fuller meaning after we have given it another formulation.

For an object moving along the x-axis, the relation between the uncertainty Δp_x (magnitude) in momentum and the uncertainty $\Delta\lambda$ in wavelength can be obtained by taking the differential of Equation 5–2.

$$\Delta\lambda = \frac{h}{p_x^{\,2}}\,\Delta p_x$$

Substituting this equation in Equation 5–6 and using Equation 5–2 give

† The value of the constant on the right side of Equation 5–8 depends on the precise definitions of the uncertainties ΔE and Δt. For the conservative convention used in Section 1–4, the constant is h, but if ΔE and Δt represent the root-mean-square values of several distinct measurements, then the constant becomes $h/4\pi$.

$$\Delta\lambda\,\Delta x = \frac{h\,\Delta p_x}{p_x^{\,2}}\,\Delta x \geq \lambda^2$$

$$\Delta p_x\,\Delta x \geq \frac{(\lambda p_x)^2}{h}$$

$$\Delta p_x\,\Delta x \geq h \tag{5–9}$$

which is the other formulation of the Heisenberg uncertainty principle. It was assumed that the wavelength was measured over the finite distance Δx along the direction of wave propagation; with reference to the wave-particle duality, the quantity Δx can be interpreted as the uncertainty in the position of the particle. Thus, the formulation of the uncertainty relation given in Equation 5–9 says that the product of the uncertainties in position and momentum equals or exceeds Planck's constant. Note that the momentum and position referred to in the equation are both measured along the *same* direction; according to the uncertainty relation, it is impossible to specify simultaneously and with infinite precision the linear momentum and the *corresponding* position of a particle or photon. Although $\Delta p_x\,\Delta x$ is equal to or greater than h, the product $\Delta p_y\,\Delta x$ can equal zero; there is no restriction on the simultaneous measurement of mutually perpendicular momentum and displacement.

The fundamental limitation on the certainty of measurements of energy and time or of position and momentum is compatible with the principle of complementarity. If the particle nature of, say, an electron is to be perfectly displayed, then both Δx and Δt must be zero. Therefore, when the particle aspect is chosen, the wave aspect is necessarily suppressed. All the quantities ν, E, λ, and p are then completely uncertain, which follows either from the uncertainty principle or from complementarity. On the other hand, if the wave characteristics of a material particle or of electromagnetic radiation are to be defined perfectly, that is, if $\Delta\nu = 0$ and $\Delta\lambda = 0$ (also $\Delta E = 0$ and $\Delta p = 0$), then by the principle of complementarity or by the uncertainty principle we are prevented from giving simultaneously the distinctively particle characteristics of precise location in space and in time, and x and t are completely uncertain.

Suppose that we want to represent an electron by its wave properties and yet localize it to some degree. We cannot use a single, sinusoidal wave; such a wave extends to infinity and is certainly not localized. We can, however, superpose several sinusoidal waves differing in frequency over a range of frequencies $\Delta\nu$ and have a *wave packet,* as described in Section 1–4. The component waves constructively interfere over a limited region Δx, identified as the somewhat uncertain localization of the "particle" and yield a resultant wave function ψ of the sort shown in Figure 5–11. Because there is necessarily a range in frequency $\Delta\nu$ and a range in wavelength $\Delta\lambda$, the associated momentum and energy are necessarily uncertain, and it is impossible to predict precisely where or when the wave packet will reach another point and what the momentum and energy will then be.

Since the uncertainty relation implies an uncertainty in energy, of magnitude $h/\Delta t$ over time Δt, it also implies that the law of energy conservation may actually be violated—by that amount, $\Delta E = h/\Delta t$—but only for the time interval Δt. The greater the amount of energy borrowed or discarded, the shorter the time interval over which the nonconservation of energy exists.

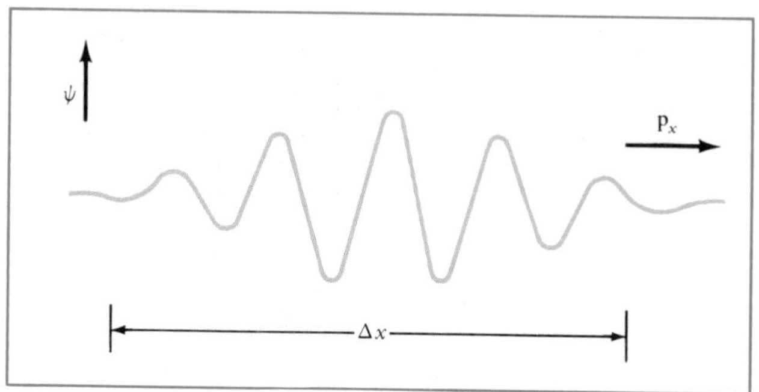

FIGURE 5–11. *The wave function of a wave packet.*

Similarly, the uncertainty in a particle's momentum $h/\Delta x$ implies that the law of momentum conservation may be violated, but only by the amount $\Delta p = h/\Delta x$ over a region Δx.

To derive the uncertainty principle in a different way, consider the diffracting of waves by a single, parallel-edged slit. A monochromatic plane wave is incident on a slit of width w, and the diffraction pattern is formed on a distant screen, as shown in Figure 5–12. The location of the points of zero intensity is given by the equation

$$\sin \theta = \frac{n\lambda}{w} \tag{5–10}$$

where λ is the wavelength and n is 1, 2, 3, ...† The total intensity within the central maximum is much greater than the intensity within any of the secondary humps, since the area under it far exceeds the area under any of the others. The area under the central hump is approximately three times the area under all the others; therefore, roughly three-fourths of the energy passing through the slit falls within this central region. The limits of the central region are given by Equation 5–10 with $n = 1$:

$$\sin \theta = \pm\frac{\lambda}{w} \tag{5–11}$$

We have not yet specified what sort of wave is diffracted by the slit. If the wave consists of electromagnetic radiation, then the intensity of the diffraction pattern is proportional to E^2, the square of the electric field at the screen. If on the other hand, the wave consists of a beam of electrons, the intensity of the diffraction pattern is proportional to ψ^2, which is the square of the wave function at the screen and gives the probability of finding an electron at any point along the screen. Whether the waves are of electromagnetic radiation or of a material particle, the diffraction effects are pronounced only when their wavelength is comparable to the slit width (see Figure 1–2); for waves much shorter than the slit width, the intensity pattern on the screen reduces to a geometrical shadow cast by the edges of the slit.

Suppose now that we reduce drastically the amount of radiation or the number of electrons, as the case may be. Then, when we observe the screen, we no longer see smooth variations along it but instead, photons or electrons arriving one by one. The intensity of the diffraction pattern is given by E^2

† See Weidner and Sells, *Elementary Classical Physics*, 2nd ed., Section 39–1.

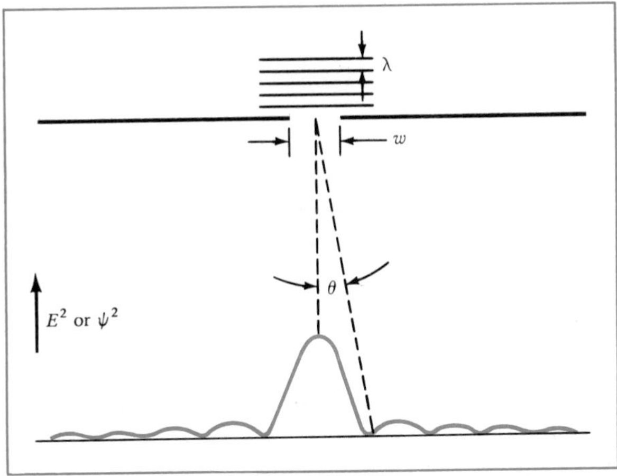

FIGURE 5–12. *Diffraction pattern of a monochromatic plane wave incident on a slit of width w.*

for photons and ψ^2 for electrons; thus, in Figure 5–12, the intensity represents the *probability* that a particle will strike a certain spot on the screen. There is a 75 percent probability that it will fall within the central region in the diffraction pattern, a smaller probability that it will fall in any other, and *zero* probability that it will fall at the zeros in the pattern. At very low illumination, bright flashes appear over a large area of the screen. As time passes, more and more particles accumulate on the screen, and the distinct bright flashes merge and form the smoothly varying intensity pattern predicted by wave theory.

There is *no* way of predicting in advance where any one electron or photon will fall on the screen. All that wave mechanics permits us to know is the probability that a particle will strike any one point. Before the particles pass through the slit, their momentum is known with complete precision both in magnitude (monochromatic waves) and in direction (vertically down in this case). When they pass through the slit, their position along a line in the x-direction, completely uncertain before they reached the slit, is now known with an uncertainty $\Delta x = w$, the slit width. What is not known, however, is precisely where any one particle will strike the screen. Any particle has approximately a three-to-one chance of falling anywhere within the central region, whose boundaries are given by Equation 5–11. There will be an uncertainty in the x-component of the momentum p that is *at least* as great as $p \sin \theta$; this can easily be seen in Figure 5–13. Therefore we write

$$\Delta p_x \geq p \sin \theta$$

Using Equation 5–11, we have

$$\Delta p_x \geq \frac{p\lambda}{\Delta x}$$

and since $p = h/\lambda$, we have

$$\Delta p_x \, \Delta x \geq h$$

the *Heisenberg uncertainty relation.*

Now suppose that Δx in our example is very large, that is, that the slit is very wide. Then the uncertainty in position is increased, and we cannot be

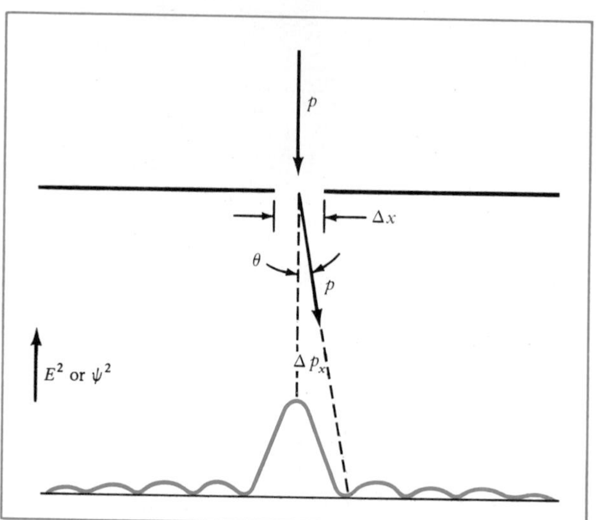

FIGURE 5–13. *An illustration of the uncertainty in the momentum of a particle that has passed through a single slit.*

certain where an electron is located along x. But the uncertainty in the momentum is reduced correspondingly; diffraction becomes less pronounced and essentially all electrons fall within the geometrical shadow (see Figure 1–2*c*). Contrarily, as the slit is reduced in width and Δx becomes very small, the diffraction pattern is expanded along the screen; and for the increase in our certainty of the electron's position, we must pay by a correspondingly greater uncertainty in the electron's momentum.

We see that when the slit width is much greater than the wavelength, particles pass undeviated through the slit to fall within the geometrical shadow. This agrees with classical mechanics, in which the wave aspect of material particles is ignored. Thus, there is a close parallel in the relation of wave optics to ray optics, and of wave mechanics to classical mechanics. Ray optics is a good approximation of wave optics whenever the wavelength is much less than the dimensions of obstacles or apertures that the light encounters; similarly, classical mechanics is a good approximation of wave mechanics whenever a particle's wavelength is much less than the dimensions of obstacles or apertures encountered by material particles. Symbolically, we can write

$$\underset{\lambda/w \to 0}{\text{Limit}} \text{ (wave optics)} = \text{ray optics}$$

$$\underset{\lambda/w \to 0}{\text{Limit}} \text{ (wave mechanics)} = \text{classical mechanics}$$

No ingenious subtlety in the design of the diffraction experiment will remove the basic uncertainty. We do not have here, as in the large-scale phenomena of classical physics, a situation in which the disturbances on the measured object can be made indefinitely small by ingenuity and care. The limitation here is rooted in the fundamental quantum nature of electrons and photons; it is intrinsic in their complementary wave and particle aspects.

Example 5–2. We compute, in illustration of the uncertainty principle, the uncertainty in the momentum of a 1000-eV electron moving along the x-axis. Assume that the position is uncertain by no more than 1 Å = 1.0×10^{-10} m, the approximate size of atoms. From $\Delta p_x \geq h/\Delta x$, it

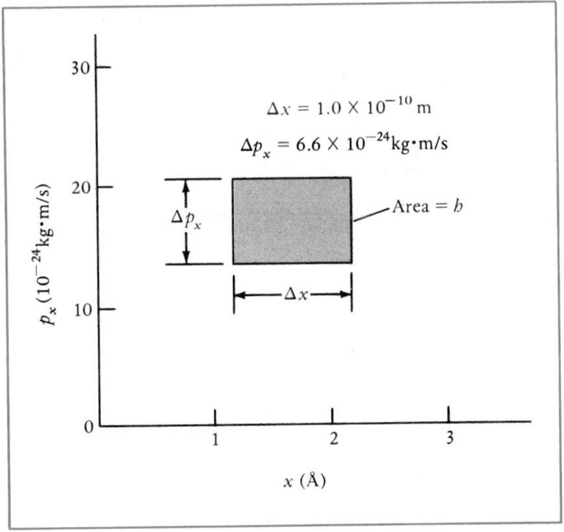

FIGURE 5–14. *Uncertainties in the simultaneous measurements of position and momentum of an electron.*

follows that $\Delta p_x = 6.6 \times 10^{-24}$ kg·m/s. Now let us compare this uncertainty in the momentum with the momentum itself, $p_x = (2mE_k)^{1/2} = 17 \times 10^{-24}$ kg·m/s. The fractional uncertainty in the momentum is $\Delta p_x/p_x = 6.6/17$, about 40 percent! Because of the uncertainty principle, it is impossible to specify with even moderate precision the momentum of an electron confined to atomic dimensions.

Consider now the uncertainty involved when a 10.0-g body moves at a speed of 10.0 cm/s; that is, an ordinary-sized object is moving at an ordinary speed. Let us further assume that the position of the object is uncertain by no more than 1.0×10^{-3} mm. We want to find the uncertainty in the momentum, and more especially, the fractional uncertainty. We find $\Delta p_x = 6.6 \times 10^{-28}$ kg·m/s and $p_x = 1.0 \times 10^{-3}$ kg·m/s; therefore, $\Delta p_x/p_x = 6.6 \times 10^{-25}$! The fractional uncertainty in the momentum arising in this example of a macroscopic body is so extraordinarily small as to be negligible compared with all possible experimental limitations. The uncertainty principle imposes a fundamental limitation on the certainty of measurements only in the microscopic domain, where the wave-particle duality is important. In the macroscopic domain, the uncertainties are, in effect, trivial (see Figure 1–1).

Figure 5–14 shows the momentum p_x of the electron in our example above, plotted against its position x. The uncertainty principle requires that the shaded area in this figure, which gives the product of the uncertainties in the momentum and the position, be equal in magnitude to Planck's constant h. If the position is known very precisely, the momentum is rendered highly uncertain; if the momentum is specified with high certainty, the position must be highly indefinite. It is therefore impossible to predict and follow in detail the future path of an electron confined to essentially atomic dimensions. Newton's laws of motion, which are completely satisfactory for giving the paths of large-scale particles, cannot be applied here; to predict the future course of any particle, it is necessary to know not only the forces that act on the particle but also its initial position and momentum. Because *both* position and momentum cannot

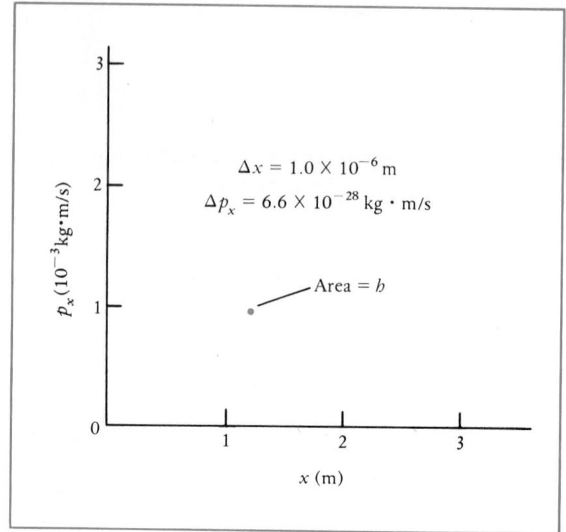

FIGURE 5–15. *Uncertainties in the simultaneous measurement of position and momentum of a 10-g body. (On the scale of this drawing, the uncertainty, area h, has been exaggerated by a factor of 10^{28}.*

be known simultaneously without uncertainty, it is not possible to predict the future path of the particle in detail. Instead, *wave mechanics* must be used to find the probability of locating the particle at any future time.

Consider again the 10.0-g body moving at 10.0 cm/s. Figure 5–15 shows its momentum and position. The area h, representing the product of the uncertainties in momentum and position, is so extraordinarily tiny in such macroscopic circumstances that it appears as an infinitesimal point on the figure. In this case, the classical laws of mechanics can be applied without entailing appreciable uncertainty.

Here we have seen still another example of the correspondence principle as it applies to relations between classical physics and quantum physics. The *finite size* of Planck's constant is responsible for quantum effects. Quantum effects are subtle because Planck's constant is very small—but not zero. Recall that the relativity effects are subtle because the speed of light is very large—but not infinite. If somehow Planck's constant were zero, the quantum effects would disappear. Thus classical physics is the correspondence limit of quantum physics as h is imagined to approach zero. Symbolically,

$$\text{Limit}_{h \to 0} \text{ (quantum physics)} = \text{classical physics}$$

5-7 Wave Packets and the De Broglie Wavespeed

From the wave point of view, the speed of a wave of frequency $\nu = E/h$ and wavelength $\lambda = h/p$ is given by

$$v_{\text{ph}} = \nu\lambda = \frac{E}{h}\frac{h}{p} = \frac{E}{p} \tag{5–12}$$

where the subscript denotes that the speed is the *phase* speed, the rate at which a point of constant phase travels through space. Using the angular frequency $\omega = 2\pi\nu$ and the wave number $k = 2\pi/\lambda$, we can write

$$E = \hbar\omega \qquad p = \hbar k \tag{5–13}$$

$$v_{\text{ph}} = \frac{\omega}{k}$$

where $\hbar = h/2\pi$.†

From the particle viewpoint, the relativistic energy E and momentum p of a particle of rest mass m_0 traveling at the speed v are given by

$$E = \gamma m_0 c^2 \qquad p = \gamma m_0 v \tag{5–14}$$

In assigning a speed v to a particle, we imply that its energy and momentum are to some degree localized and are transported at this speed. We can compare the particle speed v with the phase speed v_{ph} of the associated wave by substituting Equation 5–14 in Equation 5–12:

$$v_{\text{ph}} = \frac{E}{p} = \frac{\gamma m_0 c^2}{\gamma m_0 v} = \frac{c^2}{v} \tag{5–15}$$

First consider a photon, a particle with zero rest mass and speed $v = c$. From Equation 5–15, we have $v_{\text{ph}} = c$ for $m_0 = 0$. The associated electromagnetic wave travels at the *same* speed c as the photon does.

For a particle with a finite rest mass, the speed v must always be less than c, and Equation 5–15 becomes

$$v_{\text{ph}} > c \qquad \text{for } m_0 > 0$$

The phase speed of the associated wave of a nonzero rest mass particle must always *exceed* the speed of light. Consequently, a monochromatic wave associated with a material particle is unobservable—a circumstance that is not disturbing for wave mechanics, since observations of ψ always involve the probability ψ^2 of finding a particle, not the speed with which the phase of ψ advances.

If we are to think of some object bearing energy and momentum as a particle, its energy and momentum confined to a small region of space, then the amplitude of the associated wave function must be concentrated in a relatively small region, as shown in Figure 5–11. The figure does not depict a wave of a single frequency, for such a wave would be strictly sinusoidal and have infinite extension; it shows a wave packet, composed of waves of different frequencies. The component waves interfere constructively at the location of the particle, in the region in which the resultant wave function is large; and they interfere destructively in all other regions, as shown in Figure 1–8. Since the particle's location is that for which the wave function and the probability are large, the speed v at which the particle moves is the speed at which the region of constructive interference advances through space.

Recall now some general results given in Section 1–5 about phase and group velocities. When a group, or packet, of individual sinusoidal waves differing in frequency *and in phase speed* combine to produce a region of strong constructive interference, the speed at which that region advances, the *group speed* v_{gr}, is related to the angular frequency ω and wave number k of the component waves by the relation

† The symbol $\hbar$, equal to $h/2\pi$, is referred to as "aitch bar."

$$v_{gr} = \frac{d\omega}{dk} \tag{1–17}$$

As the arguments have suggested, the speed v_{gr} of the *group* of waves in the packet should equal the particle speed v. Let us prove it.

Since $\omega = E/\hbar$ and $k = p/\hbar$ Equation 1–32 can be written

$$v_{gr} = \frac{d\omega}{dk} = \frac{dE}{dp} \tag{5–16}$$

The particle's total energy E is related to its relativistic momentum p by

$$E = \sqrt{E_0^{\,2} + (pc)^2} \tag{3–15}$$

Taking the derivative, we have

$$v_{gr} = \frac{dE}{dp} = \frac{pc^2}{\sqrt{E_0^{\,2} + (pc)^2}} = \frac{pc^2}{E} = \frac{\gamma m_0 v c^2}{\gamma m_0 c^2}$$

$$v_{gr} = v$$

The particle's speed *is* just the group speed of the particle's wave packet.

5-8 Quantum Description of a Confined Particle

For a particle that is completely free from any external influence, the classical description is that the particle moves in a straight line with a constant momentum (Newton's first law). In the language of wave mechanics, on the other hand, a particle having a constant, well-defined momentum must be represented by a monochromatic sinusoidal wave with a well-defined wavelength. Therefore according to the uncertainty principle, the position of the particle is altogether uncertain and indeterminate. Most particles in nature, because of interactions with their surroundings, cannot be taken to be free, and a more general quantum law is needed. We shall now extend the quantum description to a confined particle.

In Figure 4–2c, we saw an example from classical physics of waves having perfectly defined wavelengths and yet confined to a limited region of space, namely, resonant standing waves on a string fixed at both ends. The wave on the string is repeatedly reflected from the boundaries, the fixed ends, and it constructively interferes with itself. The wave can persist and resonance is achieved only when the string length is some integral multiple of half-wavelengths and the standing-wave pattern fits between the boundaries.

An elementary wave-mechanical problem analogous to that of standing waves on a string follows. Let us have the wave associated with a particle confined in the same way that a transverse wave on a string is confined. We assume that the particle moves freely back and forth along the x-axis but encounters an infinitely hard wall at $x = 0$ and another at $x = L$; it is, then, confined between these boundaries. The infinitely hard walls correspond to an infinite potential energy V for all values of x smaller than zero and larger than L. Because the particle is free between zero and L, its potential energy V in this region is constant. The situation we have described fits a *particle in a one-dimensional box,* or a particle in an infinitely deep potential well. Because the walls are infinitely hard, the particle imparts none of its kinetic energy to them, its total energy remains constant, and it continues to bounce back and forth between the walls undeterred.

From the point of view of wave mechanics we can say that if the particle is confined within the limits stated and its potential energy between zero and L taken as zero for convenience, then the probability of finding it outside these limits is zero. Therefore, the wave function ψ, whose square represents this probability, must be zero for $x \leq 0$ and $x \geq L$.

Mathematically, the conditions of our problem are:

$$V = \infty \qquad \text{for } x < 0, x > L$$

$$V = 0 \qquad \text{for } 0 < x < L$$

$$\psi = 0 \qquad \text{for } x \leq 0, x \geq L$$

Only the wave functions that satisfy these conditions are allowed. Since the particle's momentum is constant in the entire region between the walls, it is represented, we know, by a sinusoidal wave. To satisfy the conditions at the boundaries, the only wavelengths that are allowed are those that will permit an integral number of half-wavelengths to be fitted between $x = 0$ and $x = L$. The condition for the existence of *stationary,* or standing, waves is then

$$L = n\frac{\lambda}{2} \qquad (5\text{–}17)$$

where λ is the wavelength and n is the *quantum number* having the possible values 1, 2, 3, ...

Figure 5–16 shows the wave function ψ and the probability ψ^2, plotted against x, for the first three possible *stationary states* of the particle in the box. Note that whereas ψ can be negative as well as positive, ψ^2 is always positive.

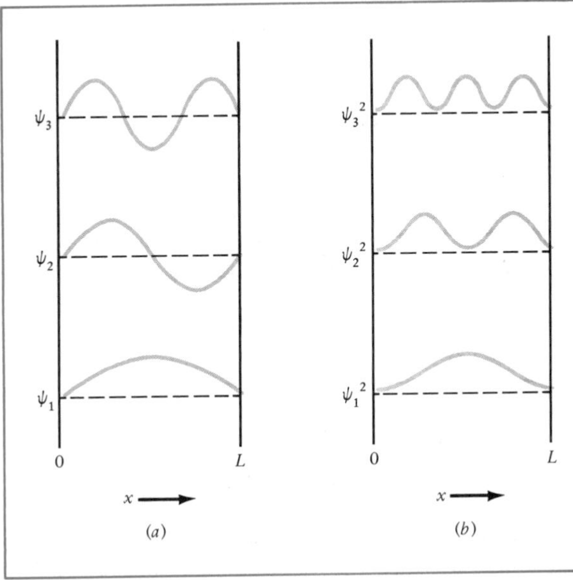

FIGURE 5–16. *The first three stationary states of a particle in a one-dimensional box: (a) wave functions and (b) probability distributions.*

The probability distribution is such that ψ^2 is always zero at the boundaries. For the first state, $n = 1$, the most probable location of the particle is the point midway betwen the two walls, at $x = L/2$; for the second state, $n = 2$, however, the least probable location ($\psi^2 = 0$) is at $x = L/2$, which is to say that it is impossible for the particle to be located there!

Imposing the boundary conditions on ψ, that is, fitting the waves between the walls, has restricted the wavelength of the particle to the values given by Equation (5–17). If only certain wavelengths are permitted, the magnitude of the momentum also is restricted to certain values, since $p = h/\lambda$. Therefore, the permitted momenta are given by

$$p = \frac{h}{\lambda} = \frac{hn}{2L} \qquad (5\text{–}18)$$

Finally, the kinetic energy E_k (and therefore the total energy E of the particle, since the potential energy is zero) is given by

$$E_k = E = \tfrac{1}{2}mv^2 = \frac{p^2}{2m} = \frac{(hn/2L)^2}{2m}$$

$$E_n = n^2 \frac{h^2}{8mL^2} \qquad (5\text{–}19)$$

where m is the particle's mass (this equation holds only for nonrelativistic speeds). The subscript n signifies that the possible values of the energy depend only on the quantum number n (1, 2, 3, . . .) for fixed values of m and L.

It is convenient to write the possible energy values E_n in terms of the smallest value (for $n = 1$). Since

$$E_1 = \frac{h^2}{8mL^2} \qquad (5\text{–}20)$$

Equation 5–19 becomes

$$E_n = n^2 E_1 \qquad (5\text{–}21)$$

The quantum state of lowest possible energy E_1 is called the *ground* state and the corresponding energy E_1 the *zero-point energy.*

By Equations 5–18 and 5–19, we see that both the *momentum* (magnitude) and the *energy* of the particle in the one-dimensional box are quantized. The first three allowed energies are shown in the *energy-level diagram* of Figure 5–17. The particle cannot assume any momentum or energy except those particular momenta and energies that satisfy the boundary conditions placed on the wave function. The quantization of the energy is analogous to the classical quantization of the frequencies of waves on a string fixed at both ends.

Note that the particle's lowest possible energy ($n = 1$) is not zero, but

$$E_1 = \frac{h^2}{8mL^2} \qquad (5\text{–}20)$$

Although this result is surprising, it is in accord with the uncertainty principle. If the particle's energy were zero, with the particle then at rest somewhere within the box ($\Delta x = L$), both the momentum p and the uncertainty in the momentum Δp_x would be zero. But this would violate the uncertainty relation, since the product $\Delta p_x \Delta x$ would be $(0)(L) = 0$, not h. Since the

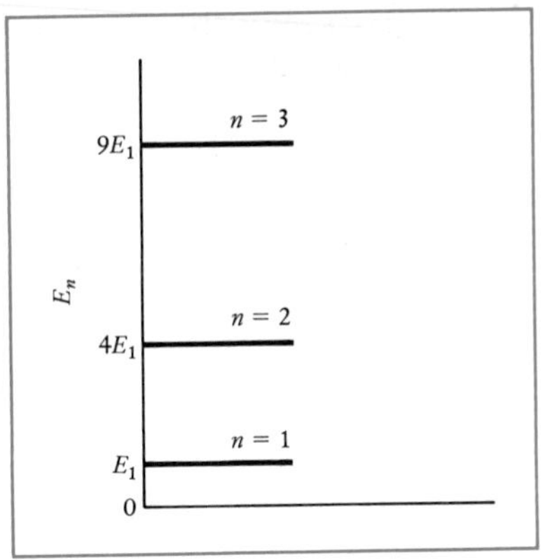

FIGURE 5–17. *First three allowed energies of a particle confined to a one-dimensional box, where $E_n = n^2E_1 = n^2(h^2/8mL^2)$.*

particle is known to be somewhere within the box, $\Delta x = L$, and by the uncertainty principle (Equation 5–9), $\Delta p_x \geq h/\Delta x = h/L$.

For the lowest state, the particle's momentum p_x in one direction must be at least as great as the momentum uncertainty Δp_x. Moreover, we cannot know whether the particle is traveling left or right, so that all told, $\Delta p_x = 2p_x$. Under these circumstances, the particle's lowest energy is

$$E = \frac{p_x^{\,2}}{2m} = \frac{(\Delta p_x/2)^2}{2m}$$

With $\Delta p_x = h/L$, we have $E = h^2/8mL$, exactly the energy of the first allowed state, given by Equation 5–20.

The problem of the particle in the box is somewhat artificial; there is no such thing as an infinitely great potential energy, and a particle cannot be made completely free from all external influences while inside a box. Nevertheless, the problem is important because it reveals the quantization of the energy. Fundamentally, energy quantization occurs because only certain discrete values of the wavelength can be fitted between the boundaries.

Example 5–3. Find the possible values of the energy and speed for the following particles confined to a one-dimensional box: (*a*) an electron of mass 9.1×10^{-31} kg constrained to move back and forth within an atomic distance $L = 4.0$ Å $= 4.0 \times 10^{-10}$ m; (*b*) a relatively large object of mass 9.1 mg $= 9.1 \times 10^{-6}$ kg constrained to move along the x-axis with $L =$ 4.0 cm $= 4.0 \times 10^{-2}$ m.

(*a*) Substituting in Equation 5–22 gives the ground state energy

$$E_1 = \frac{h^2}{8mL^2}$$

$$= \frac{(6.6 \times 10^{-34}\ \mathrm{J\cdot s})^2}{8(9.1 \times 10^{-31}\ \mathrm{kg})(4.0 \times 10^{-10}\ \mathrm{m})^2}$$

$$= 3.7 \times 10^{-19}\ \mathrm{J}$$

$$= 2.3\ \mathrm{eV}$$

The next possible energies follow from Equation 5–21:

$$E_n = n^2E_1 \qquad \text{or } 4E_1, 9E_1, 16E_1, \ldots$$

The permitted energies of the electron in a 4.0-Å box are shown on the energy-level diagram of Figure 5–18a. It is significant, in relation to atomic structure, that when an electron is confined to a distance approximately the diameter of an atom, its possible energies are in the range of a few electron volts, comparable to the binding energy of electrons in atoms.

The electron's possible speeds can be found from Equation 5–18:

$$p_n = mv_n = \frac{nh}{2L}$$

or

$$v_n = n\frac{h}{2mL} = nv_1$$

with the lowest speed v_1 given by

$$v_1 = \frac{h}{2mL} = \frac{6.6 \times 10^{-34}\ \text{J}\cdot\text{s}}{2(9.1 \times 10^{-31}\ \text{kg})(4.0 \times 10^{-10}\ \text{m})}$$

$$= 9.1 \times 10^5\ \text{m/s}$$

Note that the ground-state speed is large, but still within the nonrelativistic region, since

$$\frac{v_1}{c} = \frac{9.1 \times 10^5}{3 \times 10^8} = 0.003$$

(b) Again, using Equation 5–20 for the relatively large object, we get

$$E_1 = \frac{(6.6 \times 10^{-34}\ \text{J}\cdot\text{s})^2}{8(9.1 \times 10^{-6}\ \text{kg})(4.0 \times 10^{-2}\ \text{m})^2} = 2.3 \times 10^{-41}\ \text{eV}$$

a fantastically small amount of energy!

The energy-level diagram for these circumstances, shown in Figure 5–18b, is plotted to the *same* scale as in Figure 5–18a. The spacing between adjacent energies for such macroscopic conditions is so very small that energy is effectively continuous. That is why we never see any obvious manifestation of the quantization of the energy of a macroscopic particle;

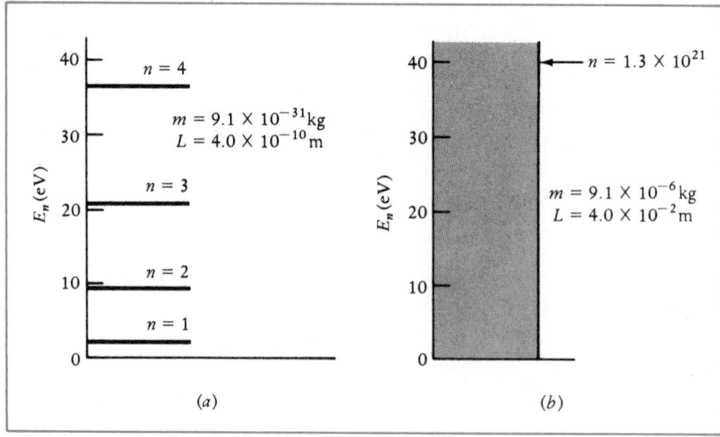

FIGURE 5–18. *(a) Allowed energies of an electron confined to a one-dimensional box of atomic dimensions. (b) Allowed energies of a 9.1-mg particle confined to a 4-cm one-dimensional box.*

the quantization is there, but too fine to be discerned. This result agrees with the correspondence principle, which requires that the discrete energies of a bound system appear continuous in large-scale phenomena.

The particle's speed follows from Equation 5–18:

$$v_n = \frac{p_n}{m} = n\frac{h}{2mL}$$

The lowest speed is

$$v_1 = \frac{h}{2mL} = \frac{(6.6 \times 10^{-34}\ \mathrm{J\cdot s})}{2(9.1 \times 10^{-6}\ \mathrm{kg})(4.0 \times 10^{-2}\ \mathrm{m})}$$

$$= 9.1 \times 10^{-28}\ \mathrm{m/s}$$

or a mere 10^{-7} Å per millennium. The particle is effectively at rest.

5-9 The Schrödinger Equation

We now examine the quantum description of a bound particle in a potential-energy field $V(x)$. From the classical one-dimensional wave equation, it is possible to arrive at the Schrödinger equation, whose solution yields the permitted wave functions and quantized energies of bound systems.

Any wave propagated along the x-direction obeys the general wave equation,

$$\frac{\partial^2 F}{\partial x^2} = \frac{1}{v_{\mathrm{ph}}^2}\frac{\partial^2 F}{\partial t^2} \tag{5–22}$$

where F is the wave function, depending on both the coordinate x and the time t, and v_{ph} is the wavespeed (or phase speed). When transverse waves are propagated along a taut string, the function F is the transverse displacement of the string from equilibrium and v_{ph} is the speed of the wave along the string; when electromagnetic waves are propagated through a vacuum, F is the electric or magnetic field and v_{ph} is the speed of light; when sound waves are propagated through a gas, F is the pressure difference and v_{ph} is the speed of sound. In the wave-mechanical behavior of particles the wave function F is that quantity whose square gives the probability of locating a particle at any point in space; we shall now find it convenient to call it Ψ.

We consider only nonrelativistic systems whose total energy E is constant and whose particles move along the x-axis and are bound.† Then, by Equation 5–1, the frequency $\nu = E/h$ associated with the bound particle is also constant. Assume that the wave function $\Psi(x, t)$ is separable into a spatial-dependent term $\psi(x)$ and a time-dependent term $f(t)$:

$$\Psi(x, t) = \psi(x)f(t)$$

Since it is assumed that the frequency is precisely defined, the time-dependent term $f(t)$ must vary sinusoidally with time; we can take it to be given by

$$f(t) = \cos 2\pi\nu t$$

† Here we can ignore the rest energy E_0, assuming E to be the sum of the kinetic and potential energies only.

The second partial derivatives in Equation 5–22 then become

$$\frac{\partial^2 \Psi}{\partial x^2} = f(t)\,\frac{d^2\psi}{dx^2}$$

and

$$\frac{\partial^2 \Psi}{\partial t^2} = \psi(x)\,\frac{d^2 f}{dt^2} = -4\pi^2\nu^2 f(t)\psi(x)$$

Setting these results in Equation 5–22 yields

$$f(t)\,\frac{d^2\psi}{dx^2} = -\frac{4\pi^2\nu^2}{{v_{\mathrm{ph}}}^2} f(t)\psi(x)$$

$$\frac{d^2\psi}{dx^2} = -\left(\frac{2\pi}{\lambda}\right)^2 \psi = -\left(\frac{p}{\hbar}\right)^2 \psi \tag{5–23}$$

where by Equation 5–12, the wavelength is $\lambda = v_{\mathrm{ph}}/\nu$, and by Equation 5–2, the momentum of the particle is $p = h/\lambda$.

We take the particle of mass m to be interacting with its surroundings (of infinite mass) through a potential-energy function $V = V(x)$. The total energy E of the system is then given by

$$E = E_{\mathrm{k}} + V = \frac{p^2}{2m} + V$$

where E_{k} is the particle's kinetic energy. Then we have

$$p^2 = 2m(E - V)$$

and Equation 5–23 becomes

$$\frac{\hbar^2}{2m}\frac{d^2\psi}{dx^2} + (E - V)\psi = 0 \tag{5–24}$$

This equation is the one-dimensional, time-independent, nonrelativistic Schrödinger equation.† To get the Schrödinger equation that applies to particles moving in three dimensions, we merely replace $d^2\psi/dx^2$ in this equation with $\partial^2\psi/\partial x^2 + \partial^2\psi/\partial y^2 + \partial\psi/\partial z^2$.

In getting the Schrödinger equation, we assumed that the "particle" propagates as a wave; we used a *wave* equation. We also assumed, however, that it interacts with its surroundings as a particle; the potential energy V is given as a function of *points* in space. The complementarity principle is built into the Schrödinger equation.

If we know the force acting on a bound particle—that is, if we know the potential energy as a function $V(x)$ of the particle's position—we can find the allowed wave functions and allowed energies of the system. Because the square of the wave function is proportional to the probability of finding the particle at position x, an acceptable solution $\psi(x)$ must be finite, continuous, and single-valued; and most especially, it must be consistent with the boundary conditions imposed by the character of the potential

† Although the correct form of the time-independent Schrödinger equation (Equation 5–24) can be arrived at from the general wave equation (Equation 5–22), the general equation is not appropriate for probability wave functions. For one thing, the correct Schrödinger time-dependent wave equation contains only the first derivative with respect to time, not the second. Furthermore, the total wave function $\Psi(x, t)$ must in general be complex; as a consequence, the probability of finding a particle is given by $\Psi^*\Psi$, where Ψ^* is the complex conjugate of Ψ.

energy V. Indeed, it is imposing boundary conditions on the wave function that leads to the quantization of the energy of a bound system. In non-analytical terms, we must regard the particle as a wave reflected back and forth within the confines of a bound system, forming standing waves; it is fitting stationary waves to boundary conditions that leads to the quantized values of a system's permitted energies. If the potential energy depends on position, but the system's total energy E is constant, then the particle's kinetic energy, momentum, and wavelength all must depend on position. More specifically, the wavelength is given by

$$\lambda = \frac{h}{p} = \frac{h}{\sqrt{2m(E - V)}} \tag{5–25}$$

For any given energy state, the energy E (and frequency ν) is constant, but the wavelength λ (and momentum p) is not; λ and p vary with position x according to Equation 5–25. Some examples of the application of Equation 5–24 follow.

PARTICLE IN A ONE-DIMENSIONAL POTENTIAL WELL OF INFINITE HEIGHT. The simplest problem to solve with the Schrödinger equation concerns a single particle confined to a one-dimensional box of width L (Section 5–8); that is, the particle is imagined to be in an infinitely deep potential well, as shown in Figure 5–19. The potential energy $V(x)$ is constant zero within the box and rises to infinity at the boundaries, $x = 0$ and $x = L$. For the interval $0 < x < L$, the Schrödinger equation, Equation 5–24, then becomes, with $V = 0$,

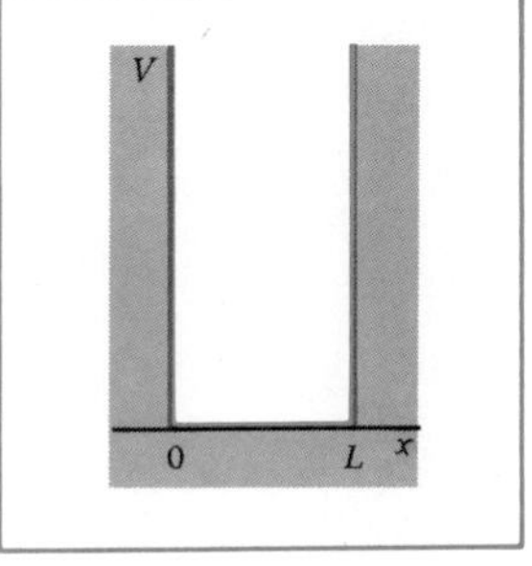

FIGURE 5–19. *Potential energy of an infinitely deep well of width L.*

$$\frac{\hbar^2}{2m}\frac{d^2\psi}{dx^2} + E\psi = 0$$

or

$$\frac{d^2\psi}{dx^2} = -B^2\psi \tag{5–26}$$

where $B^2 \equiv 2mE/\hbar^2$. Since the walls are infinitely high, the particle cannot be found outside the box. This implies that $\psi(x)$ must be zero for all points at the boundaries of the box and beyond. Thus, the allowed solutions must be consistent with the boundary conditions, which are $\psi(0) = 0$ and $\psi(L) = 0$. A suitable solution is

$$\psi(x) = A \sin Bx$$

which can be verified by substitution in Equation 5–26. The first boundary condition $\psi(0) = 0$ is satisfied for the sine function; it would *not* be satisfied for a cosine function, so the cosine solution is discarded. The second boundary condition $\psi(L) = 0$ is satisfied only if $BL = n\pi$, where n is an integer (since $\sin n\pi = 0$). Substituting for B, we then have

$$n\pi = BL = \sqrt{2mE}\,\frac{L}{\hbar} \tag{5–27}$$

The energies and wave functions of a free particle confined to a box of infinitely high walls are therefore

Energies: $$E_n = \frac{n^2\pi^2\hbar^2}{2mL^2}$$

Wave functions: $$\psi_n(x) = A_n \sin \frac{n\pi x}{L} \tag{5–28}$$

These results agree with those obtained in Section 5–5. The sinusoidally varying wave functions and the quantized energies (varying with the square of the quantum number n) are shown in Figures 5–16*a* and 5–17.

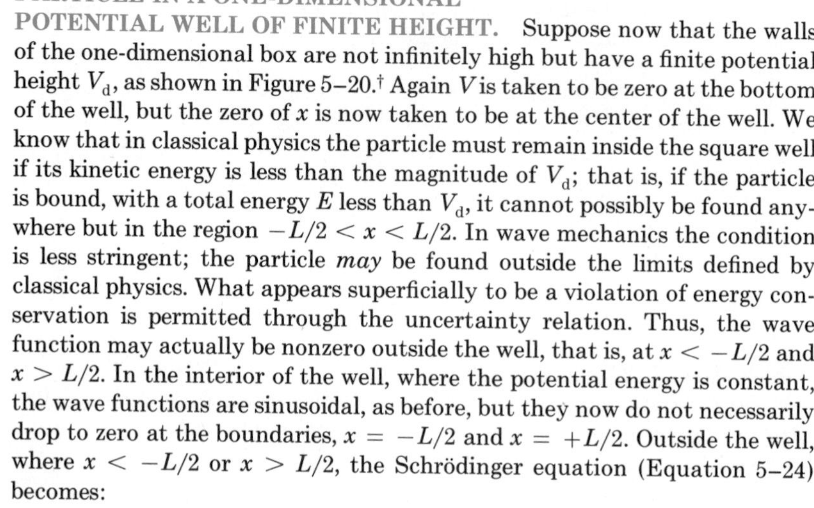

PARTICLE IN A ONE-DIMENSIONAL POTENTIAL WELL OF FINITE HEIGHT. Suppose now that the walls of the one-dimensional box are not infinitely high but have a finite potential height V_d, as shown in Figure 5–20.[†] Again V is taken to be zero at the bottom of the well, but the zero of x is now taken to be at the center of the well. We know that in classical physics the particle must remain inside the square well if its kinetic energy is less than the magnitude of V_d; that is, if the particle is bound, with a total energy E less than V_d, it cannot possibly be found anywhere but in the region $-L/2 < x < L/2$. In wave mechanics the condition is less stringent; the particle *may* be found outside the limits defined by classical physics. What appears superficially to be a violation of energy conservation is permitted through the uncertainty relation. Thus, the wave function may actually be nonzero outside the well, that is, at $x < -L/2$ and $x > L/2$. In the interior of the well, where the potential energy is constant, the wave functions are sinusoidal, as before, but they now do not necessarily drop to zero at the boundaries, $x = -L/2$ and $x = +L/2$. Outside the well, where $x < -L/2$ or $x > L/2$, the Schrödinger equation (Equation 5–24) becomes:

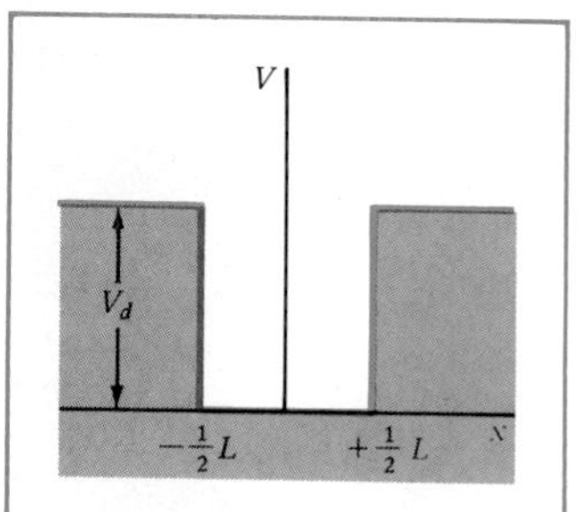

FIGURE 5–20. *Potential energy of a well of finite depth V_d and width L.*

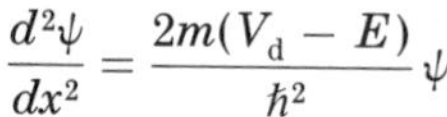

$$\frac{d^2\psi}{dx^2} = \frac{2m(V_d - E)}{\hbar^2}\psi$$

The potential energy of a bound particle exceeds the total energy E, so that $V_d - E$, appearing on the right-hand side of the equation, is *positive*. If we use $C^2 \equiv 2m(V_d - E)/\hbar^2$, this equation becomes

$$\frac{d^2\psi}{dx^2} = C^2\psi$$

where C^2 is a positive quantity for the regions outside the potential well.

The possible solutions of this equation are $\psi_+ = A_+e^{+Cx}$ and $\psi_- = A_-e^{-Cx}$, which can be verified by substitution. First we concentrate on the exterior region on the right, where $x > L/2$. Here x is always positive and we can rule out $\psi_+ = A_+e^{+Cx}$ on the following grounds. Although an acceptable wave function may be nonzero beyond the walls of the one-dimensional box, it cannot be infinite; and if ψ_+ were a solution, we should find it to be infinite at infinite distances from the box along the positive of the x-axis, implying an infinite probability of finding the particle at the greatest distances from the box. Clearly, this is physically untenable, and we are then left with the solution $\psi_- = A_-e^{-Cx}$, which means that the wave function decays exponentially in a direction away from the box on the positive of the x-axis and becomes zero at infinitely great distances. Similar arguments show that the wave function also decays exponentially along the negative of the x-axis;

[†] This situation approximates the potential between a neutron and a proton that are interacting by the nuclear force.

since x is always negative to the left of the box, the appropriate form there is $\psi_+ = A_+e^{+Cx}$.

To sum up, the wave function is a sinusoid inside the box and a decaying exponential outside it. The inside and outside wave functions must, however, join smoothly at the points $x = +L/2$ and $x = -L/2$. At both these points

$$\psi_{\text{inside}} = \psi_{\text{outside}}$$

$$\frac{d\psi_{\text{inside}}}{dx} = \frac{d\psi_{\text{outside}}}{dx}$$

The complete wave functions of the first two allowed states for particular values of V_d and L are shown in Figure 5–21. The corresponding allowed energy states are somewhat lower than the allowed energy states for a box with infinitely high walls. This is a consequence of a greater wavelength inside the box—greater since the wave functions do not drop to zero at the walls; hence there is a lower momentum, a lower kinetic energy, and a lower total energy. Furthermore, for a finite potential well V_d, the number of bound energy states is finite. Bound states can exist only for $E < V_d$.

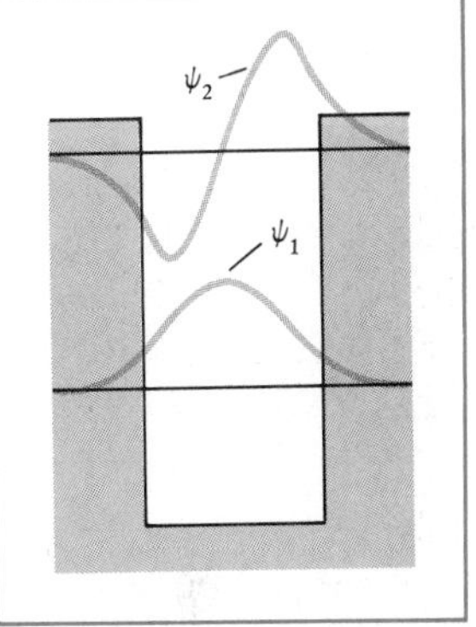

FIGURE 5–21. *Wave functions for the first two states of a particle in a potential-energy well of finite depth.*

THE TUNNEL EFFECT. Since according to wave mechanics, a particle can spill over its classical confines, we have a curious possibility when a particle encounters a potential wall of finite width as well as height (see Figure 5–22). In classical physics, a particle with too little kinetic energy to get over the wall would never get through to the other side by either climbing or passing through; it simply could not, because of the conservation of energy. In wave mechânics getting to the other side *is* a possibility. As we have seen, the wave function is nonzero and decays exponentially when the potential energy exceeds the total energy. Thus, the wave function of a particle approaching the wall from the left is sinusoidal to the left of the wall, exponential through the wall, and sinusoidal, but of much smaller amplitude, to the right of the wall. Since the amplitude of the wave function gives a measure of the probability of locating a particle, there is a small but finite probability that the particle that has approached from the left will be on the *right*. In other words, there is a high probability that the particle will be found on the left, a smaller probability that it will be found within the wall, and a still smaller probability that it will be found on the right. The particle, or more properly, the wave, can penetrate, or tunnel through,

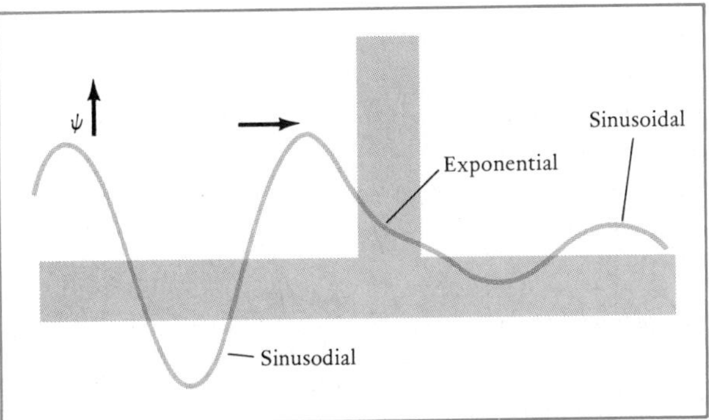

FIGURE 5–22. *Wave function of a particle incident from the left upon a barrier of finite height and width.*

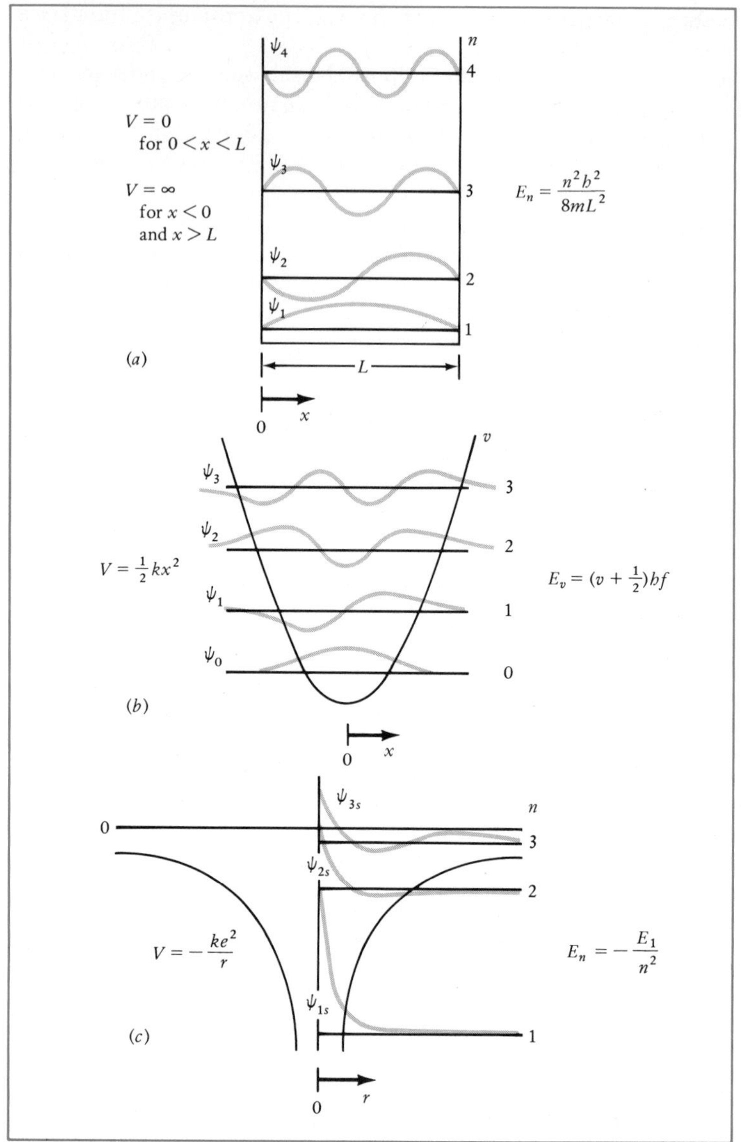

FIGURE 5–23. *Potential-energy functions, wave functions, and allowed energies of three simple potential energies: (a) an infinitely deep well, (b) a simple harmonic oscillator, and (c) a particle subject to an inverse-square attractive force (hydrogen atom).*

the classically insurmountable barrier. The probability of this *tunnel effect* is vanishingly small except at the atomic and nuclear level. The effect is observed, however, in the behavior of certain semiconducting devices (the so-called tunnel diodes) and in the emission of α particles from unstable, heavy nuclei.

OTHER EXAMPLES OF BOUND PARTICLES. The environment for particles in atoms and nuclei cannot be approximated by the simple one-dimensional potential-well models discussed earlier. Two other examples of bound particles having potential energies more representative of physical situations are: (1) a particle subject to an inverse-square force (with $V(r) =$

$-ke^2/r$), like the electron in the hydrogen atom; (2) the one-dimensional simple harmonic oscillator, a particle subject to a linear restoring force (with $V(x) = \frac{1}{2}Kx^2$), somewhat like an atom in a diatomic molecule, or an atom in a crystalline solid. The mathematical solutions of the Schrödinger equation (Equation 5–24) for the hydrogen atom and the simple harmonic oscillator are too involved to be given here, but Section 6–6 concerns their permitted energies and Problem 5–45 their wave functions.

For comparison, Figure 5–23 shows the permitted energies and associated wave functions for the infinitely deep square well, the simple harmonic oscillator, and the attractive inverse-square force. As you can see in the examples given in Figure 5–23, some general features of wave-mechanical solutions are the following.

1. The wave function of a particle in a box of infinite height is exactly zero at the boundaries and at all exterior points. The wave functions of the other two potentials, however, are finite at the classical boundaries and extend beyond them.
2. Integral multiples of half-wavelengths, $1(\lambda/2)$, $2(\lambda/2)$, $3(\lambda/2)$, . . . , successively are fitted between the boundaries at states of progressively higher energies.
3. The wavelength is constant (is independent of x) and the wave function consequently sinusoidal at a constant potential energy, like the energy inside the infinitely deep well. But the wavelength is not constant (is dependent on x) and the wave function consequently not sinusoidal for a potential energy that varies with x, like the potential energies of the simple harmonic oscillator and of the hydrogen atom. From Equation 5–25, $\lambda = h/[2m(E - V)]^{1/2}$, it follows that the wavelength depends on the potential-energy function $V(x)$ and therefore on x. For this reason it is small whenever the kinetic energy $E - V$ is large.
4. The lowest energy is *not* zero.
5. For a particle in the infinitely deep well, the energy levels are crowded at the bottom; for the simple harmonic oscillator, they are equally spaced; and for the hydrogen atom, they are crowded at the top. This behavior is related to the shape of the curve of the potential energy. In Figure 5–23*a* the potential energy is bent toward the vertical, compared with that of the simple harmonic oscillator, Figure 5–23*b*, whereas in Figure 5–23*c* it is bent toward the horizontal.

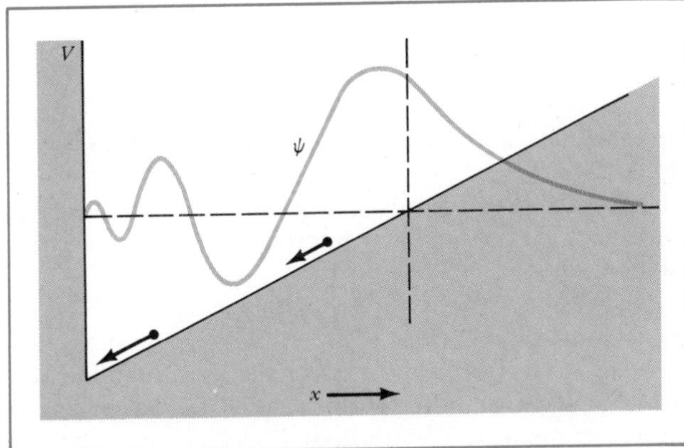

FIGURE 5–24. *The wave-mechanical inclined plane: potential-energy function and wave function of the fifth allowed state.*

Example 5–4. Sketch the wave function of an excited state of a particle moving in the potential shown in Figure 5–24. This is nothing more than the wave-mechanical version of a standard example in elementary mechanics: a particle sliding on a frictionless inclined plane (here with an infinitely hard, massive block at the bottom of the incline). Solving the problem in complete analytical detail would involve finding solutions to the Schrödinger equation for a potential energy that rises to infinity at $x = 0$ and increases steadily with x, according to $V = ax$, for $x > 0$. We can show, however, even without detailed analysis, the general features of the wave-mechanical solution.

Since the potential rises to infinity at $x = 0$, the wave function must be zero there. At the other extreme, where the particle can be found outside the classical upper limit, the wave function must decay to zero. In between, the particle has some nonzero kinetic energy, and so the wave function is undulatory there. Further, for a relatively highly excited state, the features of the wave-mechanical solution must, by the correspondence principle, approach the classical features of a particle sliding on an inclined plane and colliding with a hard wall at the bottom. Since the particle moves fast at the base of the incline and less so at the top, the wavelength of the wave function must be smallest near the base and increasingly large toward the top. Moreover, the classical high speed at the base and low speed at the top imply that the particle is more likely to be found at the top than at the bottom. Thus, the amplitude of the wave function must be greatest at the top and least at the bottom. Given all these features, we know that the wave-mechanical solution of an excited state is as shown in Figure 5–24.

In the next energy state up we should find one additional half-wavelength fitted between the left and right extremes, and in the next energy state down we should find one less half-wavelength; in the lowest state, we should find a single hump in ψ between the zeros in ψ at the left and right extremes.

SUMMARY

Every particle, whether of finite or zero rest mass, has associated with it a frequency and wavelength given by

$$\nu = \frac{E}{h} \quad \text{and} \quad \lambda = \frac{h}{p} \qquad (5\text{–}1), (5\text{–}2)$$

The wavelengths of electrons of 150 eV, and of thermal neutrons ($\frac{1}{25}$ eV), typical x-rays, and the interatomic spacings of a solid are all of the order of 1 Å.

When a wave is incident on a set of parallel Bragg planes separated by a distance d, constructive interference occurs, according to the Bragg relation, when

$$n\lambda = 2d \sin\theta \qquad (5\text{–}4)$$

X-rays and electrons can be diffracted by crystalline solids. The x-ray and electron diffraction effects can be used for measuring x-ray and electron wavelengths and interatomic spacings and for producing monochromatic beams.

According to Bohr's principle of complementarity, the wave and particle aspects of electromagnetic radiation and of material particles are complementary; the use of one description precludes the use of the other in a given circumstance.

The wave-mechanical description of material particles parallels the description of electromagnetic radiation as follows:

Electromagnetic radiation (photons)		*Material particles*
Wave function: electric field **E** (or magnetic field)		The wave function ψ
Probability of observing particle in interval dx	$E^2\,dx$	$\psi^2\,dx$

The Heisenberg uncertainty principle imposes a limit on the certainty of simultaneous measurements of energy E and time t and of momentum p_x and position x:

$$\Delta E\,\Delta t \geq h \qquad (5\text{–}8)$$

$$\Delta p_x\,\Delta x \geq h \qquad (5\text{–}9)$$

A particle can be represented by a packet of waves; the group velocity of the wave packet is the velocity of the particle.

By the correspondence principle,

$$\underset{h\to 0}{\text{Limit}}\,(\text{quantum physics}) = \text{classical physics}$$

A free particle confined to a one-dimensional box L of infinitely high walls is restricted to those states in which an integral multiple of the particle's wavelength is fitted between the boundaries. The particle's energy is thereby quantized and given by

$$E_n = \frac{n^2h^2}{8mL^2} \qquad (5\text{–}19)$$

where n is the quantum number $n = 1, 2, 3, \ldots$.

The wave functions ψ and allowed energies E of the stationary states of a particle of mass m confined by a potential-energy function V are determined by the time-independent Schrödinger equation; for one dimension this equation has the form

$$\frac{\hbar^2}{2m}\frac{d^2\psi}{dx^2} + (E - V)\psi = 0 \qquad (5\text{–}24)$$

The wave function is single-valued and continuous and chosen to be consistent with the boundary conditions imposed by the nature of the potential-energy function $V(x)$.

PROBLEMS

De Broglie waves, §5–1

● **5–1.** At what kinetic energy (eV) will electrons produce the same diffraction pattern as red light (6200 Å) diffracting through the same grating?

● **5–2.** Show that the wavelength of a charged particle moving in a uniform magnetic field is inversely proportional to the particle's radius of curvature in the magnetic field.

● **5–3.** A 150-eV electron has a wavelength of 1.0 Å. What is the wavelength of a 0.60-keV electron?

5–4. (*a*) Show that the wavelength of a particle of rest energy E_0 and kinetic energy E_k is given by

$$\lambda = \frac{hc}{\sqrt{E_k(E_k + 2E_0)}} = \frac{12\,400\ \text{Å}\cdot\text{eV}}{\sqrt{E_k(E_k + 2E_0)}}$$

where λ is in angstroms and all energies are in electron volts. (*b*) Show that the equation in part (*a*) reduces to Equation 4–7 for photons. (*c*) Show that the equation in part (*a*) reduces to $\lambda = 12\,400\ \text{Å}\cdot\text{eV}/\sqrt{2E_kE_0}$ for material particles of low energies ($E_k \ll E_0$).

5–5. Find the wavelength of electrons that have been accelerated from rest through a potential difference of (*a*) 100 V; (*b*) 100 kV; (*c*) 100 MV.

5–6. A 10-eV monoenergetic beam of electrons is projected perpendicularly into a uniform magnetic field **B** = 0.010 T. (*a*) What is the radius of the circular path of electrons in the magnetic field? (*b*) Find the wavelength of the particles. (*c*) For what energy (eV) and (*d*) radius of curvature (Å) would the electron beam have a radius equal to the wavelength?

5–7. A polyenergetic beam of electrons enters crossed electric and magnetic fields, as shown in Figure 5–25. The uniform magnetic field of 0.10 T is directed into the paper; the electric field **E** is directed from the positively charged plate to the negatively charged plate. (*a*) For what magnitude of the electric field will electrons of wavelength 5000 Å pass through the crossed fields undeflected? (*b*) What energy do particles in the emerging monoenergetic beam have? (*c*) Why is an experiment under the given conditions not feasible?

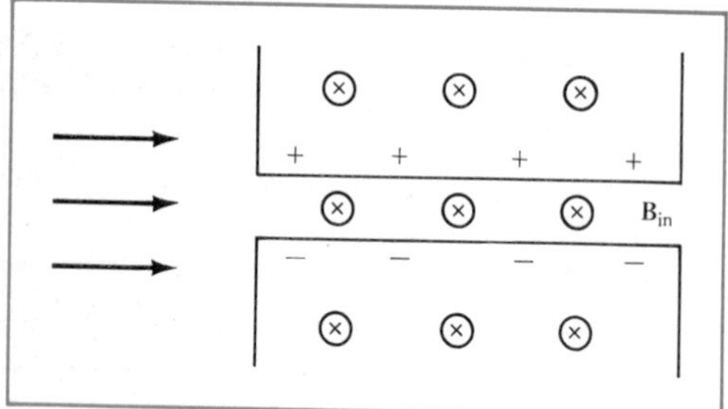

FIGURE 5–25.

5–8. The surface of solids can be investigated by observing the diffraction of extremely low energy gas atoms by the surface. In one such experiment, neon atoms (atomic mass, 20.0) having a speed of 1.22×10^2 m/s were used. (*a*) What is the kinetic energy (eV) of each neon atom? (*b*) Find the de Broglie wavelength (Å) of this beam and compare it with the distance between the surface atoms in a solid (~1 Å).

The Bragg law, §5–2

● **5–9.** At what scattering angles (2θ) do diffraction peaks occur when the incident monochromatic waves have a wavelength equal to the lattice spacing of the crystal?

5–10. Find the energies (eV) of (*a*) photons, (*b*) electrons, and (*c*) neutrons having wavelengths comparable to the lattice spacing of a single crystal of potassium chloride, 3.14 Å.

5–11. For a given set of parallel Bragg planes of an MnO crystal, the first two diffraction peaks for the Bragg diffraction of thermal neutrons (λ = 1.4 Å) at room temperature are observed at scattering angles (2θ) of 18° and 36°. (*a*) What is the lattice spacing *d*? (*b*) How many additional diffraction peaks are there for this set of planes?

5–12. (*a*) What is the wavelength of 50-keV electrons used in a typical electron microscope? (*b*) What is the smallest size that can be observed with this microscope if one takes the minimum resolvable separation distance to be 100λ?

Diffraction, §5–3

● **5–13.** Soon after the wave properties of electrons were verified experimentally in 1927, Thomson's experiment was modified by applying a uniform magnetic field in the region after the foil, as shown in Figure 5–5. (*a*) What effect would this have on the electron diffraction pattern of Figure 5–8? (*b*) What effect would it have on the x-ray diffraction pattern of Figure 5–7?

● **5–14.** A 150-eV electron has a wavelength of 1.0 Å. What is the kinetic energy of a neutron (1839 times the mass of an electron) that has the same wavelength?

● **5–15.** At what energy (MeV) will an electron have a wavelength of 10^{-15} m (approximately the size of an atomic nucleus)?

● **5–16.** A polychromatic beam of x-rays is incident on a KCl crystal whose lattice spacing is 3.14 Å. What wavelengths will be predominantly diffracted at a scattering angle of 30°?

5–17. A monochromatic beam of 2.0×10^4 eV electrons is incident on a foil, and diffraction circles are observed on a photographic plate 10 cm behind the foil (see Figure 5–5). (*a*) Find the wavelength (in angstroms) of the electrons. (*b*) For diffraction from parallel planes that are 1.5 Å apart, what are the radii of the first three diffraction rings?

5–18. A beam of electrons accelerated through a potential difference of 40 kV passes through a thin foil of polycrystalline iron. The radii of two first-order diffraction rings observed on a screen 20 cm beyond the foil (see Figure 5–5) are measured to be 0.60 cm and 0.85 cm. What is the distance between the parallel Bragg planes that gives rise to each of these rings?

■ **5–19.** A 50.0-eV beam of electrons is incident on a metal surface at an angle of 30.0° with the surface. Because of the work function φ at the metal surface the electrons refracted into the metal gain an energy of $e\varphi$. Assume the work function of the metal to be 4.00 eV. (*a*) Determine the wavelength (Å) and speed of the incident electron beam. (*b*) Determine the speed and wavelength of the refracted beam. (*c*) Find the angle of refraction of the beam within the metal.

5–20. In an interferometer experiment, an incident beam of monoenergetic neutrons of wavelength 1.445 Å was split into two separate beams using the Bragg diffraction of neutrons by a thin slab of a crystal of silicon.† Silicon is known to have a diamond structure in which the locations of the silicon atoms (projected on a cubic face) are as shown in Figure 5–26. When the beam of 1.445-Å neutrons was incident at an angle of 22.1° to the Bragg planes shown dotted in the figure, first-order Bragg diffraction was observed. Therefore the incident beam, in passing through the thin silicon slab, was divided into the undeviated beam and the first-order diffracted beam at an angle of 44.2° with the undeviated beam. (These two emerging beams were then brought back together to show interference effects.) (*a*) Find the spacing between the parallel Bragg planes in Figure 5–26. (*b*) Show that the lattice parameter a (see Figure 5–26) for silicon is 5.43 Å.

Probability interpretation, §5–5

5–21. An FM radio station emits 100-MHz radio waves at a power of 20 MW. (*a*) How many photons per second are emitted? (*b*) Assuming the radiation to be isotropic, find the photon flux 100 km away from the transmitter.

5–22. A 2.0-mW laser emits light of wavelength 6300 Å in a narrow beam of 1.0-mm² cross section. (*a*) Find the photon flux through the beam cross section. (*b*) In comparing this photon flux with the photon flux from the much more powerful 20-MW radio transmitter of Problem 5–21, explain why the laser-beam photon flux is larger.

† See R. Collela, A. W. Overhauser, and S. A. Werner, "Observations of Gravitationally Induced Quantum Interference," *Phys. Rev. Lett.*, *34*, 1472 (1975). This experiment showed that the phase of the waves associated with neutrons can be changed by the gravitational interaction between the neutrons and the earth.

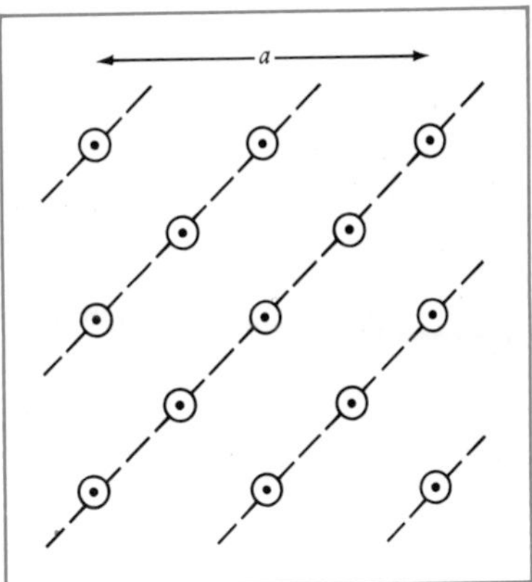

FIGURE 5–26. *Head-on view of cubic crystal of silicon with cube edge length a.*

5–23. A point source radiates 50 W of 6200-Å light uniformly in all directions. A camera located 1.8 km away takes a photograph of the source. (*a*) If the camera aperture has a diameter of 2.0 mm and is open for only 2.0×10^{-5} s, how many photons will enter the camera? (*b*) What is the average number of photons striking the film at the back of the camera? (*c*) Where are the photons most likely to strike on the film?

■ **5–24.** Each of two parallel slits 5.0×10^{-7} m apart has width 1.0×10^{-8} m and length 1.0 cm. Monochromatic light of wavelength 5000 Å is incident on the two slits and the light diffracted through the double slit falls on a screen 1.0 m behind the double slit. (*a*) How many photons strike the screen each second? (*b*) What is the intensity (W/m²) of the incident beam if on the average there is one photon between the double slit and the screen?

Uncertainty principle, §5–6

● **5–25.** An electron microscope is to resolve objects having a size of 10 Å. What minimum value of the kinetic energy (eV) of the electrons is needed for resolving objects of this size?

● **5–26.** The location of an electron is uncertain along the x-axis by 10.0 Å. What is the minimum uncertainty (in eV/c) in the electron's momentum component along x?

● **5–27.** Estimate the exaggeration of Δx and Δp_x on the graph of Figure 5–15, and verify that the area h shown on the graph is approximately 10^{28} times too large.

● **5–28.** An electron is known to have a speed lower than 1.0 cm/s. What is the minimum uncertainty in its location?

5–29. The diameter of a spherical virus of density 1.2 g/cm³ is measured to be 50 Å. What is the virus's minimum speed?

■ **5–30.** A small steel ball of mass 3.0×10^{-6} kg is dropped onto a horizontal surface from a height of 1.0 m. After each bounce the ball rises to 0.90 of the height before the bounce. How many bounces take place before the uncertainty principle becomes important, assuming that this occurs when the uncertainty in speed Δv is comparable to the speed of the ball at the surface?

5–31. In an electron microscope a beam of electrons replaces a light beam, and electric and magnetic focusing fields replace refracting lenses. The smallest distance that can be resolved by a microscope under optimum conditions (in other words, its resolving power) is approximately equal to the wavelength used in the microscope. (*a*) A typical electron microscope might use 50-keV electrons. Compute the minimum distance that can be resolved in such a microscope. (*b*) By what factor does the actual resolving power of about 10 Å, attained in well-designed electron microscopes, exceed the ultimate resolving power, which is limited by the wave properties of electrons? (The ultimate resolving power is the minimum separation between two point objects distinguishable as two distinct objects.)

5–32. Another formulation of the uncertainty principle is given by angular momentum L and angle θ; it is $\Delta L\, \Delta\theta \geq h$. The quantity ΔL (in units of $\text{kg}\cdot\text{m}^2/\text{s}$) is the uncertainty in a particle's angular momentum relative to some chosen point and $\Delta\theta$ (in radian units) is the uncertainty in the particle's angular position relative to the same point. (*a*) For a particle moving in a circular arc of radius r, use Equation 5–9 and relations between angular and linear quantities to get $\Delta L\, \Delta\theta \geq h$. (*b*) As we shall find in Chapter 6, the electron in the hydrogen atom has a ground-state orbital angular momentum of exactly $h/2\pi$. What does the uncertainty relation imply about the angular location of the electron in this orbit?

5–33. The uncertainty relation $\Delta E\, \Delta t \geq h$ can be derived from the uncertainty relation $\Delta p_x\, \Delta x \geq h$ through the following considerations. The position of a particle traveling along the x-axis is uncertain by the value Δx. Therefore, the uncertainty in the time at which the particle will arrive at the more distant point along the x-axis is $\Delta t = \Delta x/v$, where v represents the average speed of the particle. (*a*) Find the uncertainty in the particle's kinetic energy in terms of the particle's average speed v. (*b*) Find the uncertainty in its momentum Δp_x in terms of Δv_x, the uncertainty in the particle's speed. (*c*) From the results in (*a*) and (*b*), show that $\Delta E\, \Delta t = \Delta p_x\, \Delta x \geq h$.

Wave packets, §5–7

5–34. Find the phase speeds of the (*a*) photon, (*b*) electron, and (*c*) proton that have kinetic energy 1.0 MeV.

5–35. In a room at constant temperature, what is the ratio of the phase speed of an oxygen molecule's de Broglie wave to the phase speed of a hydrogen molecule's de Broglie wave? Assume average molecule speeds for each molecule.

5–36. A neutron has a radius of approximately 10^{-15} m. An early interpretation of the neutron had it consist of a proton and an electron bound by the electrostatic interaction $V = ke^2/r$. (*a*) Find what the electrostatic binding energy (eV) of an electron-proton system is when the particles are separated by 10^{-15} m. (*b*) Calculate the minimum kinetic energy (eV) of an electron confined to a size of 10^{-15} m and show that the electron-proton model for a neutron is untenable.

Confined particle §5–8

● **5–37.** Find the lowest energy state (eV) of a photon confined to a one-dimensional box of atomic dimension (10^{-10} m).

● **5–38.** A 0.50-kg ice puck slides back and forth on a horizontal, frictionless ice surface, the puck making perfectly elastic collisions at both ends, 2.0 m apart. The allowed quantum energy states for the puck are given by Equation 5–21, and the probability distributions for the three lowest states are shown in Figure 5–16*b*. Estimate the probability ratio of finding the puck (within 1.0 cm) midway between the ends, and at the location 0.50 m from an end for (*a*) the $n = 1$ quantum state; (*b*) the $n = 10^6$ quantum state. (*c*) Estimate the quantum number for a state in a typical experimental arrangement, and verify that there is equal probability that the puck will be found in any 1.0-cm length between the two ends.

■ **5–39.** Assume that an electron confined to a one-dimensional box of length $L = 1.0$ Å is in the lowest energy state E_1. To locate the electron within a small region of the box, say within a length of 0.10 Å, one might use photons of short wavelength ($\lambda \sim 0.1$ Å). (*a*) Photons of what energy would be needed? (*b*) After the measurement, what would be the uncertainty in energy of the electron? (*c*) In what energy states might the electron then be found?

5–40. A particle confined to a one-dimensional box of width L is in a stationary state for which the probability of finding the particle at $\frac{1}{3}L$ from one side of the box is zero. What is the minimum kinetic energy the particle can have to be in such a state in terms of the ground-state energy E_1?

5–41. A particle is confined to a one-dimensional box so narrow that even in the ground state the particle's kinetic energy greatly exceeds its rest energy. Show that the allowed states, as portrayed on an energy-level diagram, are equally spaced.

5–42. Assume that 1.0×10^{23} identical particles, each of mass 9.1×10^{-31} kg, are confined in the same one-dimensional, infinitely deep box of length 1.0 cm. (*a*) Find the lowest energy (eV) of the system if there is no limit on the number of particles in any state. (*b*) Find the approximate value of the lowest possible energy of the system if there is a limit of no more than two particles in any state.

Schrödinger equation §5–9

5–43. How many allowed energy states are there for a particle bound (*a*) in an infinitely deep well? (*b*) by a linear restoring force? (*c*) by an inverse-square attractive force?

5–44. An electron subject to an inverse-square attractive force with $V(r) = -ke^2/r$ is in the $n = 3$ quantum state (see Figure 5–23*c*). (*a*) How much energy (in terms of the ionization energy, E_I) is needed to free this electron from the attractive force? (*b*) Using the graph for the ψ_{3s} state in Figure 5–23*c*, locate the points of high probability of finding the electron in this state.

■ **5–45.** Consider a mass m in simple harmonic motion, with $V(x) = \frac{1}{2}kx^2$, where k is the force constant. The classical angular frequency of oscillation is $\omega = 2\pi f = \sqrt{k/m}$. The ground-state wave function ($n = 0$ state in Figure 5–23*b*) for this potential is $\psi_0 = Ce^{-ax^2}$. (*a*) Show that ψ_0 satisfies the Schrödinger equation (Equation 5–24) if $E = \frac{1}{2}\hbar\omega$ and $a = m\omega/2\hbar$. (*b*) Compare the range of values of position x for a particle with energy $\frac{1}{2}\hbar\omega$ treated classically with the range for a like particle treated quantum-mechanically.

5–46. A small bead of mass 2.0 g slides freely around a circular wire of radius 1.0 cm. (*a*) What are the allowed speeds and energies of the bead (in electron volts)? (*b*) What quantum number corresponds to the condition in which the bead revolves at 1 rps? (*c*) At what speed does the bead travel around the wire when the bead is in its lowest energy state?

5–47. For each of the bound particles in Figure 5–23: (*a*) Find the difference in energy between adjoining states $\Delta E_n = E_{n+1} - E_n$. (*b*) For each bound system in part (*a*), show that the fractional difference $\Delta E_n/E_n$ between adjoining energy levels approaches zero as the quantum number n approaches infinity. Thus, the discrete energy levels in quantum physics approach the apparent continuous distribution of energies in classical physics when n becomes large.

5–48. (*a*) Using the Schrödinger equation (Equation 5–24), show that the change in slope of the wave function $\Delta(d\psi/dx)$ over a small spatial interval Δx is given by $-(2m/\hbar^2) \times (E - V)\psi\,\Delta x$. (*b*) For what kinetic energy is the change in slope $\Delta(d\psi/dx)$ equal to zero? (*c*) Show that the wave function generally curves toward the x-axis when the kinetic energy of the particle is positive ($E > V$), and away from the x-axis when the kinetic energy is negative ($E < V$).

5–49. (*a*) For the wave function ψ_2 shown in Figure 5–21, sketch the slope ($d\psi/dx$) versus x, and show that its behavior is consistent with the curvature properties of Problem 5–48. (*b*) Show that for energies slightly greater than (or less than) the allowed energy E_2 corresponding to the wave function ψ_2 of part (*a*), the wave functions will not remain finite as $x \rightarrow \pm\infty$ and therefore are not physically acceptable solutions to the Schrödinger equation.

■ **5–50.** The ground state energy E_1 and associated wavefunction ψ_1 of a particle confined to a potential-energy well of finite depth V_d is shown in Figure 5–27*a*. (*a*) Will the ground-state energy of this same particle, now confined to the potential energy well shown in Figure 5–27*b*, be larger or smaller than E_1? Why? (*b*) Sketch the ground-state wave function.

5–51. An electron of mass $m = 9.11 \times 10^{-31}$ kg is bound in a potential-energy well of finite depth $V_d = 3.00$ eV and width 9.60 Å. Quantum-mechanical calculations show that there are only three bound states: $E_1 = 0.27$ eV, $E_2 = 1.05$ eV, and $E_3 = 2.24$ eV above the well bottom. (See Figures 5–20 and 5–21). (*a*) Sketch the wave functions for these three states. (*b*) Find the three lowest energy states in an infinitely deep well of this width, compare the corresponding energies with the well of part (*a*), and show that the lower energies for a well of finite depth are consistent with the discussion of the finite-depth well in Section 5–9.

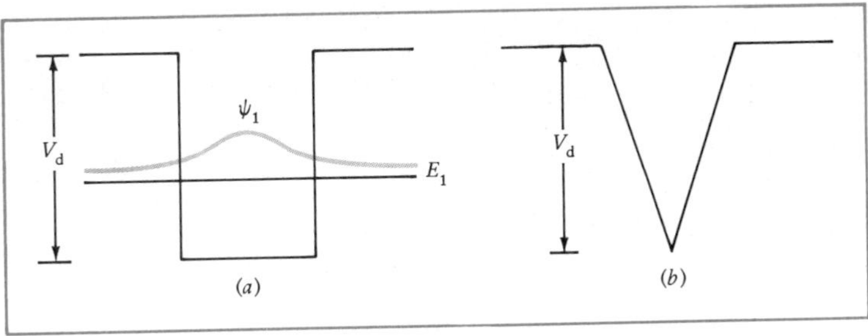

FIGURE 5–27.

6
The Structure of the Hydrogen Atom

Niels H. D. Bohr (1885, Denmark–1962). Student of J. J. Thomson and of Ernest Rutherford. 1913, quantum theory of hydrogen atom. 1920, named director, Institute of Theoretical Physics, Copenhagen, world capital of atomic physics. Developed complementarity principle, Copenhagen interpretation of quantum mechanics, liquid-drop model of atomic nucleus. 1922, Nobel Prize, atomic structure and radiation. World leader in advocating peaceful applications of atomic energy; 1957, first recipient, U.S. Atoms for Peace award. (Photo courtesy of the American Institute of Physics, Niels Bohr Library.)

6–1 α-Particle Scattering

Our modern concept of an atom postulates as essential features (1) a nucleus, which occupies a very small region of space and in which all the positive charge and practically all the atom's mass is concentrated, and (2) negatively charged electrons, which surround the nucleus. Ernest Rutherford first proposed this concept of atomic structure in 1911.

By 1900 it was known that the negative electric charge of the atom is carried by electrons, whose mass is but a small fraction of the total mass of the atom. Because atoms are ordinarily electrically neutral, it follows that if we were to remove all the electrons from an atom, then what would remain would contain all the positive electric charge and essentially all the mass. The question is, then, how are the mass and positive charge distributed within the atom?

Atoms are known, from a variety of experiments, to have a "size" (diameter) of the order of 1 Å, and because the positive charge and mass are confined to at least this small a region, it is impossible to directly observe any details of the atomic structure. Indirect measurement must be resorted to. One of the most powerful methods of studying the distribution of matter or of electric charge is *scattering*. After the α-particle scattering experiments Hans Geiger and Ernest Marsden performed in Rutherford's laboratory in 1909, Rutherford was led to propose, two years later, the existence of a massive, positively charged nucleus at the center of every atom. In 1913, Geiger and Marsden, in further α-particle scattering experiments from various thin foils, confirmed Rutherford's nuclear hypothesis.

We can best grasp the strategy of scattering by considering first a simple example. Suppose that we are confronted with a large black box, the mass of whose contents is known. Although we are not allowed to look inside to examine its internal structure, we must determine how the mass is distributed throughout the interior of the box. The box might, for instance, be filled completely with some material of relatively low density, such as wood, or it might be only partly filled with a material of high density. How can we find out whether one of these two possibilities represents the distribution of material within the box? We can use a very simple expedient—shoot bullets into the box and see what happens to them. If all the bullets emerge in the forward direction with reduced speeds, then we can infer that the box is filled with some material like wood, which deflects the bullets only slightly as they pass through. On the other hand, if we find a few bullets greatly deflected from their original paths, we can assume that they have collided with small, hard, and massive objects dispersed throughout the box. It is possible then, by studying the distribution of the scattered bullets, to learn a lot about the arrangement of material within the box. Note that it is not necessary to aim the bullets; the shots may be fired randomly over the front of the box. This is the essence of particle-scattering experiments in atomic and nuclear physics.

The essentials of a scattering apparatus are shown in Figure 6–1. A collimated beam of particles strikes a thin foil of scattering material, and a detector counts the number of particles scattered at some scattering angle θ from the incident direction within an angle $d\theta$. The experiment consists in measuring the relative number of scattered particles at various scattering

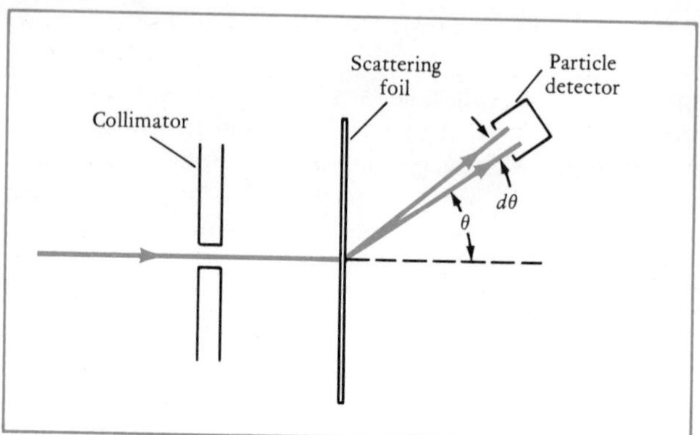

FIGURE 6–1. *Arrangement for a simple scattering experiment: collimator, scattering material, and detector.*

angles θ. Geiger and Marsden, in one of their early experiments, used α particles of kinetic energy 5.5 MeV from a radioactive source (radium C, later identified as an isotope of bismuth, ^{214}Bi); the particles struck a gold foil of thickness 6.00×10^{-5} cm. The rotatable detector consisted of a zinc sulfide screen, viewed through a microscope; the particles striking the screen produced bright flashes, or scintillations, that could be observed and counted for any angle θ.

A surprising result of the many experiments Geiger and Marsden carried out was the unexpected scattering of a small percentage of the incident α particles in the backward direction. They found, for example, approximately 1 in 8000 scattered through an angle larger than 90°. Such a small but significant number of α particles scattered through large angles could not be understood on the basis of the atomic model then current—the so-called plum-pudding model J. J. Thomson had proposed. After Thomson discovered the negatively charged electron with its small mass relative to the atom, he proposed that an atom's positive charge was uniformly distributed throughout the atomic volume, with light, point electrons embedded in it. According to Thomson's model, only an extremely small fraction of incident particles were scattered through angles larger than a few degrees—a mere 1 in 10^{3500} for 90° or larger—as against the 1 in 8000 from experiment. The experimental scattering results disturbed Rutherford, and in 1911, in a new model for atomic structure, he proposed that all positive charge and most of the atomic mass were concentrated in a very small region (essentially a point) at the atom's center, the *nucleus.* He held that the light electrons surrounded the nucleus and moved throughout the entire volume of the atom. Rutherford then showed that this nuclear model of the atom indeed predicted the observed scattering. We now turn to a more detailed account of Rutherford scattering.

Consider the behavior of the α particles as they traverse the interior of a scattering foil composed of gold. Rutherford had shown that an α particle is a doubly ionized helium atom: a positive electric charge twice that of an electron and a mass several thousand times the electron mass but considerably smaller than the mass of heavy atoms such as gold. We can dismiss as inconsequential any encounters an α particle may have with electrons within the material; the reason is that the α particle's mass is much greater

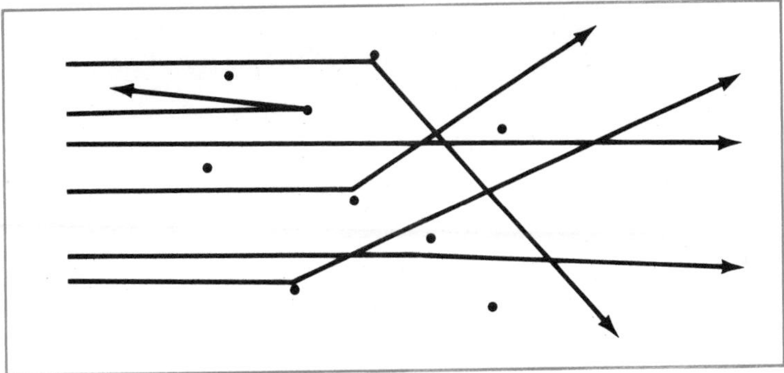

FIGURE 6–2. *Scattering of α particles by the nuclei of a material (the number scattered through sizable angles is greatly exaggerated).*

than the electron's and therefore the α particle is essentially undeflected in such collisions. Consequently, a negligible fraction of its energy is transferred to any one electron. Thus, the α particles are appreciably deflected and scattered only by close encounters with the nuclei. The nucleus of a gold atom has a mass that is considerably larger (50 times) than that of the α particle. Therefore, the gold nucleus does not recoil appreciably and can be assumed to remain at rest. Since the α particles and the nuclei are both positively charged, they repel each other. Rutherford assumed that the *only* force acting between a nucleus and an α particle (each regarded as a point charge) was the Coulomb electrostatic force. As we know, that force varies inversely with the square of the distance between the charges. Therefore, although the Coulomb force is never zero (except at infinite separation between the charges), it is strong only when an α particle comes close to a nucleus.†

Figure 6–2 shows several α-particle paths as they traverse the interior of a scattering foil. Most particles pass through with only a slight deviation from their original course; the chances of a close encounter with a nucleus, or scattering center, are fairly remote. On the other hand, the few α particles that barely miss head-on collisions are deflected at sizable angles. And the ones that are involved in the extremely rare head-on collisions are deflected through 180°—that is, are brought to rest momentarily and then returned along their paths of incidence.

Now consider the collision in detail. The incident particle is scattered by an angle θ, as shown in Figure 6–3. It moves in a nearly straight line, until it comes fairly close to the scattering center, which is the dot in the figure (the shading signifies a circular area about the nucleus, seen almost edge on). The particle is deflected and continues in a nearly straight line; its path is a hyperbola because of the repulsive inverse-square electric force. The incident particle would have passed the nucleus at a distance b had there been *no* force between them. This distance is called the *impact parameter*.

† Since by quantum theory a beam of monoenergetic particles is in effect a beam of monochromatic waves (Section 5–2), the scattering consists of the diffraction of incident waves by scattering centers. It is remarkable that a thoroughgoing wave-mechanical treatment of scattering by an inverse-square force yields precisely the same result as that yielded by the strictly classical particle analysis, discussed here. In cases of other types of force, however, the classical and wave-mechanical results differ.

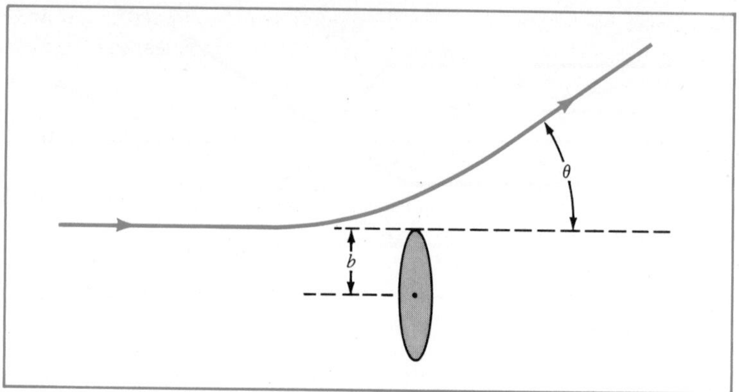

FIGURE 6–3. *Scattering of an α particle that approaches a heavy nucleus with an impact parameter b.*

We see from Figure 6–3 that all incident particles headed in a direction so as to strike at the circumference, or rim, of a circle drawn about the nucleus and having a radius b, will be deflected by an angle θ. Furthermore, any particles headed so as to strike anywhere within the shaded area πb^2 will be deflected by an angle larger than θ. The target area πb^2 is called the *cross section* σ:

$$\sigma = \pi b^2 \tag{6–1}$$

Thus, associated with every scattering center is an area σ of such a nature that an incident particle heading for it will be scattered an angle θ or larger. When a particle is deflected an appreciable angle, the cross section is extraordinarily small; that is, in large-angle scattering each nucleus presents a very small target to an incident particle.

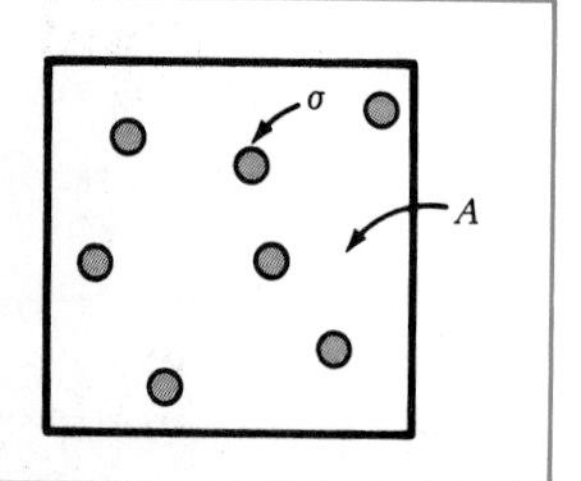

FIGURE 6–4. *Target areas presented by the scatteirng centers of a thin foil resulting in the scattering of incident particles through an angle θ or larger.*

Let us now calculate the total target area presented by all the scattering centers within a foil of area A and thickness t. We take the foil to be so thin that the cross-sectional area presented by any one nucleus does not overlap that of any other nucleus; see Figure 6–4, in which the total area represents that of the foil. With the total shaded area very small compared with the area of the foil, there is a very low probability that an incident particle will be perceptibly deflected by more than one nucleus; that is, if the foil is sufficiently thin and the cross section σ is small, "*single scattering*" rather than "multiple scattering" will occur. If there are n nuclei per unit volume, then in a foil of volume At, there are nAt scattering centers, and the total target area for single scattering is $\sigma(nAt)$. With one nucleus in each atom, the number of scattering centers per unit volume can be computed from Avogadro's number N_A, the mass density ρ, and the atomic mass M_a of the scattering foil:

$$n = \left(\frac{\text{atoms}}{\text{vol}}\right) = \left(\frac{\text{atoms}}{\text{mol}}\right)\left(\frac{\text{mol}}{\text{vol}}\right)$$

$$= N_A\left(\frac{\text{mass}}{M_a}\cdot\frac{1}{\text{vol}}\right) = \frac{N_A\rho}{M_a}$$

We wish to calculate the fraction of incident particles scattered by an angle θ or larger. The incident beam cannot be aimed to strike any one nucleus in the foil; the beam is spread over an area very large compared with σ. Therefore, the probability that any one incident particle will be deflected an angle greater than θ is simply the ratio of the shaded target area σnAt to the total area A of the foil. Of many incident particles, the fraction scattered is then given by $\sigma nAt/A$, or

$$\frac{N_s}{N_i} = \sigma nt \tag{6–2}$$

where N_i is the number of particles incident on the foil and N_s is the number of these incident particles scattered an angle θ or larger.

The relation between b and θ in the Coulomb scattering of particles of charge Q_1 and kinetic energy E_k by an infinitely massive point charge Q_2 is given by

$$b = \frac{k}{2}\frac{Q_1 Q_2}{E_k}\cot\frac{\theta}{2} \tag{6–3}$$

where $k = 1/(4\pi\epsilon_0)$, the constant in Coulomb's law.† For the scattering of α particles by a massive nucleus of atomic number Z, $Q_1 = +2e$ and $Q_2 = +Ze$. This equation is derived from the laws of conservation of energy, linear momentum, and angular momentum and by recognizing that the Coulomb force is a conservative, central, inverse-square force.

Example 6–1. Let us apply the theory of the scattering of α particles to the conditions of the Geiger and Marsden experiment, in which 5.5-MeV α particles were scattered by a 6.0×10^{-5} cm gold foil. We use the following known values, choosing θ to be 90°:

$$\theta = 90° \qquad k = 9.0 \times 10^9\ \mathrm{N\cdot m^2/C^2}$$

$$t = 6.0 \times 10^{-7}\ \mathrm{m} \qquad \rho = 1.9 \times 10^4\ \mathrm{kg/m^3}$$

$$E_k = 5.5\ \mathrm{MeV} \qquad M_a = 200$$

$$Z = 79 \qquad N_A = 6.0 \times 10^{26}\ \mathrm{atoms/kg\cdot mol}$$

$$e = 1.6 \times 10^{-19}\ \mathrm{C}$$

Equations (6–3), (6–1), and (6–2) then yield, respectively,

$$b = 2.1 \times 10^{-14}\ \mathrm{m}$$

$$\sigma = 1.3 \times 10^{-27}\ \mathrm{m^2}$$

$$\frac{N_s}{N_i} = 4.5 \times 10^{-5}$$

Under the conditions of this experiment, any incident particle that originally moves so as to miss a head-on collision with a nucleus by no more than 2.1×10^{-14} m will be deflected by at least 90°. Note that this impact parameter is much less than the distance between gold nuclei, approximately 3×10^{-10} m, which is also the size of a gold atom. The cross section, or target area, for the scattering of α particles through an angle larger than 90° is $1.3 \times 10^{-27}\ \mathrm{m^2}$, a very small area. Finally, we see that N_s/N_i is 4.5×10^{-5}. According to theory, in the stated conditions, only one out of every 22,000 incident particles will be deflected 90° or more.

† See Weidner and Sells, *Elementary Classical Physics,* 2nd ed., Section 22–6.

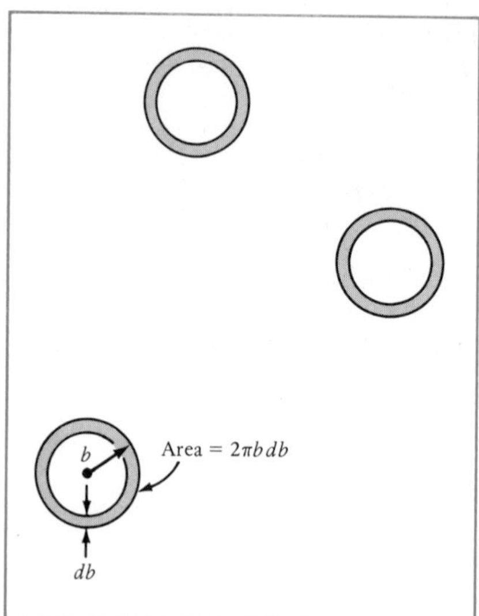

FIGURE 6–5. *Nuclear target area for the differential cross section.*

The assumption that the shaded area of Figure 6–4 is much less than the total area of the foil is amply justified, and therefore only single scattering is important in the case of such a large scattering angle. If the computation is repeated for $\theta = 1°$, one finds $N_s/N_i = 0.60$; that is, 60 percent of the incident particles are expected to be scattered 1° or more, again with single scattering assumed.

A particle detector (see Figure 6–1) actually measures the number dN_s of scattered particles within the angle $d\theta$ subtended by the detector at a scattering angle θ. The corresponding target area that will scatter particles into angles from θ to $\theta + d\theta$ is the *differential cross section* $d\sigma$, which by Equation 6–1 is

$$d\sigma = 2\pi b \, db \tag{6–4}$$

This follows as well from Figure 6–5, in which we see the differential cross section of each nucleus as a ring of radius b, circumference $2\pi b$, and width db. An incident particle heading toward the shaded ring now is scattered between θ and $\theta + d\theta$.

The number dN_s of particles scattered between the angles θ and $\theta + d\theta$ is, then, by Equations 6–2 and 6–4,

$$\frac{dN_s}{N_i} = nt \, d\sigma = 2\pi bnt \, db \tag{6–5}$$

Equation 6–5 is the basis of all scattering experiments with thin foils. The numbers of incident and scattered particles N_i and dN_s can be directly measured by experiment, and their ratio may be compared with the ratio computed from this equation.

For Coulomb scattering, db follows from Equation 6–3:

$$db = \frac{-kQ_1Q_2}{4E_k} \csc^2 \frac{\theta}{2} \, d\theta$$

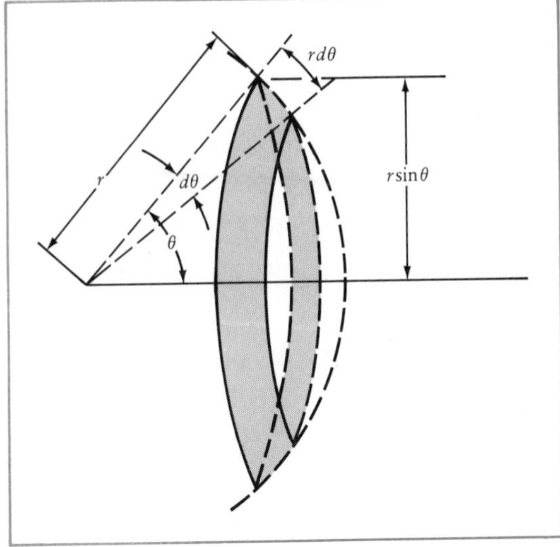

FIGURE 6–6. *The solid angle $d\Omega$ subtended at the scattering angle θ by the incremental angle $d\theta$. By definition, $d\Omega/4\pi$ is the shaded area divided by the entire area of spherical surface; the shaded area is equal to $2\pi(r \sin \theta)(r\, d\theta)$. Then $d\Omega/4\pi = (2\pi r^2 \sin \theta\, d\theta)/4\pi r^2$; therefore $d\Omega = 2\pi \sin \theta\, d\theta$.*

The minus sign implies that the scattering angle θ increases as the impact parameter b *decreases;* we shall not, however, use it hereafter. Putting the relations for b and db into Equation 6–5, we have

$$\frac{dN_s}{N_i} = \frac{\pi nt}{4}\left(\frac{kQ_1Q_2}{E_k}\right)^2 \cot\frac{\theta}{2} \csc^2\frac{\theta}{2}\, d\theta = \frac{\pi nt}{4}\left(\frac{kQ_1Q_2}{E_k}\right)^2 \frac{\cos(\theta/2)\, d\theta}{\sin^3(\theta/2)}$$

Using the trigonometric identity $\sin \theta = 2 \sin(\theta/2) \cos(\theta/2)$, we can rewrite this relation:

$$\frac{dN_s}{N_i} = \frac{nt}{16}\left(\frac{kQ_1Q_2}{E_k}\right)^2 \frac{(2\pi \sin \theta\, d\theta)}{\sin^4(\theta/2)} \tag{6–6}$$

The quantity $2\pi \sin \theta\, d\theta$ gives, as Figure 6–6 shows, the entire solid angle $d\Omega$ subtended by the incremental scattering angle $d\theta$. The scattering detector, rotated at a fixed distance r about the scattering foil, always subtends a solid angle proportional to $d\Omega$; that is, the factor $d\Omega = 2\pi \sin \theta\, d\theta$ does not change as the detector is rotated to count particles scattered at various scattering angles θ. From Equation 6–6 it then follows that

$$\frac{dN_s}{N_i} \propto \left(\frac{Q_1Q_2}{E_k}\right)^2 \frac{nt}{\sin^4(\theta/2)} \tag{6–7}$$

For α-particle scattering, $Q_1 = 2e$ and $Q_2 = Ze$, and this equation becomes

$$\frac{dN_s}{N_i} \propto \frac{Z^2e^4nt}{E_k{}^2 \sin^4(\theta/2)}$$

Since the number of scattered particles varies inversely with $\sin^4(\theta/2)$, the number dN_s falls off drastically as θ increases. For example, for every million particles observed by a detector at $\theta = 10°$, there are only 231 deflected at $\theta = 90°$ and a mere 58 deflected at $\theta = 180°$.

If nuclear point charges are assumed, most of the incident particles should be deflected only slightly and a small but significant number deflected at large angles. The nuclear hypothesis of Rutherford agreed with the experiments of Geiger and Marsden in that the measured distribution of the scattered α particles agreed with the distribution predicted when scattering under a Coulomb force by point charges was assumed—that is, a scattering proportional to $1/[\sin^4 (\theta/2)]$. The Rutherford theory was confirmed for a variety of α-particle energies, foil materials, and foil thicknesses.

Although Equation 6–7 agrees well with experiments for incident α particles having a few MeV kinetic energy, there are discrepancies between Coulomb-scattering theory and experiment for higher-energy α particles. When the distance between the α particle and the target nucleus becomes somewhat less than 10^{-14} m, the force between the nucleus and an α particle is not given merely by Coulomb's law; it must be assumed that a distinctive, additional force acts between the particles. This force has been given the name the *nuclear force*. These deviations from the Coulomb force will be critical in our considerations of nuclear physics. For the moment, however, we shall be content to note that the mass and the positive electric charge in any atom are located within a region no larger than 10^{-14} m. It is certainly proper to assume that in a typical atom, having a size of approximately 10^{-10} m, the electrons are subject to a strictly electrostatic Coulomb force of attraction originating from a point charge, the nucleus.

Example 6–2. A beam of hydrogen atoms is projected into a gas composed of xenon atoms; xenon has a much larger atomic mass than hydrogen does. Assume that the atoms of both hydrogen and xenon behave as hard spheres of radii r_H and r_X and that the xenon atoms are initially at rest (see Figure 6–7). (*a*) Find the impact parameter b between an incident hydrogen atom and a xenon atom in terms of r_H, r_X, and the scattering angle θ of the hydrogen atom. (*b*) Determine the fraction of hydrogen atoms scattered through angles larger than 90° if the number density of xenon atoms is 5×10^{17} atoms/m³, and if $r_H + r_X = 2$ Å and the length of the hydrogen beam through xenon is 10 cm. (*c*) What fraction of incident hydrogen atoms pass through the xenon gas without colliding with xenon atoms?

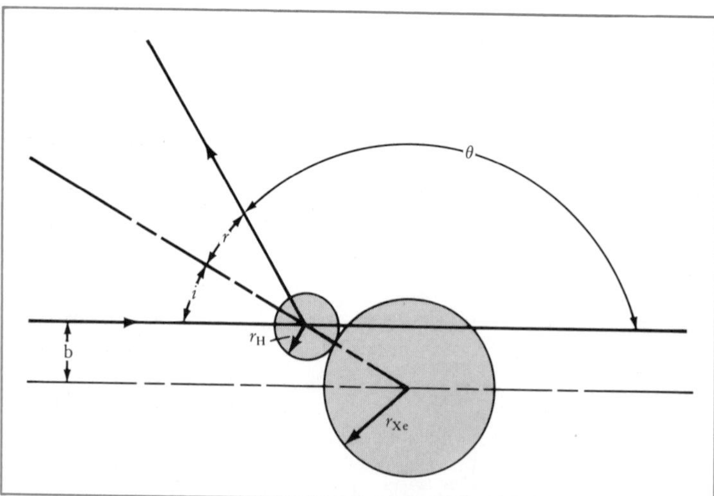

FIGURE 6–7. *Scattering of a hydrogen atom by a fixed xenon atom.*

(*a*) The perfectly elastic collision between a hydrogen atom and a xenon atom requires that the incident angle i of the incoming hydrogen atom be equal to its reflected angle r, as shown in Figure 6–7. Therefore, the angle θ through which the hydrogen atom is scattered is given by

$$\theta = \pi - (i + r) = \pi - 2i$$

At impact, the center-to-center distance between the two atoms is $r_H + r_X$, and the impact parameter b is, from the figure,

$$\begin{aligned} b &= (r_H + r_X) \sin i \\ &= (r_H + r_X) \sin \left(\frac{\pi}{2} - \frac{\theta}{2}\right) \\ &= (r_H + r_X) \cos \frac{\theta}{2} \end{aligned}$$

(*b*) The impact parameter for $\theta = 90°$ is

$$b = (r_H + r_X) \cos \tfrac{90}{2} = (2.0 \text{ Å}) \cos 45° = 1.4 \text{ Å}$$

Therefore, the fraction of hydrogen atoms scattered through angles larger than 90° is, by Equation 6–2,

$$\begin{aligned} \frac{N_s}{N_i} &= \sigma n t \\ &= \pi(1.4 \times 10^{-10} \text{ m})^2(5 \times 10^{17} \text{ atoms/m}^3)(1 \times 10^{-1} \text{ m}) \\ &= 3 \times 10^{-3} \end{aligned}$$

(*c*) The impact parameter for making any collision is found by letting $\theta = 0$; therefore,

$$b = (r_H + r_X) \cos \tfrac{0}{2} = 2 \text{ Å}$$

and the fraction of atoms making a collision is then

$$\begin{aligned} \frac{N_s}{N_i} &= (\pi b^2) n t \\ &= \pi(2 \times 10^{-10} \text{ m})^2(5 \times 10^{17} \text{ atoms/m}^3)(1 \times 10^{-1} \text{ m}) \\ &= 6 \times 10^{-3} \end{aligned}$$

The fraction of hydrogen atoms making no collisions is then

$$1 - (6 \times 10^{-3}) = 0.994$$

6–2 The Classical Planetary Model

The nucleus is the center of all the positive charge and most of the mass in the atom, and the force between charged particles at atomic dimensions is the Coulomb electrostatic force. These elements can be used as the basis of a very simple, classical model of atomic structure. We consider hydrogen, the simplest of all atoms, by this classical model.

Ordinary hydrogen consists of a nucleus with a single positive charge (a proton) and one electron. The proton, 1836 times more massive than the electron, attracts the electron with a Coulomb electrostatic force that varies inversely with the square of the distance between them. (The gravitational force, 10^{39} times smaller than the electric force, can be neglected.) The situation is like our solar system; the sun has a much greater mass than any of the planets, and a planet and the sun attract one another by an inverse-square gravitational force. When a planet is bound to the sun, it moves in an elliptical or a circular path about the sun. The atomic planetary model assumes that an atom is like a miniature solar system; the nucleus replaces the sun, an electron replaces a planet, and the Coulomb force replaces the gravitational force. The model is strictly classical; no wave aspects are ascribed to the electron, and all quantum effects are excluded. It is assumed, for simplicity, that the electron moves in a circular orbit about the hydrogen nucleus, the nucleus remaining at rest.

The system's total energy E (excluding the rest energies of the electron and proton) is the sum of the electron's kinetic energy E_k and the electric potential energy E_p between the electron and proton. An electron with mass m is imagined to move at a speed v (nonrelativistic) in a circle of radius r; the electron and the proton have each an electric charge of magnitude e. Therefore, the total energy of the system is

$$E = E_k + E_p = \tfrac{1}{2}mv^2 + \left(\frac{-ke^2}{r}\right) \qquad (6\text{–}8)$$

where $k = 1/(4\pi\epsilon_0) = 8.99 \times 10^9\ \text{N}\cdot\text{m}^2/\text{C}^2$. The inward radial force maintaining the electron in its circular orbit is supplied by the electric force due to the nucleus. Thus,

$$F = ma$$

$$\frac{ke^2}{r^2} = \frac{mv^2}{r}$$

$$mv^2 = \frac{ke^2}{r} \qquad (6\text{–}9)$$

We see from Equations 6–8 and 6–9 that when an orbit is circular, the *kinetic energy is one-half the magnitude of the potential energy.* Putting Equation 6–9 in Equation 6–8 gives

$$E = \frac{1}{2}\frac{ke^2}{r} - \frac{ke^2}{r}$$

$$E = -\frac{ke^2}{2r} \qquad (6\text{–}10)$$

Equation 6–10 shows that the total energy of the system is negative. As the radius of the electron orbit increases, E approaches zero. This means two things: (1) the electron is most tightly bound to the nucleus when it moves in a small, circular orbit; and (2) the electron is free of the binding to the nucleus only when they are separated by an infinite distance. When E is negative, the electron and proton form a bound system.

It is known that a hydrogen atom has a diameter of approximately 1 Å and that the electron in hydrogen is bound to the nucleus with an energy of 13.6 eV. Putting $E = -13.6$ eV in Equation 6–10 gives $r = 0.53$ Å; thus far the hydrogen planetary model agrees with experimental facts.

Now consider electromagnetic radiation from a classical planetary hydrogen atom. Classically, electromagnetic waves are produced by any accelerated electric charge, and the frequency of the emitted wave is precisely the frequency of oscillation of this charge.† Since the electron in a planetary hydrogen atom moves in a circle, it is continuously accelerated; therefore, the atom radiates continuously. The frequency of the radiation, moreover, is expected to equal the frequency f of the electron's orbital motion. The orbital frequency is given by

$$f = \frac{v}{2\pi r} \tag{6–11}$$

Substituting v from Equation 6–9 in Equation 6–11 gives

$$f = \frac{1}{2\pi}\sqrt{\frac{ke^2}{mr^3}} \tag{6–12}$$

If one puts $r = 0.5$ Å into this equation, the orbital frequency f is found to be 7×10^{15} Hz, in the ultraviolet region of the electromagnetic spectrum. If the atom radiates, its total energy E must decrease and become even more negative. We see from Equation 6–10 that if the total energy E decreases, the orbital radius r must also decrease, and from Equation 6–12 that f increases as r decreases. In short, when energy is radiated, E decreases, r decreases, the orbital frequency f increases, and hence the radiated frequency continuously increases.

The classical planetary theory predicts then that the electron, starting from some initial orbit, will spiral inward toward the nucleus, the atom will radiate a *continuous* spectrum, and the frequency will increase as the electron's radius decreases (see Figure 6–8). Calculations show that the electron in this classical model reaches the nucleus and combines with it in less than 10^{-8} s. The atom collapses! The classical planetary atomic model is clearly untenable on two important counts. It predicts that atoms are unstable, and it predicts a continuous radiation spectrum. This is completely at odds with the experimental facts. Atoms *are* stable, and atoms radiate *discrete* spectra of frequencies. In addition, the classical model predicts no fundamental atomic length, one that can be identified with the size of an atom.

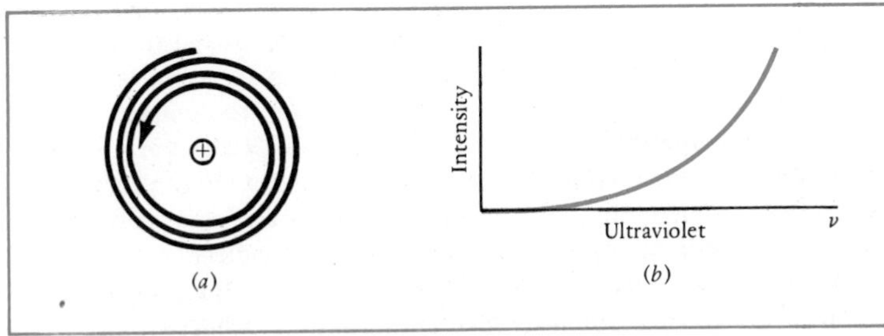

FIGURE 6–8. *Classical collapse of an atom due to continuous radiation by the orbiting electron: (a) radius decreases; (b) classical intensity distribution of electromagnetic radiation as a function of frequency.*

† See Weidner and Sells, *Elementary Classical Physics,* 2nd ed., Section 35–7.

Example 6–3. In classical radiation theory, an accelerated charged particle emits radiation of an intensity proportional to the square of the particle's acceleration. Show that this implies that an orbiting electron in the hydrogen atom emits a radiation intensity that increases as the $\frac{8}{3}$ power of the frequency ν (see Figure 6–8*b*).

Assuming that the electron of mass m revolves in a circle of radius r about the fixed proton, we have from Newton's law,

$$F = ma$$

$$\frac{ke^2}{r^2} = \frac{mv^2}{r} = \frac{m(2\pi rf)^2}{r}$$

or

$$r^3 \propto f^{-2}$$

Therefore the electron's acceleration a can be expressed as

$$a = \frac{v^2}{r} \propto \frac{r^2 f^2}{r} \propto f^{-2/3} f^2$$

or

$$a \propto f^{4/3} \tag{6–13}$$

Since the radiated intensity is proportional to a^2 and since the classical radiation frequency ν equals the frequency f of the orbiting electron, the radiation intensity is proportional to $\nu^{8/3}$.

6–3 The Hydrogen Spectrum

Since the planetary model of the atom is fundamentally defective, a correct model is needed—one that can account in detail for the observed spectra and the stability. Before we discuss the model that adequately describes the structure of hydrogen atoms, the natural starting point for any theoretical description of atomic structure, let us set down what is known about the hydrogen spectrum. To observe the spectrum of isolated hydrogen atoms, one must use gaseous atomic hydrogen, because in it the atoms are so far apart that each one behaves as an isolated system. (Molecular hydrogen, H_2, and solid hydrogen radiate spectra that reflect some aspects of hydrogen atoms bound together.) The visible spectrum emitted by hydrogen can be studied with a prism spectrometer; one is shown schematically in Figure 6–9. (A diffraction grating can be used instead of a prism for dispersing the radiation.) The hydrogen gas is excited by an electrical discharge or by extreme heating and emits radiation. The dispersed radiation, separated into its various frequency components, falls on a screen or photographic plate, making it possible to measure the frequencies and intensities of the *emission spectrum.*

Any instrument that disperses and measures the various wavelengths of a beam of electromagnetic radiation is called a *spectrometer.* An instrument that disperses the light and photographs the spectrum is known as a *spectrograph;* one that makes the spectrum visible directly to the eye is called a *spectroscope.* Spectrometers are designed for studying each of the several regions of the electromagnetic spectrum, such as radiofrequency, x-rays, and γ rays. The branch of physics that deals with the electromagnetic radiation emitted or absorbed by substances is called *spectroscopy.* Spectroscopy is a very powerful method of inquiry into atomic, molecular, and nuclear struc-

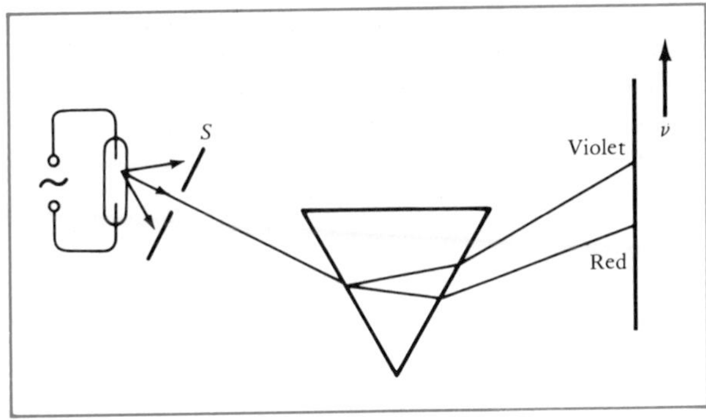

FIGURE 6–9. *Schematic diagram of a prism spectrometer.*

ture; it is characterized by very great precision (frequencies or wavelengths are easily measurable to 1 part in 10 million) and very high sensitivity (emission or absorption from samples as small as fractions of a microgram can be observed).

The spectrum from atomic hydrogen given by a prism or diffraction-grating spectrometer consists of numerous sharp, discrete, bright lines on a black background; the lines are images of the slit, shown by S in Figure 6–9. The spectra of all chemical elements in monatomic gaseous form are composed of such bright lines, each spectrum characteristic of a particular element. The spectrum is known as a *line spectrum.* The *emission spectrum* from atomic hydrogen, then, is a bright-line spectrum characteristic of hydrogen. Since each chemical element has its own characteristic line spectrum, spectroscopy is a particularly sensitive method of identifying the elements.

The line spectrum from atomic hydrogen in the visible region is shown in Figure 6–10. The lines are labeled H_α, H_β, and so on in the order of increasing frequency (decreasing wavelength). Ordinarily the H_α line is much more intense than the H_β, which is in turn more intense than the H_γ, and so on. The spacings between adjacent lines become smaller as the frequency increases, and the discrete lines approach a *series limit,* above which there appears a weak continuous spectrum. This group of hydrogen lines, which appears in the visible region of the electromagnetic spectrum, is known as the *Balmer series;* in 1885, J. J. Balmer arrived at a simple empirical formula, called the Balmer formula, from which all the observed wavelengths in the group could be computed. This formula, giving the wavelength λ for all spectral lines in this series, can be written

$$\frac{1}{\lambda} = R\left(\frac{1}{2^2} - \frac{1}{n^2}\right) \qquad (6\text{–}14)$$

where $R = 1.096\,775\,8 \times 10^7\ \mathrm{m}^{-1} \approx 1.0968 \times 10^{-3}\ \text{Å}^{-1}$ and n is an integer having the values 3, 4, 5, ...

Putting $n = 3$ in this formula gives $\lambda = 6564.7$ Å, the H_α line; similarly, putting $n = 4$ gives $\lambda = 4862.7$ Å, the H_β line. The wavelength of the series limit is given by the Balmer formula when $n = \infty$. The constant R is known as the *Rydberg constant;* its experimental value is chosen by trial to give the best fit for the measured wavelengths. In atomic spectroscopy, spectral lines typically are specified by their wavelengths rather than their frequencies, because it is the wavelength that is measured.

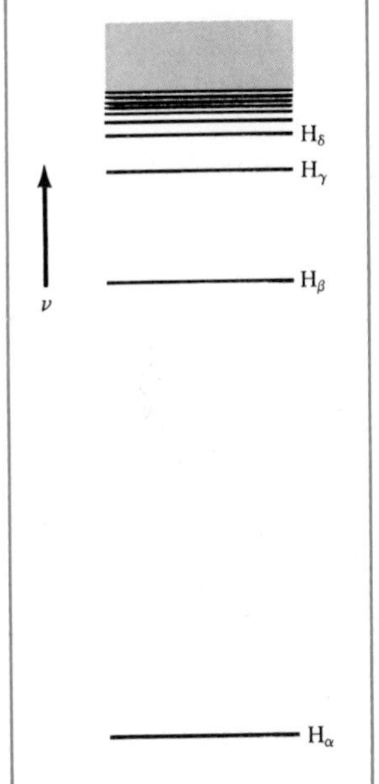

FIGURE 6–10. *Frequency distribution of radiation from atomic hydrogen in the visible region. This particular group of spectral lines is the Balmer series.*

Besides the Balmer series in the visible region, hydrogen radiates a series of lines in the ultraviolet region and several series of lines in the infrared. Each series can be represented by a formula like the Balmer equation. In fact, one general formula can be written, from which *all* the spectral lines of hydrogen can be computed. It is known as the *Rydberg equation:*

$$\frac{1}{\lambda} = R\left(\frac{1}{n_l^2} - \frac{1}{n_u^2}\right) \qquad (6\text{–}15)$$

where $n_l = 1$ and $n_u = 2, 3, 4, \ldots$ *Lyman* series, ultraviolet region
$n_l = 2$ and $n_u = 3, 4, 5, \ldots$ *Balmer* series, visible region
$n_l = 3$ and $n_u = 4, 5, 6, \ldots$ *Paschen* series, infrared region

and so on, to further series lying in the far infrared. The value of the Rydberg constant in this equation is *precisely* the same as its value in Equation 6–14; Equation 6–15 becomes 6–14 when n_l is set equal to 2. The choice of *u* ("upper") and *l* ("lower") as subscripts for the integers in the Rydberg formula will become obvious in Section 6–4. The several series of hydrogen lines are named after their discoverers. Although the Rydberg formula is remarkably successful in summarizing the wavelengths radiated by atomic hydrogen, it is merely an empirical relation; in itself, it supplies no information about the structure of hydrogen. On the other hand, a truly successful theory of the hydrogen atom must be capable of predicting the spectral lines; that is, it must yield the Rydberg formula as a result.

We have been concerned with the spectrum given by hydrogen when it is excited by an electrical discharge or by extreme heating. Atomic hydrogen at room temperature does *not*, by itself, emit electromagnetic radiation, but at room temperature it can selectively absorb electromagnetic radiation, giving an *absorption spectrum*. The absorption spectrum of atomic hydrogen is observed when a beam of white light (all frequencies present) is passed through atomic hydrogen gas and the spectrum of the transmitted light is examined in a spectrometer. What is found is a series of dark lines superimposed on the spectrum of white light; this is known as a *dark-line spectrum*. The gas is transparent to waves of all frequencies except those corresponding to the dark lines, for which it is opaque; that is, the atoms absorb only waves of certain discrete, sharp frequencies from the continuum of waves passing through the gas. The absorbed energy is very quickly radiated by the excited atoms, but in *all directions,* not just in the incident direction. Therefore, these lines will appear dark when one observes the spectrum of the light beam emerging from the gas. The dark lines in the absorption spectrum of hydrogen and the bright lines in the emission spectrum occur at precisely the same frequencies, as shown schematically in Figure 6–11, where the intensity is plotted as a function of frequency. Hydrogen is a radiator of electromagnetic radiation only at the specific frequencies or wavelengths given by the Rydberg formula; it is an absorber of radiation only at the same frequencies.

What holds for the emission and absorption spectra of atomic hydrogen holds equally well for the line spectra of all elements. A characteristic set of frequencies is emitted when the atoms radiate energy; the same set of frequencies is absorbed by the atoms when a continuous frequency band of electromagnetic radiation is sent through the gas.

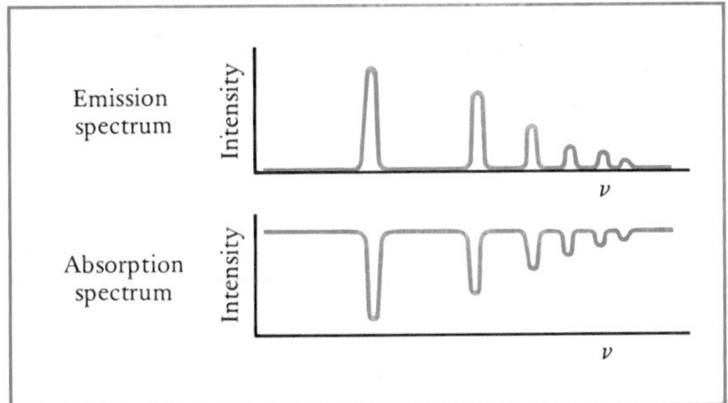

FIGURE 6–11. *Variation of intensity with frequency in emission and absorption spectra of hydrogen.*

6–4 The Bohr Theory of Hydrogen

The planetary model of the hydrogen atom cannot account for known experimental properties of hydrogen—an atomic size of the order of angstroms, the stability of hydrogen, and the discrete line spectra emitted and absorbed by a hydrogen gas. The model fails because it does not include quantum effects—the particle aspects of radiation and the wave aspects of material particles.

The first quantum theory of the hydrogen atom was proposed in 1913 by Niels Bohr, who was then working under Rutherford at the University of Manchester. Einstein had earlier established the photon nature of electromagnetic radiation, but the wave aspects of material particles were not recognized until de Broglie's work in 1924. Nevertheless, Bohr was able to make the first long step toward a thoroughgoing wave mechanical treatment of atomic structure. Although the Bohr model has limited applicability, it introduces enough quantum features to describe fairly accurately the atomic spectrum of hydrogen. Bohr's model is transitional between classical mechanics and the more complete wave-mechanical model developed during the 1920s.

Following the strictly classical model of the hydrogen atom, Bohr assumed that the more massive proton remained at rest and the electron moved about it in a circular orbit because of the electrostatic force between two charges. Thus, Equation 6–9, $mv^2 = ke^2/r$, and Equation 6–10, $E = -ke^2/2r$, still pertain in his model.

Bohr realized that no fundamental length unit comparable in size to an atom could be got from any combination of the quantities mass m, Coulomb force constant k, and electric charge e. Recalling the earlier discovered photon nature of radiation and the need for Planck's constant h in describing the properties of radiation, he then inquired whether a fundamental length might exist when h was included. By dimensional-analysis arguments, he found that there was indeed a combination of m, k, e, and h having the unit of length, namely, h^2/ke^2m (see Problem 6–21), and that its magnitude was of the order of atomic dimensions.

To account for stable states of an atom, Bohr boldly assumed that despite the predictions of classical electromagnetism, the atom could exist in certain discrete bound states with energies E_1, E_2, . . . without radiating (Figure 6–12). For unbound states ($E > 0$), all energies would be allowed. He further assumed that radiation occurred when the atom made a transition

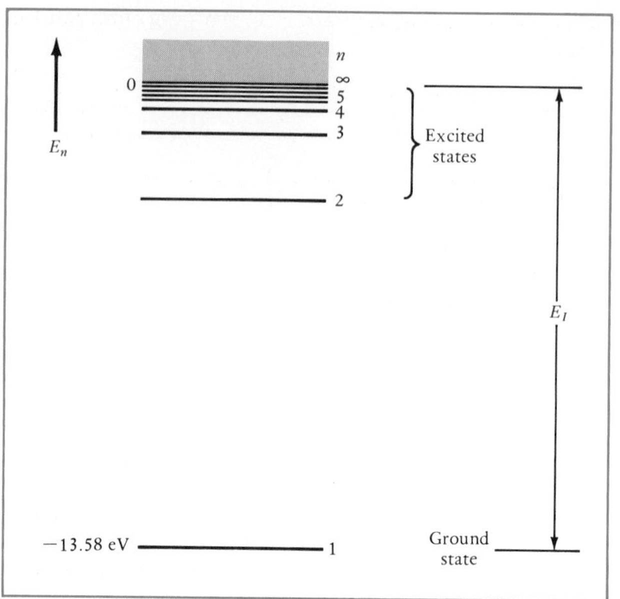

FIGURE 6–12. *Energy-level diagram of a hydrogen atom.*

between stationary energy states. He incorporated Planck's constant in the model by applying total energy conservation; this implies that the energy lost (or gained) when the atom goes from one energy state to another is converted, through the creation (or annihilation) of a single photon, into radiation energy $h\nu$. For example, an atom, initially in an upper discrete energy E_u, makes a transition to a lower discrete energy E_l with the creation and emission of a single photon of energy $h\nu$. By energy conservation,

$$h\nu = E_u - E_l \tag{6–16}$$

Whereas the Bohr theory incorporates the particle nature of electromagnetic radiation by assuming that a single photon is created whenever the atom makes a transition to a lower energy, it gives no details of the electron's quantum jump or of the photon's creation. The situation is like that in photon-electron interactions (photo-electric effect, Compton effect, and so on) in that we did not concern ourselves with the details of the interactions but merely applied the conservation laws to the states before and after the interaction.

Each of the permitted, or quantized, energies of Figure 6–12 corresponds to a stationary state in which the atom can exist without radiating. The lowest stationary state ($n = 1$) is called the *ground state.* Stationary states above the ground state ($n = 2, 3, 4, \ldots$) are called *excited states;* an atom in an excited state tends to make a transition to some lower stationary state. It can be imagined that in a downward transition, the electron jumps suddenly from one orbit to a smaller orbit. When an atom is in some excited state and has an energy E_n, the amount by which this energy exceeds the energy of the ground state is called the *excitation energy.* The term *binding energy* denotes the energy that must be added to an atom in any state to free the bound particles and thereby make $E_n = 0$. The binding energy of the atom in the ground state is the energy necessary to ionize the atom; it will be designated E_I. Obviously, $E_1 = -E_I$.

Comparing Equation 6–16 with the experimental Rydberg equation for hydrogen Equation 6–15,

$$\frac{1}{\lambda} = \frac{\nu}{c} = R\left(\frac{1}{n_l^2} - \frac{1}{n_u^2}\right) \tag{6–15}$$

Bohr obtained an expression for the allowed bound energy states

$$E_n = \frac{hcR}{n^2} \tag{6–17}$$

with $n = 1, 2, \ldots$. With $n = 1$ in Equation 6–17, we have

$$E_1 = -hcR = -E_I \tag{6–18}$$

Therefore, Equation 6–17 can be written

$$E_n = -\frac{E_I}{n^2} \tag{6–19}$$

Thus, the only possible energies of the bound electron-proton system constituting the hydrogen atom are $-E_I$, $-E_I/4$, $-E_I/9$, ...

The radius r_n of the circular orbits follow from Equations 6–10 and 6–19:

$$r_n = -\frac{ke^2}{2E_n} = \frac{n^2ke^2}{2E_I} = n^2r_1 \tag{6–20}$$

where the ground state orbit is given by

$$r_1 = \frac{ke^2}{2E_I} \tag{6–21}$$

Bohr's final task was to find theoretical values (in terms of k, e, h, c, and m) for E_I, r_1, and R, and to compare these with the known experimental values. This he accomplished by comparing the classical frequency f_n of the orbiting electron, Equation 6–12, with the quantum frequency ν_n of the emitted photon when the atom makes a transition from a stationary energy state E_{n+1} to the adjacent lower energy state E_n. Using Equation 6–20 in Equation 6–12 allows us to write the orbital frequency

$$f_n = \left(\frac{2}{\pi^2k^2e^4m}\right)^{1/2}\frac{E_I^{3/2}}{n^3} \tag{6–22}$$

The orbital frequency decreases rapidly with increasing quantum number n.

On the other hand, when an atom makes a transition from state $n + 1$ to the n state, the frequency of the emitted photon is, by Equations 6–16 and 6–19,

$$\nu = \frac{E_{n+1} - E_n}{h} = \frac{E_I}{h}\left[\frac{1}{n^2} - \frac{1}{(n+1)^2}\right] \tag{6–23}$$

The orbital frequencies, given classically, and the radiation frequencies, given by quantum considerations, are obviously not the same for low quantum numbers, since f_n and ν are different functions of n. By the Bohr correspondence principle, however, the results of quantum physics must agree with classical physics for the large quantum numbers; that is, for large n, the orbits become so large that the atom can be regarded as a macroscopic system for which classical physics yields a correct description. Requiring, then, that

$$f_n = \nu \qquad \text{for } n \to \infty$$

we have, from Equations 6–22 and 6–23,

$$\left(\frac{2}{\pi^2 k^2 e^4 m}\right)^{1/2} \frac{E_I^{3/2}}{n^3} = \frac{E_I}{h}\left[\frac{2n+1}{n^2(n+1)^2}\right] \approx \frac{E_I(2n)}{hn^4}$$

For large n, both frequencies vary with $1/n^3$, and solving the equation for E_I gives

$$E_I = \frac{k^2 e^4 m}{2\hbar^2} \tag{6–24}$$

where $\hbar = h/2\pi$.

The numerical value of the ionization energy for the hydrogen atom can now be computed from the known values of the atomic constants on the right side of Equation 6–24:

$$E_I = \frac{(8.987\,55 \times 10^9\ \text{N}\cdot\text{m}^2/\text{C}^2)^2(1.602\,19 \times 10^{-19}\ \text{C})^4(9.109\,53 \times 10^{-31}\ \text{kg})}{2(1.054\,59 \times 10^{-34}\ \text{J}\cdot\text{s})^2(1.602\,19 \times 10^{-19}\ \text{J/eV})}$$

$$= 13.6058\ \text{eV}$$

This is in excellent agreement with the experimental value, 13.60 eV.

The Rydberg constant expressed in terms of atomic constants follows from Equation 6–18:

$$R = \frac{E_I}{hc} = \frac{k^2 e^4 m}{4\pi\hbar^3 c} \tag{6–25}$$

Setting the known values of the physical constants in this equation, we find that $R = 1.097\,37 \times 10^{-3}$ Å^{-1}, which closely agrees with the experimental spectroscopic value of $1.096\,78 \times 10^{-3}$ Å^{-1} given at Equation 6–14.

The theoretical value of the electron's orbital radius r_1 in the lowest energy state, $n = 1$, follows from Equations 6–21 and 6–24:

$$r_1 = \frac{ke^2}{2E_I} = \frac{\hbar^2}{kme^2} \tag{6–26}$$

Substituting the known atomic constants in Equation 6–26 gives the theoretical value for the smallest allowed orbital radius, the so-called radius of the first Bohr orbit, $r_1 = 0.528$ Å. The Bohr model predicts that the size of a hydrogen atom with the smallest stationary orbit is of the order of 1 Å, agreeing well with experimental determinations.

The orbital speed of the electron in the stationary states follows from Equations 6–9, 6–20, and 6–26:

$$mv_n^2 = \frac{ke^2}{r_n} = \frac{ke^2 mke^2}{n^2\hbar^2}$$

or

$$v_n = \frac{1}{n}\frac{ke^2}{\hbar} \tag{6–27}$$

This equation can be written

$$v_n = \frac{v_1}{n} \qquad (6\text{–}28)$$

where $v_1 = ke^2/\hbar$. The permitted orbital speeds are v_1, $v_1/2$, $v_1/3$, . . . , the electron having a maximum speed v_1 in the first Bohr orbit. The ratio of this speed to the speed of light, v_1/c, is represented by the symbol α. From Equation 6–27, we have

$$\alpha = \frac{v_1}{c} = \frac{ke^2}{\hbar c} \qquad (6\text{–}29)$$

Inserting the known values of the constants in the right-hand side of this equation shows that $\alpha = 1/137.0360$. Thus, the electron in the first Bohr orbit moves at $\frac{1}{137}$ the speed of light. The quantity α, a dimensionless quantity that appears frequently in the theory of atomic structure, is known as the *fine-structure constant.*† To treat the Bohr hydrogen atom as a nonrelativistic problem is not unreasonable; but because of the very high precision of wavelength measurements in spectroscopy, relativistic effects can be observed and so must be included in a more complete theory.

The allowed values of the orbital angular momentum $L = mvr$ for circular orbits of the electron in the hydrogen atom are particularly simple. Substituting for v_n and r_n from Equations 6–27 and 6–26, we get

$$L_n = mv_n r_n = m\left(\frac{ke^2}{n\hbar}\right)\left(\frac{n^2\hbar^2}{kme^2}\right)$$

$$L_n = n\hbar \qquad (6\text{–}30)$$

The orbital angular momentum of the electron has only those values that are integral multiples of $\hbar$.‡ Bohr recognized the importance of the quantization of angular momentum; and both energy and angular-momentum quantization are very important in the quantum description of atomic and nuclear systems.

The observed spectral lines of hydrogen can now be interpreted by the energy-level diagram, Figure 6–13. The vertical lines represent transitions between stationary states; the lengths of these lines are proportional to the respective photon energies and therefore to the frequencies. The lines of the Lyman series correspond to those photons produced when hydrogen atoms in

† The fine-structure constant is very important in quantum electrodynamics, because it gives a relation involving the fundamental constants of electromagnetism (k and e), of the quantum theory ($\hbar$), and of relativity (c).

‡ The angular-momentum quantization relation, $mvr = nh/2\pi$, can be alternatively written by using the de Broglie equation, $\lambda = h/mv$. We have

$$\frac{h}{\lambda}r = \frac{nh}{2\pi}$$

or

$$n\lambda = 2\pi r$$

The last equation implies that an integral number n of electron waves of wavelength λ fit around the circumference $2\pi r$ of an allowed orbit in the Bohr theory. Since an electron's wave function extends into three dimensions, not merely around a circular orbit, the integral wavelength relation must be regarded merely as a useful mnemonic and not an accurate portrayal of the electron's wave behavior.

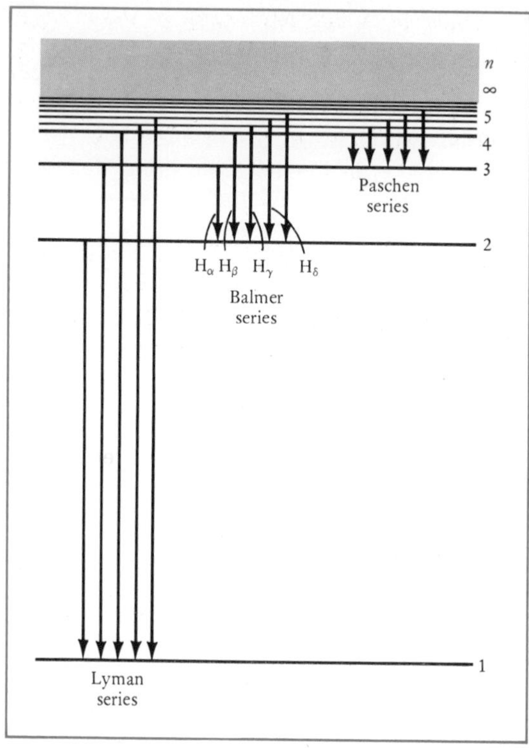

FIGURE 6–13. *Some possible energy transitions in atomic hydrogen.*

any of the excited states in which $n_u = 2, 3, 4, 5, \ldots$ undergo transitions to the ground state, in which $n_l = 1$. Transitions from the unbound states ($E > 0$) to the ground state account for the observed continuous spectrum lying beyond the series limit. We account in a similar way for the Balmer series, which is produced by transitions from the excited states in which $n_u = 3, 4, 5, \ldots$ to the first excited state, in which $n_l = 2$. Still further emission series involve downward transitions to $n_l = 3$, $n_l = 4$, and so on, these series falling progressively toward longer wavelengths.

We have examined the emission from a single hydrogen atom. The atom can exist in only *one* of its quantized energy states at any one time, and when it makes a transition from one state to a lower state, it emits a *single* photon. When the entire emission spectrum from an excited hydrogen gas, a collection of very many hydrogen atoms, is observed in a spectroscope, we see the simultaneous emission of many photons produced by downward transitions from each of the excited states. Therefore, to observe the entire emission spectrum, we must have very many hydrogen atoms in each of the excited states making downward transitions to all lower states.

Now we also have a basis for understanding absorption spectra. As shown in Figure 6–11, an absorption spectrum shows dark lines on a white background with the same wavelengths as the bright lines on the black background of the corresponding emission spectrum. When white light, consisting of photons having all possible frequencies, or energies, passes through a gas, those particular photons that have energies equal to the energy difference between stationary states can be removed from the beam. They are annihilated, thereby giving their radiant electromagnetic energy to the internal excitation energy of the atoms. The same set of quantized energy levels participates in both emission and absorption; for this reason the frequencies of the emission and absorption lines are identical. (Because atoms remain in an excited state for only a very short time, the Lyman series is the only one observed in absorption.)

FINITE NUCLEAR MASSES AND HYDROGENIC ATOMS. It has been assumed thus far that the nuclear mass, compared with the electron mass, is effectively infinite and that the nuclear charge has the magnitude e. It is a simple matter to extend the Bohr theory to include (1) *finite* nuclear masses and (2) *hydrogenic* atoms, or ionized atoms in which a single electron moves about a nucleus of charge $+Ze$.

1. In an isolated atom, both the electron of mass m and the nucleus of mass M are in motion relative to the atom's center of mass, which remains at rest. To take into account the motion of the nucleus, we can merely replace the electron's mass m, wherever it appears in a relation, with the so-called reduced mass, $\mu = m/(1 + m/M)$; see Example 6–4. Thus, the Rydberg constant for an atom with nuclear mass M becomes, from Equation 6–25, $R_M = R_\infty/(1 + m/M)$, where R_∞ represents the Rydberg constant in Equation 6–25. For example, for hydrogen ^{1}H, with a single proton of nuclear mass $M_H = 1.672\ 65 \times 10^{-27}$ kg, it is $R_H = 1.096\ 77 \times 10^{-3}$ Å^{-1}, in excellent agreement with the experimental value of Rydberg's constant, $1.096\ 78 \times 10^{-3}$ Å^{-1}. On the other hand, for deuterium, ^{2}H, or heavy hydrogen, whose nucleus of mass $M_D = 3.343\ 36 \times 10^{-27}$ kg consists of a proton and a neutron bound together, the Rydberg constant is $R_D = 1.097\ 07 \times 10^{-3}$ Å^{-1}. Although the difference in the Rydberg constants of ^{1}H and ^{2}H is small, it leads to corresponding changes in the wavelengths of the emitted and absorbed radiation. For ^{1}H, for example, the H_α line is 6564.70 Å; for ^{2}H it is 6562.94 Å. This difference in wavelength in the H_α line can be seen. Indeed, deuterium was first discovered, in 1932, when closely spaced pairs of hydrogen spectral lines were seen.
2. A hydrogenic atom has a single electron bound to a nucleus whose charge is $+Ze$. Here Z represents the atomic number, or the number of protons in the nucleus. For example, doubly ionized lithium, Li^{2+}, with $Z = 3$, is a hydrogenic atom. To find the energies, frequencies, and emitted wavelengths of hydrogenic atoms, one merely replaces the quantity e^2, wherever it appears in a relation given by the Bohr theory, with Ze^2.

Example 6–4. Derive the relation for the reduced mass as follows. Two particles interact with one another through a central force, as shown in Figure 6–14. (a) The location of particle 1 is $\mathbf{r}_1$ and the mass is m_1; and the force of particle 2 on particle 1 is $F_1\ (\mathbf{r}/r)$, where $\mathbf{r} = \mathbf{r}_1 - \mathbf{r}_2$ is the displacement of particle 1 relative to particle 2 and F_1 is the magnitude of $\mathbf{F}_1$. Write Newton's second law in vector form as it applies to the motion of particle 1 in terms of m_1, $\mathbf{r}_1$, and $F_1(\mathbf{r}/r)$. (b) The location and mass of particle 2 are $\mathbf{r}_2$ and m_2. Write Newton's second law as it applies to this particle in terms of $\mathbf{r}_2$, m_2, and $F_2(\mathbf{r}/r)$. (c) Combine the two equations of (a) and (b) to arrive at the relation $d^2\,\mathbf{r}/dt^2 = (1/m_1 + 1/m_2)F_1(\mathbf{r}/r)$. ($d$) Show that a single equivalent particle with a *reduced mass* μ, given by $1/\mu = 1/m_1 + 1/m_2$, can be thought to move under the influence of a

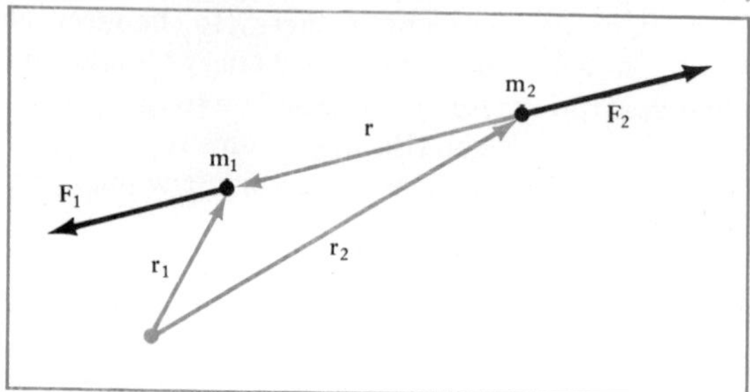

FIGURE 6–14.

force dependent only on the relative displacement between the two interacting particles.

(*a*) Newton's equation of motion for particle 1:

$$\sum \mathbf{F}_1 = \frac{m_1\, d^2\mathbf{r}_1}{dt^2}$$

$$F_1\frac{\mathbf{r}}{r} = \frac{m_1\, d^2\mathbf{r}_1}{dt^2}$$

where F_1 is the magnitude of the force $\mathbf{F}_1$.

(*b*) For particle 2, Newton's law gives

$$\sum \mathbf{F}_2 = \frac{m_2\, d^2\mathbf{r}_2}{dt^2}$$

$$-F_2\frac{\mathbf{r}}{r} = \frac{m_2\, d^2\mathbf{r}_2}{dt^2}$$

where F_2 is the magnitude of $\mathbf{F}_2$.

(*c*) By Newton's third law $\mathbf{F}_2 = -\mathbf{F}_1$ and $F_2 = F_1$. Using this in (*b*) and subtracting the accelerations of (*a*) and (*b*) gives

$$\frac{d^2\mathbf{r}_1}{dt^2} - \frac{d^2\mathbf{r}_2}{dt^2} = \left(\frac{F_1}{m_1} + \frac{F_2}{m_2}\right)\left(\frac{\mathbf{r}}{r}\right)$$

$$\frac{d^2}{dt^2}(\mathbf{r}_1 - \mathbf{r}_2) = \left(\frac{1}{m_1} + \frac{1}{m_2}\right)F_1\left(\frac{\mathbf{r}}{r}\right)$$

or with $\mathbf{r} = \mathbf{r}_1 - \mathbf{r}_2$ (see Figure 6–14),

$$\frac{d^2\mathbf{r}}{dt^2} = \left(\frac{1}{m_1} + \frac{1}{m_2}\right)F_1\left(\frac{\mathbf{r}}{r}\right)$$

This equation involves only the displacement $\mathbf{r}$ of one particle *relative* to the other.

(*d*) Replacing $(1/m_1 + 1/m_2)$ by $1/\mu$, part (*c*) takes the simple form

$$\frac{\mu\, d^2\mathbf{r}}{dt^2} = F_1\left(\frac{\mathbf{r}}{r}\right)$$

This is the equation of motion of a particle of mass $\mu = m_1m_2/(m_1 + m_2)$ under a force that depends only on the relative displacement vector $\mathbf{r}$ between the two particles.

6-5 Successes and Failures of the Bohr Theory

The fundamental postulates of the Bohr theory, succinctly expressed, are:

1. *A bound atomic system can exist without radiating only in certain discrete stationary states.*
2. *The stationary states are those in which the orbital angular momentum, mvr, of the atom is an integral multiple of $\hbar$, Planck's constant divided by 2π.*
3. *When an atom undergoes a transition from an upper energy state E_u to a lower energy state E_l, a photon of energy $h\nu$ is emitted, the conservation of energy requiring that $h\nu = E_u - E_l$; if a photon is absorbed, the atom will make a transition from the lower to the higher energy state, according to the same relation.*

These postulates are retained in their essential forms in the more complete wave-mechanical treatments of atomic and nuclear structure. The Bohr theory accounts for (1) the stability of atoms, (2) the wavelengths of the emission and absorption spectra of hydrogenic atoms, and (3) the measured ionization energies of one-electron atoms.

The Bohr atomic theory has, however, certain serious shortcomings: (1) It is nonrelativistic. (2) It gives no method of calculating the intensities of the spectral lines. (3) It can't explain the spectra of atoms having more than one electron. (4) It does not explain the binding of atoms in molecules, liquids, and solids. (5) Even for hydrogen, it fails to account for the fine details of the spectrum. (High-resolution spectrographs show that each "line" predicted by the Bohr theory consists of two or more very closely spaced lines, or fine structure.) (6) It gives for orbital angular momentum—while taking into account its quantization—a rule less complicated than what physical systems actually follow.

All these defects are corrected in a relativistic wave-mechanical treatment. A fundamental reason that the Bohr theory is defective is that it overemphasizes the classical particle nature of the electron. The electron is considered to move in a well-defined circular path, the radii, speeds, and orbital frequencies being precisely defined. In a wave-mechanical treatment, the electron is a three-dimensional wave that extends and moves throughout the whole region of space surrounding the nucleus. We shall shortly consider some simple and special examples of fitting three-dimensional electron waves, by applying the Schrödinger equation, to find the quantized stationary states of hydrogen (Section 6–7).

6-6 The Hydrogen Atom and Its Wave Functions from the Schrödinger Equation

The problem is finding allowed wave functions of the time-independent (nonrelativistic) Schrödinger equation,

$$\frac{\partial^2\psi}{\partial x^2} + \frac{\partial^2\psi}{\partial y^2} + \frac{\partial^2\psi}{\partial z^2} + \frac{2m}{\hbar^2}(E - V)\psi = 0 \qquad \text{(5–24), (6–31)}$$

for a particle of mass m (the electron) under the influence of an inverse-square attractive force (the Coulomb force of the nucleus).

We wish to find solutions to the Schrödinger equation for the Coulomb electric potential energy,

$$V = -\frac{ke^2}{r}$$

between two point charges, each of magnitude e, separated by r. Again, the positive point charge is assumed to remain fixed at the origin. To arrive at the most general solutions of the Schrödinger equation, we might transform the rectangular coordinates of Equation 6–31 into spherical coordinates (r, θ, ϕ). We should thereby be expecting the solutions to take on their simplest mathematical form when the coordinate system matches the potential energy (involving the coordinate r only). We shall, however, follow a still simpler course.

Since the potential energy depends only on the radial distance r, there must be a class of wave functions satisfying Equation 6–31 that are spherically symmetrical, that is, that depend only on the coordinate r:

$$\psi = F(r)$$

As written in Equation 6–31, the Schrödinger equation contains partial derivatives involving the rectangular coordinates x, y, and z. Since we seek solutions depending only on r, we must first transform this equation into one involving derivatives of r only. Clearly,

$$r = \sqrt{x^2 + y^2 + z^2}$$

Thus,
$$\frac{\partial r}{\partial x} = \frac{1}{2}\left(\frac{2x}{(x^2 + y^2 + z^2)^{1/2}}\right) = \frac{x}{r}$$

Now consider $\partial^2\psi/\partial x^2$. Remembering that ψ is to depend only on r, we have

$$\frac{\partial\psi}{\partial x} = \frac{d\psi}{dr}\frac{\partial r}{\partial x} = \frac{x}{r}\frac{d\psi}{dr}$$

The second derivative is then

$$\frac{\partial^2\psi}{\partial x^2} = \frac{\partial}{\partial x}\left(\frac{\partial\psi}{\partial x}\right) = \frac{\partial}{\partial x}\left(\frac{x}{r}\frac{d\psi}{dr}\right)$$

$$= \frac{1}{r}\frac{d\psi}{dx}\left(\frac{\partial x}{\partial x}\right) + x\frac{d\psi}{dx}\frac{\partial}{\partial x}\left(\frac{1}{r}\right) + \frac{x}{r}\frac{\partial}{\partial x}\left(\frac{d\psi}{dr}\right) \qquad (6\text{–}32)$$

Therefore,
$$\frac{\partial^2\psi}{\partial x^2} = \frac{1}{r}\frac{d\psi}{dr} - \frac{x^2}{r^3}\frac{d\psi}{dr} + \frac{x^2}{r^2}\frac{d^2\psi}{dr^2} \qquad (6\text{–}32)$$

The derivatives for $\partial^2\psi/\partial y^2$ and $\partial^2\psi/\partial z^2$ are just as given here except that y and z replace x. Adding the three equations, we have

$$\frac{\partial^2\psi}{\partial x^2} + \frac{\partial^2\psi}{\partial y^2} + \frac{\partial^2\psi}{\partial z^2} = \frac{3}{r}\frac{d\psi}{dr} - \left(\frac{x^2 + y^2 + z^2}{r^3}\right)\frac{d\psi}{dr} + \left(\frac{x^2 + y^2 + z^2}{r^2}\right)\frac{d^2\psi}{dr^2}$$

$$= \frac{2}{r}\frac{d\psi}{dr} + \frac{d^2\psi}{dr^2}$$

Equation 6–31 then becomes

$$\frac{d^2\psi}{dr^2} + \frac{2}{r}\frac{d\psi}{dr} + \frac{2m}{\hbar^2}\left(E + \frac{ke^2}{r}\right)\psi = 0 \qquad (6\text{–}33)$$

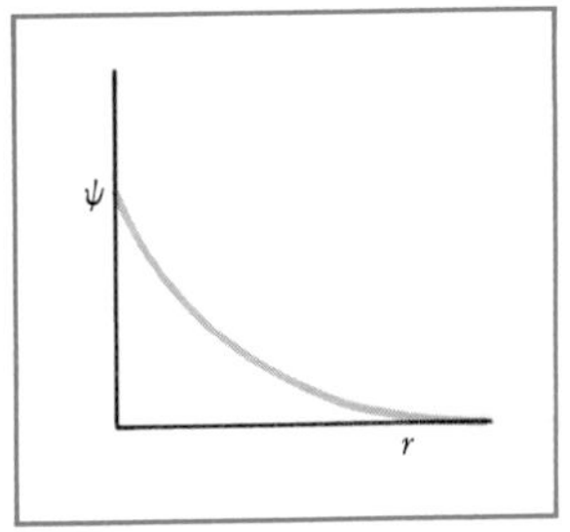

FIGURE 6–15. *The wave function $\psi = e^{-ra}$ for the ground state of hydrogen.*

Rather than try to solve this analytically in detail, we shall choose what seems to be a reasonable form of the wave function for the ground state and then test whether it is a solution. We know that the electron is more likely to be found near the nucleus rather than very far from it and the probability of finding the electron at infinity is zero. This implies that the wave function ψ is relatively large for small r and is zero for $r = \infty$. A wave function satisfying these requirements is

$$\psi = e^{-ra} \qquad (6\text{–}34)$$

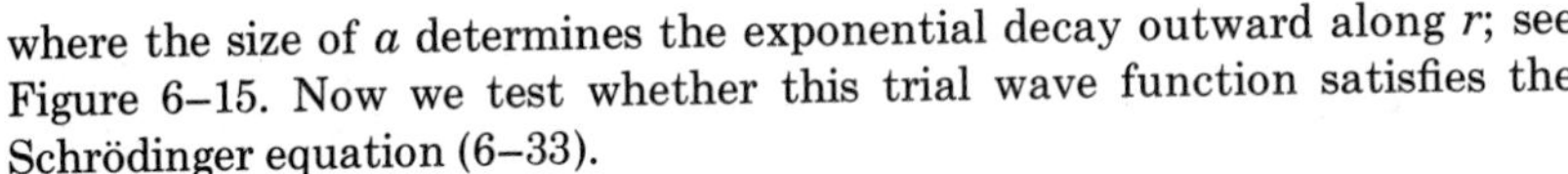

where the size of a determines the exponential decay outward along r; see Figure 6–15. Now we test whether this trial wave function satisfies the Schrödinger equation (6–33).

From Equation 6–34, we have

$$\frac{d\psi}{dr} = -ae^{-ra} = -a\psi$$

$$\frac{d^2\psi}{dr^2} = a^2e^{-ra} = a^2\psi$$

Setting ψ, $d\psi/dr$, and $d^2\psi/dr^2$ in Equation 6–33, we have

$$\left[a^2 - \frac{2}{r}a + \frac{2m}{\hbar^2}\left(E + \frac{ke^2}{r}\right)\right]\psi = 0 \qquad (6\text{–}35)$$

One possible solution is $\psi = 0$ everywhere, but this is unacceptable. A nonzero solution for ψ occurs when the quantity within the brackets equals zero. Regrouping the terms in Equation 6–35 gives

$$\left(a^2 + \frac{2m}{\hbar^2}E\right) + \frac{1}{r}\left(\frac{2mke^2}{\hbar^2} - 2a\right) = 0 \qquad (6\text{–}36)$$

This equation must hold for all values of r from zero to infinity. For r small, $1/r$ is large; therefore, the second parenthetical term in Equation 6–36 must equal zero. For r large, $1/r$ is small, and the first term must equal zero. With the second term equal to zero, we have

$$a = \frac{mke^2}{\hbar^2} \qquad (6\text{–}37)$$

With the first term in Equation 6–36 zero,

$$E = -\frac{a^2\hbar^2}{2m} = -\left(\frac{mke^2}{\hbar^2}\right)^2\frac{\hbar^2}{2m}$$

$$E = -\frac{mk^2e^4}{2\hbar^2} = -13.6\text{ eV} \qquad (6\text{–}18), (6\text{–}24)$$

The energy of the atom is $E = -13.6$ eV, just the energy of the hydrogen ground state, according to the Bohr theory. The wave function assumed

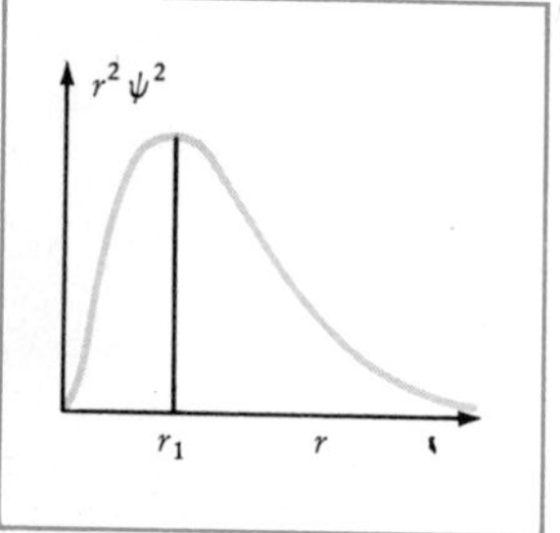

FIGURE 6–16. *The probability of finding an electron in the hydrogen ground state between r and r + dr (proportional to $r^2\psi^2$) as a function of distance r from the nucleus.*

in Equation 3–34 is, in fact, the one corresponding to the ground state in the Bohr theory.

From the wave description, Equation 6–34, of the hydrogen atom in its ground-state energy, what can we infer about the size of the atom? Recall that $\psi^2\,dv$ is the probability of finding a particle in the volume element dv. For a hydrogen atom in the ground state, the probability of the electron's being within a small volume element dv is proportional to

$$\psi^2\,dv = e^{-2ra}\,dv$$

For this state, we are more likely to find the electron in a volume element dv at the nucleus ($r = 0$) than in the *same* volume element anywhere else. We now ask what the probability is of the electron's being *between* r and $r + dr$, that is, within a spherical shell of radius r, thickness dr, and volume $4\pi r^2\,dr$.

For *equal volumes* dv, the probability $\psi^2\,dv$ is the maximum at the origin. For spherical shells of *equal thickness* dr, however, the volume $dv = 4\pi r^2\,dr$ of the shell is very small at the origin and very large at large distances from the origin; therefore, there must be a maximum probability of finding the electron in the spherical shell between $r = 0$ and $r = \infty$.

For a shell of radius r and thickness dr, the volume element is $dv = 4\pi r^2\,dr$; the corresponding probability density is proportional to r^2e^{-2ra}. This is shown plotted in Figure 6–16. The curve rises initially (because of r^2), reaches the maximum, and then goes to zero for large r (because of e^{-2ra}). At what distance r from the nucleus, within the range dr, is the electron most likely to be located? This value of r is, of course, the value corresponding to the maximum in Figure 6–16. The peak of r^2e^{-2ra} occurs at the following location:

$$\frac{d}{dr}(r^2e^{-2ra}) = 0$$

$$2re^{-2ra} - 2ar^2e^{-2ra} = 0$$

$$r_{\text{max}} = \frac{1}{a} = \frac{\hbar^2}{mke^2}$$

Comparing this with Equation 6–26, we find that the maximum is exactly at the radius of the first Bohr orbit. Although the electron may be found anywhere, when in the ground state it is more likely to be a distance r_1 from the nucleus than any other distance from it.

Still other spherically symmetrical wave functions exist. It is easy to show by substitution in Equation 6–33 that the wave function $\psi_2 = e^{-ra/2}(2 - ra)$ also is a solution. The corresponding energy is found to be $E_2 = -13.6\text{ eV}/4 = -3.4\text{ eV}$, the energy of the first excited state of hydrogen. Indeed, the energies E_n of the hydrogen atom for all spherically symmetrical wave functions are given by

$$E_n = -\left(\frac{mk^2e^4}{2\hbar^2}\right)\frac{1}{n^2} = -\frac{E_I}{n^2} \qquad (6\text{–}19), (6\text{–}38)$$

with $n = 1, 2, 3, \ldots$, again agreeing with the Bohr theory.

Since the potential energy $V = -ke^2/r$ is spherically symmetrical, there are spherically symmetrical wave functions, as we have seen. But there are also nonspherically symmetrical wave functions. The simplest of these is

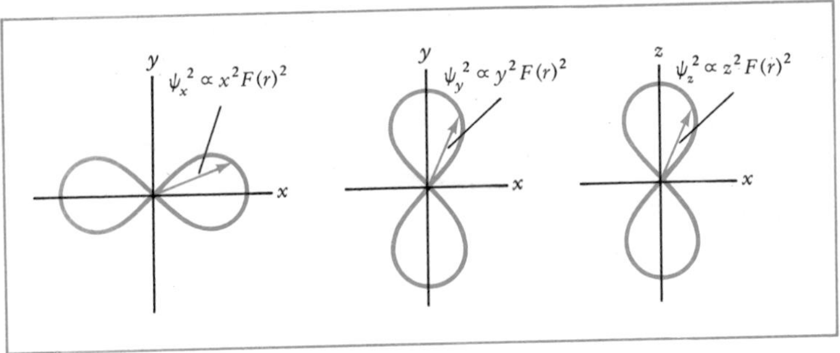

FIGURE 6–17. *Polar plot of the variation of ψ^2 with angle for p states.*

$$\psi = xF(r) \tag{6–39}$$

where $F(r)$ is again a function of r only. That this is a solution is tested by computing the derivatives $d\psi/dr$ and $d^2\psi/dr^2$ and setting them in Equation 6–33. The corresponding energy with $F(r) = e^{-ar/2}$ is found to be $E_2 = -13.6\text{ eV}/4 = -3.4\text{ eV}$, that of the first excited state of hydrogen.

If $\psi = xF(r)$ is a solution, then so are $\psi = yF(r)$ and $\psi = zF(r)$, where $F(r)$ is the same for all three states. The energy is the *same* for all three states. Nonspherically symmetrical wave functions of the type

$$\psi = xF(r)$$
$$\psi = yF(r)$$
and
$$\psi = zF(r)$$

are called *p* states (the spherically symmetrical solutions are called *s* states). Figure 6–17 shows probability distributions for a *p* state (or a *p* orbital) in a polar diagram.

All spherically symmetrical states have in common that the atom's orbital angular momentum is zero. For nonspherically symmetrical states, such as the *p* states of Figure 6–17, it is *not* zero; indeed, the orbital angular momentum (along any one direction in space) can be shown always to be an integral multiple of $\hbar$.

6–7 Atomic Excitation by Collision: The Franck-Hertz Experiment

Atomic systems, like the hydrogen atom, are quantized. The allowed energies are discrete, and as a consequence, a photon can be absorbed only if its energy $h\nu$ matches the energy difference $E_u - E_l$ between two allowed states. For example, only photons with an energy of 10.2 eV will cause hydrogen atoms in the ground state to change to the first excited state. We might well ask whether the energy of a quantized system can change not only through collisions with photons but also through collisions with particles of nonzero rest mass, such as an electron. The experiment of J. Franck and G. Hertz in 1914 first demonstrated that exciting atoms by particle bombardment is possible and that the process is also governed by the quantization of energy.

First consider atomic hydrogen. Suppose that hydrogen atoms in the ground state are bombarded by a monoenergetic beam of electrons whose kinetic energy is less than 10.2 eV, the excitation energy of the first excited state of hydrogen. Because a hydrogen atom in the ground state cannot in-

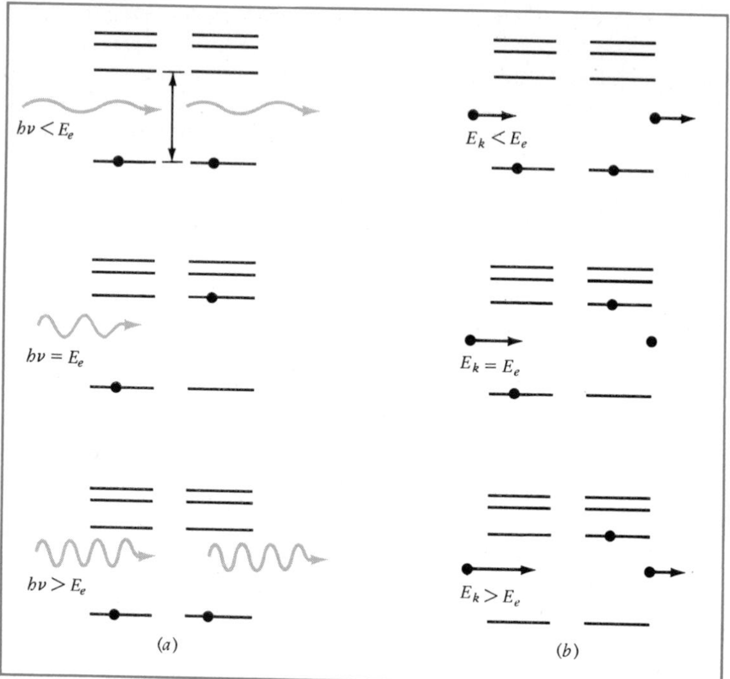

FIGURE 6–18. *Excitation of a quantum transition: (a) by photon bombardment, where E_e is the atomic excitation energy; (b) by particle bombardment, where E_k is the particle kinetic energy.*

crease its energy by any smaller amount, the electrons hit the hydrogen atoms in *perfectly elastic* collisions; the total kinetic energy of the particles emerging from a collision is precisely the total kinetic energy going into it. On the other hand, if monoenergetic electrons with a kinetic energy of exactly 10.2 eV strike hydrogen atoms in the ground state, the collisions can be inelastic, and the electron's initial kinetic energy is transformed into internal energy in the hydrogen atom as the latter makes an upward transition from the ground state to the first excited state.† Since some atoms are in this way promoted to an excited state, they can subsequently decay to the ground state with the emission of a photon of 10.2 eV.

When bombarding electrons have a kinetic energy greater than 10.2 eV, the collisions are again inelastic; only 10.2 eV is transformed into internal atomic excitation energy. The remaining kinetic energy cannot be absorbed by the hydrogen atom, and it necessarily appears as the kinetic energy of the electron emerging from the collision (and to a lesser extent, as kinetic energy of the struck atom). With a further increase in the bombarding particles' energy, atoms can be promoted to the second excited state and to still higher ones. In each such inelastic collision, the atom absorbs only the energy that will produce a transition from one quantized energy level to a higher one; see Figure 6–18. In short, if an atom's excitation energy is denoted by E_e and the light bombarding particle's kinetic energy by E_k, then inelastic collisions with atomic excitation occur only if

$$E_k \geq E_e \tag{6–40}$$

† Because momentum must be conserved in every collision, the struck atom's momentum after the collision equals the momentum of the electron before, but its kinetic energy is negligible compared with the change in its internal energy due to the collision, since its mass far exceeds the mass of the electron.

In the original Franck-Hertz experiment, electrons were made to collide with mercury atoms in a vapor. The wavelength of radiation corresponding to a transition between the ground state and the first excited state of mercury is 2536 Å; the equivalent photon energy, equal to the excitation energy, is 4.88 eV. Franck and Hertz found that electrons of at least that kinetic energy were required to produce an excitation of mercury atoms. This was inferred since the collisions were perfectly elastic when the electrons' energy was less than 4.88 eV but some inelastic collisions occurred when it was more. At the same time, it was found that mercury atoms emit radiation of 2536 Å if and only if electrons having at least the excitation energy of 4.88 eV collide with them.

The Franck-Hertz experiment was significant in showing that atomic systems are quantized, from the evidence not only of photon absorption and emission but also of particle bombardment. In practice, the inelastic collisions of electrons are observed by measuring the electric current arising from electrons in motion through a gas of molecules. If the speed of the electrons is reduced drastically by inelastic collisions that result in atomic excitation, then the current registered drops sharply when the kinetic energy of the electrons corresponds to the excitation energy. Indeed, if the electrons are energetic enough, any one of them can make numerous inelastic collisions in passing through the gas, losing an energy E_e in each.

Ionization also can result from collisions. If the bombarding particles are electrons, the atoms are ionized by those having a kinetic energy equal to the atom's ionization energy.

At room temperature, essentially all the atoms of a hydrogen gas are in the ground state, and there can't be any noticeable emission. Let us see why this is true. The average kinetic energy per molecule $\frac{3}{2}kT$ of a gas at room temperature is 0.04 eV. Thus very few atoms have a translational kinetic energy of 10.2 eV, the minimum energy necessary to raise a hydrogen atom from the ground state, in which $n = 1$ and $E_1 = -13.6$ eV, to the first excited state, in which $n = 2$ and $E_2 = -3.4$ eV. Thermal excitation of atoms occurs when some of the translational kinetic energy of two colliding atoms is transformed into *internal* excitation energy of one or both of the atoms; *translational* kinetic energy is not conserved in such a collision, and thus the collision is *inelastic*. When the gas temperature is raised to the point at which the average translational kinetic energy of the atom $\frac{3}{2}kT$ approximately equals some possible excitation energy, appreciable numbers of atoms can absorb enough energy in inelastic collisions to be raised to this higher state. To excite atoms by heating requires very high temperatures; for instance, $\frac{3}{2}kT$ is 10 eV for a temperature of 75,000 K.

A simpler and more common method of exciting atoms suggested by the Franck-Hertz experiment involves an electric discharge. Electrons and ions are accelerated to very high kinetic energies by an external electric field; this can be done by applying a potential difference between two electrodes placed in a glass chamber containing the gas. Thermal excitation and electrical excitation thus are ways of producing emission spectra.

We can now see why in the kinetic theory of gases, gas molecules and atoms can be regarded as inert particles having no internal structure and making perfectly elastic collisions with each other when the gas is at moderate temperatures. Unless the average translational kinetic energy per atom is comparable to the difference in energy between the ground state and the first excited state, the internal structure of the atom cannot change, the total translational kinetic energy in a collision is conserved, and the collision is perfectly elastic. If the gas temperature is high enough, inelastic collisions occur. Some atoms are thereby excited, and they can no longer be considered inert particles, incapable of undergoing internal change.

SUMMARY

Rutherford's experiments in the scattering of α particles showed that all the positive charge and essentially all the mass of an atom are confined to a very small region of space (no greater than 10^{-14} m), called the nucleus.

The fraction of incident particles scattered by a thin foil of thickness t containing n scatterers per unit volume is

$$\frac{N_s}{N_i} = \sigma n t \qquad (6\text{–}2)$$

where σ is the cross section associated with each scattering center.

The classical planetary model predicts that atoms are unstable and that they emit a continuous spectrum.

Hydrogen atoms are observed to emit and absorb discrete spectral lines, whose wavelengths λ are given by

$$\frac{1}{\lambda} = R\left(\frac{1}{n_l^2} - \frac{1}{n_u^2}\right) \qquad (6\text{–}15)$$

where R is the Rydberg constant and n_l and n_u are integers.

The Bohr quantum atomic theory assumes (1) stationary states, (2) orbital angular momentum $n\hbar$, and (3) $h\nu = E_u - E_l$ in transitions.

The energies and radii predicted by the Bohr atomic theory for hydrogen are given by

$$E_n = -\frac{E_I}{n^2} \qquad \text{where } E_I = \frac{k^2e^4m}{2\hbar^2} = 13.61 \text{ eV} \qquad (6\text{–}19), (6\text{–}24)$$

$$r_n = n^2 r_0 \qquad \text{where } r_1 = \frac{\hbar^2}{kme^2} = 0.5292 \text{ Å} \qquad (6\text{–}20), (6\text{–}26)$$

Wave-mechanical solutions of the Schrödinger equation for hydrogen yield the same energies; the radius r_1 of the first Bohr orbit gives that distance from the nucleus at which the electron in the ground state is most likely to be found.

The Franck-Hertz experiment confirmed that the excitation and ionization of atoms by particles in inelastic collisions follow the quantum condition.

PROBLEMS

α-particle scattering, §6–1

● **6–1.** (*a*) Find the de Broglie wavelength of the 5.5-MeV α particles used in the Rutherford scattering experiment (Example 6–1). (*b*) What is the closest separation distance between a 5.5-MeV α particle and a gold nucleus in a head-on collision? (*c*) Compare the wavelength of the α particle with the closest separation distance.

● **6–2.** A beam of 5.5-MeV α particles strikes a target composed of helium atoms. Show that none of the particles is scattered through an angle larger than 90°.

● **6–3.** Alpha particles of 5.5-MeV energy are incident on a uranium target ($Z = 92$). What is the closest distance an α particle can come to a uranium nucleus?

6–4. Ten thousand small hard spheres, each 2.0 mm in diameter, are dispersed randomly throughout the interior of a cubical container 1.0 m along an edge. One million dust particles, each very small compared with the spheres within the box, are shot randomly over a broad face of the container. (*a*) Find the probability that a dust particle will be scattered once before traversing the container. (*b*) Estimate the probability that a single dust particle will be scattered twice before emerging from the container. (*c*) How many of the dust particles will emerge from the box undeflected?

6–5. A 10.0-MeV beam of α particles is incident on a thin gold target ($\rho = 1.93 \times 10^4$ kg/m^3; $M_a = 197$; $Z = 79$) of thickness t, and a detector counts 100 particles per minute at a scattering angle of 45°. What will be the count rate if only the following quantity is changed in the experiment?

(*a*) The incident beam energy is increased to 20.0 MeV.
(*b*) A 10.0-MeV proton beam is used.
(*c*) The detector is turned to a scattering angle of 135°.
(*d*) A gold foil of thickness $2t$ is used.

6–6. A beam of 2.0-MeV protons is incident on a gold foil (atomic number $Z = 79$). (*a*) Use energy conservation to find the closest distance a proton can come to a gold nucleus. (*b*) A gold nucleus has radius $r_g = 8.1 \times 10^{-15}$ m; a proton $r_p = 1.4 \times 10^{-15}$ m. Compare $r_g + r_p$, which is the distance the two nuclei must come toward one another to produce nuclear reactions, with the distance of closest approach. Can 2-MeV protons on gold targets produce a nuclear reaction? (*c*) Estimate the radius of the innermost electron in the gold atom by finding the radius of a hydrogenic atom with $Z = 79$. Can a 2.0-MeV proton come within the orbit of this inner electron?

6–7. In Example 6–1, compare the ratio of the scattering cross section σ to the cross-sectional area of a gold atom with the ratio of the target area producing scattering greater than 90° to the total area of the target. Why are they different?

6–8. An α particle, in traveling through a gold foil, collides with some of the electrons in the gold atoms. What is the maximum energy (eV) a 4.0-MeV α particle can transfer to an electron (assumed free) in an elastic collision?

■ **6–9.** A beam of 10-MeV protons is incident on a copper ($Z = 29$; $M_a = 65$) foil. For protons scattered 90°: (*a*) find the impact parameter; (*b*) using angular-momentum and energy conservation, determine the closest distance of approach of a proton to a copper nucleus for 90° scattering, and compare this with the impact parameter.

6–10. In 1913, Geiger and Marsden confirmed Rutherford's Coulomb scattering relation (Equation 6–7) by observing the scattering of a beam of α particles through thin foils of gold, of thickness 1.86×10^{-4} cm, and silver, of thickness 2.82×10^{-4} cm. Listed below are the observed number of counts per minute at three scattering angles:

	Gold	*Silver*
$\theta = 45°$	1435	989
$\theta = 75°$	211	136
$\theta = 135°$	43	27.4

For each angle, compare the ratio of scattered α particles per minute by gold and by silver with that predicted by the Rutherford formula (Equation 6–7). For gold and silver, the nuclear charges are 79 and 47, the mass densities 1.93×10^4 and 1.05×10^4 kg/m^3, and the atomic masses 197 and 108, respectively.

Hydrogen spectrum, §6–3

6–11. A diffraction grating of 5000 lines/cm replaces the prism spectrometer of Figure 6–8 to observe the spectrum of emitted radiation from an excited hydrogen gas. On a screen 1.0 m from the slit, what distance separates the H_α and H_β spectral lines in the first-order spectrum?

6–12. Using the characteristics of the diffraction-grating spectrometer given in Problem 6–11, find the separation of the H_α lines in the first-order spectrum for hydrogen (6564.70 Å) and deuterium (6562.94 Å).

6–13. The human eye is sensitive to electromagnetic radiation between 3800 Å and 7600 Å. Show that many of the Balmer series lines, but none of the Lyman series lines or Paschen series lines, fall within this visible region.

Bohr theory, §6–4

● **6–14.** What is the energy of the photon emitted when a hydrogen atom makes a transition from the $n = 10$ state to the $n = 2$ state?

● **6–15.** What is the ionization energy of a doubly ionized lithium atom?

● **6–16.** What are the energies of the photons emitted by a collection of hydrogen atoms, all of which are initially in the second excited state?

● **6–17.** Show that the energy of a hydrogen atom in its ground state can be written $E = -\frac{1}{2}\alpha^2 E_0$, where α is the fine-structure constant and E_0 is the rest energy of the electron.

● **6–18.** What transition in a singly ionized helium atom will emit a wavelength very close to the H_α line in hydrogen?

● **6–19.** When an atom makes a transition from one energy state to another, which of the following quantities are conserved: total electric charge; total number of electrons; total number of photons; energy of atom; total energy; angular momentum of atom; linear momentum of atom; total linear momentum?

6–20. Suppose that the numerical value of Planck's constant were ten times larger than its present value and that the hydrogen atom still behaved according to quantum mechanics. What would then be (*a*) the size and (*b*) the ionization energy of the hydrogen atom?

6–21. (*a*) Use dimensional analysis to find the exponents a, b, and c such that the quantity $R = m^a(ke^2)^b\hbar^c$ has the unit of length. (*b*) Taking m to be the mass of the electron, determine the magnitude of the length R and compare it with the radius of the first Bohr orbit of the hydrogen atom.

6–22. Assume that 1200 hydrogen atoms are initially in the $n = 4$ state. The atoms then make transitions to lower energy states. (*a*) How many distinct spectral lines will be emitted? (*b*) Assuming for simplicity that all possible downward transitions are equally probable from a given excited state, determine the total number of photons emitted.

6–23. Calculate the classical orbital frequencies f_n and f_{n+1} for adjacent states of a hydrogen atom and compare them with the frequency ν of radiation emitted when the atom goes from state $n + 1$ to state n for (*a*) $n = 1$; (*b*) $n = 100$; and (*c*) $n = 1000$.

6–24. Using the fact that the average energy of a gas atom is $\frac{3}{2}kT$, estimate the minimum temperature of a gas of atomic hydrogen that will produce appreciable ionization of the atoms through collisions so that the hydrogen atoms will be broken up into protons and electrons (a plasma).

6–25. The phenomena of the aurora borealis (northern lights) and the aurora australis (southern lights), the luminous displays in the sky near the earth's poles, are produced when charged particles thrown out by the sun collide with oxygen and nitrogen 100 km or more above the earth's atmosphere, thereby exciting and ionizing these atoms. (*a*) The charged particles are known to travel the 1.5×10^{11} m from the sun to the earth in about 24 h; if we assume that they are protons, what is their average kinetic energy? (*b*) Why do charged particles from the sun produce appreciable ionization of oxygen and nitrogen only in the vicinity of the earth's poles?

6–26. An isolated hydrogen atom initially at rest emits a photon in a transition from the first excited state to the ground state. What is (*a*) the energy of the photon? (*b*) the momentum of the photon? (*c*) the momentum with which the hydrogen atom recoils upon emitting the photon? (*d*) the kinetic energy of the recoiling hydrogen atom?

6–27. A hydrogen atom in the $n = 6$ state makes a transition to the ground state with the emission of a photon. (*a*) What is the kinetic energy of the recoil atom? (*b*) How does this recoil energy compare with the average thermal energy $\frac{3}{2}kT$ of hydrogen atoms at 300 K?

6–28. According to the Bohr model, what does a hydrogen atom lose in orbital angular momentum when it makes a transition (*a*) from the $n = 2$ to $n = 1$ state? (*b*) from the $n = 5$ to the $n = 1$ state? (*c*) If total angular momentum is conserved for the system of atom and emitted photons, what angular momentum would each of the above two photons have? This prediction of the Bohr model, namely, that all photons do not have the same value of angular momentum, is corrected in the complete quantum-mechanical theory. The angular momentum of any photon, $\sqrt{2}\hbar$, is an intrinsic property of all photons.

■ **6–29.** A hydrogen gas is ionized by projecting an incident beam of particles through the gas. Use the conservation laws of energy and linear momentum to explain why such ionization requires a smaller incident-particle energy when (*a*) photons are used instead of electrons; (*b*) when electrons are used instead of protons.

6–30. (*a*) What atomic transition to the ground state of singly ionized helium will produce a photon with the same energy as the Lyman alpha ($2 \to 1$) photon of hydrogen? (*b*) Compare the loss in angular momentum of the two atoms in such a transition according to the Bohr model. (*c*) Assume that the angular-momentum loss by the atom equals the angular momentum the photon carries away. What would happen if the Lyman alpha photons emitted by hydrogen were to pass through an ionized helium gas with the He^+ ions in the ground state?

6–31. Suppose that a relatively heavy atom such as tungsten (atomic number $Z = 74$) has all but one of its electrons removed. (*a*) How much energy (eV) is required to remove this last electron completely if the ionized atom is initially in the ground state? (*b*) What would be the wavelength of the photon emitted if this single remaining electron were to make a transition from $n = 2$ to $n = 1$? (*c*) In what portion of the electromagnetic spectrum is such a photon found?

6–32. A monochromatic beam illuminates a collection of hydrogen atoms in the ground state, and this results in the radiation by the hydrogen atoms of three distinct wavelengths. What is the minimum wavelength of the incident monochromatic light?

6–33. A beam of 11.21-eV electrons collides head-on with a beam of hydrogen atoms in their ground state. The kinetic energy of each atom in the hydrogen beam is such that the center of mass of the system consisting of a colliding electron and hydrogen atom is at rest. (*a*) Find the kinetic energy (eV) and momentum (eV/c) of the incoming hydrogen atoms. (*b*) Assume that an inelastic collision occurs between an electron and a hydrogen atom and that the atom is excited to the $n = 2$ energy state. What is the total kinetic energy of the electron and the excited hydrogen atom after the collision?

6–34. Because of the random thermal motion of atoms in a gas at temperature T, the observed frequency ν' of the light emitted by an excited atom will depend on the atom's velocity away from or toward the observer; for low velocities, $v \ll c$, this *Doppler effect* results in an observed frequency $\nu' = \nu(1 \pm v/c)$, where v is the approach (plus sign) or recessional (minus sign) velocity of the atom relative to the observer. (*a*) Show that the magnitude of the fractional change in frequency or wavelength is $\Delta\nu/\nu = \Delta\lambda/\lambda = v/c$. (*b*) The *Doppler effect* is one reason that spectral lines are not infinitely sharp, since the radiating atoms have different velocity components along the propagation direction of the light that enters the spectrometer. Assuming that excited hydrogen gas in a spectrum tube is at a temperature of 300 K, and taking the maximum velocities of recession and approach to be given approximately by $\frac{1}{2}mv^2 = \frac{3}{2}kT$, where k is Boltzmann's constant, calculate the approximate width (Å) of the hydrogen H_α line arising from *Doppler broadening*.

6–35. When a hydrogen atom is excited to the first excited state, it remains in this state for an average time of 10^{-8} s before making a transition back to the ground state. (*a*) Use the uncertainty principle to find the uncertainty in the energy (eV) of the atom for this excited state. (*b*) What will be the fractional uncertainty and the uncertainty (Å) in the wavelength of the emitted photon resulting from the downward transition of the hydrogen atom? (*c*) Compare the *natural linewidth* arising from the uncertainty principle in part (*b*) with the *Doppler line width* arising from the random motion of the atoms (Problem 6–34).

6–36. Because of the Doppler effect (Problem 6–34), light of frequency ν emitted by a source receding from the earth at speed v will have an observed frequency $\nu' = \nu\sqrt{(c - v)/(c + v)}$. (*a*) At what recessional speed will the Lyman H_α line of hydrogen (1216 Å) have an observed wavelength equal to that of the Balmer H_α line (6565 Å)? (*b*) Show that the observed Lyman series for a source moving away from the earth at this speed will fall in the visible region of the electromagnetic spectrum.

■ **6–37.** Two hydrogen atoms, one in the first excited state and one in the ground state, approach one another at a relative speed $v = 0.1683c$. The excited atom emits a photon toward the atom in the ground state. (*a*) Using the Doppler effect (see Problem 6–36), show that if the ground-state atom absorbs this photon, the atom will be left in the *second* excited state. (*b*) Why is the process in (*a*) not a violation of energy conservation?

6–38. Compare the experimental value of the wavelength of the H_α line for hydrogen, 6564.70 Å, (*a*) with the calculated value using $R_\infty = 1.097\,37 \times 10^{-3}$ Å^{-1}, and (*b*) with $R_M = 1.096\,78 \times 10^{-3}$ Å^{-1} for the ^{1}H atom.

6–39. The negative muon is an unstable elementary particle with properties like those of the electron. It has a negative electric charge of magnitude e and a mass 206.8 times the electron mass. Muonic atoms, in which the muon replaces an electron, have been produced and studied in the laboratory. If a muon is captured by a proton to form a bound muonic hydrogen atom, what is (*a*) the wavelength of the H_α line ($n = 3$ to $n = 2$); (*b*) the radius of the first Bohr orbit?

6–40. Compute the Rydberg constants for ordinary hydrogen, ^{1}H, with nuclear mass $1.672\,65 \times 10^{-27}$ kg, and heavy hydrogen, ^{2}H, with nuclear mass $3.343\,38 \times 10^{-27}$ kg. From these, show that the difference in wavelengths of their H_α lines is 1.79 Å. The electron's mass is $9.109\,08 \times 10^{-31}$ kg.

6–41. An atom of positronium is formed when a positron (same mass as electron but positive electric charge) and an electron combine to form a bound system. (*a*) Compute the ionization energy of the positronium atom. (*b*) Calculate the wavelength of the $n = 2$ to $n = 1$ line for this system and compare it with the experimentally measured value of 2430 Å. [See S. Berko, K. Canter, and A. Mills, *Phys. Rev. Lett., 34,* 177 (1975).]

6–42. Show that the ionization energy of a negative muon bound to a deuteron is larger than the ionization energy of a negative muon bound to a proton by 135 eV. The deuteron is the nucleus of heavy hydrogen ^{2}H, and the muon has a charge of -1.6×10^{-19} C and a mass of 207 electron masses. Evidence of these muonic atoms has been found experimentally. [See L. W. Alvarez et al., "Catalysis of Nuclear Reactions by μ Mesons," *Phys. Rev., 105,* 1127 (1957).]

Hydrogen wave functions, §6–6

6–43. The energy of a hydrogen atom can be computed approximately by using the uncertainty relation. For a hydrogen atom in the ground state, the uncertainty in the location of the electron is approximately one Bohr radius (see Figure 6–16); consequently, the electron's momentum in the ground state is uncertain by at least $\hbar/r_1$. The absolute value of the electron's momentum must be at least as great as the uncertainty in its momentum. Compute the uncertainty in the electron's kinetic energy on this basis and compare it with the energy of hydrogen in the ground state.

6–44. A hydrogen atom is in its ground state. For this state, find the ratio of the probability that the electron will be at the second Bohr radius to the probability that it will be at the first Bohr radius.

6–45. A hydrogen atom is in its first excited state, $n = 2$, with the probability density $r^2\psi^2$ proportional to $r^2(2 - ra)^2e^{-ra}$. For this state, find the ratio of the probability that the electron will be at the second Bohr radius to the probability that it will be at the first Bohr radius.

7
Many-Electron Atoms

Wolfgang Pauli (1900, Austria—1958). Aged 20, 200-page authoritative encyclopedia article on relativity. 1925, exclusion principle. 1931, proposed neutrino (first observed 1956) as an alternative to abandoning conservation laws. 1945, Nobel Prize, exclusion principle. (Photo, CERN—courtesy of the American Institute of Physics, Neils Bohr Library.)

Although the Bohr atomic theory can't describe the structure and spectra of atoms in detail, some of its essential quantum features hold for many-electron atomic systems. These features include the existence of stationary states, the quantization of energy, and the quantization of angular momentum. A correct treatment of the many-electron atom is strictly wave-mechanical; it is mathematically difficult and does not contribute to a simple visualization of the atomic structure. In wave mechanics, an atomic electron must be regarded as a three-dimensional wave surrounding the nucleus; therefore it is incorrect, indeed impossible, to assign a well-defined path to the electron's motion. Instead, wave mechanics yields through the wave function only the probability that an electron is at a particular location. Nevertheless, within these limitations we can gain certain insights into the results of wave mechanics by using results applicable to a completely classical particle model. First we briefly review the classical problem of a particle moving under an inverse-square attractive force. Next we examine, without proof, a few important results of wave mechanics. It will then be possible to interpret these results by analogy with the corresponding classical model.

7–1 Constants of the Motion in a Classical System

When a particle moves under the influence of a central, attractive, inverse-square force, like a planet about the sun or a classical electron about a positive massive particle, the isolated, bound system is characterized by several *constants of the motion.* The constants of the motion are physical quantities that do not change with time and include the system's total energy, total angular momentum, and total electrical charge. It is useful to review the constants of the motion of a classical planetary system as a prelude to discussing the wave-mechanical aspects of atomic systems, because each one corresponds in wave mechanics not merely to a quantity that is constant in time but also one that is *quantized.* Total electric charge, for example, a constant in classical physics, is quantized, having only integral values of the fundamental charge e. Wave mechanics predicts a similar behavior for the properties of energy and angular momentum.

We know that the path traced out by a bound particle moving under the influence of a central inverse-square force from a fixed point is an ellipse whose force center is at one focus. The *first* constant of the motion, the total energy (kinetic plus potential) of a classical planetary system, such as a hydrogen atom, is given by $E = -ke^2/2a$, where a is the semimajor axis of the ellipse (Figure 7–1). The energy E depends only on a. Thus, the system's total energy is the same for each of the two orbits shown in Figure 7–1, one a fairly eccentric ellipse and the other a circle of radius a, with the force center at its center. Under an inverse-square force, the path of a bound particle is an ellipse, the total energy is the same at all points along the path, and all elliptical orbits having the same semimajor axis have the same total energy.

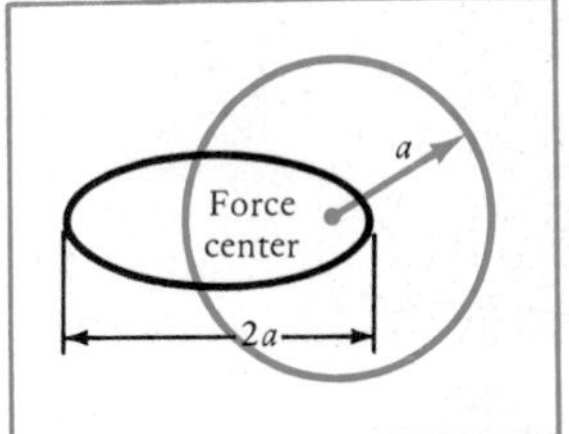

FIGURE 7–1. *For an inverse-square attractive force, a system's total energy depends only on the semimajor axis a.*

Although all orbits with the same value of a have the same energy, they differ in the magnitude of the *second* constant of the motion: the system's total orbital angular momentum. The angular momentum of an orbiting particle, measured relative to a point at the fixed force center, is given by

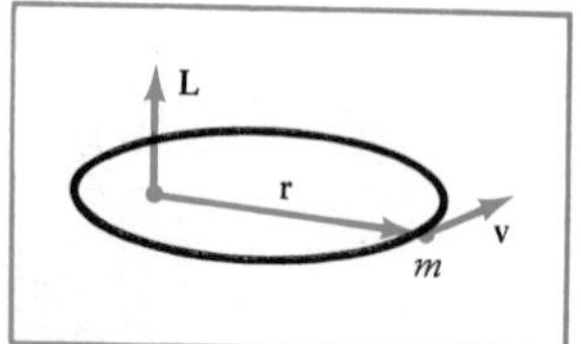

FIGURE 7–2. *The angular momentum* $\mathbf{L} = \mathbf{r} \times m\mathbf{v}$ *of an orbiting particle.*

$\mathbf{L} = \mathbf{r} \times m\mathbf{v}$; see Figure 7–2. So long as the force on the moving particle is *central* (along the line connecting it with the force center), the system's orbital angular momentum **L** is constant in both magnitude and direction. The direction of **L** is at right angles to the plane of the orbit, and it is related to the sense of rotation of the particle by the right-hand rule.

Now consider many elliptical orbits, all with the same semimajor axis a, and therefore the same total energy, but differing in eccentricity. The least eccentric orbit is that in which the particle moves in a circular path, always at the same distance a from the force center; the most eccentric is an ellipse so collapsed as to be a straight line with a focus near a turning point. The circular orbit represents the state of maximum orbital angular momentum, whereas the collapsed orbit represents the state of zero orbital angular momentum. The magnitude of **L** ranges continuously between these extremes, as shown in Figure 7–3; so for a given total energy, there are a variety of orbital angular momenta ranging continuously from zero to the maximum.

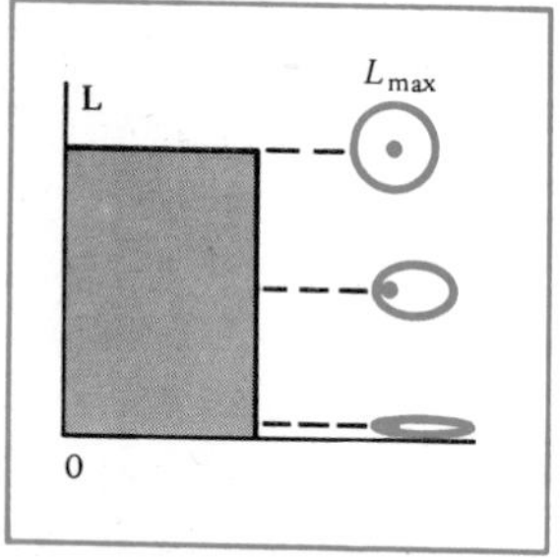

FIGURE 7–3. *Classical allowed values of the orbital angular momentum of elliptical orbits of the same major axis, or energy.*

Since the isolated system's orbital angular momentum **L** is constant both in direction and in magnitude, its component L_z along any axis in space is also constant; this is the *third* constant of the motion. Consider, for example, an arbitrary z-axis (see Figure 7–4a). We see that $L_z = L\cos\theta$, where θ is the angle between **L** and the positive z-axis. In classical physics there is no restriction on the choice of the z-direction, so L_z may range continuously from $+L$ to $-L$, as shown in Figure 7–4b. In other words, depending on the choice of z-direction, the angle θ can take on any value from 0 to 180°. In short, there is no classical restriction on possible directions of the orbital angular momentum vector **L**. This seemingly trivial consideration has important consequences in the wave-mechanical analog.

Finally, if an orbiting object also has some finite extension in space and is spinning about an internal axis of rotation, then besides the system's orbital angular momentum there is *spin angular momentum*. If no net external torque acts on the spinning object, its spin angular momentum remains constant. This is, then, a *fourth* classical constant of the motion. The spin angular momentum of an orbiting object (the daily rotation of the earth about its center of mass, for example) is computed by finding the contribution of each part of the spinning object, again using the vector relation $\mathbf{L} = \mathbf{r} \times m\mathbf{v}$ for each part. The remarkable property of spin angular momentum is that its value (magnitude and direction) is *independent of the choice of the axis for computing angular momentum,* provided only that the spinning object is symmetrical and rotates about an axis of symmetry.

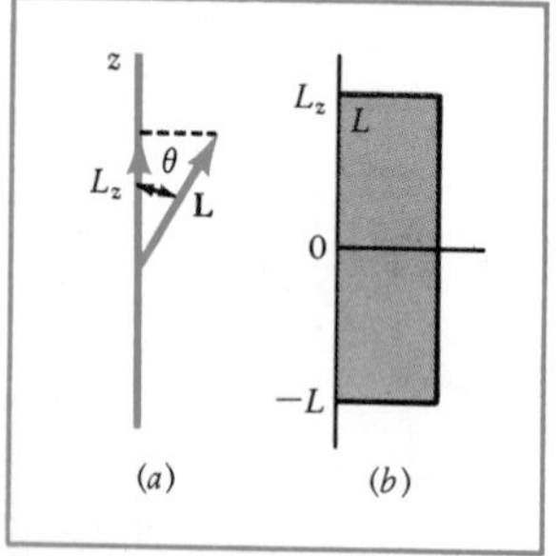

FIGURE 7–4. *Component of the orbital angular momentum in some arbitrary* **z***-direction: (a) orientation with respect to* **L***; (b) classically allowed values for a given value of L.*

The proof is straightforward. Consider the two particles, each of mass m, moving in opposite directions at the same speed v in Figure 7–5. They are symmetrically situated relative to the spin axis through the center of mass, each a distance r from the center of the circle in which they travel. We compute the total angular momentum of this pair of particles relative to the arbitrarily chosen point P. Recognizing that the angular momentum of one particle is positive while the other is negative, we have for the total angular momentum of the pair

$$L_{\text{pair}} = mv(r_{\perp} + 2r) - mvr_{\perp} = 2mvr$$

The angular momentum $2mvr$ of this symmetrically located pair of particles

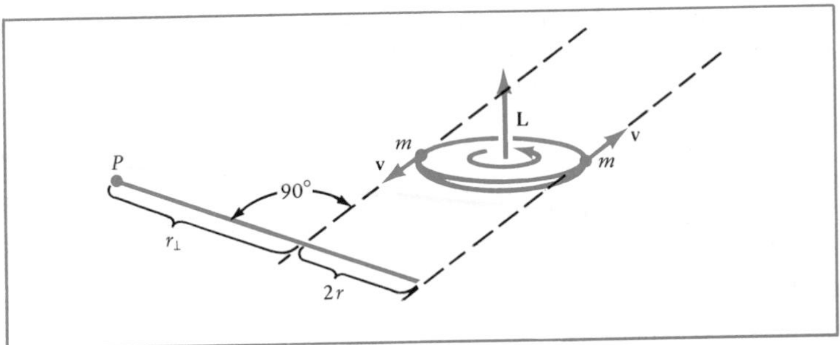

FIGURE 7–5. *The total spin angular momentum* **L**, *relative to an arbitrary point P, of two particles symmetrically located with respect to the spin axis.*

is *independent* of the location of P (independent of $r_\perp$). Since the spinning object is taken to be symmetrical about the rotation axis, we can imagine it to be composed of such pairs of particles, each contributing an angular momentum independent of the choice of axis. Thus, the object's total spin angular momentum is independent of axis. It is customary to locate the spin-angular-momentum vector along the spin axis, as shown in Figure 7–5; from the proof given above, it could be located *anywhere*. It is not hard to show that spin angular momentum is independent of the inertial frame, too. The angular momentum of a spinning symmetrical object is, then, an intrinsic property of the object; it is sometimes referred to as *intrinsic angular momentum.*

The total angular momentum of a system of objects consists of the vector sum of the orbital and spin angular momenta; when the system is isolated, its total angular momentum is constant. We shall see that such particles as electrons must be assigned intrinsic angular momenta in addition to their orbital angular momenta.

7–2 Quantization of Orbital Angular Momentum

The Bohr theory of a one-electron atom introduces the principal quantum number n, whose integral value determines the total energy of the atom according to the relation $E_n = -E_I/n^2$, where E_I is the ionization energy. The quantum number n also specifies the magnitude of the angular momentum **L**. This value comes from the electron's orbiting the nucleus in a circular path, according to $L = n\hbar$, where $\hbar$ is Planck's constant divided by 2π. It is, however, not proper from the point of view of wave mechanics to visualize the electron as moving in a well-defined path, circular or otherwise.

Although the Bohr theory agrees with wave mechanics on the quantized energy values E_n, the Bohr rule for the quantization of the magnitude of the orbital angular momentum is *not* correct. Wave mechanics does show that there is quantization of the orbital angular momentum for a one-electron atomic system, but the rules are more complicated than what the simple Bohr model provides. Because the mathematical analysis of this quantum problem is lengthy and involved, we shall state only the results here.†

First, one finds that the magnitude of the orbital angular momentum **L** of an atomic system is quantized; the possible values are given by

$$L = \sqrt{l(l+1)}\hbar \qquad (7\text{–}1)$$

† See any textbook on elementary quantum mechanics.

where l is an integer called the *orbital-angular-momentum quantum number.* Furthermore, for a given value of the principal quantum number n, the only permitted values of l are integers ranging from zero to $(n - 1)$:

$$l = 0, 1, 2, 3, \ldots, n - 1$$

Thus, for $n = 1$ (ground state), the only possible value of l is 0, and from Equation 7–1, the value of L is 0. For $n = 2$, the value of l is restricted to 0 or 1, and the corresponding values of L are 0 and $\sqrt{2}\hbar$, respectively. Generally for a given n, there are n possible values of l, and therefore n possible values of the orbital angular momentum. The integral values of the quantum number l are often represented by letter symbols (the reasons are historical), as follows:

$$l = 0,\ 1,\ 2,\ 3,\ 4,\ 5,\ \ldots$$
$$\text{Symbol} = \text{S, P, D, F, G, H, } \ldots$$

Whereas in the Bohr theory, the state of an atom is specified completely by the one quantum number n (hence, the radius of the circular orbit, or the total energy), in wave mechanics the state of an atom is specified by the values of *all* the appropriate quantum numbers. To every state there corresponds a distinctive wave function ψ, differing from the others in the way in which it depends on the spatial coordinates. Those states for which, say, n is 3 and l is 0, 1, and 2 are called 3S, 3P, and 3D states, respectively. From Equation 7–1, the corresponding magnitudes of the orbital angular momentum **L** are 0, $\sqrt{2}\hbar$, $\sqrt{6}\hbar$; see Figure 7–6. Since the 3S, 3P, and 3D states have a common value of the principal quantum number, $n = 3$, then for a single electron under the influence of a Coulomb force from a nucleus assumed to be a point charge, the three states have identical energies E but different angular momenta; they also differ in the spatial dependence of the wave function. Such states, which are identical in total energy but different in some other respect, are said to be *degenerate.*

Recall that in the classical planetary model, a bound system's total energy depends only on the magnitude of the major axis of the ellipse and not on the eccentricity of the orbits or on the orbital angular momentum.

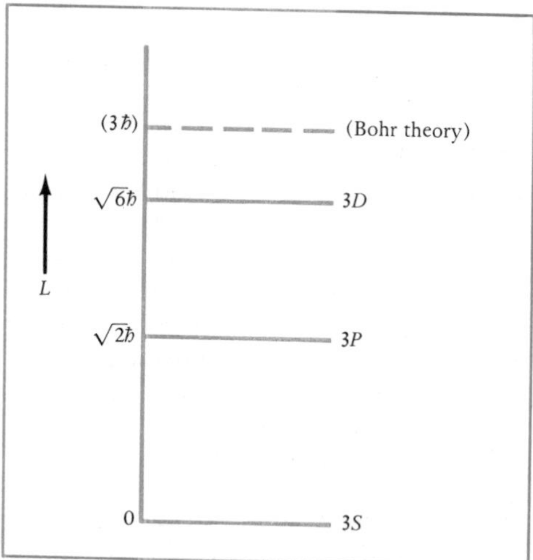

FIGURE 7–6. *Allowed values of the magnitude of the orbital angular momentum for $n = 3$.*

A similar situation obtains in the quantum theory; for a given value of n, which specifies the energy of the atom, there are n possible values of l, each l specifying a different possible value of the magnitude of the orbital angular momentum. An important difference is that whereas the classical theory places no restriction on the possible values of orbital angular momentum, the quantum theory limits them to discrete, quantized values.

In the classical theory, the orbits corresponding to the small values of angular momentum are those of high eccentricity; the circular orbit has the largest angular momentum for a given major axis, or energy (Figure 7–1). One may paraphrase this by saying that for a given major axis, or energy, an orbit of small angular momentum is one in which the orbiting particle spends an appreciable amount of time in each cycle close to the force center, whereas an orbit of large angular momentum is one in which the circulating particle is always far from the force center. The corresponding wave-mechanical situation is analogous. We find, by examining wave functions derived from the Schrödinger equation, that for a given value of n, or total energy, the probability that an electron will be at or close to the nucleus is greater for a state of low angular momentum (small l) than for a state of high angular momentum (large l).

Consider the hydrogen atom's wave function ψ plotted as a function of the distance r of the electron from the nucleus for the states in which n is 1, 2, and 3 (see Figure 7–7; also see Section 6–6 for the solution of the Schrödinger equation leading to ψ for the $n = 1$ state). We see that ψ is a maximum at $r = 0$ for an S state ($l = 0$) at *any* value of n. On the other hand, ψ is zero at $r = 0$ for all states with $l > 0$ and having nonzero angular momentum. The probability that the electron will be within any small volume element dv of *fixed* size is proportional to ψ^2. It follows that when the angular momentum is zero, the electron is more likely to be within a given dv near the nucleus than away from it. On the other hand, for states of higher angular momentum, the electron is more likely to be away from the nucleus than near it.

Now consider a related but different probability—that the electron is between r and $r + dr$, or within a spherical shell of radius r, thickness dr, and volume $dv = 4\pi r^2\,dr$. The probability that it is within the volume element dv, *not* fixed in size, is proportional to $\psi^2\,dv = \psi^2(4\pi r^2)\,dr$; and therefore the probability that it is within a spherical shell of *fixed thickness dr* is proportional to $r^2\psi^2$. The graphs on the right in Figure 7–7 show $r^2\psi^2$ as a function of r for the several wave functions plotted on the left. We see that the peaks shift to progressively larger values of r as the quantum number n, and the total energy, increases. This corresponds to the classical increase in orbit size with energy.

Figure 7–7 shows hydrogen wave functions as a function of the radial distance r only. A complete knowledge of the wave function in three dimensions involves, of course, its dependence on two other spatial coordinates. For the S states, with $l = 0$, the wave function is spherically symmetrical and depends *only* on the radial distance r. From the right-hand graphs of the figure, we see that, roughly speaking, an electron in an S state can be thought of as a ball of electric charge around the nucleus for the $n = 1$ state, as a ball of charge surrounded by a shell of charge for the $n = 2$ state (because $r^2\psi^2$ has a zero between the two peaks), and as a ball of charge surrounded by shells of successively larger sizes for higher and higher values of n. Wave

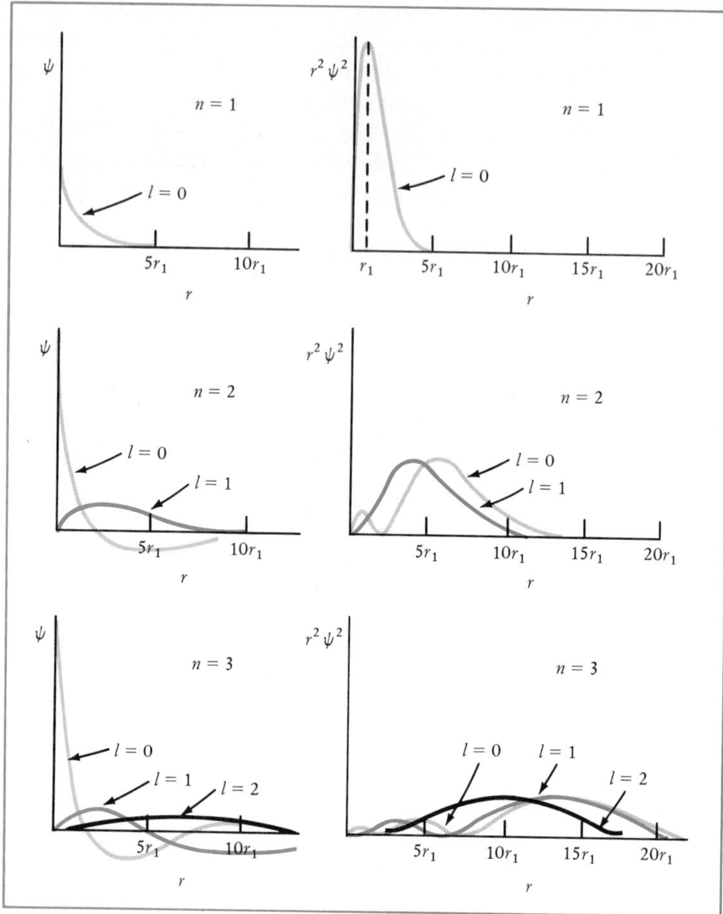

FIGURE 7–7. *Wave functions ψ (left-hand graphs) and probabilities of particles being between r and r + dr (proportional to $r^2\psi^2$, right-hand graphs) for n = 1, 2, and 3.*

functions for $l = 1$ states are not spherically symmetrical. For example, the P wave functions have the general form $xF(r), yF(r), zF(r)$, or linear combinations thereof, where $F(r)$ is a function of r only. The D wave functions, also not spherically symmetrical, have the general form of $x^2F(r)$, $y^2F(r)$, $xyF(r)$, . . . , involving the coordinates x, y, z to the second power, where again $F(r)$ is a function of r only. Figure 6–15 shows the probability distribution in space for a P wave function with $n = 2$.

Consider again an energy-level diagram of hydrogen, Figure 7–8, in which the states are identified according to the principal quantum number n and also the orbital-angular-momentum quantum number l. The n-fold degeneracy for each energy is shown. The diagonal lines connecting states represent possible transitions between stationary states leading to the emission of photons; such transitions are those in which l changes by only one unit. Wave mechanics selects from the all the combinations of stationary states only those that make for appreciable radiation (emission, or absorption). Transitions for which l changes by 1, that is, for which the change Δl is $+1$ or -1, are called *allowed transitions.* The *selection rule* for allowed transitions is, then,

$$\Delta l = \pm 1 \tag{7–2}$$

forming a relatively inert, closed shell, about which the eleventh electron moves. Similarly, the neutral atom of magnesium, an alkaline earth, with atomic number 12 and valence +2, can be regarded as consisting of an inner shell of 10 inactive electrons surrounded by two chemically active electrons. When magnesium is singly ionized, only one valence electron remains outside the closed shell. An atom of sodium or of singly ionized magnesium thus resembles the hydrogen atom in that the chemical properties are due chiefly to a single electron held to an inert core whose radius is approximately 1 Å.

If the valence electron were to stay completely outside the inner core of nucleus and electrons, it would "see" an electric charge of $+Ze$ from the nucleus, a charge of $-(Z - 1)e$ from the inner electrons, and therefore a net electric charge of $+Ze - (Z - 1)e = +1e$, just the electric charge of the nucleus of the hydrogen atom. The electron's "position" is given by its wave function, which extends over space. We recall some general features of the hydrogen wave function and the inferences concerning probability, illustrated in Figure 7–7:

1. For a given value of n, the smaller the value of l, the larger the probability that the electron is within the inert core radius, 1 Å.
2. As n increases, the probability that the electron is within the inert core decreases.

Applying these results to a hydrogenlike atom, we see that those states in which the valence electron is far from the nucleus (and inner electron core) can be expected to be most like corresponding states of hydrogen. Thus, the energy of states with large n values should be the same as for hydrogen. Moreover, for any given n value, the state of the largest possible l value should be most nearly hydrogenlike. For example, of the possible states 3S, 3P, and 3D, the last should have an energy closer to that of hydrogen than the first two. In the 3S state, on the other hand, there is a high probability that the valence electron will be *inside* the core of inner electrons. If it is inside the core, however, the nuclear charge is less well shielded by electrons in the closed shell. The valence electron will then experience a force arising from an effective charge *greater* than $+1e$. A more strongly attractive force implies a more tightly bound system, one whose energy is more negative and whose energy level is displaced downward. This follows also from the relation $E_n = -Z^2E_I/n^2$, which gives the energy of a single-electron atom with a nucleus of electric charge Ze; when Z is greater than 1, as it is when a valence electron penetrates the core, then the atom's energy E_n becomes more negative.

These effects are seen in the energy-level diagram of the hydrogenlike atom sodium, shown in Figure 7–9. The hydrogen levels are shown for comparison. First note that energy levels corresponding to $n = 2$ and $n = 1$ are *not* found (the reasons that the valence electron is excluded from those states will be explored in Section 7–8). Each of the $n = 3$ states (3S, 3P, and 3D) has less energy than the corresponding state in hydrogen. Furthermore, the 3S state is lower than the 3P, which is in turn lower than the 3D. These three states in sodium all have different energies and are therefore *not* degenerate, even though the valence electron has the same quantum number n. The energy of the sodium atom is *not* independent of the value of the orbital angular momentum, but is, instead, lowest for the smallest orbital

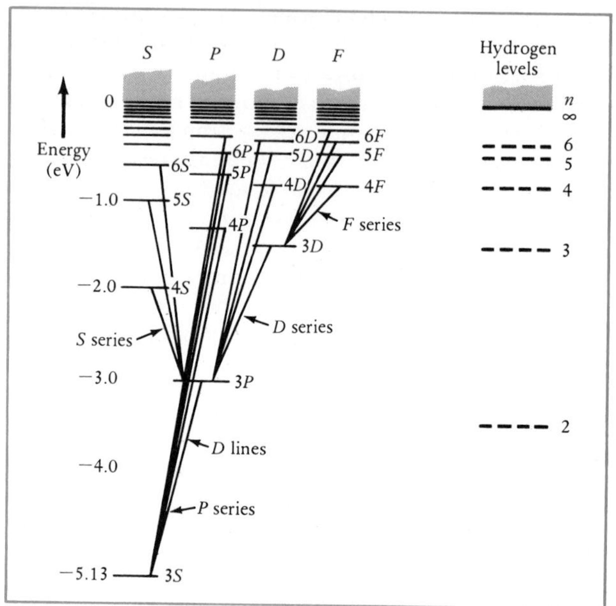

FIGURE 7–9. *Energy-level diagram of sodium. For comparison, the hydrogen energy levels are shown on the right.*

angular momentum. At higher values of n, the energy levels more closely approach those of hydrogen.

The allowed transitions, according to the selection rule $\Delta l = \pm 1$, are classified as S, P, D, and F series, as in Figure 7–8 for hydrogen, but the corresponding emitted photons are of *different* frequencies. For example, the 4S → 3P and 4P → 3S transitions in sodium yield two *different* spectral lines. The most prominent line in the spectrum of sodium is the so-called sodium *D* line which arises from the 3P → 3S transition, the first line of the P series.†

The other alkali metals (see Table 7–1) have similar emission and absorption spectra to those of sodium. The singly ionized alkaline earths (Table 7–1) are *isoelectronic* with the alkali metals, meaning that two adjacent elements have the same number of electrons; they differ principally in the size of the nuclear charge. In the $_{12}Mg^+$ atom, for example, there are ten electrons in a closed shell like that of $_{10}Ne$, and a single valence electron, as in $_{11}Na$; when

† The labels S, P, D, and F were assigned to these and similar series early in the history of spectroscopy for the following reasons. The lines of the S series were relatively "sharp." The lines of the P series were the "principal" lines in the emission or absorption spectra in that they were found even for relatively small excitation of the source (the P lines result from transitions from the *first* and higher excited P states to the *ground* state). The lines of the D series were "diffuse." And the lines of the F series, lying in the infrared, had frequencies the lowest of any of the series, corresponding to the "fundamental."

The label D for the strong yellow lines of sodium has no connection with the symbol D designating $l = 2$. Fraunhofer discovered in 1809 that the spectrum from the sun contains a number of dark absorption lines (*Fraunhofer lines*) that arise from the absorption of radiation from the interior of the sun by elements in the sun's atmosphere. These lines were labeled *A*, *B*, *C*, *D*, etc. The *D* Fraunhofer line corresponds to absorption by sodium vapor in the 3P → 3S transition. Close observation shows that this transition actually consists of two closely spaced yellow lines having wavelengths of 5890 and 5896 Å. Other lines in the sodium spectrum show a similar fine structure, whose origin will be treated in Section 7–6. Also appearing in the sun's absorption spectrum were lines identified with the element helium and named for the sun (*helios*); helium was later isolated on the earth and identified.

the valence electron is outside the inert core, it sees a net positive charge of $+2e$. Thus, the "nonpenetrating" energy states of $_{12}Mg^+$ will correspond closely to the states of the one-electron atom $_2He^+$, whose nuclear electric charge also is $+2e$. States with small n values, and especially with small l values, will be displaced downward relative to corresponding states of $_2He^+$.

We have seen that it is possible to understand qualitatively the energy levels and spectra of hydrogenlike atoms by assuming that the excited states are due to the last, or the valence, electron of the atom. We shall see that the inertness of the closed shells arises in a natural way from fundamental principles. The energy levels and spectra of atoms containing more than one active valence electron are much more complex.

7-4 Quantization of Angular Momentum Component

In a classical planetary model, the total energy, the magnitude of the orbital angular momentum, and the component of the orbital angular momentum along any direction in space are constants of the motion. In wave mechanics, all constants of motion should be quantized. Thus far, we have found that the energy of a one-electron atom is quantized and identified by the principal quantum number n, and that the atom's orbital angular momentum is quantized; the possible values depend on the value of the orbital-angular-momentum quantum number l. The third classical constant of the motion, the component of the orbital angular momentum along a fixed direction in space, also is quantized and specified by a quantum number, called m_l.

According to wave mechanics, the component of the orbital-angular-momentum vector **L** along an arbitrary direction *cannot* assume *any* value; rather, this component is restricted to *integral* multiples of $\hbar$. For example, along any axis, say the z-axis, the possible values of the z-component of **L** (see Figure 7–10) are given by the rule

$$L_z = m_l\hbar \tag{7–3}$$

where m_l, the *orbital-angular-momentum quantum number,* can assume, for a given value of l, only the integral values

$$m_l = l, l-1, l-2, \ldots, 0, \ldots, -l \tag{7–4}$$

As Equation 7–4 shows, for each value of the quantum number l, there are $2l + 1$ distinct quantum states resulting from the quantization of L_z. For example, in a D state (with $l = 2$) the possible values of m_l are $+2, +1, 0, -1$, and -2. In this state, then, the component L_z can assume only the values $2\hbar$, $1\hbar, 0\ -1\hbar, -2\hbar$, while the magnitude of **L** from Equation 7–1, is $\sqrt{6}\hbar$. Figure 7–11 shows the five possible orientations of the orbital-angular-momentum vector with respect to the arbitrary z-axis (compare with Figure 7–4). Because the angular-momentum vector is restricted to certain discrete orientations in space, it is said to be *space-quantized.* Further, since L_z is $L \cos\theta$, the rule governing the orientation of the **L** vector, that is, the rule for space quantization, is

$$\cos\theta = \frac{m_l}{\sqrt{l(l+1)}} \tag{7–5}$$

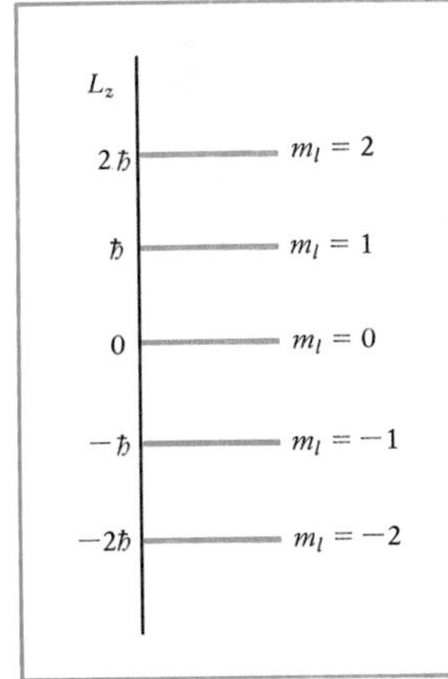

FIGURE 7–10. *Permitted quantum values of the component of the orbital angular momentum along the direction of a magnetic field for a D state.*

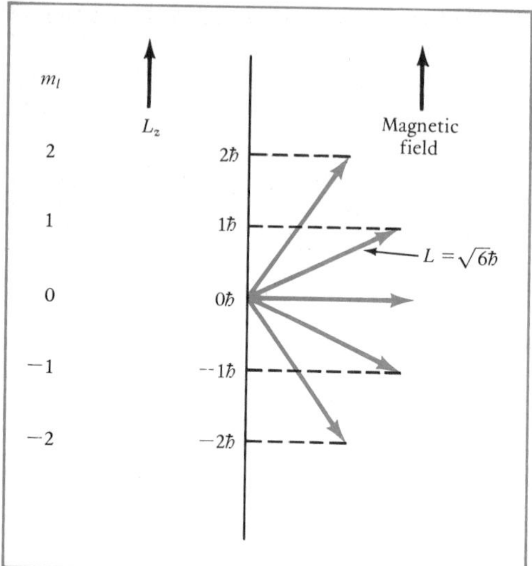

FIGURE 7–11. *Space quantization of the orbital-angular-momentum vector* **L** *for a D state.*

Note that **L**'s maximum component in the direction of space quantization, $L_z = l\hbar$, is always *less* than its magnitude, $L = [l(l + 1)]^{1/2}\hbar$. Thus, the orbital-angular-momentum vector can never be perfectly aligned along the z-direction. Wave mechanics permits the magnitude of **L** and its z-component to be precisely specified, but paradoxically does not allow its x- and y-components to be known simultaneously. It is customary to regard the **L** vector as *precessing* around the z-axis at a constant angle θ, thereby tracing out a cone, for any particular allowed value of m_l; this model correctly implies that the magnitude and the z-component of **L** are known but the x- and y-components are unknown.†

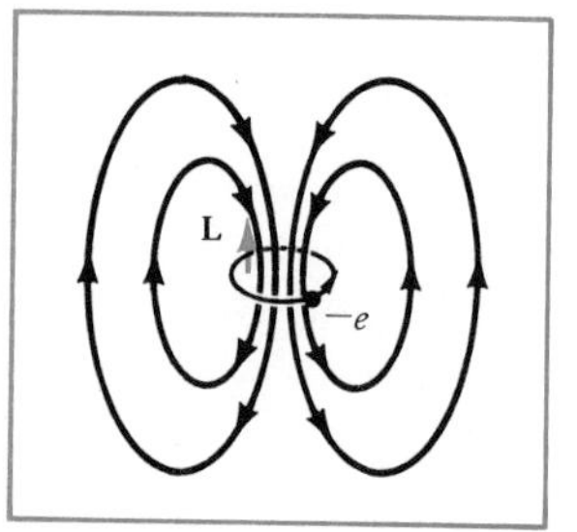

FIGURE 7–12. *Magnetic field of a magnetic dipole consisting of a circulating negative electric charge.*

The quantization of the orbital-angular-momentum components into $2l + 1$ distinct states should be closely related to magnetic properties of atoms, since a circulating negative charge produces a magnetic field. To find this relation, first consider the magnetic effects associated with a classical orbiting charged particle. The orbital angular momentum **L** of a particle moving in closed orbit is represented by a vector oriented at right angles to the plane of the orbit. A circulating negative electric charge constitutes an electric current loop and has associated with it a magnetic field (see Figure 7–12) proportional to the magnitude of the current. The magnetic field configuration is like that of a small permanent magnet, and we can associate a *magnetic dipole moment* $\boldsymbol{\mu}$ with the circulating electron. The direction of $\boldsymbol{\mu}$ is perpendicular to the plane of the electron's loop and related to the sense of a rotating *positive* charge through the right-hand screw rule. Thus, for a negatively charged particle, the angular momentum **L** and the magnetic moment $\boldsymbol{\mu}$ point in opposite directions, as shown in Figure 7–13. We wish to show that $\boldsymbol{\mu}$ is proportional to **L** and find the proportionality constant.

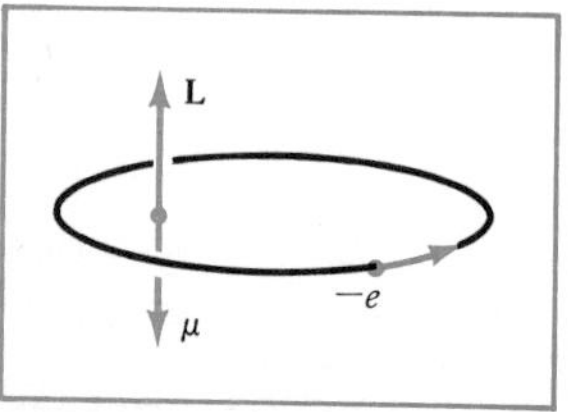

FIGURE 7–13. *Orbital angular momentum* **L** *and magnetic moment* μ *of an orbiting electron.*

The magnitude of the magnetic moment of an electric current i enclosing a loop in a plane of area A is given by‡

$$\mu = iA$$

† The indefiniteness of the x- and y-components can be shown to be a necessary consequence of the uncertainty principle. See Problem 5–32.

‡ See Weidner and Sells, *Elementary Classical Physics,* 2d ed., Equation 29–19.

When a particle of charge e completes one loop in the time T, the current is $i = e/T$; then the equation can be written

$$\mu = \frac{eA}{T} \tag{7–6}$$

The orbital angular momentum L relative to the force center for a particle of mass m moving under the influence of a central force remains constant and can be written

$$L = 2m\frac{dA}{dt}$$

where dA/dt represents the rate at which a radius vector from the force center to the moving particle sweeps out area and m is the mass of the electron (this is merely Kepler's second law of planetary motion[†]). Over the time T of one complete cycle the area swept out is A; then $dA/dt = A/T$. Substituting this result in the last equation, we have

$$L = \frac{2mA}{T} \tag{7–7}$$

Finally, combining Equations 7–6 and 7–7, we have

$$\boldsymbol{\mu} = -\frac{e}{2m}\mathbf{L} \tag{7–8}$$

The minus sign has been introduced because the magnetic-moment and angular-momentum vectors point in opposite directions. We see that $\boldsymbol{\mu}$ is directly proportional to $\mathbf{L}$. The proportionality constant, $-e/2m$, is the constant we set out to find; it is customarily called the *magnetogyric ratio*.

Wave mechanics gives precisely the same relation for the magnetogyric ratio of an electron in an atom that classical physics gives, despite that it is impossible to visualize the connection between the magnetic effects and the angular momentum in terms of a well-defined electron orbit. Since L depends on the orbital-angular-momentum quantum number l, the magnetic moment μ must, too. Using Equation 7–1 in Equation 7–8, we have

$$\mu_l = \frac{e}{2m}\sqrt{l(l+1)}\hbar \tag{7–9}$$

where the subscript denotes the magnetic moment associated with l.

Consider now the magnetic potential energy change ΔE_{m} of an atom that occurs when the atom with magnetic moment $\boldsymbol{\mu}$ is placed in an external magnetic field of flux density $\mathbf{B}$. This is given by[‡]

$$\Delta E_{\mathrm{m}} = -\boldsymbol{\mu}\cdot\mathbf{B} = -\mu B\cos\theta \tag{7–10}$$

where θ is the angle between $\boldsymbol{\mu}$ and $\mathbf{B}$, as shown in Figure 7–14*a*. When the dipole is aligned with the external field and θ is zero, then ΔE_{m} equals $-\mu B$, a minimum. When work is done on the dipole to turn it so that it is aligned in a direction opposite to that of $\mathbf{B}$ and θ is 180°, then ΔE_{m} equals $+\mu B$, a maximum. We know that in classical physics, all orientations of the dipole between 0 and 180° and therefore all energies between $-\mu B$ and $+\mu B$ are

† See Weidner and Sells, *Elementary Classical Physics,* 2d ed., Section 11–4.

‡ See Weidner and Sells, *Elementary Classical Physics,* 2d ed., Section 29–8.

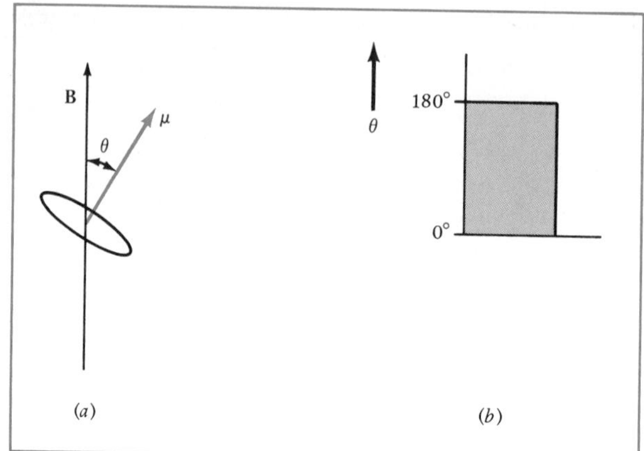

FIGURE 7–14. *Magnetic dipole in a magnetic field: (a) relative direction of* **B** *and* μ*; (b) orientations permitted in classical physics.*

allowed, as shown in Figure 7–14*b*. Wave mechanics, on the other hand, predicts only discrete values for the atom's orbital magnetic moment and therefore discrete values for the magnetic energy. Combining Equations 7–9 and 7–10 and now taking θ to be the angle between **L** and **B** instead of μ and **B**, we have

$$\Delta E_{\mathrm{m}} = \frac{e\hbar}{2m}\sqrt{l(l+1)}\, B\cos\theta \tag{7–11}$$

Thus, when the atom is immersed in an external magnetic field, its energy depends on the angle θ between the orbital-angular-momentum vector and the external field. If there were no restriction on the angle θ, the component of the orbital angular momentum in the direction of the magnetic field could assume any value between positive and negative $[l(l+1)]^{1/2}\hbar$, and similarly, ΔE_{m} could assume any value between positive and negative $(e\hbar/2m)B[l(l+1)]^{1/2}$, according to Equation 7–11. In short, if there were no rule restricting, that is, quantizing, the values of L in the field direction, a *continuum* of possible energies would exist, quite unlike what has heretofore been found to hold in bound atomic systems. Then the emission lines from atoms with magnetic moments and in magnetic fields would be continuously broadened and not split into discrete lines.

The emission lines from atoms placed in a strong external magnetic field were studied in 1896 by P. Zeeman, who found that on application of the field, the lines become split and consist of two or more closely spaced, sharp lines. Such splitting of a single spectral line into discrete components by a magnetic field is known as the *Zeeman effect.*

7–5 The Normal Zeeman Effect

If the orientation of the angular-momentum vector is quantized, so are the possible orientations of the associated magnetic dipole moment μ_l, and so too is the magnetic potential energy ΔE_{m} of the state. Using Equation 7–5, the rule for space quantization, in Equation 7–11, the equation for the change in energy of a state with quantum numbers l and m_l, we have

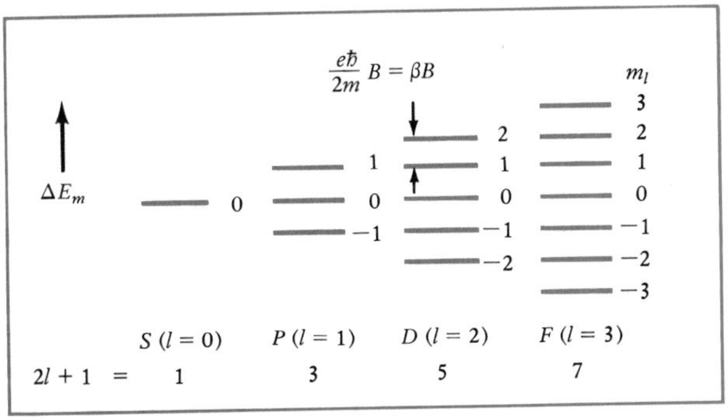

	$\frac{e\hbar}{2m}B = \beta B$			m_l
ΔE_m				3, 2, 1, 0, −1, −2, −3
	0	1, 0, −1	2, 1, 0, −1, −2	
	S ($l = 0$)	P ($l = 1$)	D ($l = 2$)	F ($l = 3$)
$2l + 1$ =	1	3	5	7

FIGURE 7–15. *Energy splitting of S, P, D, and F states of an atom in a magnetic field.*

$$\Delta E_m = m_l \frac{e\hbar}{2m} B \tag{7–12}$$

Figure 7–15 shows the *changes* in the energies of S, P, D, and F states having, respectively, 1, 3, 5, and 7 *magnetic sublevels;* in general, the number of Zeeman components for a given l equals $2l + 1$. The difference in energy between adjacent magnetic sublevels equals $(e\hbar/2m)B$ and is independent of the value of l. The quantity $e\hbar/2m$ has the units of magnetic moment; it is known as the *Bohr magneton* β, because β is the magnetic moment of a classical electron orbiting the hydrogen nucleus at the radius of the first Bohr orbit (see Equation 7–8 with $L = \hbar$):

$$\text{Bohr magneton } \beta = \frac{e\hbar}{2m} = 0.9273 \times 10^{-23}\text{ J/T}$$

Consider the spectrum of lines emitted from excited atoms in transitions between a D state and a P state in the presence of a magnetic field; see Figure 7–16. When B is zero, the energy of the D state is E_D (for all five m_l values), the energy of the P state is E_P (for all three m_l values), and photons having the single frequency ν_0 are emitted according to $h\nu_0 = E_D - E_P$. When the field is turned on, the D state splits into five equally spaced magnetic sublevels, and the P state splits into three equally spaced magnetic sublevels; the difference in energy between any two adjacent magnetic sublevels is $(e\hbar/2m)B$. The transitions between the D state ($l = 2$) and the P state ($l = 1$) obey the selection rule $\Delta l = \pm 1$; wave mechanics also requires a selection rule for transitions between magnetic sublevels:

$$\Delta m_l = 0 \quad \text{or} \quad \pm 1 \tag{7–13}$$

That is, only those transitions are allowed in which the magnetic quantum number m_l is unchanged or in which it changes by one unit. The permitted transitions and the spectrum of the emitted lines are shown in Figure 7–16. You can see that the differences $h\nu$ for allowed transitions take one of three possible values:

$$\begin{aligned} \Delta m_l &= -1: & h\nu &= h\nu_0 - \frac{e\hbar}{2m}B \\ \Delta m_l &= \ \ 0: & h\nu &= h\nu_0 \\ \Delta m_l &= +1: & h\nu &= h\nu_0 + \frac{e\hbar}{2m}B \end{aligned} \tag{7–14}$$

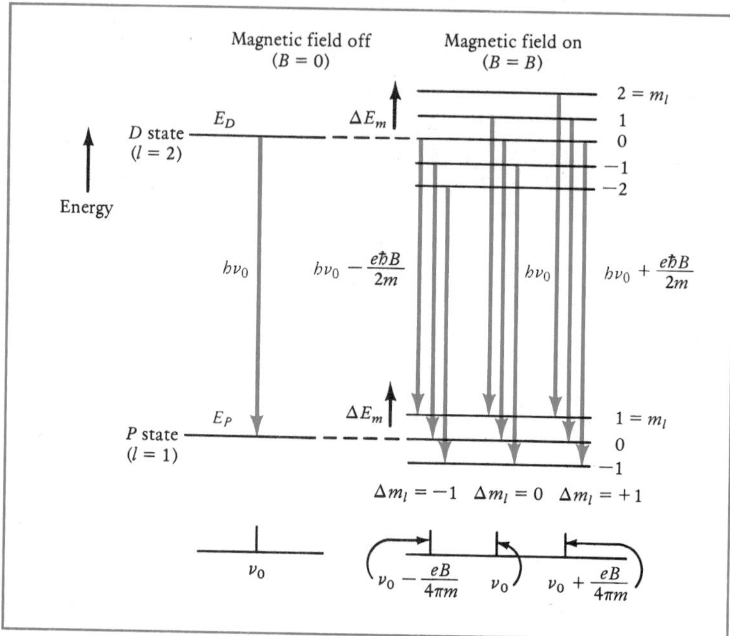

FIGURE 7–16. *Energy levels and spectra of a D → P transition: (left) zero magnetic field; (right) nonzero magnetic field, normal Zeeman effect.*

Dividing both sides of these equations by h gives the frequencies of emitted radiation:

$$\nu = \nu_0 - \frac{e}{4\pi m}B$$

$$\nu = \nu_0 \tag{7–15}$$

$$\nu = \nu_0 + \frac{e}{4\pi m}B$$

Thus, a single line in the spectrum is split by an external magnetic field into three equally spaced components: the original line of frequency ν_0 and two equally spaced satellite lines, whose separation $(e/4\pi m)B$ from ν_0 is proportional to the magnetic field B. Using Equation 7–14, we see that for a relatively strong magnetic field, say B = 1.0 T (10,000 G), the difference in energy between adjacent Zeeman levels is only 9.3×10^{-24} J, or 5.8×10^{-5} eV. Since the typical energy difference between levels giving rise to emission in the visible region of the spectrum is a few electron volts, the energy, frequency, or wavelength is changed by less than 1 part in 1000 when a strong magnetic field is applied. For this reason one needs spectrometers of moderately high resolution to see the Zeeman effect.

Figure 7–16 shows the energy levels and allowed transitions between a D and P state. The value of ΔE_m and the selection rules governing m_l are both independent of l; therefore:

All transitions for which Δl is ± 1 will give rise to the same Zeeman effect, three equally spaced Zeeman component lines.

This is called the *normal Zeeman effect,* and the observed splittings of some lines from *some* elements, such as calcium and mercury, agree completely with the spectrum shown in the figure. On the other hand, the spectra of most elements do *not* show a normal Zeeman effect, in that the magnitude of the splittings and the number of Zeeman components is *not* in accord with the theory presented here. Such Zeeman spectra are said to be *anomalous,* since the emitted radiation cannot be accounted for simply by the space quantization of the *orbital*-angular-momentum vector and the associated magnetic effects.

Note that the frequency difference $(e/4\pi m)B$ in Equation 7–15 does not involve the quantum constant h. This suggests that it may not be a distinctly quantum effect. Indeed, as Problem 7–21 shows, the so-called normal Zeeman effect can be derived with a strictly classical computation.

The space-quantization rule, which limits the value of the component of the orbital angular momentum in any direction in space to integral multiples of $\hbar$, holds whether or not a magnetic field is applied. When a magnetic field is applied, its direction specifies the direction of space quantization, and the energies of the several states differ according to the value of m_l. When the magnetic field is turned off, the space quantization persists; now, however, the energies of the states corresponding to the several possible values of m_l are all identical. Therefore, when there is no magnetic field, any state with orbital-angular-momentum quantum number l has $2l + 1$ substates, which are identical both in their energy, $\Delta E_{\text{m}} = 0$, *and* in the magnitude of the orbital angular momentum, $[l(l + 1)]^{1/2}\hbar$; they differ, however, in the component of the angular-momentum vector, $m_l\hbar$, along any direction in space. Thus, when there is no magnetic field, there is a $(2l + 1)$-fold degeneracy in the energy of the states for any particular value of l.

7–6 Electron Spin

In the quantum theory, three classical constants of the motion of a particle subject to an inverse-square force of attraction—the energy, the magnitude of the orbital angular momentum, and the component of the orbital angular momentum along any direction in space—are quantized. In classical mechanics the energy of a particle in an elliptical orbit is determined by the *size* of the orbit, that is, by the major axis of the ellipse; the magnitude of the orbital angular momentum is determined for a given major axis by the *shape* of the elliptical orbit, that is, by the eccentricity of the elliptical path; the component of the orbital angular momentum along a direction in space is determined by the *orientation* of the elliptical orbit. To these constants of the motion, there correspond in quantum mechanics the quantum numbers n, l, and m_l. There is a fourth and final quantum number s, which is associated with the concept of electron spin.

We have remarked that the strongest emission from sodium comes from the $3\text{P} \rightarrow 3\text{S}$ transition. When this radiation is examined with a spectrometer of moderately high resolution, it is seen that the transition corresponds to *two* closely spaced yellow lines (at 5890 and 5896 Å), called the sodium D lines (see Section 7–3). In fact, each of the spectral lines of sodium exhibits such *fine structure;* for each transition shown in Figure 7–9, there are two or three distinct lines separated from one another by no more than a very few angstroms of wavelength. The fine structure is *anomalous* in that it occurs without the application of an external magnetic field and so cannot be accounted for as a *normal* Zeeman effect. Fine structure in emission and absorption spectra is common to all atomic line spectra. Apparently, a distinctive additional feature of atomic structure is manifest in fine structure, one that cannot be accounted for in terms of the quantum numbers n, l, and m_l.

It is suggestive to attribute fine structure to an *internal* Zeeman effect, within the atom. Such an effect would require an internal atomic magnetic field and a new source of magnetic moment and angular momentum within the atom. The orbital angular momentum of the atom has already been taken into account. What other contribution to the angular momentum can be imagined?

S. A. Goudsmit and G. E. Uhlenbeck suggested in 1925 that an intrinsic angular momentum quite apart from orbital motion was associated with an electron. It is named *electron spin,* for it may be visualized as analogous to the intrinsic angular momentum that any extended object has by virtue of rotation, or spin, about its center of mass. It is not proper in wave mechanics to regard an electron as a simple sphere of electric charge; but for the sake of identifying the electron-spin angular momentum with some sort of model that can be visualized, it is useful to imagine the electron as having an extension in space and as continuously spinning around an axis of rotation. The electron spin, then, is the intrinsic angular momentum $\mathbf{L}_s$ arising from the rotation of the charge cloud about a rotation axis fixed with respect to the electron. Furthermore, because negative electric charge is imagined to be rotating, a magnetic field will be produced by the spinning electron, and a magnetic moment $\boldsymbol{\mu}_s$, opposite in direction to that of the spin angular momentum $\mathbf{L}_s$, can be attributed to the electron spin; see Figure 7–17.

If an electron, with its permanent spin magnetic moment, were in a magnetic field, we should expect its spin to be space-quantized. The axis of spin magnetic moment, and spin angular momentum would then be restricted to certain quantized orientations, and the energy of the atom would differ according to the particular orientation.

The internal atomic magnetic field that acts on an electron with a spin angular momentum $\mathbf{L}_s$ and a spin magnetic moment $\boldsymbol{\mu}_s$ can be thought of as originating the following way. If a spinning electron orbits a nucleus, then an observer fixed with respect to the electron sees the nucleus orbiting it. The orbiting positive charge produces at the site of the electron a magnetic field, whose magnitude and direction depend on the magnitude and direction of the electron's *orbital* angular momentum. This magnetic field acts on the spin magnetic moment $\boldsymbol{\mu}_s$. The interaction between electron spin and orbital angular momentum is aptly called *spin-orbit interaction.* The interaction exists in all orbital states except S states ($l = 0$).

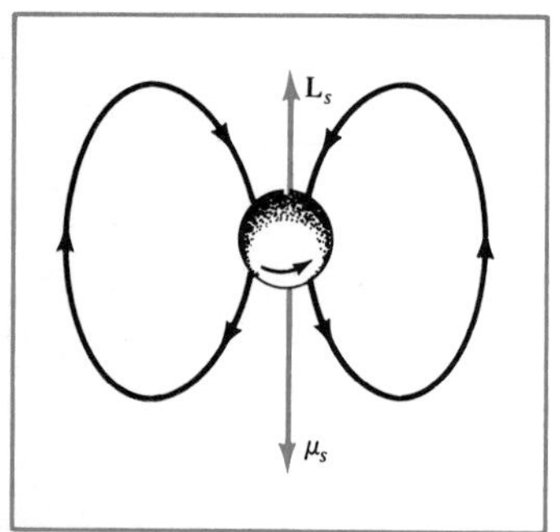

FIGURE 7–17. *Electron-spin angular momentum* $\mathbf{L}_s$ *and magnetic moment* μ_s *with associated magnetic field.*

We now turn to spectroscopic evidence to find the allowed values of the spin angular momentum L_s and the spin magnetic moment μ_s. A study of the spectral lines from an atom with a *single* valence electron, such as sodium, in the *absence* of an *external* magnetic field indicates that each of the orbital energy levels (except the S state) is split into two components (a doublet), the S state remaining unsplit (a singlet). This is why the $3P \rightarrow 3S$ transition in sodium consists of the two closely spaced D lines; the 3S state is a singlet, the 3P state a doublet. How can the doubling of all states (except the S states) be interpreted by the view that an internal magnetic field is space-quantizing the electron-spin angular momentum? In the normal Zeeman effect, any state having an orbital quantum number l is split, under the influence of an external magnetic field, into $2l + 1$ sublevels. We assume that, similarly, a state having a *spin-angular-momentum quantum number s* is split into $2s + 1$ components under the influence of an internal magnetic field. Because the number of components of all fine-structure states with a nonzero orbital angular momentum is always 2, the quantity $2s + 1$ must equal 2 and the spin quantum number s has the *single value* $\frac{1}{2}$:

$$2s + 1 = 2 \qquad \text{or} \qquad s = \tfrac{1}{2}$$

Since the spin is an intrinsic characteristic of the electron, every electron has a spin quantum number with the unique value $\frac{1}{2}$. The spin, or intrinsic, angular momentum of such a particle as an electron is as basic a characteristic as its charge and mass. The magnitude of the spin angular momentum $\mathbf{L}_s$ is given by a relation analogous to that for orbital angular momentum [Equation (7–1)]:

$$L_s = \sqrt{s(s+1)}\hbar = \tfrac{1}{2}\sqrt{3}\hbar \tag{7–16}$$

It is easy to see why the S state of sodium is a singlet. The internal magnetic field due to orbital motion is zero ($l = 0$); hence, the two spin states are unsplit and therefore *degenerate*. This degeneracy can, however, be removed by an *external* magnetic field.

In the presence of a magnetic field, the electron spin is space-quantized, such that the component $L_{s,z}$ of the spin angular momentum in the direction of the magnetic field is

$$L_{s,z} = m_s\hbar \tag{7–17}$$

where the *spin magnetic quantum number* m_s has two possible values, $+\frac{1}{2}$ and $-\frac{1}{2}$. The space quantization of the electron-spin angular momentum in a magnetic field, as shown in Figure 7–18, restricts the orientation of the electron-spin vector $\mathbf{L}_s$ to those two possible states in which its component

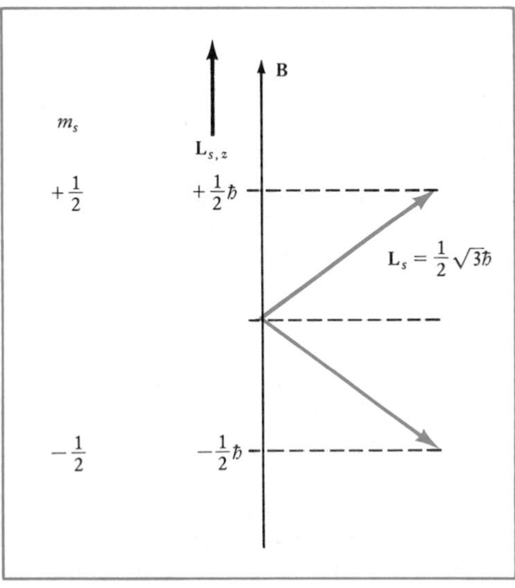

FIGURE 7–18. *Space quantization of electron-spin angular momentum.*

along z is $+\frac{1}{2}\hbar$ or $-\frac{1}{2}\hbar$. For $m_s = +\frac{1}{2}$ the spin-angular-momentum vector points more nearly in the direction of the magnetic field than away from it, and the magnetic moment $\boldsymbol{\mu}_s$ is aligned more nearly against the field than with it. Thus, the magnetic potential energy arising from the orientation of the electron-spin magnetic moment is higher for the state with $m_s = +\frac{1}{2}$ than for the state with $m_s = -\frac{1}{2}$.

As in orbital motion, in spin motion the magnetic moment lies along the same line as the angular-momentum vector but points in the opposite direction. Both quantum theory and detailed study of the Zeeman effect for atoms having fine structure show that the magnetogyric ratio associated with the electron spin is given by

$$\frac{\mu_s}{L_s} = 2(1.001\ 159\ 615)\frac{e}{2m} \qquad (7\text{–}18)$$

where e and m are the electron charge and mass. The electron spin has a magnetogyric ratio very closely *two* times that of the electron orbital motion; that is, for a given angular momentum, a spinning electron is twice as effective in magnetic effects as an orbiting one.

The magnetic potential energy change ΔE_s of an electron-spin magnetic moment in a magnetic field B is given by

$$\Delta E_s = m_s\left(2\frac{e\hbar}{2m}\right)B \qquad (7\text{–}19)$$

which is much like Equation 7–12.

Example 7–1. Find the approximate magnitude of the internal magnetic field that splits the 3P state in sodium.

Figure 7–19 shows the energy levels in the sodium atom responsible for the predominant yellow sodium D spectral lines observed in an ionized sodium gas. (The 3S and 3P states are identified by spectroscopic notation to be described in Section 7–7.) The two 3P → 3S transitions give rise to two lines differing in wavelength by 5.97 Å. Therefore, the energy difference between the two 3P states follows from $E = hc/\lambda$.

$$\Delta E = \frac{-hc}{\lambda^2}\Delta\lambda = \frac{-(12{,}400\ \text{Å}\cdot\text{eV})(-5.97\ \text{Å})}{(5893\ \text{Å})^2}$$

$$= 2.13 \times 10^{-3}\ \text{eV} \approx 2 \times 10^{-3}\ \text{eV}$$

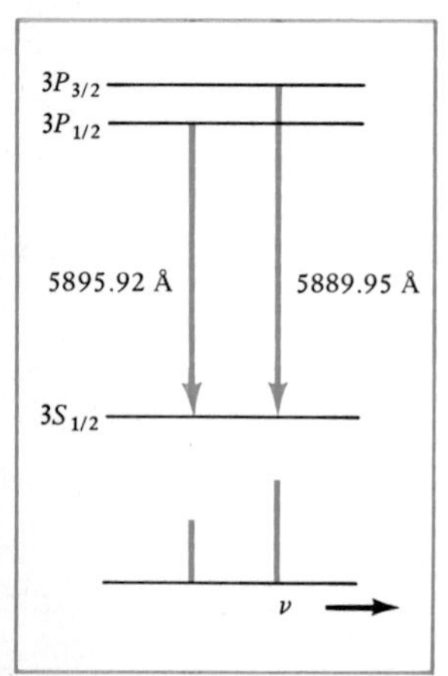

FIGURE 7–19. *Energy levels and spectrum of the sodium D lines (not to scale).*

To estimate the internal magnetic field necessary to produce an energy difference of 10^{-3} eV, we use Equation 7–19:

$$\Delta E = \Delta E_{1/2} - \Delta E_{-1/2} = 2\left(\frac{1}{2}\right)\left(\frac{e\hbar}{m}\right)B$$

$$B = \frac{m\,\Delta E}{e\hbar} = \frac{(1 \times 10^{-30}\ \text{kg})(2 \times 10^{-3}\ \text{eV})(2 \times 10^{-19}\ \text{J/eV})}{(2 \times 10^{-19}\ \text{C})(1 \times 10^{-34}\ \text{J}\cdot\text{s})}$$

$$= 20\ \text{T} = 2 \times 10^5\ \text{G}$$

This is a very strong magnetic field, typical of the internal atomic magnetic fields arising from the spin-orbit interaction in atoms.

7–7 Quantum Theory of "Single-Electron" Atoms

All atoms are composed of a small, massive nucleus containing Z positive charges and surrounded by clouds of an equal number of negative charges. The atom will have a total angular momentum primarily contributed by some vector combination of all the electrons' orbital-angular-momentum and spin-angular-momentum values.† It is possible, by the quantum theory, to determine the quantization rules for angular momenta. These rules will depend on how the Z intrinsic-spin vectors and Z orbital-angular-momentum vectors all couple together to give the atom's total angular-momentum vector.

The *"single-electron" atom* includes not only the hydrogen atom and hydrogenic atoms (for example, $_2He^+$) but also all atoms having a single-valence electron. (As we shall find in Section 7–9, all the electrons within a closed shell are so arranged that the total angular momentum of the electrons within the closed shell is *zero*).

To specify a specific quantum state of a single-electron atom requires four quantum numbers corresponding to the four classical constants of motion in Section 7–1. Two of these quantum numbers are the principal quantum number n and the orbital-angular-momentum quantum number l (restricted to the integral values $0, 1, \ldots, n-1$). The other two quantum numbers are related to the atom's angular-momentum values and will depend on the coupling interaction between the orbital angular momentum L_l and spin angular momentum L_s.

NO SPIN-ORBIT INTERACTION. First, we assume that there is *no* spin-orbit interaction between $\mathbf{L}_s$ and $\mathbf{L}_l$ ($\mathbf{L}_s$ and $\mathbf{L}_l$ are "uncoupled.") Subsequently we consider the quantum description for spin-orbit coupling. When the interaction between the spin and orbital angular momenta is ignored, the spin–angular-momentum vector $\mathbf{L}_s$ and the orbital-angular-momentum vector $\mathbf{L}_l$ precess each independently about an arbitrary z-axis, as shown in Figure 7–20. For this uncoupled system, the components of $\mathbf{L}_l$ along the z-axis are quantized according to Equation 7–4, with integral magnetic quantum numbers $m_l = l, \ldots, 0, \ldots, -l$; and the components of $\mathbf{L}_s$ along the z-axis are quantized according to Equation 7–17, with magnetic quantum numbers $m_s = +\frac{1}{2}, -\frac{1}{2}$. The four quantum numbers specifying a particular state of an *uncoupled single-valence atom* are then

Quantum numbers for *no L-S* coupling: (n, l, m_l, m_s)

Table 7–2 lists, as an example, all possible quantum states for the principal quantum number $n = 3$ of the hydrogen atom *if* spin-orbit interaction is ignored. For convenience, the states have been grouped according to different l values, and the two states differing only in the magnetic spin quantum number ($m_s = +\frac{1}{2}$ or $-\frac{1}{2}$) are combined in single terms.

† The nucleus of an atom, too, may have spin, or intrinsic angular momentum, called *nuclear spin,* which is added to the angular momentum of the electrons. *Hyperfine structure,* consisting of very closely spaced spectral lines (typically less than 10^{-3} Å), has its origin in the interaction of the spin and magnetic moment of nuclei with their counterparts of the electrons. The magnetic moments associated with nuclei are always smaller than the Bohr magneton by a factor of approximately 10^3, and the hyperfine splitting is correspondingly less than the fine splitting. In a classical planetary model, the nuclear spin is analogous to the spin angular momentum of the sun.

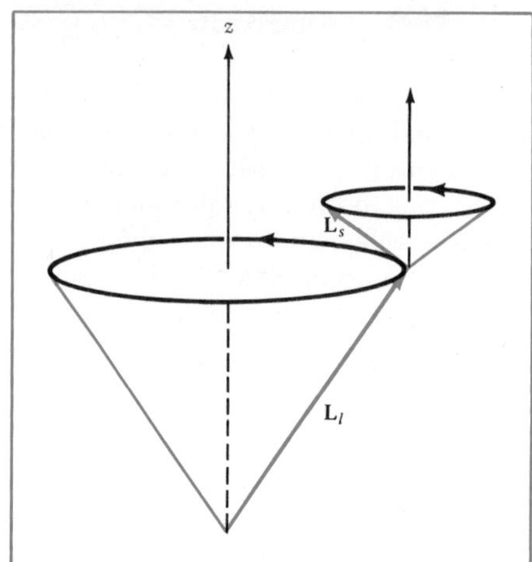

FIGURE 7–20. *With no spin-orbit coupling, each of $\mathbf{L}_l$ and $\mathbf{L}_s$ precesses independently about the z-axis.*

Since each entry in Table 7–2 corresponds to two distinct states, there are two S states, six P states, and ten D states, a total of 18 quantum states for $n = 3$. For single-valence atoms, the two S states would be degenerate, having the same energy when there is no external magnetic field; and likewise for the six P and the ten D states. Applying an external magnetic field would, of course, remove this degeneracy. The selection rules for transitions between states are:

$$\Delta l = \pm 1 \text{ and } \Delta m_s = 0 \tag{7–20}$$

Although this model ignores spin-orbit interaction, it is still useful in classifying the quantum states of an atom, and we shall apply it in the discussion of the periodic table (Section 7–9).

TABLE 7–2. *The 18 quantum states $(3, l, m_l, m_s)$ for a single-valence atom, with L-S coupling ignored*

$l = 0$	$l = 1$	$l = 2$
$(3, 0, 0, \pm\frac{1}{2})$	$(3, 1, 1, \pm\frac{1}{2})$	$(3, 2, 2, \pm\frac{1}{2})$
	$(3, 1, 0, \pm\frac{1}{2})$	$(3, 2, 1, \pm\frac{1}{2})$
	$(3, 1, -1, \pm\frac{1}{2})$	$(3, 2, 0, \pm\frac{1}{2})$
		$(3, 2, -1, \pm\frac{1}{2})$
		$(3, 2, -2, \pm\frac{1}{2})$

SPIN-ORBIT INTERACTION COUPLING. We have established that there is spin-orbit interaction for all orbital-angular-momentum states of an atom having a single-valence electron (except the S state, $l = 0$). When this interaction is included in the Schrödinger equation, even in the absence of an external magnetic field two quantization rules hold for the atom's angular momentum. These rules are different from those found to hold when the z-components of $\mathbf{L}_l$ and $\mathbf{L}_s$ were separately quantized and characterized by the quantum numbers m_l and m_s. For the spin-orbit interaction, the theory now shows that the *total angular momentum,* symbolized by $\mathbf{L}_j$, of an atom with a single-valence electron is characterized by two new quantum numbers:

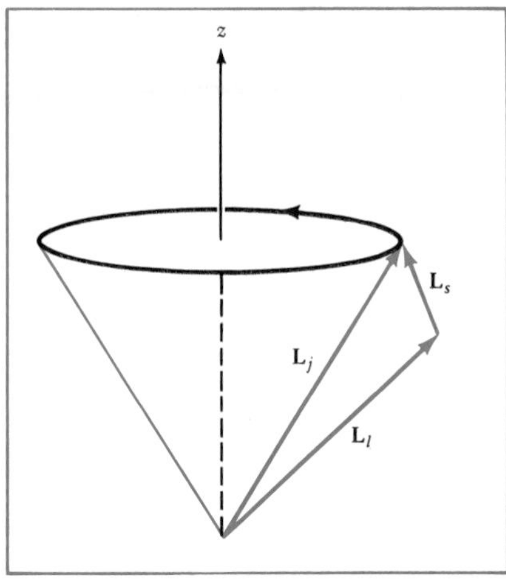

FIGURE 7–21. *Spin-orbit coupling.* $\mathbf{L}_l$ *and* $\mathbf{L}_s$ *vectors are first added to form* $\mathbf{L}_j$*; then* $\mathbf{L}_j$ *precesses about the z-axis.*

1. The *total angular-momentum quantum number j*, which determines the two possible values of the *magnitude* of the total angular momentum $\mathbf{L}_j$, given by

$$L_j = \sqrt{j(j+1)}\hbar \tag{7–21}$$

where the j quantum number can assume only two values, either $(l + s)$ or $(l - s)$; and l is given by Equation 7–1; and $s = \frac{1}{2}$. For an S state ($l = 0$), there is only one value $j = s = \frac{1}{2}$.

2. The *magnetic quantum number* m_j, which determines the possible values of the total angular momentum's z-component $L_{j,z}$, given by

$$L_{j,z} = m_j\hbar \tag{7–22}$$

where $m_j = j, j-1, \ldots, -j$.

The vector model for this spin-orbit coupling is illustrated in Figure 7–21. The vectors $\mathbf{L}_l$ and $\mathbf{L}_s$, of magnitudes $\sqrt{l(l+1)}\hbar$ and $\sqrt{s(s+1)}\hbar$, are first added to form the total angular momentum vector $\mathbf{L}_j$. The magnitude of $\mathbf{L}_j$, by Equation 7–21, has two possible values, one for $j = l + s$ and another for $j = l - s$. (For an S state, $l = 0$, and j has only one value, $j = s = \frac{1}{2}$.) Note that the values of j, l, and s must always be positive; otherwise, the magnitude of $\mathbf{L}_j$, $\mathbf{L}_l$, or $\mathbf{L}_s$ would be imaginary.

The quantization rules, Equations 7–21 and 7–22, are depicted on the vector diagram, Figure 7–21. One imagines that the resultant angular-momentum vector $\mathbf{L}_j$ precesses about the z-axis with the restrictions of Equations 7–21 and 7–22.

For spin-orbit coupling, then, the quantum numbers for a single-valence atom are

Quantum numbers for L-S coupling: (n, l, j, m_j)

Comparing the quantum description that includes spin-orbit coupling with the description that ignores spin-orbit coupling, we find different predictions in the quantized angular momentum quantities. With spin-orbit coupling, the magnitude of the total angular momentum $\mathbf{L}_j$ and its z-component $L_{j,z}$ are quantized; whereas with no coupling, the z-components of orbital angular momentum $\mathbf{L}_l$ and spin angular momentum $\mathbf{L}_s$ are separately quantized.

Since the atom's magnetic moment is related to its angular momentum, the two schemes have different predictions about the splitting of the atom's energy levels when a magnetic field is present. Experimental spectroscopic measurements confirm the spin-orbit coupling model for single-valence atoms.

In illustration of the spin-orbit coupling scheme, Table 7–3 shows the classification of quantum states for $n = 3$, according to the four quantum numbers (n, l, j, m_j).

TABLE 7–3 *The 18 quantum states $(3, l, j, m_j)$ for a single-valence atom with L-S coupling*

$l = 0$	$l = 1$	$l = 2$
$(3, 0, \frac{1}{2}, \pm\frac{1}{2})$	$(3, 1, \frac{3}{2}, \pm\frac{3}{2})$	$(3, 2, \frac{5}{2}, \pm\frac{5}{2})$
	$(3, 1, \frac{3}{2}, \pm\frac{1}{2})$	$(3, 2, \frac{5}{2}, \pm\frac{3}{2})$
	$(3, 1, \frac{1}{2}, \pm\frac{1}{2})$	$(3, 2, \frac{5}{2}, \pm\frac{1}{2})$
		$(3, 2, \frac{3}{2}, \pm\frac{3}{2})$
		$(3, 2, \frac{3}{2}, \pm\frac{1}{2})$

Each entry corresponds to two states (m_j positive and m_j negative). By Equation 7–21, the quantum number j has only one value for $l = 0$ and only two values for both $l = 1$ and $l = 2$. There are, then, two S states, six P states, and ten D states, a total of eighteen states for the principal quantum number $n = 3$.

We have an example of the splitting of the energy levels arising from internal spin-orbit interaction in sodium, $_{11}$Na, which has ten electrons within a closed shell and a single valence electron. All those electrons within the closed shell are so arranged that the total angular momentum and total magnetic moment of the closed shell are zero, and the lowest energy state for the valence electron is the 3S state. Except for the S states, there is an internal interaction between the valence electron's orbital angular momentum and its spin angular momentum; therefore the electron's total angular momentum magnitude $\sqrt{j(j+1)}\hbar$ will be quantized, with $j = (l + s)$ or $j = (l - s)$, according to Equation 7–21. This results in two different values for the total orbital magnetic moment μ_j, and therefore two different energies (see Equation 7–10) $\Delta E_m = -\mu_j \cdot \mathbf{B}$. When there is no external field, the spin-orbit interaction separates each l energy state (except an S state) into two separate energy levels. Quantum theory correctly yields the energy difference between the state with $j = l + \frac{1}{2}$ and the state with $j = l - \frac{1}{2}$. For a single-valence electron in an atom of atomic number Z, the energy difference is found to be

$$\Delta E_{l+1/2} - \Delta E_{l-1/2} = \frac{(7.24 \times 10^{-4}\ \text{eV})Z_{\text{eff}}^4}{n^3\, l\,(l+1)} \qquad (7\text{–}23)$$

where Z_{eff} is the effective positive charge of the nucleus and closed shell acting on the valence electron, and n and l are the principal and orbital quantum numbers. The effective charge Z_{eff} depends on the probability distribution of the valence electron within the closed shell. For a given n, the probability that the valence electron will be within the closed shell decreases with increasing l (see Figure 7–7); for a given l, this probability decreases for increasing n. Obviously, $1 \leq Z_{\text{eff}} \leq Z$.

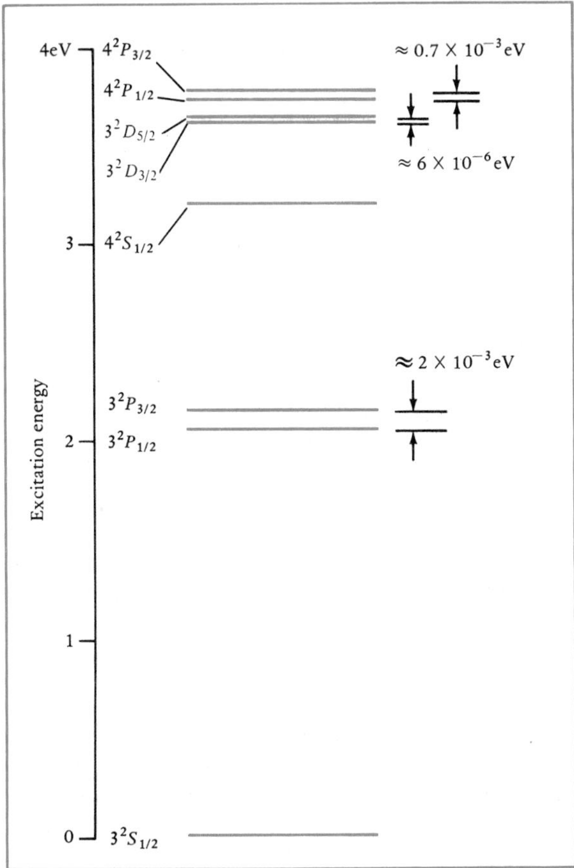

FIGURE 7–22. *Splitting of several states of sodium, due to spin-orbit interaction. (The splittings are grossly exaggerated.)*

Figure 7–22 shows the energy splitting of several states of sodium resulting from the spin-orbit interaction, where the states are identified by the common spectroscopic notation:

$$n^{2s+1}[\text{symbol for } l]_j \qquad (7\text{–}24)$$

For single-valence atoms, $s = \frac{1}{2}$ and $2s + 1 = 2$. As an example of spectroscopic notation, the P states in Table 7–3 are represented by $3^2P_{3/2}$ and $3^2P_{1/2}$ (read as "three doublet P three-halves" and "three doublet P one-half"). Similarly, for $n = 3$, we have the states $3^2S_{1/2}$, $3^2D_{5/2}$, and $3^2D_{3/2}$. This model agrees well with the emission spectral lines observed from ionized gaseous sodium vapor; for example, compare the sodium D emission lines in Figure 7–19 with the predicted lines in Figure 7–22.

In the absence of an external magnetic field, the m_j states are not resolved by the internal spin-orbit interaction and thus remain as degenerate energy states. For example, the four m_j states for $n = 3$, $l = 1$, and $j = \frac{3}{2}$ have the same energy (the $3^2P_{3/2}$ level in Figure 7–22) and the two m_j states for $n = 3$, $l = 1$, and $j = \frac{1}{2}$ have the same energy ($3^2P_{1/2}$ level in Figure 7–22). These degenerate m_j states can be resolved by applying an external magnetic field **B**, chosen to be in the $+z$-direction. The total angular momentum $\mathbf{L}_j$ of magnitude $\sqrt{j(j+1)}\hbar$ is then space-quantized along the z-axis; and following the procedure of Equation 7–3, the $2j + 1$ permitted components are

$$L_{j,z} = m_j\hbar \qquad (7\text{–}25)$$

where m_j is equal to $j, j - 1, \ldots, -j$. This is the basis of the *anomalous Zeeman effect,* which occurs whenever there is spin-orbit interaction and an external field (less than 100 T).

The selection rules for transitions between states are:

$$\Delta l = \pm 1 \qquad \Delta j = 0, \pm 1 \qquad \Delta m_j = 0, \pm 1 \qquad (7\text{–}26)$$

The computation of anomalous Zeeman splitting requires a knowledge of the magnetic moment μ_j associated with the *total* angular momentum $\mathbf{L}_j$, and computing μ_j is complicated (see Problem 7–33). Further, an analysis of the structure of atoms is increasingly complex as the number of valence electrons increases, because one must combine the spin and orbital angular momentum vectors of two or more electrons. Here we shall briefly describe the normal Zeeman effect in an atom of two valence electrons (for example, $_{12}$Mg).

The two electron-spin angular-momentum quantum numbers $s_1 = \frac{1}{2}$ and $s_2 = \frac{1}{2}$ usually combine to form the total spin quantum number S of the atom, where $S = s_1 + s_2 = 1$ (parallel spins) or $S = s_1 - s_2 = 0$ (antiparallel spins). Similarly, the two orbital-angular-momentum quantum numbers, l_1 and l_2, usually form a total orbital-angular-momentum quantum number L. The resultant spin and orbital quantum numbers S and L then combine to form a total angular-momentum quantum number J of the atom. Consider a state for which $S = 0$; such a state is a singlet state, inasmuch as $2S + 1 = 1$ (when $S = 1$ and $2S + 1 = 3$, the state is a triplet). Then $J = L$, and the total angular momentum is due only to orbital motion. In a magnetic field, the transitions between *singlet* states exhibit the *normal* Zeeman effect. In general, the *normal* Zeeman effect requires that the *total spin* angular momentum of the atom be *zero*, which is to say that the atom has an *even* number of electrons grouped in pairs of antialigned spins.

In a nonrelativistic wave-mechanical analysis of atomic structure the three quantum numbers n, l, and m_l emerge in a natural way by the fitting of three-dimensional waves, representing the electrons, into the region surrounding the nucleus. The electron spin, which has no classical analog (except for the fictitious model of a spinning sphere of charge), is *not* a consequence of nonrelativistic wave mechanics. The first relativistic wave-mechanical treatment incorporating the three space coordinates and the one time coordinate was successfully made by P. A. M. Dirac in 1928. In his relativistic quantum theory, the electron spin, having an angular momentum of $[s(s + 1)]^{1/2}\hbar$ and a magnetogyric ratio of $2(e/2m)$, emerges in a natural way with the three quantum numbers n, l, and m_l. Another consequence of the Dirac wave mechanics was the first prediction of the electron's antiparticle, the positron. The positron has a spin-angular-momentum quantum number s of $\frac{1}{2}$ and a magnetogyric ratio of $2(e/2m)$, just like the counterparts of the electron, but its spin-angular-momentum and magnetic-moment vectors point in the same direction, by virtue of the positive electric charge.

Example 7–2. Use the measured wavelengths of the sodium D lines in Figure 7–19 to find the effective charge of the inner ten electrons plus nucleus of sodium acting on the valence electron when it is in the 3P state.

The difference in energy ΔE between the two 3P states is, by Equation 7–23,

$$\Delta E = \frac{(7.24 \times 10^{-4}\ \text{eV})\, Z_{\text{eff}}^4}{n^3\, l\,(l+1)} = \frac{(7.24 \times 10^{-4}\ \text{eV})\, Z_{\text{eff}}^4}{3^3\,(1)\,(2)}$$

In Example 7–1, we found the energy difference between the two states to be $\Delta E = 2.13 \times 10^{-3}$ eV. Substituting this in the equation gives

$$Z_{\text{eff}} = \left[\frac{(27)(2)(2.13 \times 10^{-3}\ \text{eV})}{(7.24 \times 10^{-4}\ \text{eV})}\right]^{1/4} = 3.55$$

Since $Z_{\text{eff}} > 1$, the valence electron does penetrate the inner electron core, whose radius is approximately 0.3 Å.

7–8 The Stern-Gerlach Experiment

The space-quantization phenomenon, which limits the orientations of the angular-momentum and magnetic-moment vectors in a magnetic field, can be shown *directly* in an experiment that O. Stern and W. Gerlach first performed in 1921. They showed that atoms of silver are space-quantized in a magnetic field. The only contribution to the angular momentum of these atoms is the spin of a single electron.

To see the basis of the direct experimental confirmation of space quantization, let us consider how a magnetic dipole behaves in a magnetic field. A magnetic dipole moment μ in a uniform magnetic field $\mathbf{B}$ (one with uniformly spaced and parallel magnetic field lines) is subject to a torque.†

$$\tau = \mu \times \mathbf{B}$$

which tends to orient the magnet along the field lines. The dipole is not, however, subject to a *net* force tending to displace it as a whole. Consider now the behavior of a magnetic dipole in an *inhomogeneous* magnetic field, one showing a divergence of the magnetic field lines. Figure 7–23 illustrates that a *net* force pulls the magnet into the region of strong field when the magnetic dipole and the field are more nearly aligned than not, and that it pushes it out when they are more nearly antialigned.

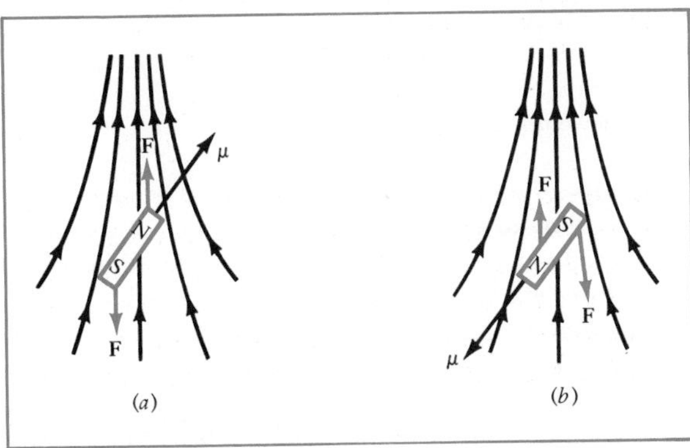

FIGURE 7–23. *Forces on a magnetic dipole in an inhomogeneous magnetic field for two orientations.*

† See Weidner and Sells, *Elementary Classical Physics,* 2d ed., Equation 29–20.

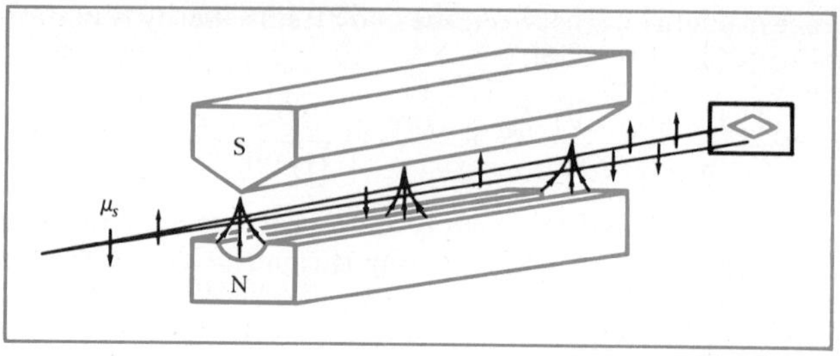

FIGURE 7–24. *Schematic representation of the Stern-Gerlach experiment.*

Let us compute the force on a magnetic moment in an inhomogeneous magnetic field. The magnetic potential energy of the electron spin is given by

$$\Delta E_s = m_s\left(2\frac{e\hbar}{2m}\right)B \tag{7–19}$$

Since a force is just the negative space derivative of the corresponding energy, the magnetic force acting on the dipole is given by†

$$F_z = -\frac{\partial}{\partial z}\Delta E_s = -m_s\frac{e\hbar}{m}\frac{\partial B}{\partial z}$$

where the z-direction is the symmetry direction of the inhomogeneous magnetic field of gradient $\partial B/\partial z$ as well as the direction of space quantization of the electron spin.

Consider now the arrangement of the original Stern-Gerlach experiment, shown in Figure 7–24. Silver atoms leave an oven at relatively high speeds, are collimated by slits, pass through an inhomogeneous magnetic field, and fall upon a photographic plate, where their final location is recorded. The electron spins are space-quantized in the magnetic field into such orientations that the component of the electron-spin angular momentum is $+\frac{1}{2}\hbar$ or $-\frac{1}{2}\hbar$ for an m_s of $+\frac{1}{2}$ or $-\frac{1}{2}$, respectively. (See Figure 7–18.) For $m_s = +\frac{1}{2}$, the atoms are deflected downward (Figure 7–23*b*); for $m_s = -\frac{1}{2}$, they are deflected upward (Figure 7–23*a*). On the photographic plate there is, then, *not* a continuous spread in the positions of the arriving atoms, as would be expected if space quantization had not occurred; rather, two distinct lines, corresponding to the silver atoms in the two allowed spin orientations, are observed. For an atom's total angular-momentum quantum number J, the number of lines appearing on the plate is $2J + 1$; thus atomic beams can be used for evaluating angular-momentum quantum numbers.

7-9 The Pauli Exclusion Principle and the Periodic Table

The state of an electron in an atom can be specified in the quantum theory by the values of each of the four quantum numbers n, l, m_l, and m_s. (The number s need not be indicated, for there is no other possible value but $\frac{1}{2}$.) By the procedures of quantum mechanics, it is possible to compute an atom's energy, its angular momentum, its magnetic moment, and its other measurable characteristics. Indeed, it is possible, at least *in principle,* to predict *all* properties of the chemical elements. In practice, such a program cannot easily be carried out, because formidable mathematical difficulties arise with sys-

† See Weidner and Sells, *Elementary Classical Physics,* 2d ed., Equation 10–7.

tems having many component particles. Only the problem of the simplest atom, hydrogen, has been solved completely by relativistic quantum theory. Work on this atom has shown that experiment and theory agree perfectly.

Even though solutions for the other atomic elements are not known exactly, the quantum theory does provide a wealth of information about their chemical and physical properties. One of its greatest achievements is furnishing a basis for understanding the ordering of the chemical elements as they appear in the periodic table. (The periodic table was first constructed merely by listing elements in the order of their atomic weights, and the remarkable periodicities in the properties of the elements were thereby revealed.) The key by which this ordering can be understood through the quantum theory is a principle W. Pauli proposed in 1924. The Pauli exclusion principle, together with the quantum theory, can be used to predict many of the known chemical and physical properties of atoms.

Consider again the energy levels available to the single electron in the hydrogen atom. Figure 7–25 shows schematically (although not to scale) the energy levels for the first three principal quantum numbers, $n = 1, 2, 3$. Each horizontal line corresponds to a particular possible set of values for the quantum numbers n, l, and m_l. For each line, there are two possible values of the electron-spin quantum number, $m_s = \pm\frac{1}{2}$. The occupancy of an available state by an electron is indicated by an arrow, whose direction indicates the electron-spin orientation, up for $m_s = +\frac{1}{2}$ and down for $m_s = -\frac{1}{2}$. For a given value of n, the S states are lowest, the P states next lowest, and so on. For a given value of the orbital-angular-momentum quantum number l, the possible values of the orbital magnetic quantum number m_l are shown horizontally arranged. Every one of the states (two for each dash) is available to the electron in the hydrogen atom. Some of the states are degenerate, having the same total energy; they are nevertheless distinguishable when a strong magnetic field or other external influence is applied to the atom.

Let us review the rules governing the possible values of the quantum numbers:

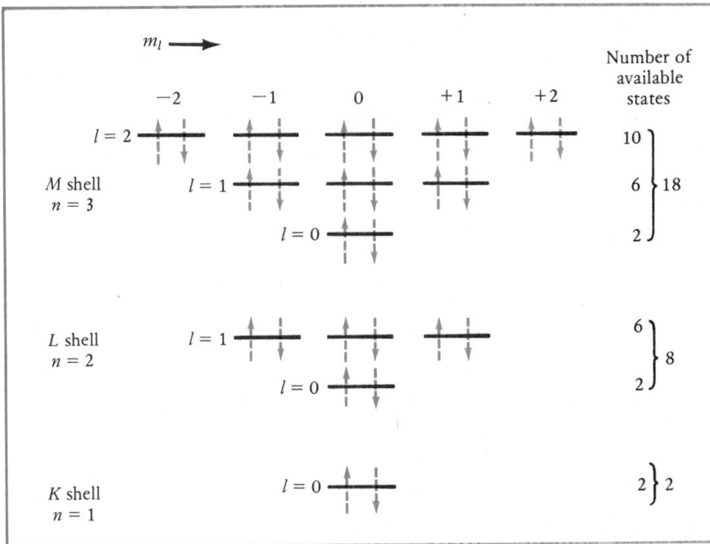

FIGURE 7–25. *Representation of the energy states available to the electron in the hydrogen atom (not to scale). There are two states for each horizontal line, corresponding to the two electron-spin orientations.*

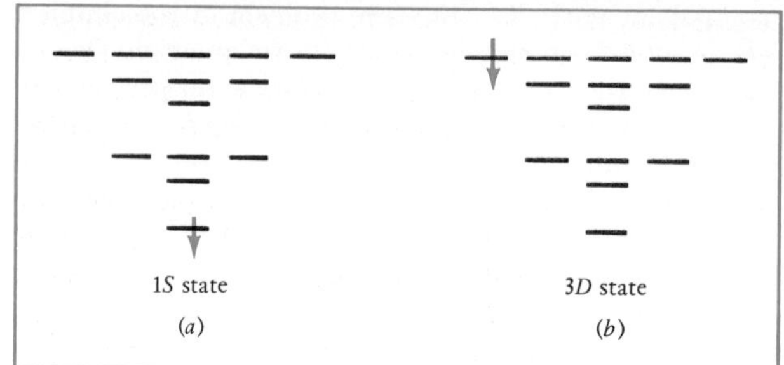

FIGURE 7–26. *Hydrogen atom in its ground state (1S) and in a 3D state.*

For a given n:	$l = 0, 1, 2, \ldots, n - 1$	(n possibilities)	(7–27)
For a given l:	$m_l = l, l - 1, \ldots, 0, \ldots, -l$	($2l + 1$ possibilities)	
For a given m_l:	$m_s = +\frac{1}{2}, -\frac{1}{2}$	(2 possibilities)	

When the hydrogen atom is in its lowest, or ground, state, the single electron is in the state in which $n = 1, l = 0, m_l = 0$, and $m_s = -\frac{1}{2}$. Excitation of the hydrogen atom may promote the electron to any of the higher-lying available states, from which the atom can then decay to the ground state by downward transitions with the emission of one or more photons. Figure 7–26 depicts a hydrogen atom in its ground states and in an excited 3D state.

Consider next the element helium, $_2$He. This atom has two electrons to be arranged among the levels shown in Figure 7–25. The separation between the levels is, however, not the same as in hydrogen, because the nuclear charge, $+2e$, is different, and also because there are three interacting particles instead of two. Nevertheless, the order of the states is the same. When the helium atom is in its ground state, both electrons are in the 1S state, with $n = 1, l = 0$, and $m_l = 0$. Then for each electron, two values of m_s are possible, $\frac{1}{2}$ and $-\frac{1}{2}$.

The possible values of m_s for each of the two electrons in helium are found in a study of its spectrum. When the various states of the helium atom are inferred from its spectrum, there is *no* 1^3S_1 state, although there is a 1^1S_0 state. A 1^3S_1 state would represent that situation in which both electron spins had the same m_s value, yielding the atomic total spin quantum number S = 1. But only the 1^1S_0 state is found in helium, a state in which the total spin quantum number S is zero and the two elements have *different* m_s values. Therefore, in the ground state the two electrons, which have identical values of n, l, and m_l, must have different spin magnetic quantum numbers $m_s = \frac{1}{2}$ and $m_s = -\frac{1}{2}$, respectively. We can interpret the non-occurrence of the 1^3S_1 state in helium in the following way—two electrons in a helium atom cannot have the same set of four quantum numbers; that is, the two electrons cannot exist in the same state.

Spectroscopic evidence from all elements is similar; it shows that atoms simply never occur in nature with two electrons occupying the same state. The Pauli exclusion principle formalizes this experimental fact:

No two electrons in an atom can have the same set of quantum numbers n, l, m_l, and m_s; or no two electrons in an atom can exist in the same state.

Exceptions to the exclusion principle, which applies also to other systems beside atoms and to other particles besides electrons, have never been found. The Pauli principle is analogous, but not equivalent, to the classical assertion that no two particles can be in the same place at the same time (the particles are regarded as impenetrable).

Thus, the two electrons in helium in the normal state occupy the two lowest available states indicated in Figure 7–25. No more electrons can be added to the $n = 1$, or K, shell; in helium the K shell is filled, or closed. Since the electron spins are oppositely aligned in the 1^1S_0 state, the helium atom has no magnetic moment and no angular momentum, either orbital or spin. Furthermore, the two electrons are tightly bound to the nucleus, and a lot of energy is required to excite one of them to a higher energy state. It is primarily for these reasons that helium is chemically inactive.

When the values of the quantum numbers of each and every electron in an atom are known, the *electron configuration* of the atom is given. A simple convention is used for specifying an electron configuration. When a helium atom is in its ground state, each of the two electrons has $n = 1$ and $l = 0$, and their configuration is represented by $1s^2$. The leading number specifies the n value, the lowercase letter s designates the orbital quantum number l of *individual* electrons, and the postsuperscript gives the number of electrons having the particular values of n and l. An energy-level diagram of neutral helium is shown in Figure 7–27. The energy levels are segregated according

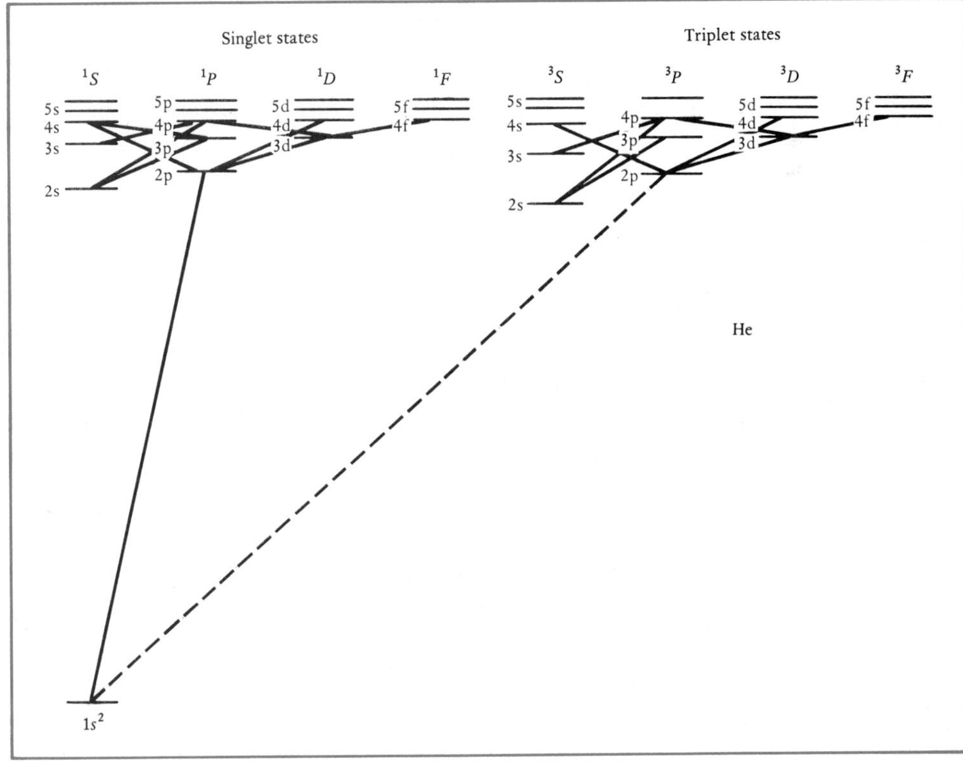

FIGURE 7–27. *Energy-level diagram of helium, showing some possible transitions. Note that the 1^3S_1 state does not exist.*

to whether they are singlets or triplets, that is, whether the two electron spins are antialigned or aligned, respectively. Note that the 1^3S state does *not* exist. When the atom is in an excited state, one of the electrons may remain a 1s electron in the K shell and the second electron occupy any of the higher excited levels. For example, the electron configuration of the first excited state in helium is $1s^1 2s^1$. Typically, transitions occur only between singlet states or only between triplet states.

The element with the next lowest atomic number, $Z = 3$, is lithium, $_3$Li. Of the three electrons in this atom, the first two occupy the two available $n = 1$ states. Therefore, as the exclusion principle requires, when lithium is in its ground state, the third electron goes to the lowest of the remaining available levels. The next-lowest available level after the K shell is $n = 2$ and $l = 0$. Then the ground-state configuration of the electrons in lithium is $1s^2 2s^1$, indicating two electrons in the closed K shell and one electron in the incomplete $l = 0$ subshell of the L shell; see Figure 7–28.

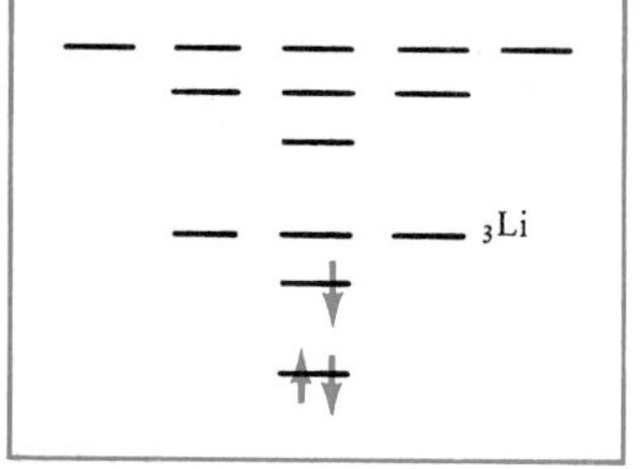

FIGURE 7–28. *Electron configuration of lithium in the ground state.*

Proceeding in this fashion—adding one electron as the nuclear charge or atomic number is increased by one unit, but always with the restriction that no *two* electrons within the atom can have the same set of quantum numbers—we can find the ground-state electron configurations of other atoms. We see from Figure 7–25 that two electrons can be accommodated in the *s* subshell of the L shell and six electrons in the p subshell, after which the L shell is completely occupied, holding its full quota of eight electrons. Table 7–4 gives the electron configurations of the elements from beryllium to sodium.

TABLE 7–4

ELEMENT	ELECTRON CONFIGURATION FOR THE GROUND STATE			
$_4$Be	$1s^2$	$2s^2$		
$_5$B	$1s^2$	$2s^2$	$2p^1$	
$_6$C	$1s^2$	$2s^2$	$2p^2$	
$_7$N	$1s^2$	$2s^2$	$2p^3$	
$_8$O	$1s^2$	$2s^2$	$2p^4$	
$_9$F	$1s^2$	$2s^2$	$2p^5$	
$_{10}$Ne	$1s^2$	$2s^2$	$2p^6$	
$_{11}$Na	$1s^2$	$2s^2$	$2p^6$	$3s^1$

We shall shortly note the chemical properties of several elements listed in Table 7–4 relative to their electron configurations and the occurrence of closed shells and subshells. First we note some of the properties of sodium that are directly related to its electron configuration, $1s^2 2s^2 2p^6 3s^1$. A single valence electron is outside the closed L shell, and the lowest state available to this valence electron is the 3s state. In the inner, closed subshells, the electrons' total angular momentum and magnetic moment is zero, since both their orbital angular momenta and spin angular momenta are paired off. These closed shells, $1s^2$, $2s^2$, and $2p^6$, are chemically inert and correspond to the electron configuration of the inert gas neon. The reason that sodium behaves approximately like a hydrogen atom is clear; a single valence electron moves about inner, inert, closed electron shells. The optical spectrum of sodium originates from the change in the state of the valence electron, while the ten electrons in the inner closed shells remain in their same states.

As the atomic number increases, there is a continuous filling of the sublevels in the expected order 1s, 2s, 2p, 3s, and 3p; the last element is argon $_{18}$Ar, whose complete electron configuration in the ground state is $1s^2 2s^2 2p^6 3s^2 3p^6$, or $3p^6$ for short. After the completion of the 3p subshell with argon, the succeeding elements, we might expect, should fill, in sequence, the ten available states of the 3d sublevel. Spectroscopic and chemical evidence indicates, however, that the 4s subshell is filled first, because its two electrons have a lower energy and are more tightly bound to the atom than 3d electrons. We can understand this apparent anomaly in the following way. The wave function of a 4s electron at the site of the nucleus is greater than the wave function of a 3d electron. Consequently, a 4s electron is more strongly attracted to the nucleus, and the atom's energy is lowered more by the addition of a 4s electron than by the addition of a 3d electron. For example, the electron configuration of the element potassium, $_{19}$K, is $1s^2 2s^2 2p^6 3s^2 3p^6 4s^1$. Note that the last electron is a 4s, not a 3d. Whereas in hydrogen the 3d state lies *below* the 4s state, in potassium, with 19 electrons, the reverse is true. According to experimental evidence, the general order in which the electron subshells are filled as the atomic number increases is as follows:

1s, 2s, 2p, 3s, 3p, 4s, 3d, 4p, 5s, 4d, 5p, 6s, 4f, 5d, 6p, 7s, 6d

One useful mnemonic for remembering this general ordering is to fill the states according to the diagonal scheme

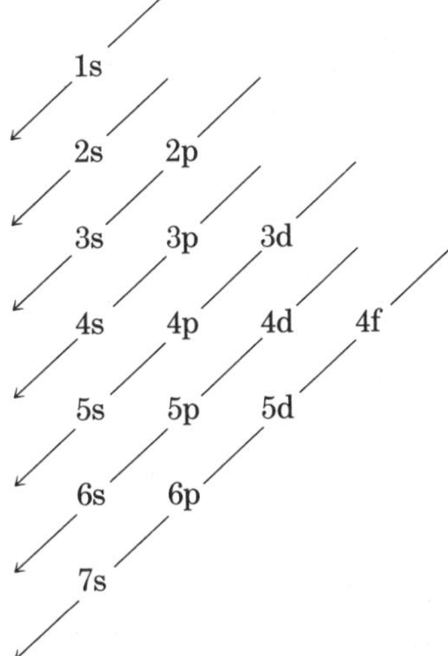

The periodic table of chemical elements is shown in Table 7–5. The electron configuration of the outer electrons in the free atom is given with the atomic number and chemical symbol of each element. The arrangement is such that any column typically contains elements with a common orbital state and also with the same number of electrons in this orbital state. For example, carbon, $_6$C, which has an outer electron configuration of $2p^2$, is found above silicon, $_{14}$Si, which has an outer electron configuration of $3p^2$; both have two electrons in an incomplete p subshell. The groups of elements corresponding to the filling of the 3d, 4d, 4f, and 5d subshells are listed

TABLE 7–5. *Periodic table of the chemical elements.*[a]

1s	1 H $1s^1$	2 He $1s^2$								
2s	3 Li $2s^1$	4 Be $2s^2$		5 B $2p^1$	6 C $2p^2$	7 N $2p^3$	8 O $2p^4$	9 F $2p^5$	10 Ne $2p^6$	2p
3s	11 Na $3s^1$	12 Mg $3s^2$		13 Al $3p^1$	14 Si $3p^2$	15 P $3p^3$	16 S $3p^4$	17 Cl $3p^5$	18 Ar $3p^6$	3p
4s	19 K $4s^1$	20 Ca $4s^2$	21–30 see (*a*)	31 Ga $4p^1$	32 Ge $4p^2$	33 As $4p^3$	34 Se $4p^4$	35 Br $4p^5$	36 Kr $4p^6$	4p
5s	37 Rb $5s^1$	38 Sr $5s^2$	39–48 see (*b*)	49 In $5p^1$	50 Sn $5p^2$	51 Sb $5p^3$	52 Te $5p^4$	53 I $5p^5$	54 Xe $5p^6$	5p
6s	55 Cs $6s^1$	56 Ba $6s^2$	57–80 see (*c*)	81 Tl $6p^1$	82 Pb $6p^2$	83 Bi $6p^3$	84 Po $6p^4$	85 At $6p^5$	86 Rn $6p^6$	6p
7s	87 Fr $7s^1$	88 Ra $7s^2$	89– see (*d*)							

(*a*) *Transition elements*: The ten 3d states follow the 4s states.

3d	21 Sc $3d^1$	22 Ti $3d^2$	23 V $3d^3$	24 Cr $4s^1\,3d^5$	25 Mn $3d^5$	26 Fe $3d^6$	27 Co $3d^7$	28 Ni $3d^8$	29 Cu $4s^1\,3d^{10}$	30 Zn $3d^{10}$

(*b*) 4d *Elements*: The ten 4d states follow the 5s states.

4d	39 Y $4d^1$	40 Zr $4d^2$	41 Nb $5s^1\,4d^4$	42 Mo $5s^1\,4d^5$	43 Tc $4d^5$	44 Ru $5s^1\,4d^7$	45 Rh $5s^1\,4d^8$	46 Pd $5s^0\,4d^{10}$	47 Ag $5s^1\,4d^{10}$	48 Cd $4d^{10}$

(*c*) *Rare earths and* 5d *elements*: The ten 5d states and fourteen 4f states follow the 6s states.

5d	57 La $5d^1$	72 Hf $5d^2$	73 Ta $5d^3$	74 W $5d^4$	75 Re $5d^5$	76 Os $5d^6$	77 Ir $5d^7$	78 Pt $6s^1\,5d^9$	79 Au $6s^1\,5d^{10}$	80 Hg $5d^{10}$				
4f	58 Ce $5d^4\,4f^1$	59 Pr $5d^0\,4f^3$	60 Nd $5d^0\,4f^4$	61 Pm $5d^0\,4f^5$	62 Sm $5d^0\,4f^6$	63 Eu $5d^0\,4f^7$	64 Gd $5d^1\,4f^7$	65 Tb $5d^1\,4f^8$	66 Dy $5d^0\,4f^{10}$	67 Ho $5d^0\,4f^{11}$	68 Er $5d^0\,4f^{12}$	69 Tm $5d^0\,4f^{13}$	70 Yb $5d^0\,4f^{14}$	71 Lu $5d^1\,4f^{14}$

(*d*) *Actinides and transuranic elements*: The 6d and 5f states follow the 7s states.

6d	89 Ac $6d^1$	90 Th $6d^2$	105 Ha $6d^3\,5f^{14}$	106	107									
5f	91 Pa $6d^1\,5f^2$	92 U $6d^1\,5f^3$	93 Np $6d^1\,5f^4$	94 Pu $6d^1\,5f^5$	95 Am $6d^1\,5f^6$	96 Cm $6d^1\,5f^7$	97 Bk $6d^1\,5f^8$	98 Cf $6d^0\,5f^{10}$	99 Es $6d^0\,5f^{11}$	100 Fm $6d^0\,5f^{12}$	101 Md $6d^0\,5f^{13}$	102 No $6d^0\,5f^{14}$	103 Lw $6d^1\,5f^{14}$	104 Ku $6d^2\,5f^{14}$

a. The ground-state configuration for the outermost electron shell is given, except when an inner shell is incomplete, in which case both shell configurations are given. For example, the element $_{42}$Mo has the complete electron configuration of $1s^22s^22p^63s^23p^63d^{10}4s^24p^65s^14d^5$.

separately. It is clear why the main body of the periodic table has a periodicity of 8—the total number of electrons completing an s subshell (two) and a p subshell (six) is eight.

The basis of the chemical properties of elements is the electron configurations of the atoms. Atoms with similar electron configurations show remarkably similar chemical behavior; the periodicity of chemical properties reflects the periodicity of the electron configurations. Let us consider some of these properties.

THE RARE GASES. The inert, or noble, rare gases are $_{2}He$, $_{10}Ne$, $_{18}Ar$, $_{36}Kr$, $_{54}Xe$, and $_{86}Rn$. We see from Table 7–5 that all these elements, except helium, have configurations in which the outermost electrons complete a p subshell. All rare gases have a ground state of $^{1}S_{0}$. The total angular momentum of the atom, from orbital motion and from electron spin, is zero; therefore, the total magnetic moment of the atom is zero. The atoms are chemically inert, or almost so, because there is no excess of electrons beyond a closed subshell and no deficiency of electrons in a subshell. Electrons within closed subshells are very strongly bound; thus, the ionization energies of elements in this group are particularly high. The atoms ordinarily fail to form chemical compounds, or nonmonatomic molecules. Furthermore, the rare gases have very low electrical conductivities and liquefaction (boiling) points.

THE ALKALI METALS. The alkali metals, in the first column of the periodic table, are $_{3}Li$, $_{11}Na$, $_{19}K$, $_{37}Rb$, $_{55}Cs$, and $_{87}Fr$. In every atom of an alkali metal, there is a single electron outside a closed rare-gas subshell; in the ground state this electron is in an s subshell. The chemical activity of the elements can be attributed to this single electron, which has a relatively low binding energy and which can be removed easily from the neutral atom leaving a singly charged, positive ion. Alkali metals, then, have a valence of +1. Clearly, they can be regarded as hydrogenlike; their spectra resemble the spectrum of hydrogen.

THE ALKALINE EARTHS. The alkaline earths, in the second column of the periodic table, are $_{4}Be$, $_{12}Mg$, $_{20}Ca$, $_{38}Sr$, $_{56}Ba$, and $_{88}Ra$. All have two s electrons outside a closed p subshell when in the normal state. The two electrons have a relatively small binding energy and are responsible for the valence +2. When the elements are singly ionized, they become hydrogenlike; their spectra are similar to the spectra of the alkali metals.

THE HALOGENS. The halogen group consists of the elements in the seventh column of the periodic table: $_{9}F$, $_{17}Cl$, $_{35}Br$, $_{53}I$, and $_{85}At$. The atoms lack one electron for completing a closed p subshell; therefore they have a valence of −1. The halogen elements are highly active chemically and form stable compounds when combined with elements of the alkali metal group; an example is the compound NaCl. When two atoms, one from each of these groups, are in close proximity, the halogen atom, lacking one electron of a closed subshell, wants an additional electron to complete its p subshell, and the alkali metal atom, having one electron outside a closed subshell, is ready to relinquish its last valence electron. When halogen and alkali metal elements unite to form compounds, each atom increases the stability of its electron configuration. Formation of the molecule by such combination of ions is called ionic binding (Section 8–1).

THE TRANSITION GROUP. The so-called transition group, in which the 3d subshell progressively is filled, contains $_{21}$Sc, $_{22}$Ti, $_{23}$V, $_{24}$Cr, $_{25}$Mn, $_{26}$Fe, $_{27}$Co, $_{28}$Ni, $_{29}$Cu, and $_{30}$Zn. The electrons in the incomplete 3d subshell are responsible for some important properties of these elements. Many of these substances are either paramagnetic (weakly magnetic) or ferromagnetic (strongly magnetic) as elements or in compounds. Their magnetism has its origin in the incomplete 3d subshell, whose total magnetic moment is *not* zero. On the other hand, their chemical activity is primarily a result of the outer, or 4s, electrons.

THE RARE EARTHS. The atoms of the 14 rare earths $_{58}$Ce to $_{71}$Lu have incomplete 4f subshells. The 4f electrons are well shielded from the 6s valence electrons. Thus, the chemical properties of the rare earths result primarily from the 6s electrons, and for this reason the elements are nearly chemically indistinguishable.

Example 7–3. Estimate the ionization energy of the single valence electron in lithium (*a*) assuming the $_3$Li atom to be hydrogenlike, with the outer electron always outside the inner two electrons; (*b*) assuming some overlap of the valence electron cloud distribution with the cloud distributions of the inner electrons. (*c*) Compare the ionization energies found in parts (*a*) and (*b*) with the experimental ionization energy of lithium, 5.390 eV.

(*a*) In the ground state of lithium, the two inner electrons are in 1s states and the valence electron in the 2s state. With no overlap of the cloud distributions of the $n = 2$ valence electron and the $n = 1$ inner electrons, the effective charge of each inner electron as "seen" by the valence electron is $-e$, and the total electric charge of the lithium nucleus ($+3e$) and inert core electrons ($-2e$) is $+3e - 2e = e$. Therefore, the ionization energy of the valence $n = 2$ electron is, by Equations 6–17 and 6–24,

$$E_{\mathrm{I}} = E_\infty - E_2 = \frac{k^2 m e^4}{n^2(2\hbar^2)}$$

$$= \frac{13.605\ \mathrm{eV}}{2^2}$$

$$= 3.401\ \mathrm{eV}$$

Because this assumption ignores the penetration of the inner core by the valence electron, with consequent larger binding energy of the outer electron, we should expect the predicted ionization energy, 3.4 eV, to be too small.

(*b*) For overlap of electron clouds, one must use the quantum theory to calculate the ionization energy of the outer electron. The results predicted by the theory can be roughly expressed by assuming that the effective charge of each electron in the adjacent closer shell is $-0.85e$ (instead of $-e$). For lithium, this implies that the effective charge of the two $n = 1$ electrons is $2(-0.85e) = 1.7e$. Therefore, the total electric charge of the lithium nucleus ($+3e$) and core electrons ($-1.7e$) is $Z_{eff} = +3e - 1.7e = 1.3e$. For hydrogenlike atoms, one replaces e^2 in Equation 6–24 with ($Z_{eff}e^2$), and

$$E_I = \frac{Z_{eff}^2 k^2 m e^4}{n^2(2\hbar^2)}$$

$$= \frac{(1.3)^2(13.605 \text{ eV})}{2^2}$$

$$= 5.7 \text{ eV}$$

(*c*) Ignoring overlap of the outer electron cloud with the inner-core electron distributions results in a low estimate (3.4 eV); approximating the overlap effect by assuming 85 percent shielding by each electron in the adjacent closer shell gives a reasonable estimate (5.75 eV) of the experimental value (5.39 eV).

7–10 Characteristic X-Ray Spectra

When atoms are bombarded by electrons having only a few electron volts of kinetic energy, the resulting excitation or ionization involves a change in the state of one or more of the weakly bound, outer electrons. The optical emission spectrum is induced by such collisions, while the more tightly bound electrons in the inner closed shells remain in their initial states. An inner electron, however, also can be excited or removed, provided that enough energy is added to the atom. When an inner electron is displaced, the *characteristic x-ray spectrum* results.

Consider the schematic diagram, Figure 7–29, which indicates the electron configuration and energy levels of copper, $_{29}$Cu, in the ground state.

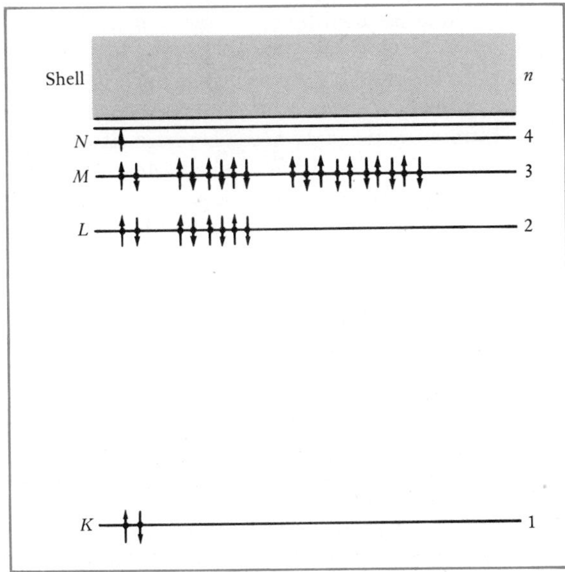

FIGURE 7–29. *Electron configuration and energy levels of $_{29}$Cu in the ground state.*

Fine-structure splittings are ignored. The K, L, and M shells are completely filled with their respective quotas of electrons (2, 8, and 18), specified by the exclusion principle. There is a single electron, the valence electron, in the N shell. Near the top of the diagram are the unfilled shells, or "optical" levels; transitions of the outer electron to these states give rise to the optical spectrum. The difference in energy between any of the unfilled levels and the onset of the continuum, $E = 0$, is very small compared with the difference between the K shell and $E = 0$.

Suppose that a very energetic electron strikes the atom, thereby removing one of the two electrons in the K shell to an available higher-energy level. The electron cannot be accepted in the filled L and M shells; it must, then, either go to one of the unfilled energy states or go out of the atom completely, thereby ionizing the atom. In any event, the removal of the electron from the K shell has increased the energy of the atom drastically (by well over 1000 eV), and a vacancy exists in the K shell. The "hole" thus created can be filled by a transition of one of the eight electrons from the L shell to the K shell, where it will then be more tightly bound. This transition, which reduces the energy of the atom by more than 1000 eV, gives rise to the emission of the so-called K_α x-ray line. Because the difference in energy is typically of the order of thousands of electron volts, the photon created in such a transition is an x-ray photon with a wavelength between 0.1 and 10 Å.

Still other transitions can occur. The vacancy in the K shell can be filled through a somewhat less probable transition, in which an electron in the M shell jumps inward to the K shell; the corresponding emitted photon is labeled K_β. The transitions from successively higher energy states, all *terminating* at the K shell, are identified in the x-ray notation by K_α, K_β, K_γ, K_δ, . . . , as shown in Figure 7–30. This group of K x-ray lines composes the K *series*.

Other series of x-ray lines having smaller energies, or longer wavelengths, also occur. When an L electron jumps to the K shell to fill the vacancy created by the removal of a K electron, a vacancy is created in the L shell. This hole can be filled by electrons that make transitions from still higher states, giving rise to the L series, L_α, L_β, L_γ, L_δ, . . . ; the symbols indicate transitions terminating in the L shell and originating in the respective higher-energy states.

The minimum energy necessary to excite a K electron is close to the ionization energy *for the K shell.* The ionization energy for the K shell, E_K, is the energy required to remove a K electron completely out of the atom so that the freed electron and the atom remain at rest; the atom is then ionized. The energy E_K is only *slightly* higher than the energy needed to bring a K electron to *any* of the unoccupied optical levels.

The energy required to produce x-ray ionization from the L shell is called E_L. Because E_K is ordinarily much greater than E_L, the x-ray lines in the K series have appreciably shorter wavelengths than those in the L series (for Z greater than 30, lines in the K series have wavelengths of less than 1 Å, and lines in the L series have wavelengths of less than 10 Å). Ordinarily the lines originating in still higher transitions, in the M, N, . . . series, are still weaker and of longer wavelength.

The x-ray ionization energies E_K, E_L, . . . are related in a simple way to the frequencies of the x-ray emission lines. Consider the K_α line. If an energy E_K removes a K electron, the atom then has an energy increased by the

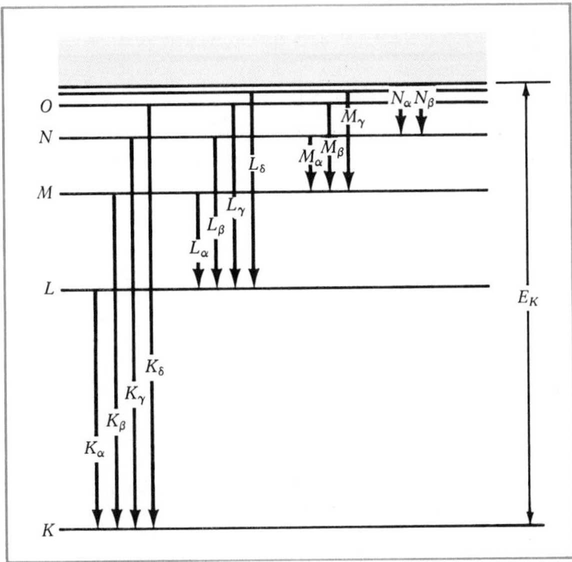

FIGURE 7–30. *Inner-electron transitions giving rise to characteristic x-ray lines.*

amount E_K; similarly, if an electron is removed from the L shell, the atom's energy is increased by E_L. These two conditions correspond, respectively, to the situations before and after the emission of the K_α line. The energy of an emitted K_α photon, $h\nu$, is simply $E_K - E_L$. Therefore,

K series:
$$\begin{aligned} h\nu_{K\alpha} &= E_K - E_L \\ h\nu_{K\beta} &= E_K - E_M \\ h\nu_{K\gamma} &= E_K - E_N \\ &\cdots\cdots \end{aligned} \qquad (7\text{–}28)$$

L series:
$$\begin{aligned} h\nu_{L\alpha} &= E_L - E_M \\ h\nu_{L\beta} &= E_L - E_N \\ h\nu_{L\gamma} &= E_L - E_O \\ &\cdots\cdots \end{aligned} \qquad (7\text{–}29)$$

We can now interpret the characteristic x-ray lines that appear when the target in an x-ray tube is struck by electrons having energies of the order of thousands of electron volts. Figure 7–31 shows the measured intensity of emitted x-radiation from a molybdenum target. There is a smoothly varying, continuous x-ray spectrum arising from bremsstrahlung collisions of electrons with the target; the continuous spectrum has a well-defined maximum frequency, determined solely by the accelerating potential of the x-ray tube (see Section 4–3). Superimposed on the x-ray continuum are sharp peaks in the intensity, characteristic of the x-ray spectrum of molybdenum. Those peaks are identified as the K_α and K_β lines; the lines of the L series, having much longer wavelengths and lower intensity, do not appear.

An x-ray spectrum is typically broken down into its component wavelengths with an x-ray crystal spectrometer. A single crystal having a known crystalline structure can be used to separate the various wavelengths, and the wavelength can be computed from the Bragg law. The intensity of the x-radiation can be measured with a chamber that is sensitive to the ionization effects produced by x-rays passing through a gas.

The frequency or wavelength of the K_α line can be calculated approxi-

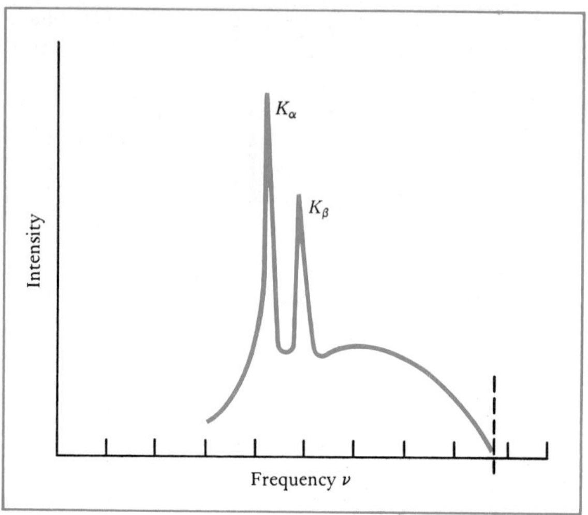

FIGURE 7–31. *X-ray spectrum of molybdenum.*

mately by a simple theoretical analysis involving the Bohr atomic theory. The wavelength λ of lines emitted by one-electron, or hydrogenic, atoms, is given by the Rydberg formula,

$$\frac{1}{\lambda} = RZ^2\left(\frac{1}{n_l{}^2} - \frac{1}{n_u{}^2}\right) \qquad (6\text{–}15), (7\text{–}30)$$

where R is the Rydberg constant,1.097×10^{-3} Å^{-1}, n_u and n_l are the principal quantum numbers of the upper and lower states of the transition, and Z is the atomic number of the one-electron atom. When a K electron, $n_l = 1$, has been removed from the atom of a heavy element, an electron in the L shell, $n_l = 2$, will "see" the nuclear electric charge Ze shielded by the charge $-e$ of the one remaining K electron. Electrons in the M, N, and higher states do not penetrate appreciably the region between the K and L shells. Therefore, an L electron will approximate closely the single electron in a hydrogenic atom, moving in the electric field of the nucleus plus the K electron, which has an *effective* atomic number, $Z - 1$. Equation 7–30 then yields, for the frequency $\nu_{K\alpha}$ of the K_α line emitted when an L electron jumps to the hole in the K shell,

$$\nu_{K\alpha} = \frac{c}{\lambda} = cR(Z-1)^2\left(\frac{1}{1^2} - \frac{1}{2^2}\right)$$

$$\nu_{K\alpha} = \frac{3cR}{4}(Z-1)^2 \qquad (7\text{–}31)$$

A plot of $\nu_{K\alpha}{}^{1/2}$ versus the atomic number Z of the emitting x-ray elements should therefore yield a straight line. H. G. J. Moseley in 1913 made the first comprehensive study of the characteristic x-ray frequencies. Moseley found that the relation shown in Equation 7–31 represented the data on the K lines very well. His measurements were the first to establish clearly the values of the atomic numbers of the elements. If the chemical elements in the periodic table are listed in the order of their atomic *weights,* the resulting list is (with a few notable exceptions) identical with a listing by atomic number. One such exception is the ordering of the elements cobalt and nickel. The work of Moseley established that although $_{27}Co$ has a greater atomic weight than $_{28}Ni$ (58.93 as against 58.71), its atomic number is smaller.

Our study of atomic structure has so far been restricted to free atoms of a gas and not to atoms strongly interacting with one another, as in molecules, liquids, or solids. Why is it proper in the theory of x-ray emission to consider the atoms essentially free when actually x-ray target materials are solids? The answer is that x-ray transitions involve the innermost, tightly bound electrons, not the outer electrons; and the latter have their configurations and energies changed when atoms are brought close together in a solid, while the former are hardly influenced by the state of the material, whether solid, liquid, or gas.

SUMMARY

For a particle subject to an inverse-square force of attraction, the classical constants of the motion (having a continuum of possible values) are quantized in the quantum theory of the atom.

The possible values of the principal quantum numbers are

$$\begin{array}{l} n = 1,\ 2,\ 3,\ 4, \ldots \\ \quad\ \ \mathrm{K, L, M, N}, \ldots \end{array}$$

For a given n:

$$\begin{array}{ll} l = 0, 1, 2, 3, 4, \ldots, n-1 & (n \text{ possible values}) \\ \quad\ s, p, d, f, g, \ldots & \end{array}$$

For a given l:

$$m_l = l, l-1, \ldots, -(l-1), -l \qquad (2l+1 \text{ possible values})$$

For a given m_l:

$$m_s = +\tfrac{1}{2}, -\tfrac{1}{2} \qquad (2 \text{ possible values})$$

When the valence electron of a hydrogenlike atom is in a state for which l is small, the energy of the atom is lowered relative to the corresponding hydrogen energy level. The selection rule giving the allowed transitions is $\Delta l = \pm 1$.

The ratio of the magnetic moment to the angular momentum, called the magnetogyric ratio, is:

$$\frac{\mu_l}{L} = \frac{e}{2m} \qquad \text{for orbital motion}$$

$$\frac{\mu_s}{L_s} = 2\frac{e}{2m} \qquad \text{for electron spin}$$

The change in the magnetic energy of a state for a magnetic moment in a magnetic field B is:

$$\Delta E_m = m_l \frac{e\hbar}{2m} B = m_l \beta B \qquad \text{for an orbital magnetic moment}$$

$$\Delta E_s = m_s \frac{2e\hbar}{2m} B = m_s(2\beta)B \qquad \text{for a spin magnetic moment}$$

$$\beta \text{ (the Bohr magneton)} = \frac{e\hbar}{2m} = 0.9273 \times 10^{-23} \text{ J/T}$$

The normal Zeeman effect arises from the interaction of the *orbital* magnetic moment with a magnetic field. The magnetic field causes each line in the spectrum to split into three equally spaced lines. The selection rule for allowed transitions is $\Delta m_l = 0, \pm 1$.

TABLE 7–6

CONSTANT OF THE MOTION	QUANTUM NUMBER	ALLOWED VALUES
Energy	n (principal quantum number)	$E_n = -Z^2E_I/n^2$ (for a single electron, ignoring spin-orbit interaction)
Magnitude of the orbital angular momentum	l (orbital-angular-momentum quantum number)	$L = \sqrt{l(l+1)}\hbar$
Component of the orbital angular momentum along z	m_l (orbital magnetic quantum number)	$L_z = m_l\hbar$ Space quantization: $\cos\theta = \dfrac{m_l}{\sqrt{l(l+1)}}$
Magnitude of the spin angular momentum	s (electron-spin quantum number)	$L_s = \sqrt{s(s+1)}\hbar = \frac{1}{2}\sqrt{3}\hbar$
Component of the spin angular momentum along z	m_s (spin magnetic quantum number)	$L_{s,z} = m_s\hbar$ Space quantization: $\cos\theta = m_s/\sqrt{s(s+1)}$

The Stern-Gerlach experiment, in which atoms with a net electron spin pass through an inhomogeneous magnetic field, shows directly the phenomenon of space quantization.

The fine structure of spectral lines has its origin in the interaction between L and L_s, spin-orbit interaction.

The spectroscopic nomenclature for atoms having one or more electrons is:

Total orbital-angular-momentum quantum number	$L = 0, 1, 2, 3, 4, \ldots$ S, P, D, F, G, . . .
Total spin-angular-quantum number	$S = \frac{1}{2}$ (for one electron) $S = 0$ or 1 (for two electrons) $J = j = l + s$, or $l - s$

The Pauli exclusion principle, the basis for understanding the periodic table of chemical elements, specifies that no two electrons in the same atom can have the same set of the four quantum numbers n, l, m_l, m_s.

PROBLEMS

Classical constants of motion, §7–1

7–1. A classical bound electron in a hydrogen atom would move in elliptical orbits whose semimajor axis a is 0.53 Å when the atom is in the ground state (Figure 7–1). According to this classical model, what is the distance from the proton at which such an electron, in an orbit of zero angular momentum, will have the same speed as an electron in a circular orbit of radius 0.53 Å?

7–2. The earth is bound to the sun by the gravitational force and has orbital angular momentum (annual rotation) about the sun. In addition, the earth has intrinsic spin angular momentum (daily rotation) and the sun has intrinsic spin angular momentum. Identify all quantities that stay constant: (a) assuming no interactions between spin-orbit and spin-spin angular momenta; (b) assuming an interaction between the earth's spin and its orbital angular momentum; (c) assuming interactions among all the angular-momentum terms.

Quantization of L, §7–2

● **7–3.** Identify in Figure 7–8 the transitions that contribute to the H_β spectral line (4863 Å) in hydrogen.

● **7–4.** For which state, the 2S or the 2P, does the electron in the hydrogen atom have the greater probability of being within a distance $5r_1$ from the proton? (See Figure 7–7.)

7–5. A diatomic molecule such as hydrogen consists of two identical point nuclei separated by an equilibrium distance d and two bound electron clouds. (a) Find the moment of inertia of the molecule about an axis through the center of mass and perpendicular to the line joining the two nuclei. (b) The orbital angular momentum of such a molecule follows the same quantum rules as those for the orbital angular momentum of an atomic electron. Using these rules, find the allowed values of the orbital angular momentum about the axis of part (a) for the lowest four rotational energy states of the molecule. (c) The selection rule for transitions between energy states is $\Delta l = \pm 2$. What photon transition frequencies are possible among these four rotational states?

Hydrogenlike atoms, §7–3

● **7–6.** (a) For any *given* element (except hydrogen), the sharp and diffuse series have a common short-wavelength series limit but the principal series limit is different (see Figure 7–9). Explain why this is so. (b) What is true for these three series limits for hydrogen?

● **7–7.** What is the change in *magnitude* of the orbital angular momentum when an atom makes a transition (a) from a P state to an S state? (b) from a D state to a P state?

● **7–8.** What ion of potassium, $_{19}$K, is isoelectronic to the sodium atom, $_{11}$Na?

● **7–9.** In general, an atom initially in an excited state remains in that state for such a short time (about 10^{-8} s on the average) that the probability of its absorbing a photon and jumping to a still higher state is very small. Explain on this basis why only the P series (*principal* series) is ordinarily observed in the *absorption* spectrum of potassium.

7–10. Assume that a thousand sodium atoms are initially in the 5S state. List all possible downward transitions that can occur (see Figure 7–9) and indicate which of these transitions result in spectral lines in the visible region (3800–7600 Å).

7–11. Calculate the ionization energy of the $_{19}$K atom, assuming that each of the electrons in the $n = 3$ shell has an effective charge of $-0.85e$, and each of those in the $n = 1$ and $n = 2$ shell has an effective charge of $-e$. The experimental value of E_I for potassium is 4.34 eV.

7–12. Calculate the total energy (eV) necessary to remove all three electrons from the $_3$Li atom, and compare this with the experimental value of 202 eV. (See Example 7–3 and assume that the effective charge of one inner electron on the other inner electron is $-0.35e$.)

7–13. (a) Show that for large l, the magnitude of the orbital angular momentum can be approximated by $l(1 + 1/2l)\hbar$. (b) How large must l be for the fractional difference between the magnitude of the orbital angular momentum and its maximum component to be less than 10^{-6}?

7–14. Transitions of the valence electrons in the alkali metal rubidium ($Z = 37$) between highly excited states (up to $n = 85$) to the ground state (electron configuration, $5s^1$) have been observed. For example, the transition frequency between the ground state ($n = 5$) and the excited state $n = 85$ is found to be $1.009\,55 \times 10^{15}$ Hz, slightly less than the ionization frequency, $1.010\,03 \times 10^{15}$ Hz. Assume that the rubidium atom is hydrogenlike and use Equation 6–15 to find the frequency difference between the $n = 85$ state and the $n = \infty$ (ionization) state and compare this with the observed frequency difference.

Quantization of L_z, §7–4

7–15. A particle with a mass of 10^{-6} kg moves in circle of radius 10^{-2} m at a speed of 10^{-6} m/s. (a) What is the orbital-angular-momentum quantum number l of this particle? (b) What is the maximum angular difference between the allowed orientations of the orbital-angular-momentum vector?

7–16. A point particle of mass m orbits in a circle of radius r and moves at speed v. Assume that this particle's orbital-angular-momentum quantum number is $l = 10$. (a) What is the magnitude of the orbital angular momentum $\mathbf{L}_l$? (b) Find what the orbital quantum number would be for a second particle of mass $2m$ moving at the same speed v and orbiting in a circle with the radius closest to r.

Normal Zeeman effect, §7–5

● **7–17.** For the principal quantum number $n = 2$, what are the allowed values of (*a*) the magnitude of the orbital angular momentum, and (*b*) the component of the orbital angular momentum along any direction?

7–18. For the 5F state of helium, what are the possible values of (*a*) the magnitude of the orbital angular momentum, and (*b*) the component of the orbital angular momentum along any direction?

7–19. One of the elements displaying the normal Zeeman splitting is calcium, $_{20}Ca$, which has no net spin angular momentum. The ground state of the two valence electrons in $_{20}Ca$ is the 4S state. Transitions from the first excited 4P state to the 4S state result in a prominent spectral line of wavelength 4226.73 Å. Show that if the calcium atoms are placed in a uniform magnetic field of 2.0 T, this line will split into three closely spaced spectral lines according to the normal Zeeman effect; determine the wavelengths of these Zeeman lines.

7–20. The separation between adjacent normal Zeeman components of the radiation emitted by calcium at 4227 Å is 0.1251 Å when the atoms are in a magnetic field of 1.500 T. Find the value of e/m from these data.

7–21. A classical gyromagnet, an object with angular momentum **L** and magnetic moment $\boldsymbol{\mu}$, is immersed in an external field **B**, as shown in Figure 7–32. One can imagine the gyromagnet to consist of rotating charged particles with a magnetogyric ratio of $\mu/L = e/2m$. The gyromagnet is subject to a magnetic torque $\tau = \boldsymbol{\mu} \times \mathbf{B}$ at right angles to its angular momentum in the same fashion that a spinning top is subject to a gravitational torque at right angles to its angular momentum. (*a*) Show that the gyromagnet precesses about the direction of **B** at the angular frequency $\omega = (\mu/L)B$, the so-called Larmor precession frequency. (*b*) Show that the precession frequency of a classical gyromagnet about an external field is the same as the difference in angular frequency between magnetic sublevels in the normal Zeeman effect.

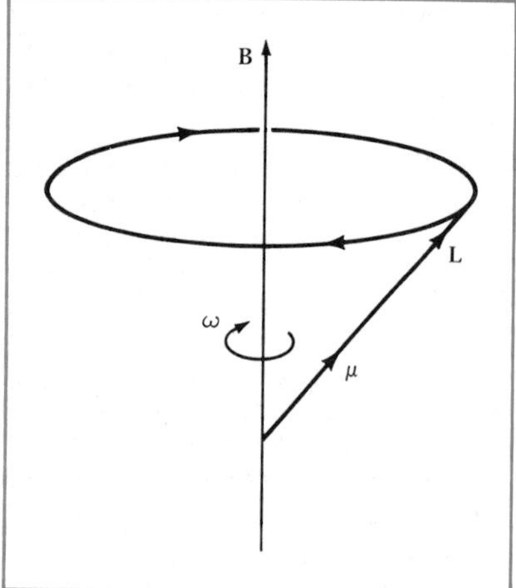

FIGURE 7–32. *Classical gyromagnet precessing about a magnetic field.*

Electron spin, §7–6

● **7–22.** A collection of electron spins is immersed in an external magnetic field **B**. Show that if there are more electrons occupying the lower energy state than the higher one, the collection of electrons is paramagnetic, that is, that more magnetic dipoles are aligned with **B** than against **B**.

● **7–23.** Show that the component of the electron's spin magnetic moment along the direction of the magnetic field is $\pm\beta$, where β is the Bohr magneton.

7–24. One classical model of the electron assumes that the electron is a sphere of radius 2.8×10^{-15}m with its mass and charge uniformly distributed throughout its volume. Assuming this model (which is fictitious), determine (*a*) at what angular speed the electron would be spinning to have spin angular momentum of $\frac{1}{2}\sqrt{3}\hbar$; (*b*) what the tangential speed of a point on the sphere at the "equator" would be.

7–25. The spin angular momentum and spin magnetic moment of the *nucleus* of the atom can give rise to a hyperfine splitting of the emitted spectral lines. Assuming $B \approx 10$ T and taking a typical nuclear-spin magnetic moment to be approximately $10^{-3}\beta$, where β is the Bohr magneton of the electron, (*a*) show that the difference between energy levels, due to nuclear spin, is of the order of 10^{-6} eV, and (*b*) show that the corresponding difference in wavelength of visible light is $\Delta\lambda \approx 10^{-3}$ Å.

■ **7–26.** In the normal Zeeman effect, the emitted light displays characteristic polarization properties. (*a*) Show that by classical theory, light propagated in the direction perpendicular to the external **B** field is linearly polarized as follows: (1) The component of unchanged frequency is linearly polarized in the direction parallel to **B**; (2) the two frequency-shifted components are linearly polarized in the direction perpendicular to **B**. (*b*) Show that all light propagated parallel to **B** is shifted in frequency and that it is circularly polarized.

7–27 (*a*) Show that the difference in energy between the two allowed electron-spin orientations in a magnetic field **B** is given by 2βB. (*b*) What is the frequency of radiation that can induce transitions (spin flips) between these two states when $B = 0.50$ T. This effect, in which photons flip electron spins, and for which $h\nu = 2\beta$B, is called *electron-spin resonance.*

7–28. An "antiatom" is one in which *all* the elementary particles are replaced with their corresponding antiparticles. Show that the lower energy state of a spin-orbit doublet is still $j = l - s$. (An "antiuniverse" cannot be distinguished from the universe simply by examining the electromagnetic radiation from it and inferring the energy levels of atoms.)

"Single-electron" atoms, §7–7

7–29. Show with a vector diagram that the vector sum of the orbital magnetic moment μ_l and spin magnetic moment μ_s does *not* lie along the same line as the sum of the orbital angular momentum $\mathbf{L}_l$ and spin angular momentum $\mathbf{L}_s$.

7–30. (*a*) Identify the two sodium *D* lines at the bottom of Figure 7–19 with the two atomic transitions shown on the energy-level diagram. (*b*) Why is the one spectral line more intense than the other?

7–31. For sodium, use Equation 7–23 to estimate the energy difference between the states $j = l + \frac{1}{2}$ and $j = l - \frac{1}{2}$ for (*a*) the 3P states and (*b*) the 3D states. (*c*) Compare the values obtained in parts (*a*) and (*b*) with those shown in Figure 7–22.

7–32. For spin-orbit coupling (see Figure 7–21), use the law of cosines to show that the angle between the orbital and total angular momenta is given by

$$\cos(l, j) = \frac{j(j+1) + l(l+1) - s(s+1)}{2\sqrt{j(j+1)}\sqrt{l(l+1)}}$$

■ **7–33.** (*a*) Knowing that the orbital magnetic moment is $\mu_l = (e\hbar/2m)[l(l+1)]^{1/2}$ and the spin magnetic moment is $\mu_s = 2(e\hbar/2m)[s(s+1)]^{1/2}$, show that the *component* of the total magnetic moment μ_j along the direction of the total angular momentum $\mathbf{L}_j$ is given by

$$\mu_j = \sqrt{j(j+1)}\,\frac{e\hbar}{2m}\left[1 + \frac{j(j+1) + s(s+1) - l(l+1)}{2j(j+1)}\right]$$

See Problem 7–32. (*b*) The quantity in the brackets is called the Landé *g* factor. Show that the Landé *g* factor gives the magnetic moment, in units of the Bohr magneton, divided by the total angular momentum, in units of $\hbar$.

Stern-Gerlach experiment, §7–8

7–34. If a beam of zinc atoms passes through the inhomogeneous magnetic field in the Stern-Gerlach experiment, how many lines will be found on the photographic plate?

■ **7–35.** Sodium atoms are normally in the $^2S_{1/2}$ state. In a Stern-Gerlach experiment, the sodium atoms emerge from an oven at a temperature of 1000°C and pass through an inhomogeneous magnetic field whose gradient is 50 T/cm for a distance of 2.0 cm. The atoms then continue through a field-free space for 10.0 cm before being deposited on a collector plate. (*a*) How many lines will be found on the collector plate? (*b*) What is the maximum distance between adjacent lines?

Pauli exclusion principle, §7–9

● **7–36.** What would be the atomic number of the next alkali metal after $_{87}$Fr?

● **7–37.** The ground-state electronic configuration of a divalent atom is 3P. What is the element?

7–38. Find the total angular momentum of (*a*) $_{31}$Ga and (*b*) $_{55}$Cs in their ground states.

7–39. An atom of $_3$Li is in its ground state. Assume no spin-orbit coupling. (*a*) What are the four quantum numbers for each of the three electrons? (*b*) What are the quantum numbers of the third electron for the two lowest excited states of this atom?

7–40. (*a*) Show, in the fashion of Figure 7–28, the occupied states of $_{13}$Al in the ground state. (*b*) Assuming *L-S* coupling, write the ground state or states in spectroscopic notation.

7–41. Which of the following elements can show a normal Zeeman effect: $_{10}$Ne, $_{21}$Sc, $_{38}$Sr, $_{80}$Hg?

X-ray spectra, §7–10

7–42. What is the approximate energy (keV) and wavelength (Å) of the K_α photon emitted (a) from $_{92}$U, and (b) from $_{13}$Al?

7–43. X-ray fluorescence is that process in which the absorption of a photon of relatively high energy in a material results in the emission of a number of x-ray photons of lower energy. Show that the complete x-ray fluorescence of a material can be produced only if the material is irradiated with characteristic x-rays produced by a target of *higher* atomic number.

7–44. One technique used for identifying trace elements in a sample such as blood serum is the following. A beam of protons of a few MeV kinetic energy is projected onto the sample. Some protons knock out inner tightly bound electrons in an atom and x-rays are emitted when electrons from high energy levels make transitions to these lower-energy vacancies. Estimate, as an example, the energies of the K_α peaks for copper and zinc, and compare them with the experimental values of 8.041 keV for copper and 8.631 keV for zinc. (For a description of trace-element analysis of blood serum, see the article by D. N. Breiter and M. L. Roush in *Am. J. Phys., 43,* 569.)

8

Molecular and Solid-State Physics

Enrico Fermi (1901, Italy—1954). Doctorate at age 21. 1926, full professor, University of Rome, development of Fermi-Dirac quantum statistics. One of the chief architects of the nuclear age; first to achieve controlled nuclear-fission reaction (December 2, 1942, basement of Stagg Field, University of Chicago). 1938, Nobel Prize, new radioactive elements and discovery of nuclear reactions affected by slow neutrons. Extraordinarily talented as both experimental and theoretical physicist. Element of atomic number 100 named fermium. 1954, first recipient of Fermi Award, U.S. Atomic Energy Commission. (Photo, Fermi Film Collection—courtesy of the American Institute of Physics, Niels Bohr Library.)

The structure of many-electron atoms can be treated only approximately because of the formidable mathematical problems involved in applying the quantum theory. It might at first sight seem to be impossible to deal with situations involving still larger numbers of particles—the molecules of a gas, the atoms of a solid, or the free electrons of an electrical conductor. But the very largeness of number of particles leads to the possibility of treating the average behavior of any one particle by applying the basic distribution laws—the laws governing how the particles of a many-particle system are distributed among the possible states of the system. We can examine some fundamental problems in molecular and solid-state physics by considering a few simple examples: molecular binding, the classical and quantum distribution laws, the quantum theory of specific heats of solids, and the free-electron theory of metals.

8-1 Molecular Binding

A diatomic molecule, consisting of two atoms held together by an attractive force, can be bound in different ways. If a diatomic molecule is to exist as a stable bound system, energy must be added to it in order to unbind, or separate, the constituent atoms; therefore the energy of the two atoms when they are close together must be lower than the energy of the two atoms when they are separated by a great distance. Thus, to show molecular binding, we have merely to establish that the total energy of the atoms is reduced when they are brought sufficiently close together.

The two important ways in which molecules are bound are *ionic*, or *heteropolar, binding* and *covalent*, or *homopolar, binding*. An example of a molecule bound almost completely by ionic binding is sodium chloride, NaCl; an example of covalent binding is the hydrogen molecule, H_2.

IONIC BINDING. We wish to show that NaCl can exist as a stable molecule; the concomitant condition will be that the total energy of the bound molecule is lower than the energy of the two atoms when they are separated. The element sodium, $_{11}$Na, is in the alkali-metal group, the first column of the periodic table (see Table 7–5); as such, it has one electron outside a closed subshell. Its electron configuration in the ground state is $1s^22s^22p^63s^1$. The single 3s electron is relatively weakly bound to the atom; it can be removed by adding 5.1 eV, thereby ionizing the atom and leaving a sodium ion with a net electric charge of $+1e$. Alkali metals are said to be *electropositive;* they are easily ionized to form positive ions, with a resulting electron configuration consisting of closed electron shells, like those of the inert gases.

The element chlorine, $_{17}$Cl, a halogen, falls in the seventh column of the periodic table; all elements in this column lack one electron of completing a closed p subshell. The electron configuration of chlorine in the ground state is $1s^22s^22p^63s^23p^5$. The neutral chlorine atom lacks one electron of filling a tightly bound, complete 3p subshell, and indeed, its energy is *lowered* by 3.8 eV when an electron is added to it, forming a negative ion of charge $-1e$. Halogen elements are said to be *electronegative*, and the *electron affinity energy* of $_{17}$Cl is 3.8 eV. It follows that 3.8 eV of energy must be added to the Cl^- ion to remove the last electron, leaving a neutral Cl atom.

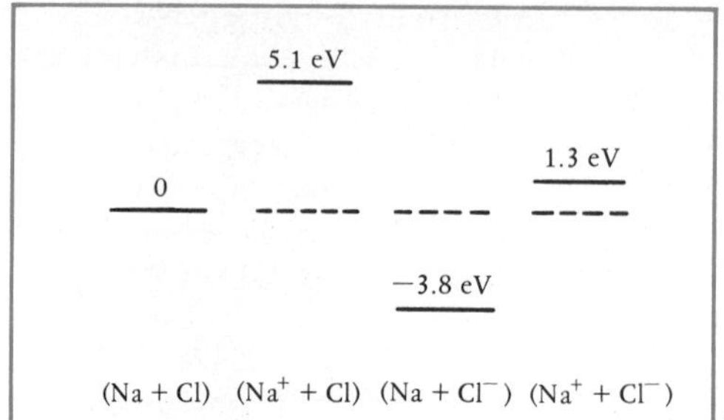

FIGURE 8–1. *Energy differences of sodium and chlorine atoms and ions when infinitely separated.*

Consider a neutral sodium atom and a neutral chlorine atom, infinitely separated. To remove one electron from Na, thereby forming Na^+, costs 5.1 eV, but if this electron were to be transferred to Cl, thereby forming Cl^-, 3.8 eV of this energy would be repaid; overall, the energy required would be only $5.1 - 3.8 = 1.3$ eV. The energy differences are shown in Figure 8–1. We should then have a positive ion and a negative ion, still separated. These ions would attract one another by the Coulomb electrostatic force; the Coulomb potential energy is $-ke^2/r$, where r is the distance between the centers of the two ions. If the sodium and chlorine ions were brought together, the energy of the system would decrease, since the force is attractive and the potential energy is negative. If r is chosen as, say, 4.0 Å, which is larger than the sum of the radii of the closed subshells of the respective ions, the Coulomb energy is easily found to be -1.8 eV. Thus, for ions separated by 4.0 Å, their total energy would be $1.3 - 1.8 = -0.5$ eV, clearly lower than the energy of a sodium atom and chlorine atom infinitely separated. The net cost of forming two ions is more than repaid by the potential-energy loss in bringing the ions to within 4 Å of one another.

Sodium and chlorine ions attract one another even when their nuclei are separated by less than 4 Å. When the separation between the nuclei is decreased still further, however, a repulsive force begins to act between the ions. The ions repel one another when the electron clouds of the two ions, each of which can be regarded as spherical, begin to overlap. The Pauli exclusion principle governs the number of electrons that can be accommodated in any given atomic electron shell. Both ions have their full electron quotas, and further electrons can be accommodated only if they occupy relatively high energy states. Consequently, as the interatomic distance is reduced, the electron shells must not overlap, by the Pauli exclusion principle. The electrons must go to higher available states, and the total energy of the molecule is then increased. When the atoms are attracting one another at large distances but repelling each other at sufficiently small distances, there exists an equilibrium interatomic separation distance r_0, at which the total potential energy of the system is a minimum (Figure 8–2).

A molecule's dissociation energy is the molecular binding energy, the energy that must be added to the molecule in its lowest energy state to separate it into its component atoms or ions. The molecule NaCl has a dissociation energy of 4.24 eV and an equilibrium separation distance r_0 of 2.36 Å. Since the molecule is held together by ionic binding, the end with the Na nucleus represents a region of positive electric charge, and the end with the Cl nucleus a region of negative electric charge. Thus, an ionic molecule

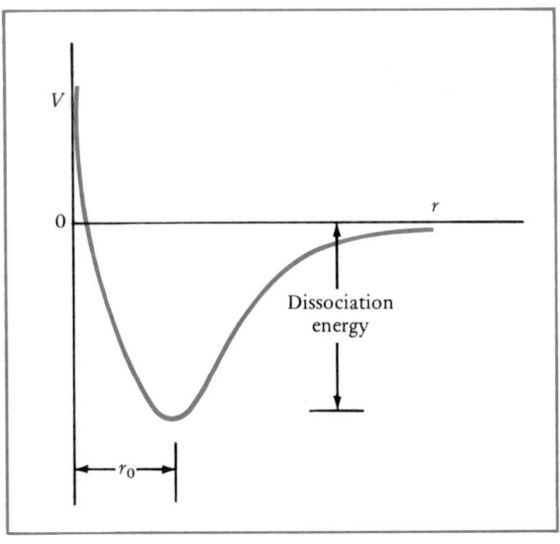

FIGURE 8–2. *Molecular potential as a function of interatomic distance.*

is a polar molecule; it has a permanent electric dipole moment—hence the alternative name, heteropolar binding (*hetero-*, from the Latin, meaning "different").

COVALENT BINDING. A simple example of covalent binding is found in the hydrogen molecule, H_2. Before considering this molecule, however, let us note a still simpler system, the hydrogen molecule ion, H_2^+. Suppose that it is formed when a neutral hydrogen atom (an electron bound to a proton) is brought together with an ionized hydrogen atom (a bare proton). When the protons are far apart, the electron will be bound to the one proton alone, but when the protons are separated by a distance of approximately 1 Å, the electron can be imagined to jump from one proton to the other, making, as it were, orbits about one or the other of the protons or about both.

Using wave mechanics, one can solve in detail for the wave function ψ of the single electron; the probability that the electron is at any position is proportional to ψ^2. The results are shown in Figure 8–3. We see that there is a relatively high probability that the single electron is in the region between the two nuclei compared with its being at either "end" of the molecule. When it is between the protons, the electron attracts both protons by the Coulomb electrostatic force; when it is at some exterior location, the protons are less strongly attracted to the electron, and their mutual repulsion becomes more important. Because the mutual repulsive force between the two nuclei increases as r (their separation) decreases, the H_2^+ molecule has a minimum

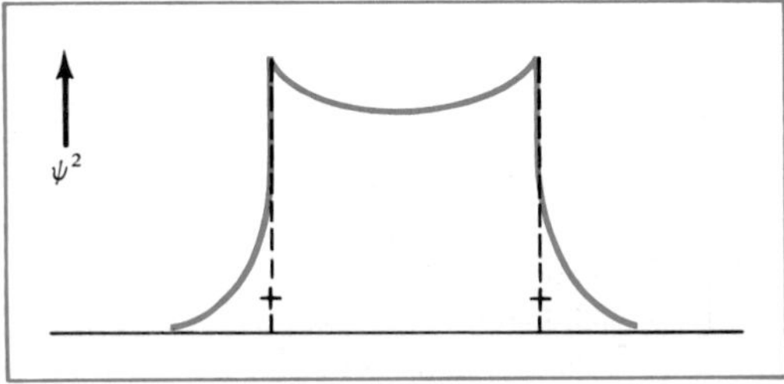

FIGURE 8–3. *Quantum-mechanical probability for finding the electron in the hydrogen molecule ion, H_2^+.*

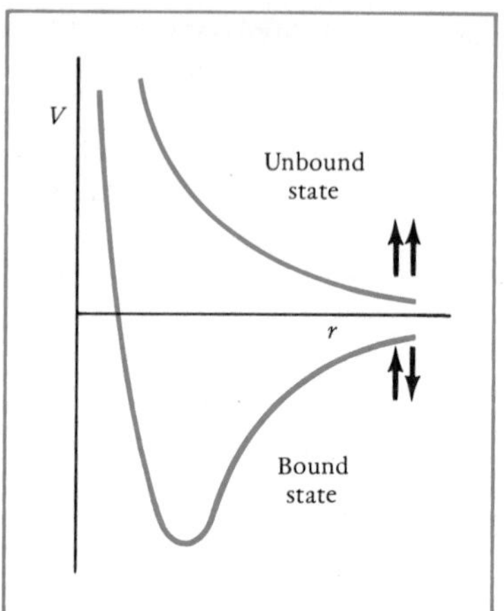

FIGURE 8–4. *Potential between two hydrogen atoms with aligned and antialigned spins.*

in the potential energy. It is found that the equilibrium separation is $r_0 = 1.06$ Å, and the dissociation energy, the energy required to separate H_2^+ into a proton and a hydrogen atom, is 2.65 eV.

Now consider the hydrogen molecule, H_2, which is formed when two neutral hydrogen atoms are brought close together. The binding together of two neutral, identical atoms is completely inexplicable in classical terms; it can be understood only by wave mechanics and the Pauli exclusion principle. When the two hydrogen atoms are separated by a distance that is large compared with the size of the first Bohr orbit (~1 Å), each electron is clearly identified with its own parent nucleus; but when the atoms are separated by a distance comparable to the size of either atom, there is, one must recognize, absolutely no way of knowing which of the two nuclei either of the electrons is identified with.

There are, however, two distinct ways in which the two hydrogen atoms can be brought together—with the two electron spins aligned in the same direction, or with the electron spins antialigned, that is, in opposite directions. Wave mechanics shows that when the hydrogen atoms are brought together, the total energy of the system increases when the electron spins are aligned but decreases when they are antialigned (Figure 8–4). This difference in energy arises from the operation of the Pauli exclusion principle. For the spin-aligned configuration, a bound state does not exist. Two antialigned electrons can both be found simultaneously in the region between the protons; therefore, the two electrons can exert attractive forces on both protons. Roughly speaking, the two nuclei of the hydrogen molecule share the two electrons, so that each nucleus can, at least for a part of the time, have both electrons filling a closed shell about it.

The total energy of two hydrogen atoms is reduced as the atoms approach one another with antialigned spins. Eventually, however, the mutual repulsion of the nuclei exceeds the binding due to exchange. Thus, a minimum energy exists, corresponding to the equilibrium configuration of the molecule. For the hydrogen molecule, the equilibrium distance is $r_0 = 0.74$ Å and the dissociation energy is 4.48 eV.

The chemical binding of two identical, nonpolar atoms is called *covalent binding,* since sharing valence electrons is primarily responsible for the attractive force. A molecule such as H_2, with no permanent electric dipole moment, is said to be nonpolar. For this reason, covalent binding is also referred to as *homopolar binding*.

A bound hydrogen molecule can exist with two atoms but not with three or more. The valence forces operating in covalent binding show *saturation,* in that the number of atoms that can be bound together is limited.

A third type of force, the only one operating in the interaction between closed-shell atoms of inert gases, is the *Van der Waals force*. This force is responsible for the cohesion in the liquid and solid state of rare gases. When two such atoms approach one another, the "center" of the negative charge is displaced from the positive nucleus. The atoms then weakly interact through the electric dipoles induced by the charge displacement.

The ionic and homopolar binding processes can hold atoms together to form crystalline solids. An example of an ionic crystal is NaCl (shown in Figure 5–1), an alkali halide, in which the ions Na^+ and Cl^- are found alternately on the corners of a cubic lattice. An example of a covalent crystal is diamond (carbon, C), in which there are carbon nuclei at the center and corners of a tetrahedron in the elementary cell structure. A third kind of crystalline binding, for which there is no counterpart in molecules, is metallic binding. This kind of binding, arising from the Coulomb interaction between the fixed positive ions and the free electrons of the metal, is of wave-mechanical origin.

8–2 Statistical Distribution Laws

Many problems in physics concern the behavior of systems composed of very large numbers of weakly interacting, identical particles. *Statistical mechanics* is the name given to the statistical methods of handling large numbers of particles whose mechanics are known. A familiar example of a system that can be treated by statistical mechanics is the ideal gas comprising numerous identical point particles that obey Newton's laws of motion. Although it is possible in principle to describe in detail the motion of every particle of such a system, the problem is so mathematically formidable as to be beyond solution. What is of interest is not the *detailed behavior* of every particle of the system but the *average behavior* of the microscopic particles and their influence on macroscopic measured quantities. It is possible, for example, to relate the pressure (a macroscopic quantity) of a gas on its container to the mass and average speed of the molecules (microscopic quantities). Furthermore, it is possible to relate another macroscopic quantity, the absolute temperature T of a gas, to a microscopic quantity, the average kinetic energy of the molecules ($\bar{\epsilon} = \frac{3}{2}kT$).†

In an ideal gas, the particles interact only weakly with one another and an equilibrium distribution results. The molecules interact with one another and with the walls by collisions, whose duration is short compared with the time between collisions. Some molecules will have small kinetic energy, and others will have large kinetic energy; briefly, there will be a distribution of the energies (and thus of the speeds) over a considerable range. If the molecules are in equilibrium and their number is very large, the relative

† See Weidner and Sells, *Elementary Classical Physics,* 2d ed., Sections 19–3 and 19–6.

number of molecules of any particular energy will be essentially constant, although the energy of any one molecule will change with time through collisions.

Statistical mechanics allows one to determine the energy distribution of *any* system of weakly interacting particles in thermal equilibrium, whether the particles obey classical or quantum mechanics. Each particle in a quantum system composed of a very large number of identical, weakly interacting particles has available to it a discrete set of quantum states, the energy of each state i being designated by ϵ_i. (In a classical system, on the other hand, there is a continuum of allowed energies, and the separation between adjacent energies can be taken to be zero.) Since the particles interact only weakly with one another, each particle has its own set of states, and if the particles are all identical, they will all have identical sets of states available to them for occupancy.

A very simple hypothetical example is shown in Figure 8–5, where each particle must at any one time exist in one of the four possible states. Note that states 2 and 3 have identical energies, $\epsilon_2 = \epsilon_3$. Whenever two or more distinct states, such as 2 and 3, have the same energy, the states are said to be *degenerate*. Now the question to be answered is, How are the particles of the system distributed among the various available states?

Statistical mechanics predicts the most probable distribution of the particles among the various possible states. Because most systems of interest have many particles, the most probable distribution becomes overwhelmingly more probable than any other distribution and thus represents (with almost complete certainty) the actual distribution. Table 8–1 lists the three types of probability distributions: Maxwell-Boltzmann, Bose-Einstein, and Fermi-Dirac statistics. Also included in the table are the characteristic properties defining the statistical behavior and examples of physical systems that obey the various distributions.

TABLE 8–1

TYPE OF DISTRIBUTION	CHARACTERISTICS	DISTRIBUTION FUNCTION $f(\epsilon_i)$	EXAMPLES OF SYSTEMS
Maxwell-Boltzmann	Identical but distinguishable particles	$f_{MB}(\epsilon_i) = Ae^{-\epsilon_i/kT}$	Essentially all gases at all temperatures
Bose-Einstein	Identical, indistinguishable particles of integral spin	$f_{BE}(\epsilon_i) = \dfrac{1}{e^{\epsilon_i/kT} - 1}$	Liquid helium (spin 0) Photon gas (spin 1) Phonon gas (spin 0)
Fermi-Dirac	Identical, indistinguishable particles of half-integral spin (therefore obeying Pauli exclusion principle)	$f_{FD}(\epsilon_i) = \dfrac{1}{e^{(\epsilon_i - \epsilon_F)/kT} + 1}$	Electron gas (spin $\frac{1}{2}$) Protons (spin $\frac{1}{2}$) Neutrons (spin $\frac{1}{2}$)

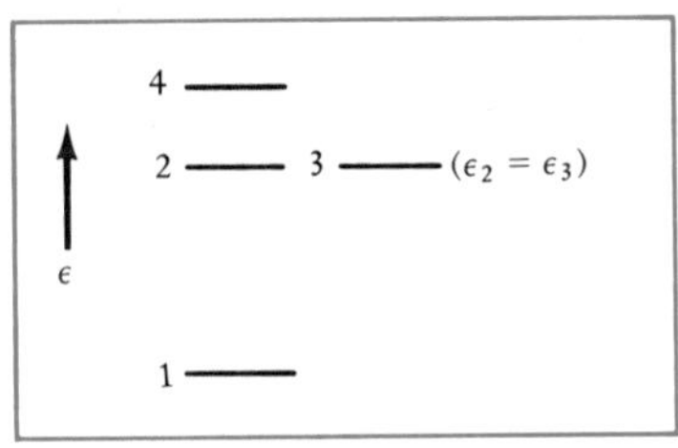

FIGURE 8–5. *Quantized energy states available to a particle.*

The *distribution function* $f(\epsilon_i)$ in Table 8–1 *represents the average number of particles in the state i.* Inasmuch as $f(\epsilon_i)$ depends only on the energy of the state, the average number of particles in states having the same energy, such as states 2 and 3 in Figure 8–5, will be the same, for example, $f(\epsilon_2) = f(\epsilon_3)$.

The distribution functions are derived from the *fundamental postulate of statistical mechanics:*

> *Any particular distribution of the particles among the various available states of a system is just as likely as any other distribution.*

Any particular distribution must be consistent with the characteristics of the particles and with such conservation laws as the conservation of energy and of particles.

MAXWELL-BOLTZMANN DISTRIBUTION. The Maxwell-Boltzmann distribution, a classical distribution, applies to a system of *identical particles* that are nevertheless *distinguishable* from one another (for example, a collection of billiard balls of identical mass and diameter but numbered 1, 2, 3, . . .). The average number f_{MB} of particles in a state i with energy ϵ_i is

$$f_{\mathrm{MB}}(\epsilon_i) = Ae^{-\epsilon_i/kT} \qquad (8\text{–}1)$$

where A is a constant for a given temperature, k is the Boltzmann constant, 1.38×10^{-23} J/K, and T is the absolute temperature of the system of particles, always assumed to be in equilibrium. The Maxwell-Boltzmann distribution law describes the *average* behavior of a gas of atoms or molecules. This distribution function, $f_{\mathrm{MB}}(\epsilon_i)$, is plotted in Figure 8–6 for two different temperatures. Note that for a given temperature, the average number of particles with energy ϵ_i drops off (exponentially) as the energy ϵ_i increases. As the temperature of the system rises, the particles are distributed more heavily in higher energy states; at the absolute zero of temperature, all particles have zero energy.

BOSE-EINSTEIN DISTRIBUTION. The Bose-Einstein distribution law applies to a system of *identical particles* that are *indistinguishable,* each particle having an *integral spin.* Such particles are called *bosons.* The average number of particles occupying a particular state i of energy ϵ_i is given by

$$f_{\mathrm{BE}}(\epsilon_i) = \frac{1}{e^{\epsilon_i/kT} - 1} \qquad (8\text{–}2)$$

The Bose-Einstein distribution is plotted in Figure 8–7. It can be seen from both Equation 8–2 and Figure 8–7 that the distribution function $f_{\mathrm{BE}}(\epsilon_i)$

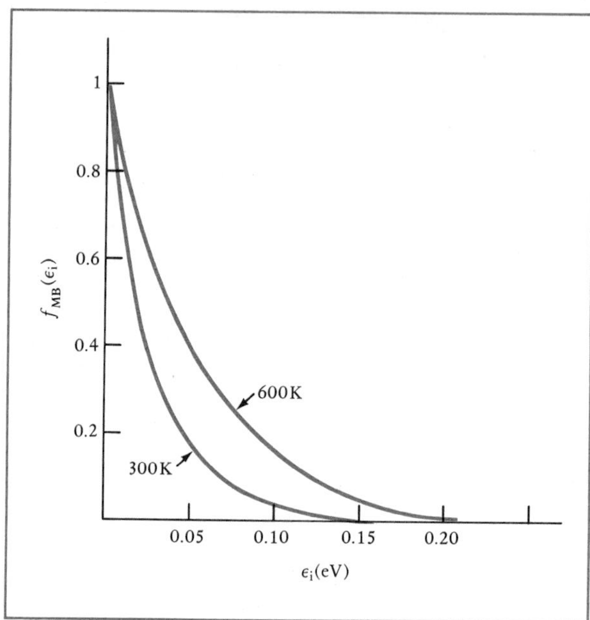

FIGURE 8–6. *Maxwell-Boltzmann distribution function.*

approaches the Maxwell-Boltzmann distribution $f_{MB}(\epsilon_i)$ when $\epsilon_i \gg kT$. At low energies, when $\epsilon_i \ll kT$, the term -1 in the denominator of Equation 8–2 becomes important and has the effect of making $f_{BE}(\epsilon_i)$ much larger than $f_{MB}(\epsilon_i)$ for the same energy. Like the classical Maxwell-Boltzmann distribution, the Bose-Einstein distribution concentrates the particles at the lowest energy states; indeed, for the Bose-Einstein distribution, the particles are more heavily concentrated in the states of lowest energy. With a rise in temperature, the particles are likewise shifted—to progressively higher energies.

FERMI-DIRAC DISTRIBUTION. The Fermi-Dirac distribution applies to a system of *identical particles* that are *indistinguishable*, each particle having a *half-integral spin*. Particles of half-integral spin (*fermions*), such as the electron, proton, or neutron, obey the Pauli exclusion principle. This keeps two or more particles from existing in the same state at the same time. The Pauli principle represents, so to speak, a very strong interaction between identical Fermi particles, keeping any two from occupying the same state. Although the Maxwell-Boltzmann and Bose-Einstein distribution laws impose no restriction on the number of particles that can occupy the same state, the Fermi-Dirac statistics allow at most only one particle in a particular state. The average number of particles in a particular quantum state i with energy ϵ_i is given by

$$f_{FD}(\epsilon_i) = \frac{1}{e^{(\epsilon_i - \epsilon_F)/kT} + 1} \qquad (8\text{–}3)$$

The quantity ϵ_F, called the *Fermi energy,* is a constant for many problems of interest and is nearly independent of temperature.

The physical meaning of the Fermi energy can be seen from Equation 8–3. For the state in which $\epsilon_i = \epsilon_F$, the average number of particles is $\frac{1}{2}$; that is, the probability that a state of energy ϵ_F is occupied is just $\frac{1}{2}$. For those

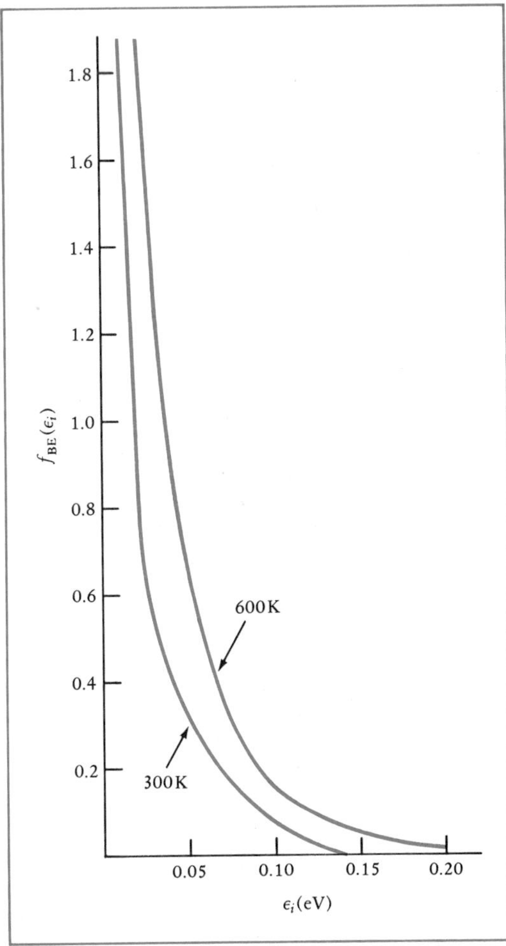

FIGURE 8–7. *Bose-Einstein distribution function.*

states with energies much lower than ϵ_F, the exponential term in the denominator of Equation 8–3 is essentially zero, and $f_{FD} = 1$; all such states have their full quota of particles, one per state, and are filled. For states with energies much higher than the Fermi energy, on the other hand, the exponential term becomes much larger than $+1$, and f_{FD} reduces to the Maxwell-Boltzmann distribution, Equation 8–1. At the absolute zero of temperature, the Fermi-Dirac distribution f_{FD} is 1 for all states up to ϵ_F and zero for all states with energies greater than ϵ_F; see Figure 8–8. Whereas particles are concentrated at the lowest energies in both the Maxwell-Boltzmann and Bose-Einstein distributions and have zero energy for $T = 0$, for the Fermi-Dirac distribution the states are filled uniformly up to the Fermi energy at absolute zero temperature. For a finite temperature, the lowest energy states remain uniformly filled, and changes in f_{FD} occur only in the vicinity of ϵ_F.

For all three types, the distribution function $f(\epsilon_i)$ gives only the *average* number of particles occupying a *state i* of energy ϵ_i and *not* the number $n(\epsilon_i)$ of particles *with the energy* ϵ_i. This is so because there may be two or more states with the same energy. Therefore, the quantity $g(\epsilon_i)$, called the *statistical weight,* is introduced. It gives the number of states with the same energy

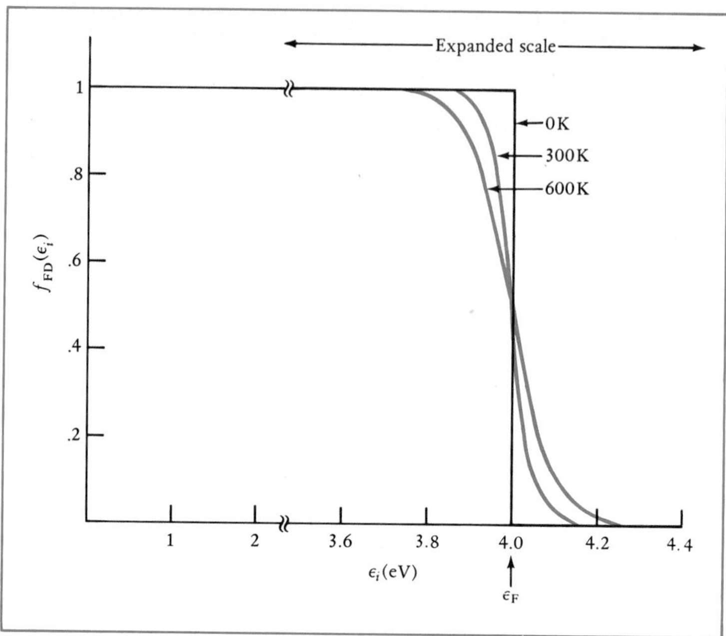

FIGURE 8–8. *Fermi-Dirac distribution function.*

ϵ_i; for instance, in Figure 8–5, $g(\epsilon_1) = 1$, $g(\epsilon_2) = 2$, and $g(\epsilon_4) = 1$. It follows that

$$n(\epsilon_i) = f(\epsilon_i)g(\epsilon_i) \tag{8–4}$$

In many situations, the energy levels are so closely spaced that they are regarded as continuous. One then wishes to know the number of particles, $n(\epsilon)\, d\epsilon$, having energies between ϵ and $\epsilon + d\epsilon$. Equation 8–4 then is written as

$$n(\epsilon)\, d\epsilon = f(\epsilon)g(\epsilon)\, d\epsilon \tag{8–5}$$

where $g(\epsilon)$, called the *density of states,* gives the number of states per unit energy.

Equation 8–5 will form the basis for the subsequent material in this chapter; for if one knows the applicable distribution function $f(\epsilon)$, and if the density of states $g(\epsilon)$ can be computed for a particular system, then one will know the (most probable) number $n(\epsilon)\, d\epsilon$ of particles within the range ϵ to $\epsilon + d\epsilon$. Given the energy distribution of the system's particles, one can compute such macroscopic properties of the system as its average energy or its specific heat.

8–3 Maxwell-Boltzmann Statistics Applied to an Ideal Gas

Consider a classical, ideal gas composed of N identical atoms or molecules assumed to be point particles obeying Newton's laws of motion. We want to calculate the number $n(\epsilon)\, d\epsilon$ of atoms within the energy range $d\epsilon$. Since this system obeys the Maxwell-Boltzmann statistics, Equation 8–5 becomes

$$n(\epsilon)\, d\epsilon = f_{MB}(\epsilon)g(\epsilon)\, d\epsilon$$

$$n(\epsilon)\, d\epsilon = Ae^{-\epsilon/kT}g(\epsilon)\, d\epsilon \tag{8–6}$$

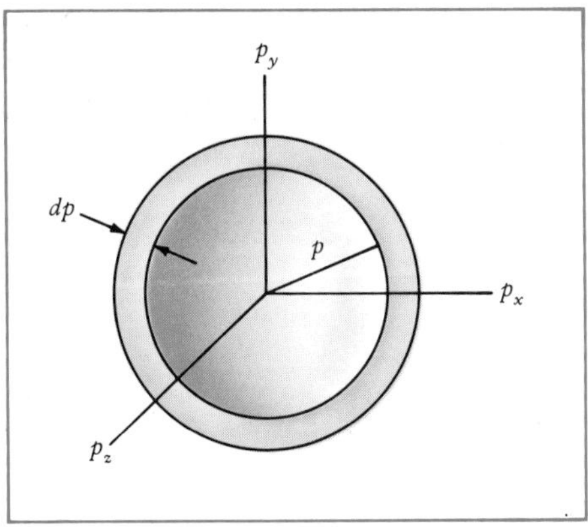

FIGURE 8–9. *Momentum states between p and p + dp.*

The quantity $g(\epsilon)$ for this system is most easily evaluated as follows. The only energy these particles possess (strictly, the only energy that can change) is translational kinetic energy, and any one particle has available to it a whole continuum of energy states ranging from zero upward. The state of such a particle can be specified by giving the three components of its momentum—p_x, p_y, p_z. Except during collisions of negligibly short duration, a particle is free and the potential energy can be taken as zero. Therefore, we need not specify a particle's location. The momentum components alone specify the particle's state completely.

It is convenient to represent the states available to a particle by points in *momentum space;* coordinates in this space are the three components p_x, p_y, and p_z of the momentum. Then each point in classical momentum space corresponds to a possible available state, and all points in momentum space are allowed.

We now have a way of finding the number $g(\epsilon)\, d\epsilon$ of states in the energy range $d\epsilon$, where ϵ represents the total energy (and here also the kinetic energy) of any one particle. We write

$$\epsilon = \tfrac{1}{2}mv^2 = \frac{p^2}{2m}$$

$$d\epsilon = \frac{p\, dp}{m} \qquad (8\text{–}7)$$

where the magnitude of the momentum $\mathbf{p}$ is $p = (p_x{}^2 + p_y{}^2 + p_z{}^2)^{1/2}$

$$p = (p_x{}^2 + p_y{}^2 + p_z{}^2)^{1/2}$$

Furthermore,
$$g(\epsilon)\, d\epsilon = g(p)\, dp \qquad (8\text{–}8)$$

where $g(p)\, dp$ gives the number of states with a momentum magnitude between p and $p + dp$. This number is proportional to the volume of a spherical shell in momentum space, that is, to $4\pi p^2\, dp$, as shown in Figure 8–9. Thus,

$$g(p)\, dp \propto p^2\, dp \qquad (8\text{–}9)$$

Using Equations 8–7 and 8–8 in Equation 8–9, we have

$$g(\epsilon)\, d\epsilon = g(p)\, dp \propto p^2\, dp \propto p\, d\epsilon$$

$$g(\epsilon)\, d\epsilon \propto \epsilon^{1/2}\, d\epsilon \qquad (8\text{–}10)$$

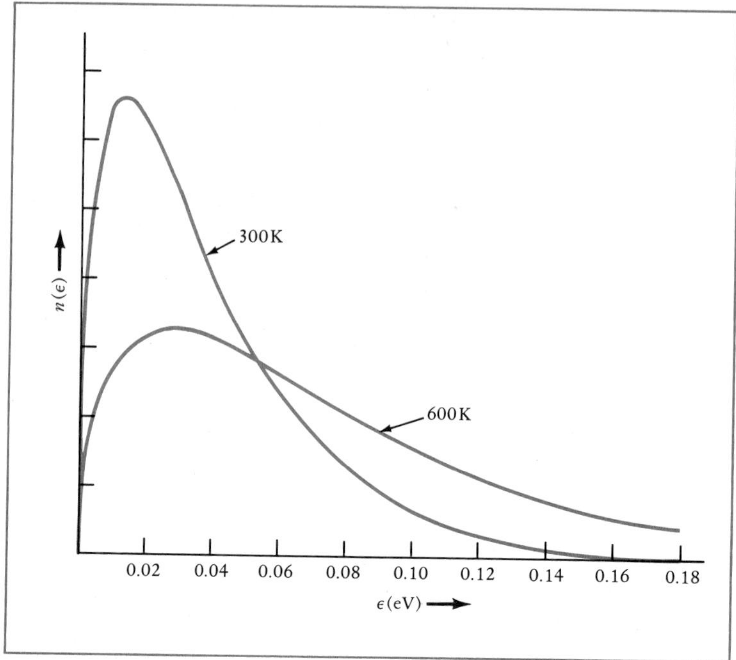

FIGURE 8–10. *Energy distribution of the molecules of an ideal gas for two temperatures.*

Finally, the most probable number of particles with energies between ϵ and $\epsilon + d\epsilon$ is, from Equation 8–6,

$$n(\epsilon)\, d\epsilon = Ce^{-\epsilon/kT}\epsilon^{1/2}\, d\epsilon \tag{8–11}$$

where C is a proportionality constant. The distribution of the particles as a function of energy is shown in Figure 8–10. The constant C can be evaluated by applying the conservation of particles; their total number N is fixed:

$$N = \int_0^\infty n(\epsilon)\, d\epsilon = C\int_0^\infty \epsilon^{1/2}e^{-\epsilon/kT}\, d\epsilon \tag{8–12}$$

Each of the $n(\epsilon)\, d\epsilon$ particles within the energy range $d\epsilon$ has an energy ϵ; therefore, the total energy E of the gas is

$$E = \int_0^\infty \epsilon n(\epsilon)\, d\epsilon = C\int_0^\infty \epsilon^{3/2}e^{-\epsilon/kT}\, d\epsilon$$

Then, integrating this equation by parts and using Equation 8–12, we have

$$E = \tfrac{3}{2}NkT$$

The average energy $\bar{\epsilon}$ per atom is then

$$\bar{\epsilon} = \frac{E}{N} = \tfrac{3}{2}kT \tag{8–13}$$

The average translational kinetic energy per atom, $\bar{\epsilon} = \frac{3}{2}kT$, can be considered equally divided among the three translational degrees of freedom. A *degree of freedom* is defined as one of the independent coordinates needed to specify the position of the particle. Therefore, the average energy per degree of freedom can be taken to be $\frac{1}{2}kT$.

The molar specific heat C_v of a gas at a constant volume is defined as the energy necessary to increase the temperature of 1 mole of gas by 1 K, with the volume of the gas fixed; that is,

$$C_v \equiv \frac{1}{n}\frac{dE}{dT} = \frac{N_A}{N}\frac{dE}{dT} \tag{8–14}$$

where n is the number of moles and N_A is the number of particles per mole (Avogadro's number). Using Equation 8–13, we get $dE/dT = \frac{3}{2}kN$; and Equation 8–14 becomes

$$C_v = \tfrac{3}{2}kN_A = \tfrac{3}{2}R \tag{8–15}$$

where the gas constant R is kN_A. Thus, the molar specific heat C_v of a classical ideal gas composed of atoms or molecules regarded as point particles (that is, as having no internal structure) is predicted to be $\frac{3}{2}R$. The measured specific heats of *monatomic* gases agree well with this theoretical value; for example, C_v is $1.50R$ for both helium and argon. On the other hand, the specific heats for gases composed of diatomic or polyatomic molecules may exceed $1.50R$. Each molecule of such gases has additional energies associated with the vibration of atoms within a molecule and the rotation of the molecule as a whole. Consequently, when a diatomic or a polyatomic molecular gas gains energy through a rise in its temperature, the total molecular energy may include, besides the translational kinetic energy, energy associated with molecular vibration and rotation; and there is also a corresponding increase of the specific heat over $1.50R$, which includes translational kinetic energy only.

Example 8–1. It is assumed that a room of typical size (~50 m³) is filled with *atomic* hydrogen gas. The total number of hydrogen atoms in the room at standard temperature (273 K) and pressure is readily computed to be 1.34×10^{27}. How many of these atoms are in the first excited state ($n = 2$) of hydrogen?

The ground state of hydrogen has an energy $E_1 = -13.6$ eV. With two degenerate spin states for 1s, two independent states are available, and the statistical weight of the ground state is $g_1 = 2$.

The first excited state has an energy $E_2 = E_1/2^2 = -13.6$ eV$/4 = -3.4$ eV. With two states for 2s and six states for 2p, eight states are available for $n = 2$, and $g_2 = 8$.

From Equation 8–5, the number of atoms in the two states is

$$n_1 = Ag_1e^{-E_1/kT} \qquad \text{and} \qquad n_2 = Ag_2e^{-E_2/kT}$$

where A is a constant. Therefore,

$$\frac{n_2}{n_1} = \left(\frac{g_2}{g_1}\right)e^{-(E_2-E_1)/kT} = 4e^{-(10.2\text{ eV}/kT)}$$

$$n_2 = n_1(4)e^{-433} \approx n_1 10^{-188}$$

Since n_1 is only 1.34×10^{27}, *not even one* atom is in the first excited state. Indeed, it would take a room larger in volume by a factor 10^{161} (far larger than the volume of the universe!) to yield just one atom in an excited state. It is therefore obvious that since no hydrogen atoms occupy excited states at room temperature, such a gas cannot emit radiation.

8-4 The Quantum Theory of the Specific Heats of Solids

Quantum statistics can be used to elucidate the observed specific heat of solids. We first review the only partially successful classical theory, which holds for relatively high temperatures, and then give the quantum theory, using the Bose-Einstein statistics.

Consider a crystalline solid composed of N atoms. Each atom is bound in the crystal lattice by forces arising from its neighboring atoms. When any one atom is displaced from its equilibrium position, it is subject, in a first approximation, to a restoring force proportional to its displacement. Thus, any atom displaced from its equilibrium position will undergo simple harmonic motion. But if one atom is displaced from its equilibrium position, so too are the neighbors with which it is coupled by interatomic binding forces. Consequently, if one atom undergoes simple harmonic motion, it causes neighboring atoms also to oscillate, and the disturbance or deformation to propagate through the crystal as an elastic wave. The total energy content of the solid, which may change with temperature, consists of the following contributions from each atom: the kinetic energy of the essentially free, outer, valence electrons and the energy of vibration of the remainder of the atom, namely the nucleus plus the tightly bound, inner electrons. At all moderate temperatures the quantum state of any of the bound electrons is unchanged. For this reason, the nucleus plus the bound electrons can be treated as a single, inert, vibrating particle. If the internal energy of the solid changes, so too does the temperature, and the change in the internal energy of the crystal per unit change in temperature is the specific heat of the solid. The total specific heat of the solid consists of the *electronic specific heat* (of *free* electrons) and the *lattice* (vibrational) *specific heat* (of the atoms' ions). Since for all temperatures except the very lowest, the electronic specific heat is negligible (Section 8–5), we shall here consider the contributions arising from the lattice vibrations only.

We first compute the specific heat of a solid using the classical theory in which the lattice-energy content is associated with N simple harmonic oscillators. For each degree of freedom of a simple harmonic oscillator, there is $\frac{1}{2}kT$ of energy associated with potential energy and $\frac{1}{2}kT$ associated with kinetic energy, or a total of kT per oscillator in one dimension. Therefore, for oscillations in three dimensions the total vibrational energy E is the number of degrees of freedom $3N$ times the energy per degree of freedom kT,

$$E = (3N)(kT) = 3NkT$$

and the classical lattice specific heat per mole C_v is

$$C_v = \frac{1}{n}\frac{dE}{dT} = 3\frac{N}{n}k = 3N_A k = 3R \qquad (8\text{–}16)$$

where n is the number of moles, N_A is Avogadro's number, and R is the constant of the general-gas law. This classical relation, known as the *Dulong-Petit law,* predicts that the molar specific heat of any solid is the same constant, $3R$, independent of the material and of the temperature. It agrees with experiment at *high* temperatures. The classical theory is incapable, however, of explaining the observed decrease in the specific heat at low temperatures, as shown in Figure 8–11.

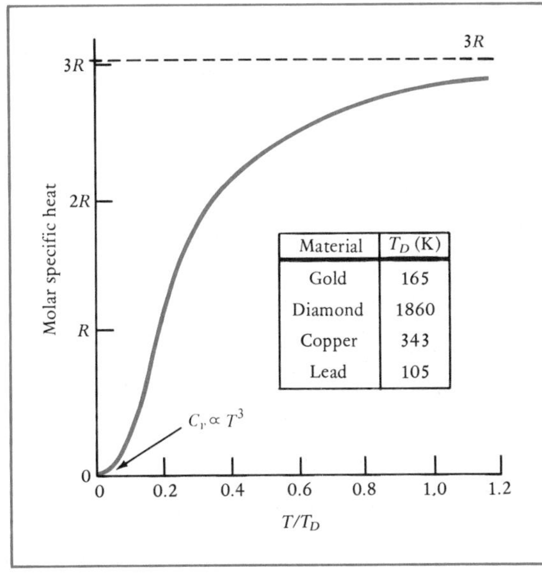

FIGURE 8–11. *Observed molar specific heat of solids as a function of temperature. The temperatures are given in units of the Debye temperature $T_D = hf_D/k$ (see Equation 8–26).*

The first successful theoretical treatment of the lattice specific heat for *all* temperatures was given by A. Einstein in 1906. This early quantum treatment was improved upon by P. Debye in 1912 and is usually referred to as the *Debye theory of specific heats.*

The essential quantum feature of lattice vibrations is the quantization of the atoms' vibrational energy. Each lattice atom is considered a simple harmonic oscillator, whose allowed energies are given by $E_v = (v + \frac{1}{2})hf$, where the quantum number v may be given the values 0, 1, 2, . . . and f is the classical vibrational frequency (Section 5–9). Each lattice atom then loses (or gains) energy in discrete amounts and transfers energy to (or from) a neighboring atom to which it is coupled; more specifically, the energy lost or gained by any one atom is hf, corresponding to a change in v of ± 1. Energy is propagated through the lattice as an elastic deformation, as one atom loses energy hf and its neighbor gains energy hf. Since the vibrational energy is propagated in quantized amounts, the elastic waves are also quantized; therefore, making an analogy with quantized electromagnetic waves, or photons, we can now speak of the propagation of quasi-particle quanta of vibrational energy, or *phonons.* Thus, a phonon is a quantized elastic wave, created and absorbed by quantized vibrators when they change their quantum states. The thermal-energy content of the crystalline lattice, in this view, consists of the total energy of the phonons.

The energy of an atomic oscillator may have any value given by $E_v = (v + \frac{1}{2})hf$; therefore, the *number* of phonons is unrestricted. This implies that the distribution of phonons, as particlelike quanta, is governed by the Bose-Einstein statistics. In analogy with a photon, a phonon has energy $\epsilon = hf$, and the phonon momentum is given by $p = \epsilon/v_s$, where v_s is the speed of the phonon propagation, the speed of sound.

Our strategy in finding the quantum specific heat for lattice vibrations is then this. We first find the total number of phonon states within a macroscopic crystalline material; then we use the Bose-Einstein statistics to

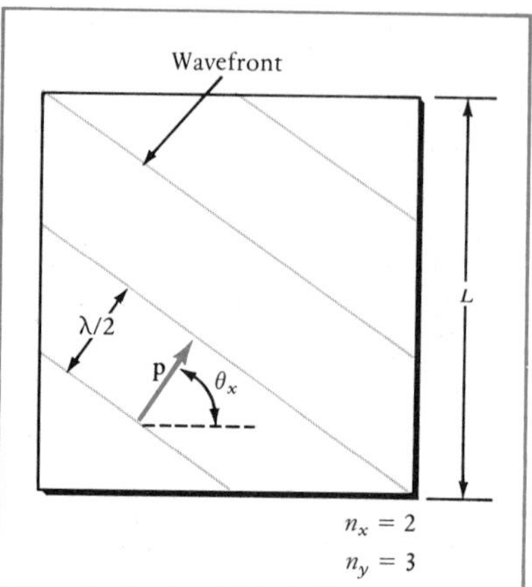

FIGURE 8–12. *An allowed stationary plane elastic wave in a cubical box.*

find the energy distribution of the phonons among the possible states. Finally, the rate at which the total phonon energy changes with temperature yields the specific heat.

The strategy for finding the number of available phonon states (the number of possible elastic waves) is the following. We imagine plane elastic waves to be confined within a cube (for simplicity) of solid of edge L. Then we count the number of stationary, or standing, wave patterns that can exist within the cube. This procedure is not unduly restrictive, for the cube can be imagined arbitrarily large, so that even the longest waves can be accommodated.

The state of a phonon is completely specified by giving the three components of its linear momentum, p_x, p_y, and p_z. The procedure for fitting stationary waves within a three-dimensional enclosure is altogether analogous to the procedure used in Section 5–8 for finding the permitted quantum states of a particle confined to a one-dimensional potential-energy well. Only certain values of p_x, p_y, and p_z will lead to stationary states.

Figure 8–12 shows one particular elastic wave, traveling obliquely relative to the sides of the box; its direction of propagation is given by the direction of the momentum vector **p**. The wavefronts shown, to which **p** is perpendicular, are one half-wavelength apart, where $\lambda = h/p$. For stationary waves to exist within the cubical box, the projection of any side along the direction of propagation must be an integral number of half-wavelengths. Only then will waves reflected at the boundaries not cancel one another. Thus, for the side parallel to the p_x direction, we must have

$$L \cos \theta_x = n_x \frac{\lambda}{2} \tag{8–17}$$

where n_x is an integer and θ_x is the angle between **p** and the p_x axis. For phonons, we have $p = h/\lambda$, and therefore,

$$p_x = p \cos \theta_x = \frac{h \cos \theta_x}{\lambda} \tag{8–18}$$

Combining this with Equation 8–17, we find, for the permitted values of the p_x component of **p**,

$$p_x = \frac{h}{2L} n_x$$

Similarly, $$p_y = \frac{h}{2L} n_y \quad \text{and} \quad p_z = \frac{h}{2L} n_z \tag{8–19}$$

Figure 8–13 shows the allowed values of p_x, p_y, p_z in momentum space. (The heavy dot in Figure 8–13 corresponds to the state illustrated in Figure 8–12.) For a macroscopic length L the separation between the adjacent points, $h/2L$, is very small compared with the momentum of all phonons except those of the longest wavelengths.

We are interested in the number of states within the small energy range ϵ to $\epsilon + d\epsilon$, where $\epsilon = pv_s$ and $p = (p_x^2 + p_y^2 + p_z^2)^{1/2}$. This number can be found by computing the number of states within a shell of radius p and thickness dp, counting only the positive values of p_x, p_y, and p_z. Then

$$g(p)\,dp = \frac{\frac{1}{8}(4\pi p^2\,dp)}{(h/2L)^3} \tag{8–20}$$

The factor $\frac{1}{8}$ is introduced because only one octant of the spherical shell may be included (only *positive* values of n_x, n_y, and n_z are permitted), and the factor $(h/2L)^3$ represents the volume associated with each point in momentum space. With $p = \epsilon/v_s$ and $L^3 = V$, the total volume of the box (the crystal), Equation 8–20 becomes

$$g(p)\,dp = g(\epsilon)\,d\epsilon = \frac{4\pi V\epsilon^2\,d\epsilon}{h^3 v_s^3}$$

or since $\epsilon = hf$,

$$g(f)\,df = \left(\frac{4\pi V}{v_s^3}\right) f^2\,df \tag{8–21}$$

Actually, two distinct types of elastic waves are propagated through an elastic medium: (1) a transverse wave traveling at a speed v_t and having two possible, mutually perpendicular polarization directions, and (2) a longitudinal wave traveling in general at a different speed, v_l. The number of

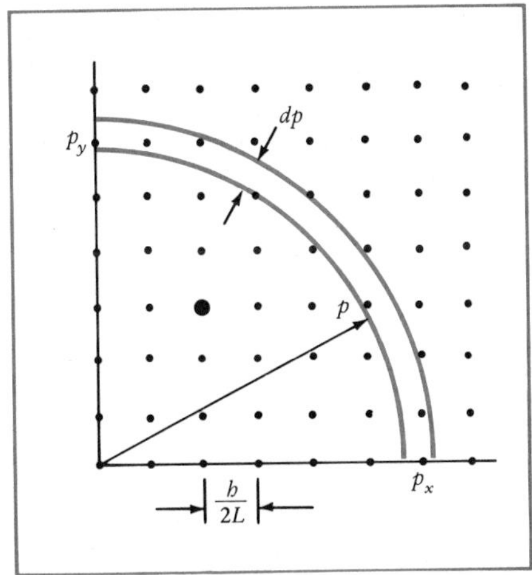

FIGURE 8–13. *Allowed values of p_x, p_y, and p_z in momentum space. The heavy dot corresponds to the state (n_x = 2, n_y = 3) illustrated in Figure 8–12.*

elastic modes of vibration, or the total number of available phonon states, in the frequency interval df is then

$$g(f)\,df = 4\pi V\left(\frac{2}{v_t^3} + \frac{1}{v_l^3}\right) f^2\,df \tag{8–22}$$

The total number of vibrational modes is limited, however; it cannot exceed the total number $3N$ of degrees of freedom of the crystal. One of Debye's contributions consisted in recognizing this restriction. The elastic vibrations are cut off at the frequency f_D, called the *Debye frequency,* as follows:

$$\text{Total number of modes} = \int_0^{f_D} g(f)\,df = 4\pi V\left(\frac{2}{v_t^3} + \frac{1}{v_l^3}\right)\int_0^{f_D} f^2\,df = 3N$$

Therefore,

$$f_D^{\,3} = \frac{9N}{4\pi V}\left(\frac{2}{v_t^3} + \frac{1}{v_l^3}\right)^{-1} \tag{8–23}$$

Equation 8–22 can now be written in terms of the Debye frequency f_D:

$$g(f)\,df = \frac{9N}{f_D^{\,3}} f^2\,df \tag{8–24}$$

The elastic vibrations are quantized, but the number of phonons in any particular state is not restricted by the exclusion principle. Therefore, the phonon distribution is given by the Bose-Einstein statistics (Equation 8–2). The number of phonons in the frequency range between f and $f + df$ is then given by

$$n(f)\,df = g(f)\frac{df}{e^{hf/kT} - 1}$$

Since the energy per phonon is hf, the total vibrational energy content of the crystal is

$$E = \int_0^{f_D} hf\frac{g(f)\,df}{e^{hf/kT} - 1} = 9N\left(\frac{kT}{hf_D}\right)^3 kT\int_0^{x_m}\frac{x^3\,dx}{e^x - 1} \tag{8–25}$$

where we have used Equation 8–24 for $g(f)$ and defined $x \equiv hf/kT$ and $x_m \equiv hf_D/kT$.

It is convenient to define a characteristic temperature, called the *Debye temperature* T_D, as that temperature for which $hf_D = kT_D$. Then,

$$T_D = \frac{hf_D}{k} \qquad \text{and} \qquad x_m = \frac{T_D}{T}$$

Equation 8–25 can then be written

$$E = 9N\left(\frac{T}{T_D}\right)^3 kT\int_0^{x_m}\frac{x^3\,dx}{e^x - 1} \tag{8–26}$$

The integral in this equation must be evaluated numerically.

The molar specific heat C_v immediately follows from the definition $C_v = (1/n)(dE/dT)$. Although one cannot easily evaluate C_v for all temperatures (because x and x_m in Equation 8–26 are functions of T), it is possible to find expressions for E and C_v for high temperatures and low temperatures.

In the high-temperature limit, we have $kT \gg hf_D$ and $x \ll 1$, and therefore, $e^x \approx 1 + x$. Equation 8–26 becomes, for high temperatures,

$$E \approx 9N\left(\frac{T}{T_{\mathrm{D}}}\right)^3 kT \int_0^{x_m} x^2\,dx = 9N\left(\frac{T}{T_{\mathrm{D}}}\right)^3 (kT)\left(\frac{T_{\mathrm{D}}^{\,3}}{3T^3}\right) = 3NkT$$

and the molar specific heat is

$$C_v = \frac{1}{n}\frac{dE}{dT} = 3N_{\mathrm{A}}k = 3R$$

In the high-temperature limit, the quantum theory of lattice specific heat gives exactly the same result, $C_v = 3R$, as the classical theory (Equation 8–16).

Consider now the low-temperature limit, where $kT \ll hf_{\mathrm{D}}$ and $x_m \to \infty$. The integral in Equation 8–26 can be evaluated in closed form to yield

$$\int_0^{\infty} \frac{x^3\,dx}{e^x - 1} = \frac{\pi^4}{15}$$

and Equation 8–26 becomes, for low temperatures,

$$E \approx \tfrac{3}{5}\pi^4 NkT\left(\frac{T}{T_{\mathrm{D}}}\right)^3$$

and the molar specific heat is

$$C_v = \frac{1}{n}\frac{dE}{dT} = \left(\frac{12\pi^4 R}{5T_{\mathrm{D}}^{\,3}}\right)T^3$$

In the low-temperature region, the lattice specific heat varies with T^3, again in accord with observation; see Figure 8–11.

The observed values for temperature-dependence of the specific heats for solids, *whether insulators or conductors*, are found to agree well with the Debye theory. See Figure 8–11. This is, at first sight, surprising, since the Debye theory takes into account the internal energy arising from the lattice vibrations but *not* the contribution to the specific heat of the conduction electrons. An electric or a thermal insulator is a material in which there are essentially no free electrons. It is to be expected, then, that the specific heat of an insulator would have contributions from the lattice vibrations alone, agreeing with experiment and the Debye theory, and that a conductor would have an added contribution to the specific heat from the free electrons.

A good conductor is imagined to have a large number of unbound, free electrons that can wander throughout the material. (These conduction electrons will show a net flow in one direction when a temperature gradient or external electric field is applied, accounting qualitatively for the high thermal and electrical conductivities of metals.) Let us compute the electronic specific heat of a solid under the assumption that each of the N atoms of the solid has one free, or conduction, electron and that the N free electrons may be regarded as *classical* particles of a Maxwell-Boltzmann gas. Three degrees of freedom are associated with the translational motion of each particle, and if these free electrons are regarded as classical particles wandering throughout the solid conductor, much as molecules in a gas, then the total electronic energy is $E_{\mathrm{e}} = N(\frac{3}{2}kT)$. The electronic contribution to the molar specific heat would be $C_{v,\mathrm{e}} = (1/n)(dE_{\mathrm{e}}/dT) = \frac{3}{2}(Nk/n) = \frac{3}{2}R$. But at the high-temperature limit of the Debye theory, the lattice specific heat is $3R$.

Thus, if the conduction electrons of a conductor were to behave as classical free particles, the conductor's *total molar* specific heat at relatively high temperatures would be $3R + \frac{3}{2}R$, or $\frac{9}{2}R$, whereas the observed value of C_v for both insulators and conductors is $3R$ (for $T \gg T_D$).

A classical treatment of electronic specific heat is clearly untenable. Electrons are *not* classical particles, but particles obeying the Pauli exclusion principle and the Fermi-Dirac statistics. The almost negligible electronic specific heat of metallic conductors, which is inexplicable with the classical free-electron theory, is understood on the basis of the quantum free-electron theory of a metal.

Example 8–2. A physical system also illustrating the Bose-Einstein distribution law, and closely analogous to the elastic solid in the quantum theory of specific heats, is a blackbody and its radiation. The successful theoretical explanation of the electromagnetic radiation from a solid that Max Planck made in 1900 marked the beginning of the quantum theory.

Solids at a finite temperature radiate a continuous electromagnetic spectrum, and blackbody radiation theory must account for how the radiation is distributed among the various frequency components and how it varies with the temperature of the emitting surface. By a *blackbody* is meant any material that absorbs *all* incident radiation, reflecting none. From the point of view of the quantum theory, a blackbody is, then, a material that has so many quantized energy levels, spaced over so wide a range of energy differences, that *any* photon, whatever its energy or frequency, is absorbed when incident on it. Since energy absorbed by a material would increase its temperature if no energy were emitted, a perfect absorber at a constant temperature (a blackbody) is also a perfect emitter.

A very good approximation of an ideal blackbody, one that can be achieved in the laboratory, is a hollow container, completely closed except for a small hole, through which radiation can enter or leave. There is very small probability that any radiation entering the container through the hole will be immediately reflected out again. Instead, the radiation is absorbed or reflected repeatedly at the inner walls, so that effectively all radiation incident through the hole is absorbed in the container. By the same token, the radiation leaking out through the hole is representative of the radiation in the interior.

When the container is maintained at some fixed temperature T, the inner walls emit and absorb photons at the same rate. Under these conditions, the electromagnetic radiation can be said to be in thermal equilibrium with the inner walls; in different language, the *photon gas* is in thermal equilibrium with the system of particles (in the walls) creating and absorbing the photons.

We can regard the equilibrium radiation within the blackbody enclosure in either of two ways: as electromagnetic waves or as particlelike photons:

1. When the radiation is treated as a collection of electromagnetic waves, the waves can be imagined to be repeatedly reflected from the walls of the container, producing standing waves.
2. When the radiation is treated as a collection of electromagnetic particles, the photons can be imagined to interact only with the container walls and to be in thermal equilibrium with the container.

This situation is, then, altogether analogous to what has just been discussed—quantized elastic waves, or phonons, imagined to exist within a rectangular box. Now we have photons of frequency ν and energy $\epsilon = h\nu$ traveling at speed c. For each photon propagation direction, however, there are two possible polarization directions for the transverse electromagnetic wave, but no longitudinal wave. Therefore, Equation 8–22, which gives the density of phonon states, yields the density of photon states with $f = \nu$, $v_t = c$, and the longitudinal term deleted:

$$g(\nu)\,d\nu = \frac{8\pi V}{c^3}\nu^2\,d\nu$$

Photons, like phonons, are bosons and follow the Bose-Einstein distribution law (Equation 8–2), so that the number of photons of energy ϵ within the energy range $d\epsilon$ is given by

$$n(\epsilon)\,d\epsilon = f_{\mathrm{BE}}g(\epsilon)\,d\epsilon = f_{\mathrm{BE}}g(\nu)\,d\nu = f_{\mathrm{BE}}\frac{8\pi V\nu^2\,d\nu}{c^3}$$

Since each photon has an energy $h\nu$, the radiation energy per unit volume $E(\nu)\,d\nu$ within the frequency range $d\nu$ is

$$E(\nu)\,d\nu = (h\nu)\frac{n(\epsilon)\,d\epsilon}{V} = f_{\mathrm{BE}}\frac{8\pi h\nu^3\,d\nu}{c^3}$$

$$E(\nu)\,d\nu = \left(\frac{8\pi h\nu^3}{c^3}\right)\left(\frac{1}{e^{h\nu/kT}-1}\right)d\nu \tag{8–27}$$

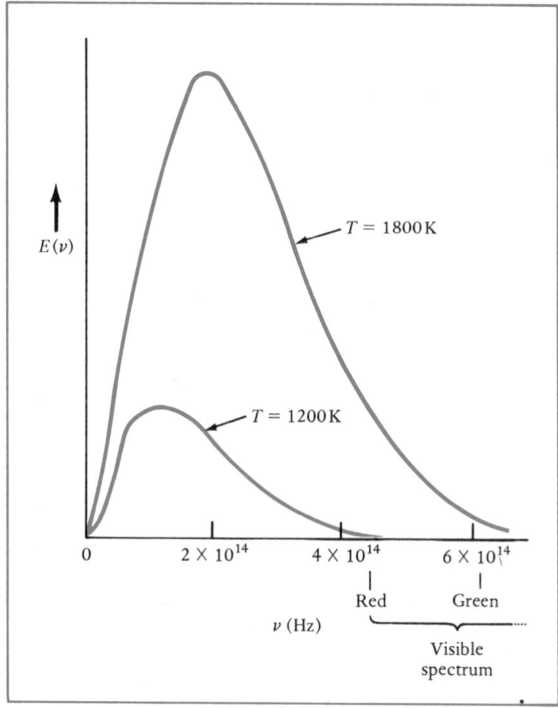

FIGURE 8–14. *Energy distribution as a function of frequency of electromagnetic radiation emitted by a blackbody for two temperatures.*

Equation 8–27 is the *Planck radiation relation;* it gives the radiated spectrum of a blackbody as a function of the photon frequency and blackbody temperature. The observed blackbody radiation, in agreement with Equation 8–27, is illustrated for two temperatures in Figure 8–14, p. 265. Note that as the temperature of the blackbody increases, the total radiation (area under curve in Figure 8–14) rapidly increases, and the radiation emitted comes from increasingly higher frequency regions of the electromagnetic spectrum. (See Problems 8–35 and 8–36.)

8-5 The Free-Electron Theory of Metals

The simple free-electron model of a metal, which W. Pauli and A. Sommerfeld first developed in 1927, is an example of a system of particles subject to the Pauli exclusion principle and obeying the Fermi-Dirac statistics.

A crystal of conducting material can be imagined to consist of two parts: (1) the nuclei, together with their tightly bound electrons, regularly spaced throughout the lattice, and (2) the weakly bound valence electrons, which can wander throughout the solid and therefore, can be considered to belong to the entire crystalline solid rather than to any one particular atom. Because a lattice atom has lost a valence electron, it bears a positive charge and attracts valence electrons. As any one electron moves throughout the solid it experiences a periodic potential, as shown in Figure 8–15; the electron feels a strong attractive force when it is close to the site of an atom, but essentially no net force when it is between atoms. We can, as a first approximation, take each electron to be truly free; that means we approximate the actual periodic crystalline potential by a potential that is constant (zero) within the interior of the crystal and rises abruptly to E_i at the surface, as shown in Figure 8–16. Nevertheless, we take the "free" electrons to interact with one another and with the lattice to the degree that the free electrons reach thermal equilibrium at the temperature of the lattice.

We want to find the number $n(\epsilon)\, d\epsilon$ of free electrons with energies in the range ϵ to $\epsilon + d\epsilon$, where $n(\epsilon) = g(\epsilon) f_{FD}(\epsilon)$. If we know the energy distribution of the free electrons, we shall be able to interpret some aspects of the macroscopic behavior of a metal.

The density $g(p)\, dp$ of states in the momentum range p to $p + dp$ is computed by considering the total number of ways in which N free electrons, regarded as waves, can be fitted within a three-dimensional box of side L. This problem is exactly analogous to finding the total number of ways in which N phonons, or quantized elastic waves, can be arranged within boundaries to form stationary wave patterns. Therefore, Equation 8–20, derived in Section 8–4, can be used directly—with, however, an additional factor 2, to account

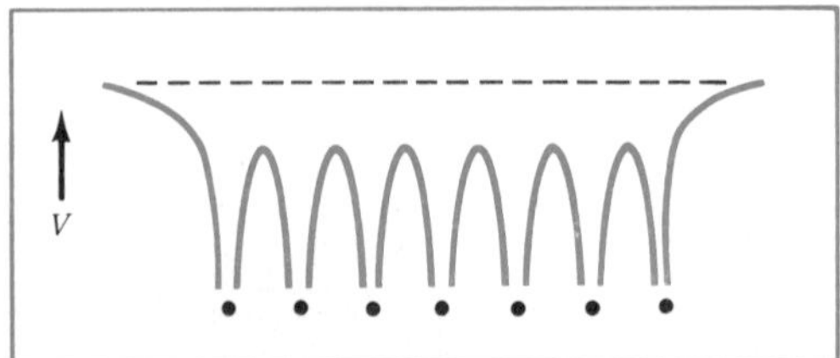

FIGURE 8–15. *Periodic potential seen by an electron in a crystalline solid.*

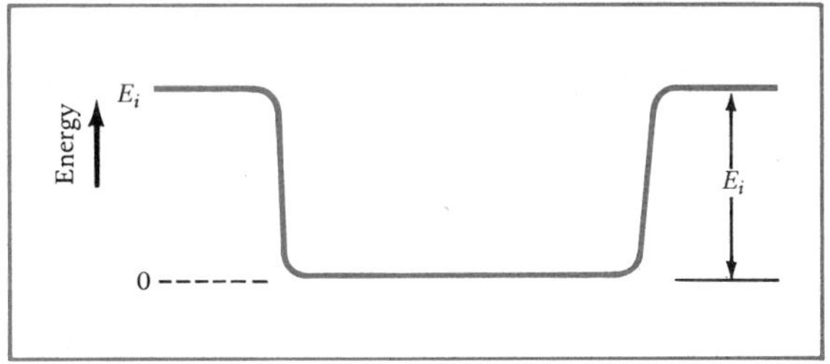

FIGURE 8–16. *Average potential energy of a free electron in a conducting solid.*

for two electrons, one with spin up and the other with spin down, having the same momentum components.

$$g(p)\,dp = \frac{8\pi V p^2\,dp}{h^3} \tag{8–28}$$

The total energy ϵ of a free electron is purely kinetic, and we can write

$$\epsilon = \frac{p^2}{2m}$$

$$d\epsilon = \frac{p}{m}\,dp = \frac{\sqrt{2m\epsilon}}{m}\,dp$$

Equation 8–28 then becomes

$$g(p)\,dp = \frac{8\pi V}{h^3}\,2m\epsilon\,\sqrt{\frac{m}{2\epsilon}}\,d\epsilon = g(\epsilon)\,d\epsilon$$

or

$$g(\epsilon)\,d\epsilon = C\epsilon^{1/2}\,d\epsilon \tag{8–29}$$

where

$$C = \frac{8\sqrt{2}\pi V m^{3/2}}{h^3} \tag{8–30}$$

Note that whereas $g(\epsilon)$ varies with ϵ^2 for photons and phonons (bosons), $g(\epsilon)$ varies with $\epsilon^{1/2}$ for molecules obeying the Maxwell-Boltzmann statistics and for electrons (fermions).

The distribution function for a collection of fermions (Equation 8–3) is given by

$$f_{\mathrm{FD}} = \frac{1}{e^{(\epsilon-\epsilon_{\mathrm{F}})/kT} + 1} \tag{8–31}$$

where $n(\epsilon)\,d\epsilon = f_{\mathrm{FD}}(\epsilon)g(\epsilon)\,d\epsilon$. Using Equations 8–29, we have

$$n(\epsilon)\,d\epsilon = \frac{C\epsilon^{1/2}\,d\epsilon}{e^{(\epsilon-\epsilon_{\mathrm{F}})/kT} + 1} \tag{8–32}$$

This equation gives the energy distribution of free electrons in equilibrium with a material at a temperature T. The significance of the quantity ϵ_{F}, called the Fermi energy, is best seen by considering the distribution of electron energies in a metal at the zero of absolute temperature.

METALS AT ZERO ABSOLUTE TEMPERATURE. A plot of the energy distribution (Equation 8–32) of an electron gas is shown in Figure 8–17 for $T = 0$. The plot is based on the product of two energy-dependent terms:

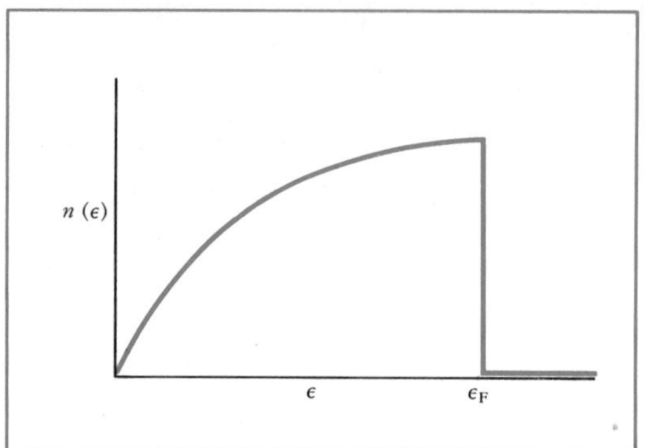

FIGURE 8–17. *Energy distribution of free electrons in a metal at $T = 0$ K.*

the factor $\epsilon^{1/2}$, which accounts for the parabolic rise in $n(\epsilon)$ from $\epsilon = 0$ upward, and the factor $1/(e^{(\epsilon-\epsilon_F)/kT} + 1)$, the Fermi-Dirac probability distribution function (Figure 8–8), which for $T = 0$ has the value of 1 from $\epsilon = 0$ to $\epsilon = \epsilon_F$ and zero for $\epsilon \geq \epsilon_F$.

Figure 8–17 shows that the free electrons do *not* all have zero kinetic energy at the zero absolute temperature, as the particles in a classical gas would; rather, there are electrons with finite energies up to a maximum energy, the Fermi energy ϵ_F. The free electrons have finite energies and are in motion, even at $T = 0$, because of the Pauli exclusion principle, to which electrons are subject—no more than two electrons, one for each of the two possible electron-spin orientations, are permitted in any particular energy state; hence, all the lowest states become filled, until the most energetic electrons reach $\epsilon = \epsilon_F$. At zero absolute temperature, the Fermi energy is the kinetic energy of the most energetic electrons; all states of lesser energy are filled and all states of greater energy are empty.

The value of the Fermi energy ϵ_F can be computed directly. The total number N of free electrons is

$$N = \int_0^{\epsilon_F} n(\epsilon)\, d\epsilon = C\int_0^{\epsilon_F} \epsilon^{1/2}\, d\epsilon = \tfrac{2}{3}C\epsilon_F^{\,3/2} \tag{8–33}$$

Using the value of C from Equation 8–30, we have

$$\epsilon_F = \frac{h^2}{2m}\left(\frac{3n}{8\pi}\right)^{2/3} \tag{8–34}$$

where n is the number of free electrons per unit volume and m is the electron mass. For copper, with one free electron per atom, the Fermi energy is computed from this equation to be 7.0 eV; for the conductor sodium, 3.1 eV. The values of ϵ_F are typically of the order of a few electron volts. Thus, the most energetic electrons of a conductor have a kinetic energy of several electron volts even at the lowest possible temperature.

The average kinetic energy $\bar{\epsilon}$ of a free electron at absolute zero can also be found directly, as follows.

$$\bar{\epsilon} = \frac{1}{N}\int_0^{\epsilon_F} \epsilon n(\epsilon)\, d\epsilon = \frac{C}{N}\int_0^{\epsilon_F} \epsilon^{3/2}\, d\epsilon = \frac{2C\epsilon_F^{\,5/2}}{5N}$$

But $C\epsilon_F^{\,3/2} = \tfrac{3}{2}N$, from Equation 8–33, and hence

$$\bar{\epsilon} = \tfrac{3}{5}\epsilon_F \tag{8–35}$$

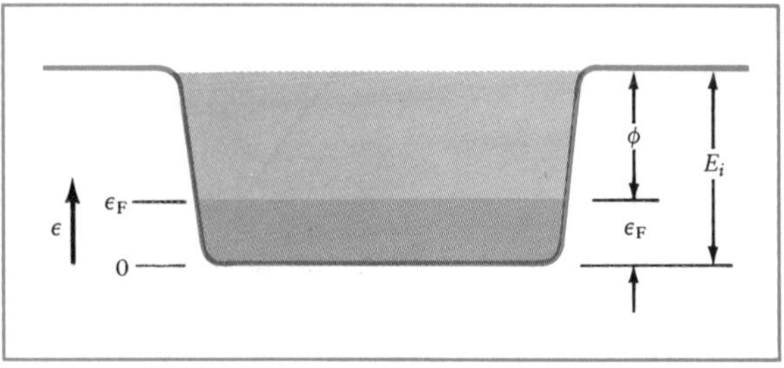

FIGURE 8–18. *Occupation of the energy levels by the free electrons of a metal at T = 0* K.

The relatively high average kinetic energy, again a few electron volts, of a free electron in a metal at $T = 0$ K may be contrasted with the average kinetic energy per classical free particle, $\frac{3}{2}kT$, which is a mere 0.04 eV at room temperature and zero at $T = 0$ K. This extraordinary behavior, in which electrons of a material at $T = 0$ K have a sizable kinetic energy, is strictly a quantum phenomenon, like the zero-point vibration of a simple harmonic oscillator.

An energy-level diagram showing the occupied energies of the free electrons of a metal at absolute zero is shown in Figure 8–18. Free electrons occupy energy states continuously up to the Fermi energy; all higher states are unfilled. The binding energy of the least tightly bound electrons of the metal (those at the Fermi surface) is the work function φ (Section 4–2); hence,

$$E_i = \epsilon_F + \varphi \qquad (8\text{–}36)$$

Since all three quantities in this equation can be determined independently, the simple features of the quantum free-electron model can be verified. For example, the conductor lithium, with one valence electron, has $E_i = 6.9$ eV, $\epsilon_F = 4.7$ eV, and $\varphi = 2.2$ eV. The values of ϵ_F and E_i are nearly temperature-independent and Equation 8–34 holds for all moderate temperatures.

METALS AT A FINITE TEMPERATURE. What are the changes that occur when the temperature of the conductor is raised? A plot of the energy distribution for a finite temperature is shown in Figure 8–19, following Equation 8–32. The energy distribution at a moderate temperature is similar to the distribution at $T = 0$ K; the only significant difference is that the corners of the plot, at $\epsilon = \epsilon_F$, are slightly rounded. The rounding of the corners at the Fermi energy follows from the change in the Fermi-Dirac distribution function $f_{FD}(\epsilon)$, shown in Figure 8–8.

At room temperature, kT is about 0.03 eV, an energy far lower than the typical Fermi energy ϵ_F, which is always a few electron volts. Therefore, the Fermi-Dirac distribution function $f_{FD}(\epsilon)$ is unchanged from $f_{FD}(\epsilon)$ at $T = 0$ except for an electron energy ϵ close to the Fermi energy ϵ_F. For $kT \ll \epsilon_F$,

$$f_{FD}(\epsilon) = \frac{1}{e^{(\epsilon-\epsilon_F)/kT} + 1} \approx 1 \qquad \text{when } \epsilon \ll \epsilon_F$$

$$\approx 0 \qquad \text{when } \epsilon \gg \epsilon_F$$

The energy distribution $n(\epsilon)$ at a finite temperature T differs from the distribution at $T = 0$ K only within the small energy range in which $|\epsilon - \epsilon_F| \approx kT$; that is, the energy distributions differ significantly only within the range kT

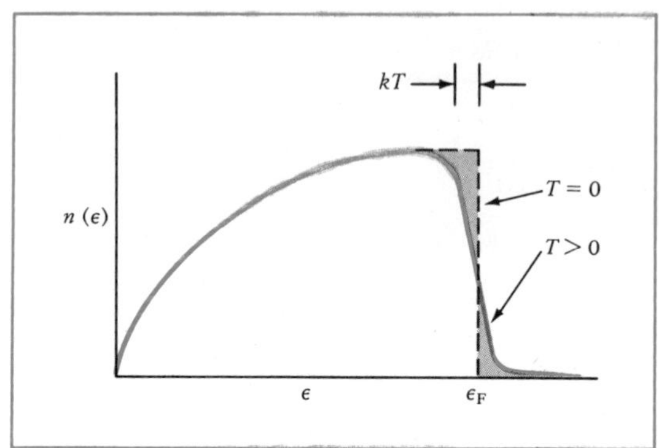

FIGURE 8–19. *Energy distribution of free electrons in a metal at a finite temperature.*

of the Fermi energy. We see from Equation 8–31 that when $\epsilon = \epsilon_F$, then $f_{FD}(\epsilon) = \frac{1}{2}$; thus, for a finite temperature, the Fermi energy corresponds to that energy for which there is a one-to-one chance that the state will be occupied. It can be shown that ϵ_F is independent of the temperature up to temperatures of a few thousand degrees.

The distribution in energy of the free electrons of a metal at a finite temperature (Figure 8–19) has an interesting interpretation. Electrons whose energy is much lower than the Fermi energy remain in the low energy states that they occupy at $T = 0$ K. Only the most energetic electrons, those within the range kT of the Fermi energy, have available to them unoccupied higher-energy states, to which they can be excited by thermal excitation. The low-energy electrons are, so to speak, locked in their energy states when the metal is excited thermally, since no unoccupied states are available to them within an energy range kT, either above or below their states at $T = 0$ K. Roughly speaking, as the temperature is raised, a fraction of only those electrons lying within the small energy range kT of ϵ_F are promoted to states on the high side of the Fermi energy, again within the range kT of the Fermi energy. Thus, as the temperature of a conductor is raised, only a very small fraction of all the free electrons can shift to higher energy states and thereby increase the electronic energy content of the solid.

We see from Figure 8–19 that the electronic energy $E_e(T)$ of the metal at some finite temperature T is larger than its energy $E_e(0)$ at $T = 0$ K because of the shifting of a small fraction of the electrons (shaded areas) from below to above the Fermi energy. The number of electrons promoted to higher energies is proportional to kT. Furthermore, the average increase in the energy of each promoted electron is approximately kT. Therefore, we can write

$$E_e(T) = E_e(0) + A(kT)^2$$

where A is a constant. The molar *electronic specific heat* $C_{v,e}$ is then given by

$$C_{v,e} = \frac{1}{n}\frac{dE_e(T)}{dT} = \gamma T \qquad (8\text{–}37)$$

where γ is a constant. The quantum free-electron theory thus predicts that the electronic contribution to the specific heat of a conductor is directly proportional to the absolute temperature. A more detailed analysis shows that $\gamma = (\pi^2/2)z(k/\epsilon_F)R$, where z is the number of valence electrons per atom.

Therefore Equation 8–37 becomes

$$C_{v,e} = \frac{\pi^2}{2} z \frac{kT}{\epsilon_F} R \tag{8–38}$$

In copper, a typical conductor, $z = 1$, $\epsilon_F = 7.0$ eV, and $kT = 0.03$ eV at room temperature. Using Equation 8–38, we find that $C_{v,e} \approx 0.02R$ for copper at room temperature. The *lattice* molar specific heat for copper is nearly $3R$ at this temperature, so that the electronic contribution to the specific heat is indeed negligible at moderate temperatures.

The electronic specific heat becomes comparable to the lattice specific heat only at the very lowest temperatures (a few degrees Kelvin), for which the lattice specific heat is itself quite small.

If the temperature of a conductor is raised to several thousand degrees, so that kT becomes comparable to the work function of the metal, some of the free electrons will have enough energy to escape from the metal surface, their kinetic energy ϵ then equaling or exceeding the internal potential energy E_i (see Figure 8–18). Thus, severe thermal excitation of free electrons can result in their emission from the metal, a process known as *thermionic emission.*

Although the quantum free-electron theory of metals can account for such properties as the electronic specific heat and thermionic emission, it cannot account for other important properties of solids, properties that depend on the fact that the electrons, even the valence electrons, in a metal are *not* completely free. In a more realistic model of solids, the electrons experience a nonconstant potential within the metal (see Chapter 9).

Example 8–3. A certain metal has a Fermi energy $\epsilon_F = 4.0$ eV and is at temperature $T = 400$ K. Find the number of free electrons per unit energy $n(\epsilon)$ for (*a*) $\epsilon = \epsilon_F + kT$, and (*b*) $\epsilon = \epsilon_F - kT$.

(*a*) By Equation 8–32, the number of electrons per unit energy is

$$n(\epsilon) = \frac{C\epsilon^{1/2}}{e^{(\epsilon - \epsilon_F)/kT} + 1}$$

For $\epsilon = \epsilon_F$, $n(\epsilon_F) = C\epsilon_F^{1/2}/2$. Therefore, to simplify, we let $C = 2n(\epsilon_F)/\epsilon_F^{1/2}$ in Equation 8–32. We then have

$$n(\epsilon) = \frac{2(\epsilon/\epsilon_F)^{1/2}\, n(\epsilon_F)}{e^{(\epsilon - \epsilon_F)/kT} + 1} \tag{8–39}$$

For $\epsilon = \epsilon_F + kT$, Equation 8–39 becomes

$$n(\epsilon_F + kT) = \frac{2[(\epsilon_F + kT)/\epsilon_F]^{1/2}\, n(\epsilon_F)}{(e^1 + 1)}$$

With $T = 400$ K, then

$$kT = (1.38 \times 10^{-23}\ \text{J/K})(400\ \text{K})(1.6 \times 10^{-19}\ \text{J/eV})^{-1}$$
$$= 0.034\ \text{eV}$$

and for $\epsilon_F = 4.0$ eV, we find that $n(\epsilon_F + kT)$ becomes

$$n(4.034\ \text{eV}) = \frac{2(4.034/4)^{1/2}\, n(4.0\ \text{eV})}{e + 1} = 0.54n(4.0\ \text{eV})$$

Above the Fermi energy, the energy distribution decreases rapidly with energy, dropping to one-half the value at ϵ_F when the energy is only 0.034 eV larger than the Fermi energy (4.0 eV).

(*b*) For $\epsilon = \epsilon_F - kT$, Equation 8–39 becomes

$$n(\epsilon_F - kT) = \frac{2[(\epsilon_F - kT)/\epsilon_F]^{1/2}\, n(\epsilon_F)}{(e^{-1} + 1)}$$

Again, with $T = 400$ K ($kT = 0.034$ eV) and $\epsilon_F = 4.0$ eV,

$$n(\epsilon_F - kT) = n(3.966 \text{ eV}) = \frac{2(3.966/4.0)^{1/2}\, n(4.0 \text{ eV})}{e^{-1} + 1}$$

$$n(3.966 \text{ eV}) = 1.5\, n(4.0 \text{ eV})$$

For energies below the Fermi energy, the energy distribution increases rapidly, rising to three-halves the value at ϵ_F when the energy is only 0.034 eV smaller than the Fermi energy of 4.0 eV. The rapid change in $n(\epsilon)$ about the Fermi energy is illustrated in Figure 8–19.

SUMMARY

The two principal types of molecular binding are ionic (heteropolar) binding, resulting from the electrostatic attraction of ions, and covalent (homopolar) binding, resulting from the sharing of valence electrons.

The three kinds of probability distribution for dealing with large numbers of weakly interacting particles are:

- Maxwell-Boltzmann—classical particles; example, molecules of ideal gas
- Bose-Einstein—particles with integral spin (bosons); example, phonons, photons
- Fermi-Dirac—particles with half-integral spin (fermions); example, electrons

The properties of these distribution functions are summarized in Table 8–1.

The number $n(\epsilon_i)$ of particles with an energy of ϵ_i is given by $n(\epsilon_i) = f(\epsilon_i)g(\epsilon_i)$, where $f(\epsilon_i)$, the distribution function, is the average number of particles in the state i, and $g(\epsilon_i)$ is the number of states with the energy ϵ_i. For very closely spaced energy levels, one may write $n(\epsilon)\, d\epsilon = f(\epsilon)g(\epsilon)\, d\epsilon$, where $g(\epsilon)$, the density of states, gives the number of states per unit of energy, and $n(\epsilon)$ is the number of particles per unit energy.

PROBLEMS

Molecular binding, §8–1

● **8–1.** The radii of the closed electron shells of the ion Na^+ is 0.97 Å and of Cl^- is 1.81 Å. Compare the separation distance when these two ions are just "touching" with the equilibrium distance r_0, and with the 4.0-Å separation distance assumed in Section 8–1 to be larger than the touching distance.

● **8–2.** Estimate the probability that the electron is somewhere between the two protons in the hydrogen molecular ion, H_2^+, shown in Figure 8–3.

● **8–3.** Sketch the probability ψ^2 for finding the electron along the line joining the two protons in the ion H_2^+ if the protons are separated by 100 Å (see Figure 8–3).

8–4 (*a*) Compare the Coulomb potential energy between the ions Na^+ and Cl^- when their closed shells are just touching (2.78 Å) with the estimated value of the potential energy from Figure 8–2. The equilibrium separation distance for the Na^+Cl^- molecule is 2.36 Å and the minimum potential energy is -4.24 eV. (*b*) Is the overlap of the electron clouds discernible at this distance?

8–5. The electron affinity of bromine is 3.6 eV; the ionization energy of sodium is 5.1 eV. (*a*) Find the difference in energy between a neutral Na atom and a neutral Br atom far separated and a Na^+ and a Br^- ion far separated. (*b*) The radius of the closed electron shell of Na^+ is 0.97 Å and of Br^- is 1.96 Å. What is the Coulomb potential energy of the Na^+Br^- molecule when the two ions are just touching? (*c*) Find the difference in energy between a neutral Na atom and a neutral Br atom just touching, and the Na^+Br^- molecule when ions touch (ignore overlap of electron clouds).

8–6. The ionization energy of a hydrogen atom is 13.6 eV. Using this information and other data given in Section 8–1, show that the energy required to separate an H_2 molecule into an ion H_2^+ and an electron is 15.4 eV.

8–7. The radii of the closed shells of ${}_{11}Na^+$ is 0.97 Å and of ${}_{17}Cl^-$ is 1.81 Å. Compare each of these with the radius of its adjacent inert gas and justify the difference. The radius of ${}_{10}Ne$ is 1.12 Å; that of ${}_{18}Ar$, 1.54 Å.

Statistical distribution laws, §8–2

● **8–8.** A system consists of three identical particles (*A*, *B*, and *C*), each one of which has available to it the quantized energy levels shown in Figure 8–5 (states ϵ_1, ϵ_2, ϵ_3, and ϵ_4, with $\epsilon_1 < \epsilon_2$, $\epsilon_2 = \epsilon_3$, and $\epsilon_4 = 2\epsilon_2$). Display the twelve distributions of the three particles among the available states that yield a total energy of the system of $2\epsilon_2$ (relative to $\epsilon_1 = 0$).

● **8–9.** Show that the Bose-Einstein and Fermi-Dirac distribution functions approach the Maxwell-Boltzmann distribution in the high-energy limit ($\epsilon \gg kT$).

8–10. Assume that the quantum energy states available to a system of weakly interacting particles are evenly spaced, with the ground state energy $\epsilon_1 = 0$. The energy difference between adjacent states is much lower than kT, which is 0.03 eV, and the weighting factor $g(\epsilon) = 1$. For the following distribution functions find the ratio of the number of particles with energy 10 eV to the number with 5 eV: (*a*) A Fermi-Dirac gas, where $\epsilon_F = 10$ eV. (*b*) A Bose-Einstein gas.

■ **8–11.** A certain gas obeys the Fermi-Dirac distribution law. For a Fermi energy of 4.00 eV and a gas temperature of 5800 K, find the probability that a state will be occupied with a particle for the following energies: (*a*) $\epsilon = 0$; (*b*) $\epsilon = 3.00$ eV; (*c*) $\epsilon = 5.00$ eV.

Maxwell-Boltzmann statistics and ideal gas, §8–3

● **8–12.** Two containers have the same volume. One contains 3.0 mol of an ideal gas at room temperature (300 K) and atmospheric pressure. The other contains 3.0 mol of ideal gas at 600 K. If one plotted $n(\epsilon)$ versus ϵ curves for these two gases, after Figure 8–10, what would be the ratio of the areas under the curves?

● **8–13.** For an ideal gas at temperature *T*, show that the distribution function $f(\epsilon + \Delta\epsilon)$, for molecules having an energy $\Delta\epsilon$ larger than those at energy ϵ, is given by $f(\epsilon)\, e^{-\Delta\epsilon/kT}$.

8–14. (*a*) Show that in an ideal gas at temperature *T*, the number of molecules at energy $(\epsilon + \Delta\epsilon)$ can be written as $n(\epsilon + \Delta\epsilon) = [1 + (\Delta\epsilon/\epsilon)]^{1/2}\, e^{-\Delta\epsilon/kT}\, n(\epsilon)$. (*b*) If $\Delta\epsilon$ is much less than both ϵ and kT, show that

$$n(\epsilon + \Delta\epsilon) \approx \left[1 + \left(\frac{\Delta\epsilon}{2\epsilon} - \frac{\Delta\epsilon}{kT}\right)\right] n(\epsilon)$$

(*c*) For what range of values of ϵ does $n(\epsilon)$ increase with ϵ? (*d*) decrease with ϵ?

8–15. The distribution in the translational kinetic energy of the molecules of an ideal gas is given, according to Equation 8–11, by $n(\epsilon)\, d\epsilon = Ce^{(-\epsilon/kT)}\epsilon^{1/2}\, d\epsilon$. Show that the most probable molecular translational kinetic energy is $\frac{1}{2}kT$.

8–16. From Problem 8–15, the maximum value of $n(\epsilon)$ for an ideal gas exists at $\epsilon_m = \frac{1}{2}kT$. Find the ratio $n(\bar{\epsilon})/n(\epsilon_m)$, where $\bar{\epsilon}$ is the average translational kinetic energy per molecule (Equation 8–13).

8–17. Consider a gas of hydrogen *atoms*. At what temperature will 99 percent of the hydrogen atoms be in the ground state ($n = 1$), essentially all the remaining atoms being in the first excited state ($n = 2$)? [Note that two levels are available in the ground state of hydrogen (electron configuration $1s^2$) while eight levels are available in the first excited state (electron configuration $1s^22s^22p^6$); therefore, the respective statistical weights for the two states are $g_1 = 2$ and $g_2 = 8$.]

The result for this question—a temperature of nearly 20,000 K—implies that for temperatures of this order or lower, atomic hydrogen cannot emit radiation, since essentially no inelastic collisions of hydrogen atoms occur, and therefore no atoms occupy excited states.

8–18. Show that the Maxwell-Boltzmann velocity distribution of an ideal gas of classical point particles is given by $n(v) = Av^2e^{-mv^2/2kT}$, where *A* is a constant. (*Hint:* $n(v)\, dv = n(\epsilon)d\epsilon$.)

8–19. Using $\int_0^\infty x^2\, e^{-ax^2}\, dx = \sqrt{\pi/16a^3}$, show that the constant *C* in Equation 8–12 has the value $2N/\sqrt{\pi(kT)^3}$.

8–20. Use Equation 8–11 and the fact that *C* in Equation 8–11 varies as $T^{-3/2}$ to find the energy ϵ (eV) in Figure 8–10 at which the two distribution curves cross one another.

8–21. Assume that the universe is a sphere of radius 1.0×10^{27} m. If this volume were filled with atomic hydrogen atoms at standard temperature and pressure, what is the probability that one of these atoms would be in the first excited state? (See Example 8–1.)

8–22. The atoms of the H_2 molecule undergo oscillations along the interatomic axis at the frequency of 1.3×10^{14} Hz (in the infrared region of the electromagnetic spectrum). The allowed energies of such a simple harmonic oscillator are, according to the quantum theory, given by $E_v = (v + \frac{1}{2})hf$, where f is the vibrational frequency and the vibrational quantum number has the allowed values $v = 0, 1, 2, \ldots$ (Figure 5–23*b*); the statistical weight $g = 1$. (*a*) Show that the ratio of the number of oscillators in any state v to those in the next lower state $(v - 1)$ is given by $e^{-hf/kT}$. (*b*) At what temperature will 1 percent of the H_2 molecules occupy the first excited vibrational state ($v = 1$), with the remaining 99 percent in the ground ($v = 0$, or zero-point oscillation) state?

The result of part (*b*) shows that for H_2 molecules at a temperature lower than 1000 K, the excited vibrational states are essentially unpopulated. This implies that the energy an H_2 molecule gains cannot appear as vibrational energy at this temperature. At temperatures at and above about 10,000 K, a significant fraction of H_2 molecules occupy excited vibrational states; this is manifest through an increase in the specific heat for H_2.

■ **8–23.** Air is at room temperature (300 K). One state for an air molecule has the energy $\epsilon_1 = \frac{3}{2}kT$ (the average translational kinetic energy per molecule). Another state has the energy $\epsilon_2 = \frac{3}{2}kT + \Delta\epsilon$. At what value of energy $\Delta\epsilon$ would the number of molecules populating the state with energy ϵ_2 be one-half the number populating the state with energy ϵ_1?

■ 8–24. A paramagnetic ion can be considered to consist of an essentially free electron relatively loosely bound to an atom. When a collection of paramagnetic ions is in thermal equilibrium, the energy distribution of the "free" electrons is governed by the temperature of the paramagnetic material and the Maxwell-Boltzmann distribution. Suppose that a paramagnetic material at temperature T is subject to an external magnetic field **B**. (*a*) Show that the excess of electron spins aligned against the magnetic field over those aligned with the magnetic field relative to the number of electron spins in the lower-energy state is given approximately by $2\beta B/kT$, where β is the Bohr magneton. (*b*) By what percentage do the electrons aligned against an external magnetic field exceed the electrons aligned with the magnetic field for a paramagnetic material at a temperature of 300 K in a magnetic field of 0.30 T?

Quantum specific heats, §8–4

● **8–25.** Diamond has a Debye temperature of 1860 K. What is the highest frequency elastic oscillation propagated in diamond?

● **8–26.** Use Figure 8–11 to estimate the molar specific heats of (*a*) gold, (*b*) diamond, and (*c*) lead at room temperature and compare with the experimental values: gold, 0.031 cal/gm·C°; diamond, 0.124 cal/gm·C°; copper, 0.092 cal/gm·C°. (The atomic masses of gold, diamond, and copper are 197, 12, and 63.5.)

● **8–27.** The molar specific heat of copper is 2.8R and of beryllium is 1.7R at room temperature. Which material has the higher Debye temperature?

8–28. The element copper has a Debye temperature of 343 K and an atomic mass of 63.5 g/mol. Using Figure 8–11 and the low-temperature expression for C_v where appropriate, estimate the lattice contribution to the specific heat of copper (in cal/g·K) (*a*) at 100 K, and (*b*) at 10 K? Compare these values with the experimental values of 0.061 and 2.1×10^{-4} cal/g·K.

8–29. Assume that the speed of elastic waves in a solid is 2.2×10^3 m/s. (*a*) Calculate the momentum and energy of phonons in a box 10 cm on a side, in the available states for which n_x, n_y, n_z is 1, 0, 0 and 1, 1, 0. (*b*) What is the difference in energy between these two states? (*c*) Repeat parts (*a*) and (*b*) for the states 100, 0, 0 and 100, 1, 0.

8–30. In a copper cube 1.0 cm on a side, what is the number of transverse elastic vibrational waves having wavelengths between 1.0×10^{-5} and 1.1×10^{-5} cm? The speed of transverse elastic waves in copper is 2.3×10^3 m/s.

8–31. Show that if the atoms of a lattice are assumed to be arranged in a cubical array (as in Figure 5–1), the Debye frequency corresponds to an elastic wave for which the distance between adjacent atoms is approximately one half-wavelength. Assume, for simplicity, that $v_t = v_l$. It is assumed in the Debye theory that the elastic waves are propagated through an essentially continuous medium, for which the wavelength is long compared with the interatomic distance. The Debye cutoff occurs when the wavelength is so short that elastic waves cannot be propagated.

8–32. The speed of the transverse elastic waves through copper is 2.32×10^3 m/s and of the longitudinal waves is 4.76×10^3 m/s. The number density of copper atoms is 8.49×10^{28} atoms/m³. (*a*) Find the Debye frequency f_D. (*b*) Calculate the Debye temperature $T_D = hf_D/k$ for copper and compare the value with that given in Figure 8–11.

8–33. For electromagnetic waves, what is the number of modes between the wavelengths 9.9 cm and 10.1 cm in a black box 1.0 m on a side?

8–34. Show that the Planck radiation relation (Equation 8–27) reduces for low frequencies ($h\nu/kT \ll 1$) to

$$E(\nu)\, d\nu = \left(\frac{8\pi\nu^2 kT}{c^3}\right) d\nu$$

This is the *classical Rayleigh-Jeans radiation formula;* it correctly describes the low-frequency radiation from a blackbody but fails for high frequencies. (The failure of the classical relation for high frequencies—a prediction of infinite radiation at high frequencies—is usually referred to as the ultraviolet catastrophe.)

8–35. The total energy radiated by a blackbody at temperature T over the entire range of frequencies is given by

$E_T = \int_0^\infty E(\nu)\,d\nu$. Using the blackbody radiation relation (Equation 8–27) and changing to the variable $x = h\nu/kT$, show that $E_T = CT^4$, where C is a constant. The proportionality of total radiated energy to the fourth power of the absolute temperature is known as the *Stefan-Boltzmann relation.*

8–36. The wavelength at which a blackbody has a maximum radiated energy depends on the temperature of the blackbody. Write the Planck blackbody relation, Equation 8–27, in terms of wavelength, $E(\lambda)\,d\lambda$. Then set $dE(\lambda)/d\lambda = 0$ to find that the wavelength λ_{max} corresponding to the peak in the energy distribution. $E(\lambda)$ depends on temperature according to the relation $\lambda_{max}T =$ constant. This relation, in which the constant has the value 2.878×10^7 Å·K, is known as the *Wien displacement law.*

8–37. Measurements in radio astronomy show that the universe is filled with blackbody radiation. This is held to survive as a vestige of the primordial fireball of electromagnetic radiation produced shortly after the creation of the universe according to the Big Bang cosmological theory. This universal background blackbody radiation has a temperature of 2.7 K. At what wavelength does the peak in this low-temperature blackbody-radiated energy fall? (See Problem 8–36).

Free-electron theory, §8–5

● **8–38.** Use Equation 8–34 to show that the Fermi energy for a typical metal is of the order of magnitude of a few electron volts.

● **8–39.** Use Equation 8–38 to verify that the electronic specific heat of a typical conductor is very small compared with the lattice specific heat at room temperature.

● **8–40.** At what temperature would zinc have kT equal to the Fermi energy of 10.9 eV?

8–41. For the element sodium, the Debye temperature is 160 K, the Fermi energy is 3.1 eV, and the number of valence electrons per atom is 1. At what temperature are the contributions from the lattice and the electronic specific heats equal?

8–42. Show that for very low temperatures ($T \to 0$), the slope of molar specific-heat-versus-temperature curve is always smaller for the lattice contribution than for the valence electron contribution.

8–43. A certain metal has a Fermi energy of 4.0 eV and is at room temperature (300 K). By how much must the electron energy exceed the Fermi energy for the electron population probability to be $\frac{1}{4}$? (*Hint:* First try $\epsilon - \epsilon_F = kT$.)

8–44. The Fermi energy for copper is 7.1 eV; therefore, a free electron in copper at the Fermi level has an energy of 7.1 eV for any temperature, including absolute zero. Calculate the temperature of a classical collection of particles—for example, the molecules of an ideal gas—for which the average energy per particle is 7.1 eV.

8–45. For silver, the work function is 4.46 eV, the internal potential energy seen by a free electron is 9.94 eV, the atomic weight is 108, and the density is 10.6 g/cm^3. Calculate the number of free electrons per atom in silver.

8–46. The mean drift speed of charge carriers in an electric conductor is typically far less than 1 m/s, even for relatively large currents through conductors of typical dimensions. Show that a free electron in a conductor with an energy of a few electron volts, comparable to the Fermi energy, has a speed that far exceeds the drift speed of an electron that is contributing to a current in a conductor.

8–47. In treating the free-electron theory of metals (Section 8–5), it was assumed for simplicity that the actual periodic potential a "free" electron experienced (Figure 8–15) was a constant potential (Figure 8–16). This approximation is valid provided that the wavelength of a free electron is large compared with the spacing d between atoms. Show that an electron at the Fermi level, one having an energy equal to the Fermi energy, has a wavelength that is approximately $2d$. This indicates that the approximation of constant potential is just barely satisfied for high-energy electrons and the free-electron model merely an approximate description of a conductor.

9
Quantum Effects and Devices

John Bardeen (born 1908, United States). Only recipient of two Nobel Prizes in physics. 1956, Nobel Prize (with W. B. Shockley and W. H. Brattain) for invention of transistor at Bell Telephone Laboratories. 1972, Nobel Prize (with L. C. Cooper and J. R. Schrieffer), for theory of superconductivity, University of Illinois. (Photo, Meggers Gallery of Nobel Laureates—courtesy of the American Institute of Physics, Niels Bohr Library.)

Some of the basic principles governing many-particle systems have important application. In particular, the semiclassical theory of conductivity is reviewed and the phenomenon of superconductivity treated briefly. Then the elements of the band theory of semiconductors is developed and applied to devices—the *p-n* junction, the solar cell, the photoemitter, and the solid-state detector. Finally, the fundamental principles operating in the design of lasers are considered, and it is shown how the special properties of a laser are exploited in holography.

9-1 Conductivity and Superconductivity

SEMICLASSICAL THEORY OF CONDUCTIVITY. In the semiclassical theory of conductivity, the conduction electrons are regarded strictly as particles in thermal equilibrium with the lattice of the conductor.† Although the conduction electrons are considered particles when interacting with the lattice, it is assumed that they obey the quantum Fermi-Dirac statistical distribution law; therefore, a conduction electron has an average kinetic energy of the order of a few electron volts and an average speed of the order of 10^6 m/s at all temperatures (see Section 8–5). The model is indeed *semi*-classical.

When no external electric field exists within a conducting material, the conduction electrons are in random motion, colliding with one another and with the lattice. Altogether, no net kinetic energy is gained or lost by the conduction electrons in such collisions. For a copper conductor at room temperature, the conduction electron speeds (near the Fermi energy) are of the order of 10^6 m/s and the mean time interval between collisions by a conduction electron is approximately 5×10^{-14} s; such electrons have a mean free path of the order of $(10^6 \text{ m/s})(5 \times 10^{-14} \text{ s}) \approx 500$ Å.

When an electric potential difference is applied across a conductor, the potential difference produces an electric field within the conductor. Under this influence the conduction electrons are accelerated and gain energy. (Strictly, only those conduction electrons with an energy close to the Fermi energy are promotable and can gain energy.) Under the influence of the driving electric field, the conduction electrons acquire an additional drift speed, one superimposed on their random thermal motion. For a typical conductor of ordinary dimensions, the average *drift* speed arising from the external electric field can be of the order of a mere 10^{-4} m/s, which is far smaller than the 10^6 m/s *thermal* speed (see Problem 9–5). After colliding with the lattice, a conduction electron is again accelerated by the electric field to acquire again an additional drift speed. The conduction electron, in its next collision with the lattice, loses the additional kinetic energy it has gained by virtue of its drift under the electric field. In short, the energy conduction electrons gain from the driving field appears finally as additional thermal energy in the lattice, and the collisions of conduction electrons with the lattice are the means of this energy transfer. Since the mean drift speed (10^{-4} m/s) is far smaller than the mean thermal speed (10^6 m/s), the external electric field has essentially *no* effect on conduction electron speeds; this means also *no* effect on the mean time between collisions and the mean free

† See Weidner and Sells, *Elementary Classical Physics,* vol. 2, 2d ed., Section 27–5.

path between collisions. This circumstance—that the mean time interval between collisions is unaffected by the driving electric field—is the microscopic basis of Ohm's law; the mean *drift* speed is proportional to the electric field, so that the current is proportional to the potential difference. Since many conductors follow Ohm's law precisely over an enormous range of current magnitudes, the semiclassical theory of conductivity has strong support.

Collisions between conduction electrons and the lattice are the origin of a finite resistivity in the semiclassical theory of conductivity. To the degree that collisions become more frequent—with a shorter time interval between collisions and a shorter mean free path—the resistivity of the conducting material rises (and its conductivity decreases). This simple model agrees with observation in several respects:

1. The resistivity of most materials rises with the temperature. This observation has a ready explanation. As the temperature rises and the thermal motions of the lattice atoms increase, the atoms are increasingly displaced from a geometrical lattice arrangement; consequently, collisions between conduction electrons and the lattice atoms become more probable.
2. When a metal is annealed, its resistivity decreases. Annealing means elevating the temperature of the material and then cooling it slowly. This procedure allows lattice atoms that had been displaced from a regular geometrical array in the crystal to return to the regular array. Reducing such "geometrical impurities" in this way allows conduction electrons to travel farther between collisions, and the resistivity is thereby reduced.
3. A conductor containing impurity atoms has a higher resistivity than a pure conductor. The impurity atoms do not fit in the geometrical array of the conducting crystal, and their presence increases the probability of collisions by conduction electrons.

In short, any change in the lattice that increases the probability of collisions by conduction electrons affects the material's resistivity. If somehow one could keep conduction electrons from colliding with lattice atoms, there would be an infinite time between collisions and an infinite mean free path; and then, according to the semiclassical theory of conductivity, the material's resistance would be zero.

SUPERCONDUCTIVITY. Superconductivity is a phenomenon that H. Kamerlingh Onnes first observed in 1911. It refers to the abrupt drop to zero of the electrical resistance of certain materials. (See Figure 9–1.) Tungsten, for example, the element with the lowest critical temperature, becomes superconducting below 0.015 K; and niobium, the element with the highest critical temperature, has a critical temperature of 9.5 K, below which it is superconducting. Certain chemical compounds and alloys are also superconductors; an alloy of niobium and germanium (Nb_3Ge) has the highest known critical temperature, 23.2 K. Those materials that have relatively high conductivities at room temperature (for example, silver) do not, however, exhibit superconductivity.

It must be emphasized that the resistance of a superconductor is truly zero, not merely very small; currents induced in macroscopic superconduct-

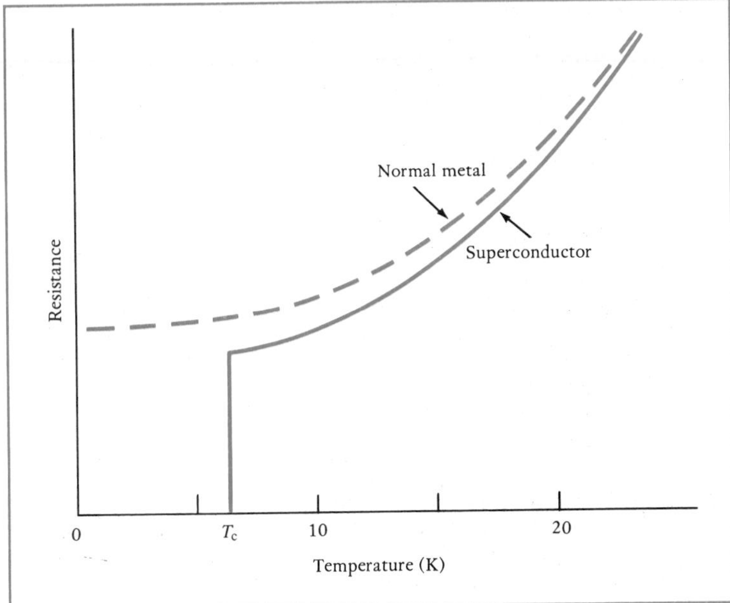

FIGURE 9–1. *Variation of resistance with temperature for a superconductor (solid line) and a normal conductor (dashed line) at very low temperatures. The resistance of the superconducting material drops abruptly to zero at the critical temperature T_c.*

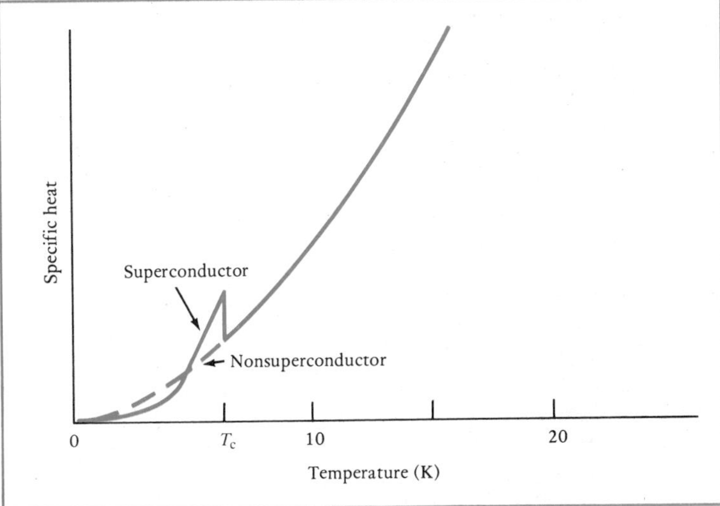

FIGURE 9–2. *Variation of the specific heat with temperature for a superconductor (solid line) and a nonsuperconductor (dashed line) at very low temperatures. As the temperature is lowered, the specific heat of the superconducting material rises abruptly at the critical temperature T_c.*

ing materials persist indefinitely without diminution. Superconductors show, besides their extraordinary electrical resistivity, other nontypical behavior in their magnetic and thermal properties. For example, a strong enough external magnetic field applied to a superconductor can make it become a normal conductor (the Meissner effect). Moreover, the electronic specific heat of a superconductor shows an abrupt change at the critical temperature. (See Figure 9–2.) The superconducting state must be regarded as a distinctive state of matter.

One might imagine that as suggested by the semiclassical model for conductivity, the superconducting electrons of a superconductor make no collisions with the lattice, or at least no collisions that transfer energy to the lattice. The successful quantum theory of superconductivity actually requires a relatively strong interaction between a superconducting electron and the lattice.

A current in any material implies that electrons are in motion around the loop of conducting material. A persistent superconducting current implies that the superconducting electrons traverse an entire macroscopic loop and continue to do so indefinitely. Superconducting electrons must, in a quantum description, be characterized through their associated wave functions; so the existence of macroscopic superconducting current loops implies that superconducting electrons have coherent wave functions, wave functions that extend not merely over some fairly limited region of atomic space but around the entire macroscopic dimensions of the current. The phenomenon of superconductivity—unlike most quantum effects, which are subtle or are manifest only microscopically—is a quantum effect on a large scale.

We have examples of coherent electron wave functions in atomic structure. For a bound state of the hydrogen atom, for example, the electron wave function must fit around the nucleus. Indeed, an allowed state is one for which the fit of the wave function is exact.

The microscopic theory of superconductivity of J. Bardeen, L. N. Cooper, and J. R. Schrieffer (the BCS theory), for which the Nobel Prize in physics was awarded in 1972, accounts in detail for the properties of superconductors. Interaction between superconducting electrons and the lattice is central to superconductivity. In any conducting material, free electrons are coupled through the Coulomb force to the positive ions of the lattice. A "collision" of a superconducting electron with the lattice produces lattice vibration; the superconducting electron, in interacting with the lattice, can be thought of as "emitting" a phonon. Now a second electron, not necessarily physically close (on a microscopic scale) to the first electron, may also interact with the lattice. Indeed, the emission of a phonon by one electron may be followed by the absorption of this phonon by the second electron, leaving the net lattice energy unchanged. In effect, the two electrons interact. This interaction lowers the energy of the system, so that the electron-electron interaction via the lattice and a phonon exchange is effectively an attractive force; and if this force exceeds in magnitude the necessarily repulsive Coulomb force between a pair of electrons, it leads to a net attractive force and an overall lowering of the energy of the system. These pairs, so-called Cooper pairs, then are superconducting electrons weakly bound to one another, with a potential energy of the order of magnitude of 10^{-3} eV. If such pairs are created, the conductor is a superconductor.

The BCS theory shows that the energy of the phonon hf_D exchanged by the electrons of a Cooper pair, where f_D is the Debye frequency, is directly proportional to the critical temperature T_c, below which the material becomes superconducting. The theory also shows that the energy 2Δ that is required to break apart the electrons of a Cooper pair (an energy Δ for each of the two electrons) is related to T_c by

$$2\Delta = 3.52\, kT_c \tag{9–1}$$

The interaction between a pair of electrons through the exchange of a phonon is most effective, from considerations based on momentum conservation, when the two electrons have oppositely directed momentum and also

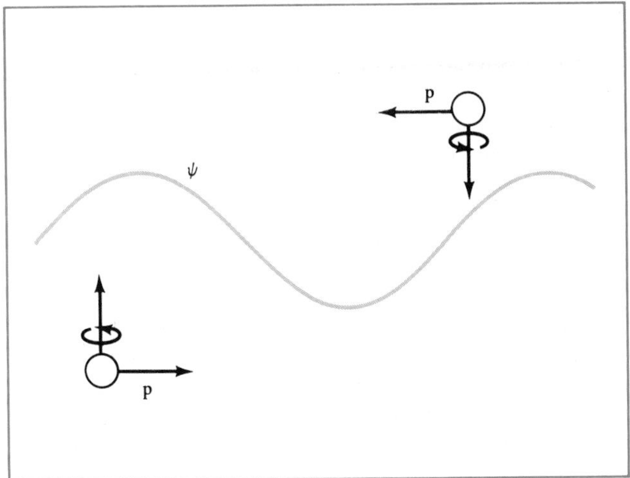

FIGURE 9–3. *Schematic representation of a Cooper pair. The electron spins are opposite. The electron momenta are opposite in direction and differ by a precisely defined amount. The pair is represented by a single coherent wave function.*

oppositely directed spins. When no superconducting current exists within a material below the critical temperature and consequently no net flow of superconducting electrons, the momenta of the two electrons of a Cooper pair are exactly equal in magnitude and opposite in direction. On the other hand, when a superconducting current exists, with a net electron flow at any one point within the material, then the momenta of the two electrons of a Cooper pair, although opposite in direction, are of unequal magnitude. This momentum difference is very precisely defined. In short, when a superconducting current exists, the two electrons of the Cooper pair have opposite spins, travel in opposite directions, have a sharply defined difference in momentum magnitude, and may be separated from one another by substantial distances on an atomic scale. (See Figure 9–3.)

Now consider the implications of having Cooper pairs for each of which the net momentum of the pair is precisely specified. If the pair's momentum is sharply defined, the uncertainty Δp_x in the momentum of the pair approaches zero. From the uncertainty principle ($\Delta x \geq h/\Delta p_x$), it follows that the uncertainty in the location Δx of the pair must become very large, approaching macroscopic dimensions. We can express this differently by saying that the wave function associated with the two electrons of a Cooper pair is not confined to a relatively small region of space; it may extend all the way around the loop in which superconducting current exists. Thus, the general requirement of a single coherent wave function for superconducting electrons is met.

Since superconductivity depends critically on the interaction between electron and lattice that leads to the existence of Cooper pairs, an interaction too weak to exceed the Coulomb repulsion between electrons will not permit superconductivity. Materials that are good conductors at room temperature have a relatively small electron-lattice interaction. Such materials are silver and gold, and other noble metals, and they do not become superconductors, even at the lowest temperatures, since the electron-lattice interaction is too small to allow the formation of bound Cooper pairs.

Example 9–1. Lead becomes superconducting at temperatures below 7.175 K. (*a*) What is the binding energy of a Cooper pair of electrons in superconducting lead? (*b*) What is the minimum frequency of electromagnetic radiation, illuminating superconducting lead just below the critical temperature, that will break a Cooper pair?

(*a*) The BCS theory relates the binding energy of a Cooper pair to the critical temperature T_c, as given by Equation 9–1:

$$2\Delta = 3.52kT_c = \frac{(3.52)(1.38 \times 10^{-23}\ \text{J/K})(7.175\ \text{K})}{(1.60 \times 10^{-19}\ \text{J/eV})}$$

$$= 2.18 \times 10^{-3}\ \text{eV}$$

(*b*) To break apart the Cooper pair, the photon energy $h\nu$ must equal or exceed the binding energy of the pair 2Δ. Therefore,

$$h\nu = 2\Delta$$

$$\nu = \frac{(2.18 \times 10^{-3}\ \text{eV})(1.60 \times 10^{-19}\ \text{J/eV})}{6.63 \times 10^{-34}\ \text{J}\cdot\text{s}}$$

$$= 5.27 \times 10^{11}\ \text{Hz}$$

This frequency corresponds to millimeter electromagnetic waves of wavelength

$$\lambda = \frac{c}{\nu} = \frac{3 \times 10^{8}\ \text{m/s}}{5.27 \times 10^{11}\ \text{Hz}} = 0.569\ \text{mm}$$

9-2 The Band Theory of Solids: Conductors, Insulators, and Semiconductors

The *band theory* of solids is the basis for understanding such phenomena as electrical and thermal conductivities and especially the distinction between conductors, insulators, and semiconductors. The band theory can account for the tremendous range in electrical resistivities from a good insulator to a good conductor (a ratio as large as 10^{30}.) Although a detailed, quantitative treatment of the band theory of solids involves a rigorous application of wave mechanics, one can understand some of the important qualitative features of this highly successful theory without mathematical analysis. There are two approaches to the band theory:

1. The theory of F. Bloch (1928) emphasizes that a valence electron in a metal does not see a constant potential in its motion through the crystal but experiences a periodic potential, corresponding to the periodicity of the crystalline structure.
2. The theory of W. Heitler and F. London (1927) considers the effects on the electron wave functions when isolated atoms are brought close together to form a crystalline solid.

A perfect crystal has nuclei located at fixed lattice positions (called *sites*) within the crystal; these nuclei form a geometrically ordered array. The electrons, whose total number is such that the crystal as a whole is electrically neutral, are of two kinds: (1) an inner electron tightly bound to an individual nucleus, and remaining at all times with this nucleus, and (2) an outer, or valence, electron only weakly bound to any one nucleus and therefore free to wander from one nucleus to another. Such a wandering valence electron is considered to belong to the entire crystal rather than to any one nucleus. Strictly, a valence electron experiences a periodic electric potential arising from the fixed nuclei and the remaining electrons (Figure 8–15).

Determining the allowed states and energies of valence electrons is fundamentally determining what electron wavelengths are possible within the crystal. Specifically, one is confronted with finding what electron wavelengths can be fitted to the periodic potential that characterizes the interaction between a valence electron and the remainder of the crystal—a complicated problem indeed.

The second approach to the band theory of solids, proposed by Heitler and London, lends itself more easily to a qualitative description. We consider N identical, isolated (noninteracting) atoms. Each free atom has its own particular set of energy levels, and the permitted states of some one atom are identical with those of any other. For an example, the energy-level diagram of a lithium atom is shown in Figure 9–4*a* together with the number of available states for each allowed energy. Since all atoms have identical energy-level diagrams, the combined energy-level diagram of the N atoms, all widely separated, is simply the diagram of the single atom, except that the number of available states for each energy level is now increased by a factor N. If a single atom can accommodate 2 electrons in an s energy level and 6 electrons in a p energy level, then N atoms have room for $2N$ electrons in the s energy level and $6N$ electrons in the p energy level. See Figure 9–4*b*.

When the N atoms are brought together so that the separation between adjacent atoms is comparable to the separation of the atoms in a crystalline solid, they interact fairly strongly with one another. A consequence of the interaction is to broaden the energy levels of the system so that those states that were degenerate earlier, with the same energy, now have slightly different energies. (We saw an example of this splitting in Figure 8–4, in the case of two hydrogen atoms that were brought together to form H_2.) The effect of bringing together very many of the originally isolated atoms (Figure 9–4*b*) to form a bound system is shown schematically in Figure 9–4*c*. The $2N$ available states for the 1s energy level are no longer coincident

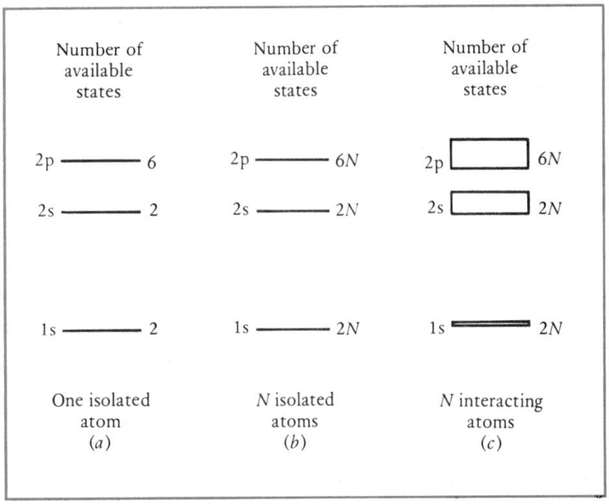

FIGURE 9–4. *Schematic representation of the energy levels and states available to (a) one isolated lithium atom, (b) N isolated lithium atoms, and (c) N interacting lithium atoms.*

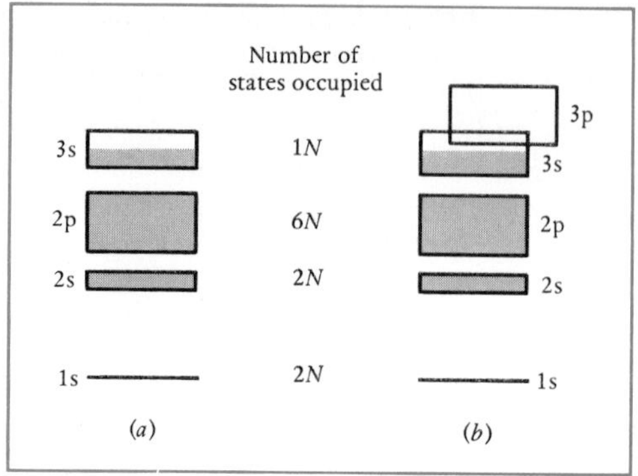

FIGURE 9–5. *Schematic representation of the energy bands and their occupancy by electrons in sodium. The dark regions correspond to occupied states.*

but spread essentially continuously throughout the 1s *energy band.* Similarly, there are $2N$ available states in the 2s energy band and $6N$ available states in the $2p$ energy band. The regions between the available energy bands cannot be occupied by any electron at all and are known as *forbidden bands.* The width and separation of the energy bands depend on the particular crystalline material with which they are associated.

So far, we have discussed only the *available* states in the energy bands of a solid and not how the electrons occupy them. To see how the electrons are distributed among the various energy bands of the crystal, we first consider the conductor sodium in the ground state at $T = 0$ K. For simplicity, we assume at first that the several energy bands do not overlap. The electron configuration of an isolated sodium atom in the ground state is $1s^2 2s^2 2p^6 3s^1$; thus, all electron shells are filled to the 3s shell, which contains only one electron. Therefore, the sodium crystal has energy bands for each of the electron shells of the atom, as shown in Figure 9–5*a*. The 1s, 2s, and 2p bands are all filled, with $2N$, $2N$, and $6N$ electrons. The 3s band, which has $2N$ available states, is only half-filled, with N 3s electrons. Note that the 1s band, corresponding to inner electrons, is narrow compared with higher bands; this 1s band is relatively narrow because the inner electrons are strongly attracted to their parent nuclei and less influenced by neighboring electrons and nuclei.

That the uppermost energy band of a conductor, such as sodium, is only partially filled is responsible for the high electrical conductivity of these materials. Consider what happens to the occupation of the energy bands when an electric field is applied to the metal. It is then possible for the electrons in the partially filled band to gain small amounts of energy by the action of an external electric field and so be promoted to the continuum of available states lying immediately above.

The distribution of the electrons among the available states at some finite temperature differs only slightly from the distribution at absolute zero. The shift in occupation of electrons is controlled in the band theory as in the simple free-electron theory by the Fermi-Dirac statistics. Consequently, the significant change in the distribution occurs only for those very few electrons that lie within a region of energy kT about the uppermost filled level (the Fermi level) at $T = 0$ K.

The energy bands of sodium shown in Figure 9–5*a* do not include the unoccupied 3p band, which comes immediately after the 3s band. Not only is the 3p band broad, but also it overlaps the 3s band, as shown in Figure 9–5*b*.

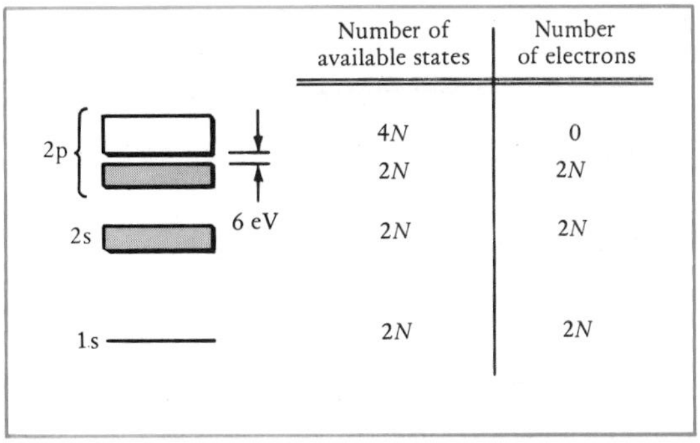

	Number of available states	Number of electrons
2p	$4N$	0
2p	$2N$	$2N$
6 eV		
2s	$2N$	$2N$
1s	$2N$	$2N$

FIGURE 9–6. *Schematic representation of the energy bands and their occupancy in diamond. Note the sizable forbidden region between the two 2p bands.*

Thus, the number of unoccupied levels available to the electrons in the 3s shell is further increased, and this leads to a correspondingly high electrical conductivity.

The very low electrical conductivity of an insulator, such as diamond, $_6$C, can also be understood from the band theory. The electron configuration of carbon in its ground state is $1s^22s^22p^2$. Because the 2p energy band is only partially, filled, with $6N$ available states but only $2N$ electrons, it might at first appear that diamond would be an electrical conductor. There are, however, two distinct 2p energy bands, separated from each other by a forbidden region of 6 eV, as shown in Figure 9–6. This separation of the 2p band arises from the specific crystalline structure of diamond. The lower 2p band is filled completely with $2N$ electrons in the $2N$ available states. At room temperature, kT is about 0.03 eV; thus, the gap width for diamond is so much larger than the thermal excitation energy kT that virtually no electrons occupy the upper 2p band. When an external electric field is applied, electrons cannot gain enough energy to be promoted to the upper, unoccupied 2p band. Thus, an external field cannot cause a net electron flow, or an electric current. In short, a substance such as diamond is a good insulator in that there is a sizable energy gap between a filled band, called the *valence band,* and the next empty (but available) energy band, the *conduction band.*

Similarly, the electrons in the valence band of an insulator cannot have their energies raised by the absorption of photons whose energy is less than the width of the energy gap. This means that in diamond, all visible-light photons are transmitted through the crystal without absorption; this is to say that diamond is perfectly transparent to visible light. By the same token, conductors are opaque; this follows, since a continuum of unfilled energy states lies immediately above the filled states, to which electrons in a conductor can be promoted by the absorption of photons over a continuum of wavelengths.

Some crystalline solids, such as silicon and germanium, have, like diamond, a filled valence band and an empty conduction band; they have, however, a much *smaller* forbidden region separating the bands, as shown in Figure 9–7. For silicon, the energy gap between the valence and conduction bands is 1.1 eV, for germanium, 0.70 eV, both an order of magnitude smaller than the insulator diamond. At very low temperatures, the thermal excitation of the valence electrons is so small that essentially none of these electrons is excited to a state in the conduction band; thus at low temperatures these materials behave as insulators.

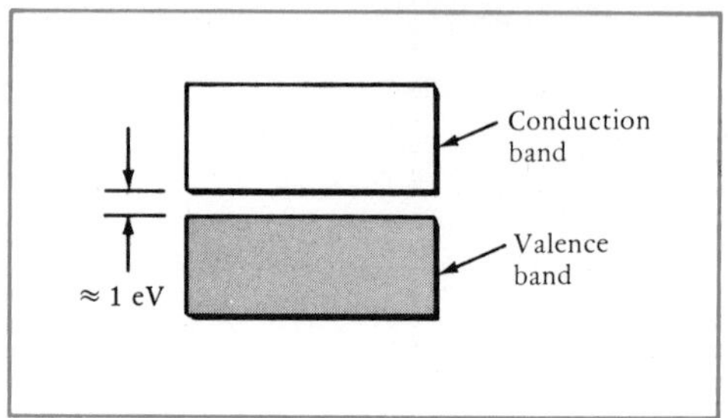

FIGURE 9–7. *Energy bands for semiconductors, such as silicon or germanium, with a small, forbidden energy gap separating the valence and conduction bands.*

Consider now, however, the occupation of states in the conduction band at higher temperatures. If the gap width is small, some electrons will occupy available states within the conduction band, and under the influence of an external electric field they can participate in a net electron flow through the material. At the same time, the unfilled states in the valence band, called *holes,* arising from missing electrons, also contribute to the electric current. The conductivity of such materials lies between the very low values for insulators and the very high values for conductors; they are known as *semiconductors.* The type of semiconductor just described, which consists strictly of atoms of a single type and depends for its semiconductivity on the electrons in the conduction band that have been thermally excited across the energy gap, is known as an *intrinsic semiconductor.*

A second type of semiconductor, called an *extrinsic,* or *impurity, semiconductor,* depends for its semiconductivity on the presence within a semiconducting crystal of a few atoms called *impurity atoms;* they are of a type different from the atoms of the crystal. Before we examine how impurity atoms influence the energy-band structure, we must first examine the bonding of atoms and impurity atoms within the crystal. Silicon and germanium both lie in the fourth column of the periodic table; the outermost shell of silicon has a $3s^2 3p^2$ electron configuration, and of germanium, a $4s^2 4p^2$. Each silicon or germanium atom in the crystalline solid is bound to its four nearest neighbors by covalent bonds, each saturated bond representing two shared valence electrons, as shown schematically in Figure 9–8. This structure is called the diamond structure, since it is like the diamond crystal ($_6C$, with $2s^2 2p^2$ electrons).

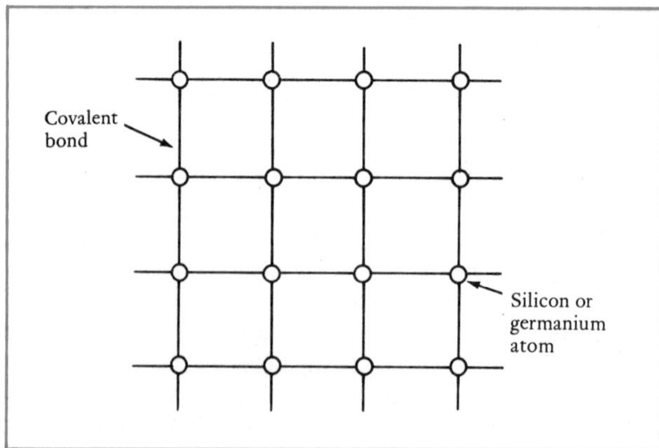

FIGURE 9–8. *Covalent bonding of the atoms in a crystalline solid of silicon or germanium. Each line between nearest neighboring atoms represents a saturated covalent bond, with the sharing of two valence electrons.*

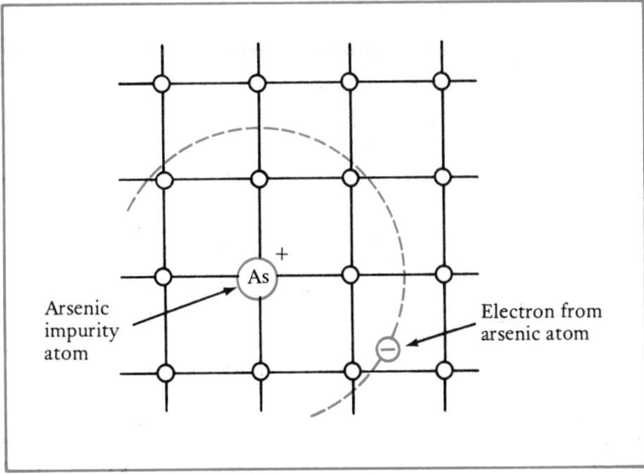

FIGURE 9–9. *An example of a donor impurity atom in silicon, producing n-type semiconductivity.*

Now consider the effect of a few atoms of arsenic in a silicon crystal. The element ${}_{33}$As lies in the fifth column of the periodic table and has the ground-state electron configuration $4s^2 4p^3$. Thus, a neutral atom of arsenic has one more electron than a neutral atom of silicon does, and when an arsenic atom replaces a silicon atom in the crystal, there is one additional electron that cannot be bonded covalently, since each covalent bond is saturated with two electrons (see Figure 9–9). The unpaired electron from the impurity atom can be imagined to move in a very large orbit (because of the high dielectric constant in the crystal's interior) about the nucleus of the arsenic atom. The electron consequently is very weakly bound, and one can regard the impurity atom as having donated a carrier of negative electric charge to the whole crystal. Therefore, impurity atoms in the fifth column of the periodic table are *donors* and produce *n-type impurity semiconductors.*

Suppose now that the impurity atoms are from elements in the third column of the periodic table, such as gallium, ${}_{31}$Ga, with an electron configuration of only *one* p electron. Then the covalent bonding around an impurity atom in the crystal is not complete. One incomplete bond is associated with each impurity atom (see Figure 9–10). The covalent bonding

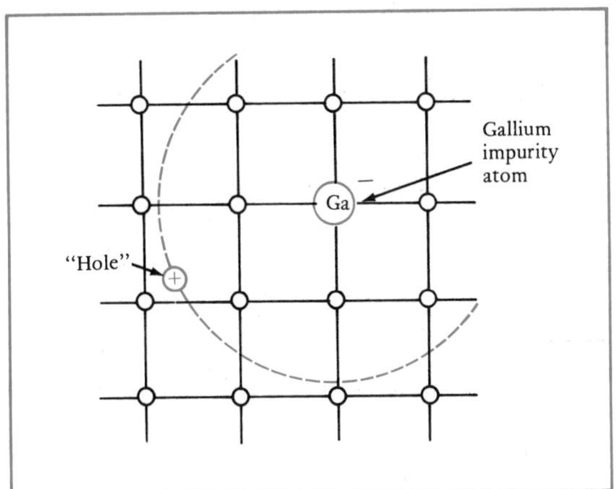

FIGURE 9–10. *An example of an acceptor impurity atom in silicon, producing p-type semiconductivity.*

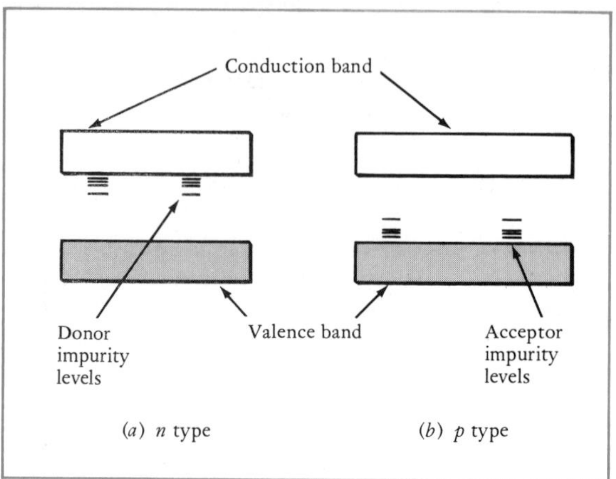

FIGURE 9–11. *Energy-level diagram of a silicon crystal modified by (a) donor impurities and (b) acceptor impurities.*

around such an impurity atom is complete when the impurity atom accepts an electron from the valence band, and one vacancy, or hole, is thereby produced in the valence band. This hole can be imagined to move in an orbit about the now negatively charged impurity ion. When one speaks of the motion of a hole, imagined as an equivalent positive charge, one is really describing the motion of electrons in the opposite direction. An impurity atom that accepts an electron for complete bonding is an *acceptor,* and a semiconductor containing such impurities (and hence, holes), is known as a *p-type impurity semiconductor.*

The effect on the energy-band structure of impurities in silicon is shown in Figure 9–11. The n-type impurities introduce additional, closely spaced energy levels lying just below the edge of the conduction band; these levels are occupied by the weakly bound electrons the donor atom has contributed. The electrons in these discrete energy levels may be excited to the available states in the conduction band that lie immediately above. The p-type impurities introduce closely spaced levels just above the valence band. Therefore, electrons from the valence band just below can be excited upward and occupy these available impurity states; the slight conductivity of the semiconductor is thereby accounted for. The conductivity of an impurity semiconductor can be controlled by the relative concentration of the impurity atoms.

9-3 Semiconducting Devices

Semiconductors, or more generally solid-state devices, have numerous technological applications. Used singly or in combination, they can form rectifiers, amplifiers (transistors), or detectors of current in electronic circuits; they can be used to convert electromagnetic radiation into an electric current, or current into radiation; and they can be used to detect high-energy particles. In what follows we discuss two-terminal devices only, not three-terminal devices, such as the transistor.

THE *p-n* JUNCTION. Consider the junction between a p-type semiconductor, with positive charge carriers, or holes, and an n-type semiconductor, with negative charge carriers, or electrons; see Figure 9–12. The boundary

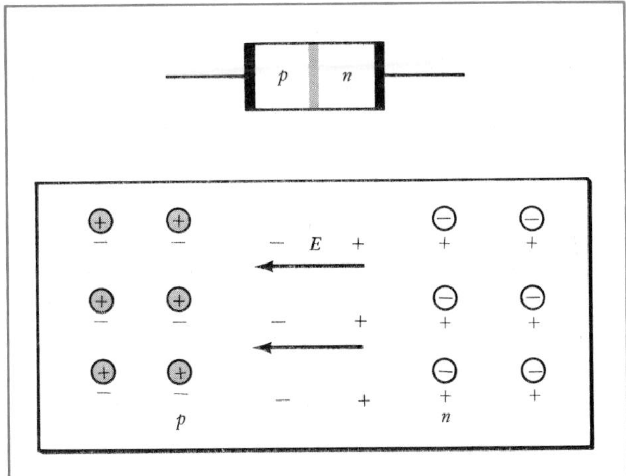

FIGURE 9–12. *A p-n junction. The charge carriers from impurity atoms, positive on the p side and negative on the n side, are shown encircled.*

is not produced merely by pressing together the two semiconductor types; rather, the boundary changes continuously from p type to n type through so-called doping procedures, in which impurity atoms are diffused in a controlled way throughout the material. A concentration of holes within the p-type material exists on the left of the figure, and a concentration of electrons within the n-type material on the right. Both sides of the junction are electrically neutral. At the junction itself, the mobile electrons and holes combine. Thus, the n-type side adjoining the junction has, through a loss of electrons, a net positive charge, and the p-type side adjoining the junction has, through a loss of holes, a net negative charge. The charge layers on either side of the junction thereby produce an internal electric field from the n-type side to the p-type (toward the left in Figure 9–12).

In its effects this internal electric field produces a force to the left on the holes on the p-type side; strictly, the force is on electrons, which move to fill vacancies. The internal electric field also produces a force to the right on the electrons on the n-type side, thereby preventing further recombination of electrons and holes at the junction.

Suppose now that an *external* electric field is applied across the junction through an electric potential difference across the electrodes attached to the p-type and n-type sides. If the p-type side is negative and the n-type side positive, the external electric field is to the left, in the same direction as the internal electric field. Then the electrons and holes are further kept from moving or recombining; under these circumstances, the electric current through the junction is small, arising solely from thermally generated electrons and holes at the junction. This condition is referred to as the *reverse bias* of a p-n junction.

When the polarity of the potential difference is reversed, so that the p-type side is positive and the n-type side negative, the external applied electric field is to the right. This condition is known as *forward bias.* The external electric field to the right exceeds the internal electric field to the left, and holes are driven to the right and electrons to the left. Both kinds of charge carrier contribute to a conventional current to the right, which may be substantial. In short, the p-n junction is a good conductor for forward bias but a poor conductor for reverse bias. The junction therefore acts as a rectifier, passing current easily in one direction but not the other. Moreover, the junction is a nonlinear, or nonohmic, circuit element; its current-voltage characteristic is not the straight line that corresponds to Ohm's law.

THE SOLAR CELL. The simplest kind of solar cell, or a photodiode, uses a p-n junction to convert radiant energy to an electric current. The main effect is that a photon absorbed in a semiconductor may release a bonded electron, to produce a mobile conduction hole and an electron. Mobile charge carriers, conduction electrons and holes, can arise in a semiconductor in three ways: (1) thermal excitation; (2) donor (n-type) and acceptor (p-type) impurity atoms; and (3) photon absorption.

Thermal excitation in a pure semiconductor produces conduction electrons and holes that are responsible for the relatively low conductivity of such materials. Thermal energy of the lattice may give an initially bonded electron enough energy to free it from its bond and allow it to wander freely through the material as a conduction electron. Freeing an initially bonded electron from its attachment to the two atoms between which it forms a bond produces a hole at that site. The hole, an equivalent positive charge, wanders relatively freely through the semiconductor as bonded electrons jump from one site to another adjacent site to fill the hole. The mobile hole also contributes a conduction current but is less mobile than its opposite, the conduction electron.

In the language of energy bands, an electron freed from a bond by thermal excitation has gained enough energy to be promoted from the valence band to the conduction band; the minimum energy is that of the energy gap. At the same time, a hole is created in the valence band. See Figure 9–13. If an external electric field is applied to the pure semiconductor, thermally excited electrons are forced to move in one direction and holes in the opposite direction, and both charge carriers contribute to a current in the same direction.

How does the conductivity of a pure semiconductor depend on temperature? The conductivity at any one temperature is proportional to the number of charge carriers. The number of charge carriers in turn depends on thermal excitation and therefore on the temperature. Indeed, the number of charge

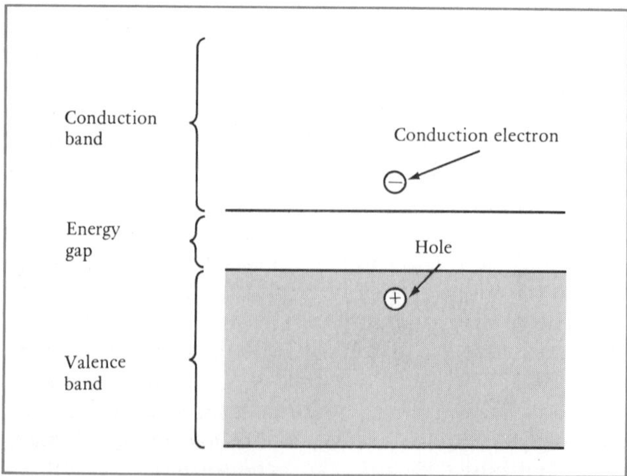

FIGURE 9–13. *A bound valence-band electron elevated to the conduction band by thermal excitation. Both the electron in the conduction band and the hole in the valence band are free charge carriers.*

carriers, it can be shown, is proportional to $e^{-E_g/2kT}$, where E_g is the gap energy. Therefore, the conductivity of a pure semiconductor is proportional to $e^{-E_g/2kT}$, and for a semiconductor the conductivity *rises* rapidly with temperature. (This is unlike ordinary conductors, for which the conductivity *decreases* slowly with temperature.)

Thermally produced conduction electrons and holes wander throughout the semiconducting material until each meets an opposite. For example, when a conduction electron meets a hole, the electron may fill the hole to become bonded again, and the number of charge carriers is thereby decreased. Thermal excitation produces electron-hole pairs continuously; these charge carriers wander through the material and eventually undergo mutual annihilation. The distance a conduction electron travels on the average between its creation and later annihilation is called the *diffusion length,* typically of the order of 10^{-6} m.

The second way by which charge carriers can be produced in a semiconductor is through impurity atoms. Impurity atoms in a semiconductor can produce *permanent* conduction electrons or holes. A donor impurity atom (n-type) contributes an electron free to wander throughout the lattice. When the unbonded electron leaves the impurity atom, the resultant positively charged ion remains *fixed* at its original lattice site. Similarly, an acceptor impurity atom (p-type) introduces a hole, free to wander throughout the material, while the acceptor atom becomes a negatively charged ion, fixed to its lattice site. Thermally produced conduction electrons and holes are both mobile, whereas donor atoms produce both fixed positive ions and mobile electrons and acceptor atoms produce both fixed negative ions and mobile holes.

Whereas the conductivity of an intrinsic, or *pure,* semiconductor can be changed by varying the material's temperature, the conductivity of an extrinsic (p- or n-type) semiconductor can be changed in that way and also by varying the number of impurity atoms.

Finally, electron-hole pairs can be produced in a semiconductor by photon absorption. If the photon energy exceeds the energy gap between the valence and conduction bands (1.1 eV in silicon)—that is, if the photon's energy equals or exceeds the minimum energy to promote an electron from the valence to the conduction band—then a mobile electron is in the conduction band and a mobile hole in the valence band. Photons in the visible region (1.8–3.1 eV) certainly have the necessary energy for creating electron-hole pairs. See Figure 9–14. (Electron-hole production in a semiconductor by photon absorption is somewhat analogous to electron-positron pair production through the absorption of a gamma-ray photon; or it can be viewed as an internal photoelectric effect in the sense that an electron initially bound to an atom is freed by photon absorption, with a positive ion left behind.)

Photon-excited charge carriers are susceptible to mutual annihilation just as thermally excited electron-hole pairs are. Furthermore, because the electron and hole paths through the semiconductor are random, such pairs in an ordinary semiconductor produce no net current. For the relatively transient electron-hole pairs resulting from photon absorption to produce a net measurable current—with the conversion of electromagnetic energy into usable energy derived from an electric current—the pairs must be created at a location that favors their acceleration under an electric field before the charge carriers disappear.

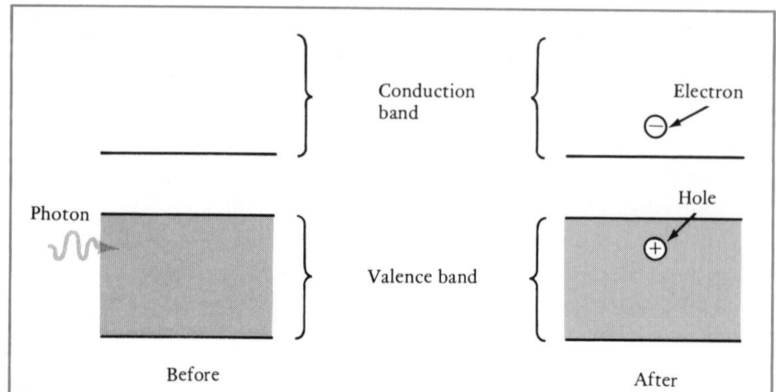

FIGURE 9–14. *A photon incident on a semiconductor surface (before) is absorbed and a mobile electron and mobile hole are created (after).*

A silicon *p-n* junction provides these circumstances. In the junction shown in Figure 9–15, the charge distribution is that for equilibrium with no applied external fields. The *p* side has mobile holes and fixed negative ions; the *n* side has mobile electrons and fixed positive ions. At the junction itself, there are relatively few electrons or holes. Mobile electrons from the *n* side and mobile holes from the *p* side annihilate each other, leaving the region of the *p-n* junction relatively free of mobile charge carriers. For this reason, the region of the junction is known as the depletion region. On the *p* side of the junction, with the removal of the mobile holes, the fixed negative ions form a negative space-charge layer; and on the *n* side, with the removal of the mobile electrons, the fixed positive ions form a positive space-charge layer. Across the junction, there is a permanent internal electric field in the

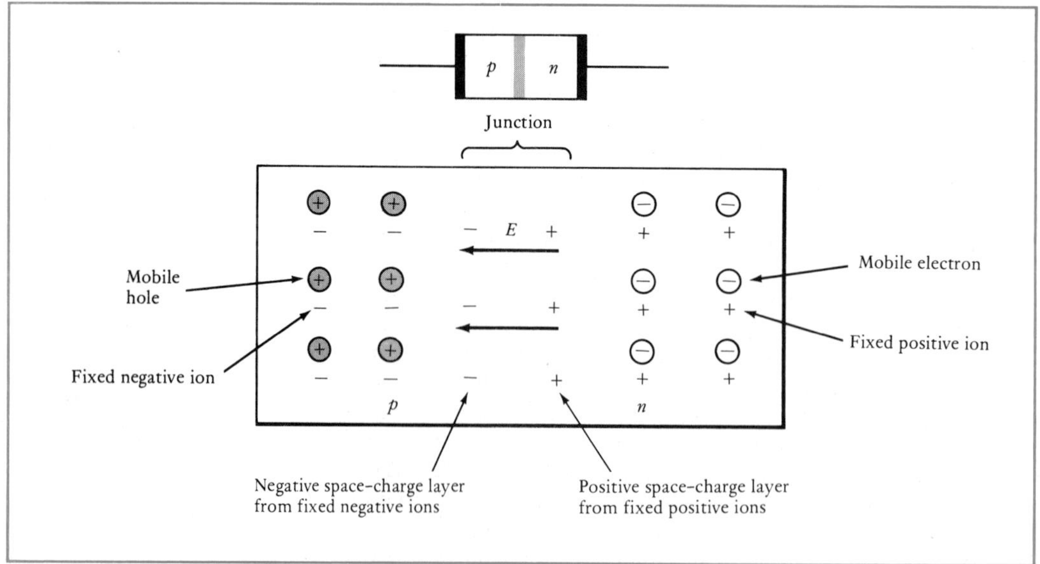

FIGURE 9–15. *The internal electric field in a p-n junction. Annihilation in the depletion region of mobile holes from the p side and mobile electrons from the n side results in a negative space-charge layer on the n side.*

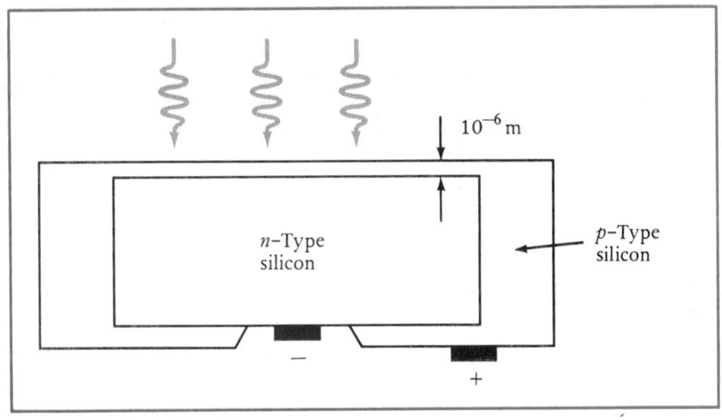

FIGURE 9–16. *Schematic diagram of a silicon solar cell.*

direction from the n side to the p side. An additional conduction electron introduced in this region would be driven by the internal electric field in a direction opposite to that of the field; holes, on the other hand, would be driven in the same direction as the field.

Suppose now that the junction is illuminated with light in just that region in which the p side merges with the junction. The freed electrons from photon-excited pairs are driven across the junction by the field to constitute a conventional net current from n side to p side. Circumstances are far less favorable for the survival of photon-created electron-hole pairs far from the junction. Only at or near the region in which the p side joins the junction will an appreciable number of photon-produced electrons be accelerated across the junction.

Figure 9–16 shows schematically the essential parts of a silicon solar cell. Exposed to the light is a very thin (10^{-6} m, approximately the diffusion length in silicon) surface of p-type silicon, which forms a junction with a larger volume of n-type silicon. The outer p surface is made large to maximize its exposure to radiation; its thickness is small so that photon-excited electrons can reach and pass through the junction before being removed through capture by a hole. With electrons flowing across the junction from the p- to the n-side, the p side acquires a positive potential and the n-side a negative potential. The photodiode becomes an energy cell.

A p-n junction operating as a light-emitting diode (LED) is, in a sense, merely a solar cell run backward. A conduction electron jumps from the conduction band to the valence band to fill a hole there, and in so doing creates a photon. Energy must be supplied to the diode by applying a potential difference in forward bias to a p-n junction to compensate for the energy emitted as light. More specifically, let us look again at Figure 9–15 but now with an external negative potential applied to the n-side and an external positive potential to the p-side. Then an external electric field is produced at the junction, and not only is it opposite to the internal electric field $\mathbf{E}$ but it also exceeds its magnitude. Therefore, the net electric field at the junction is to the right; and a mobile electron from the n side is driven by the net electric field from the n side across the junction to the p side, where the mobile electron is annihilated by a mobile hole, creating a photon. If the crystal faces on either side of the p-n junction are polished, a resonant optical cavity is formed, from which radiation can be emitted coherently in a solid-state laser (typically operating in the infrared region of the spectrum).

THE p-n JUNCTION AS DETECTOR. Another related form of photodiode is the germanium or silicon gamma-ray detector. The energy of the x-ray or gamma-ray photon far exceeds the energy gap. Each photon produces a very large number of electron-hole carriers. Strictly, the photon first produces *one* pair of a highly energetic electron and hole; these produce still other pairs, until all have an energy comparable to that of the gap. Then the number of carriers is directly proportional to the photon energy, so that the magnitude of the output current pulse is a direct measure of photon energy. The result is a *linear* detector.

A p-n junction can also be used as a so-called solid-state detector of energetic charged particles encountered in nuclear physics, for example, electrons or protons with kinetic energies measured in keV or MeV. Electron-hole pairs are created in or near the depletion region. The ionizing particle strikes the semiconductor and creates electron-hole pairs. Just as for the gamma-ray solid-state detector, numerous pairs are created by a single particle in coming to rest. The number of pairs and the pulse of electric current they produce are also directly proportional to the incident particle's initial kinetic energy. A typical material for a charged particle detector is silicon, with phosphorus as donor and boron as acceptor. Solid-state detectors, both for charged particles and gamma rays, have besides their linear characteristics, the special advantage of a very short collection time (10^{-8} s).

Example 9–2. A typical silicon solar cell has an efficiency of 12 percent in converting incident radiant energy to electric energy. The area of a single cell is 3.4 cm², and the radiant energy flux at the earth's surface with the sun overhead on a bright day is 100 mW/cm². Suppose that one were to use solar cells to produce a total output of 1 kW. (*a*) How many solar cells would be needed? (*b*) What would be the total cost at \$2 per cell? (*c*) What would be the total sensitive area exposed to sunlight?

$$(a)\qquad \text{Number of cells} = \frac{1\text{ kW}}{(12\%)(100\text{ mW/cm}^2)(3.4\text{ cm}^2/\text{cell})}$$

$$= 2.45 \times 10^4 \text{ cells} \approx 25{,}000 \text{ cells}$$

$$(b)\qquad \text{Cost} = (25{,}000\text{ cells})(\$2/\text{cell})$$

$$= \$50{,}000$$

$$(c)\qquad \text{Total cell area} = (25{,}000\text{ cells})(3.4\text{ cm}^2/\text{cell})(10^{-4}\text{ m}^2/\text{cm}^2)$$

$$= 8.5\text{ m}^2$$

9-4 The Laser

The term *laser* is an acronym for "light amplification by the stimulated emission of radiation." Such a device produces unidirectional, monochromatic, intense, and—most important—coherent visible light.† For an introduction to the laser, let us look first at several processes by which the energy

† The 1964 Nobel Prize in physics was awarded to the U.S. physicist C. H. Townes and the Soviet physicists A. M. Prokorov and N. Basov, for their fundamental work relating to the development of lasers.

For a discussion of coherent and incoherent sources of light, see Weidner and Sells, *Elementary Classical Physics,* 2d ed., Section 38–5.

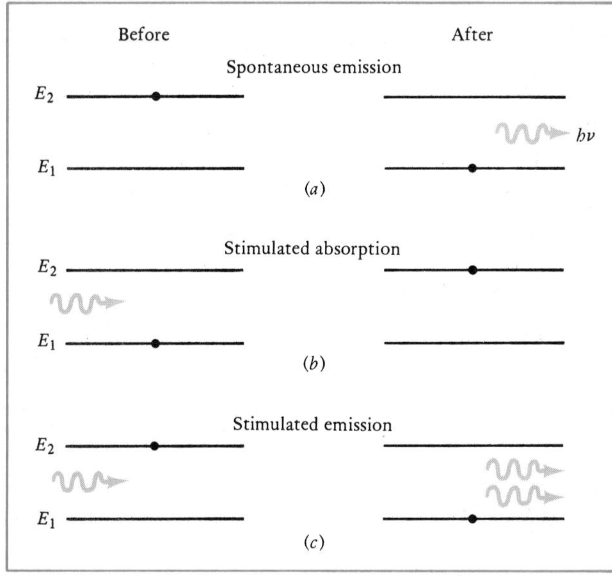

FIGURE 9–17. *Processes in which the quantum state is changed by the absorption or emission of photons. (a) Spontaneous emission. (b) Stimulated absorption. (c) Stimulated emission.*

of a free atom can change in a quantum transition with the emission or absorption of a photon. (See Figure 9–17.) These processes are (1) spontaneous emission, (2) stimulated absorption, and (3) stimulated emission.

In *spontaneous emission,* an atom is initially in an excited state and decays to a lower state as a photon of energy $h\nu = E_2 - E_1$ is emitted. Like the radioactive decay of unstable nuclei (Section 10–7), the decay of unstable atoms is governed by an exponential decay law. Typically, an excited atomic state has a lifetime of the order of 10^{-8} s; on the average, the time for an atom in an excited state to decay spontaneously with the emission of a photon is only 10^{-8} s. A few atomic transitions are, however, much slower. For such so-called metastable states, the atomic lifetime may be as long as 10^{-3} s. (The *spontaneous* transition of an atom from a low energy state to a higher is ruled out by energy conservation.)

In *stimulated absorption* an incoming photon stimulates, or induces, an atom to make an upward transition, and the photon is thereby absorbed.

In *stimulated emission,* an incoming photon stimulates an atom initially in an excited state to make a downward transition. As the atom's energy is lowered, the atom emits a photon, which is *in addition* to the photon inducing the transition. One photon approaches the atom in an excited state, and two photons leave; the atom is then in the lower energy state. Moreover, the two photons both leave in the same direction as that of the incoming photon, and they are exactly in phase relative to one another; that is, they are coherent. We can see that stimulated emission produces coherent radiation; if the two photons were out of phase by any amount, they would at least partially interfere destructively, violating energy conservation. The stimulated emission produces light amplification, or photon multiplication. The trick in constructing a laser is to make the stimulated emission dominate competing processes.

The probability of decay by spontaneous emission can be characterized by the mean life of the excited state. Similarly, one can assign probabilities

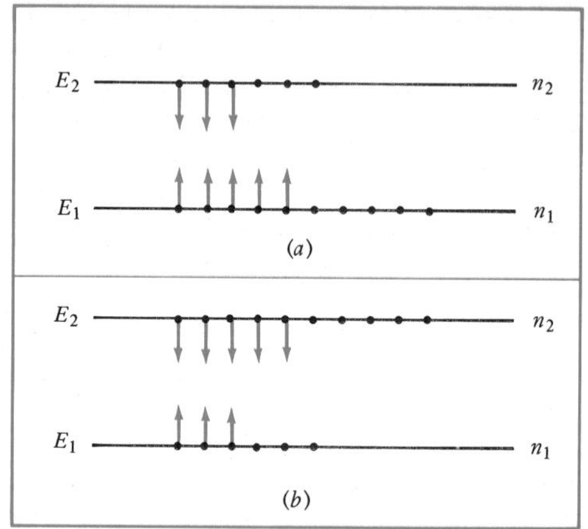

FIGURE 9–18. *Changes in the occupancy of quantized states through stimulated absorption and stimulated emission only. (a) In thermal equilibrium stimulated absorption dominates stimulated emission, and the number of photons is reduced. (b) For a population inversion, stimulated emission dominates stimulated absorption, and the number of photons is enhanced.*

P_a and P_e to the processes stimulated absorption and stimulated emission. Detailed analysis shows (Section 9–5) that

$$P_a = P_e \tag{9–2}$$

That is, for a given photon energy and type of atomic system, stimulated emission is just as probable as stimulated absorption. For example, if a certain number of photons directed at a collection of atoms all initially in a low energy state cause, say, a tenth of the atoms to undergo stimulated absorption, then the same number of photons directed at the same collection of atoms in the upper energy state will cause a tenth of the atoms to undergo stimulated emission.

The three processes—spontaneous emission, stimulated absorption, and stimulated emission—apply to free atoms interacting with photons. If a system consisting of many weakly interacting atoms is in thermal equilibrium, still other so-called relaxation processes may operate to change the quantum state of an atom without, however, emission or absorption of photons. An atom in an excited state may, for example, make a *nonradiative transition* to a lower energy state; the excitation energy goes into the thermal energy of the system rather than into creating a photon. Conversely, an atom may be raised to a higher energy state as the thermal energy of a system decreases.

Consider a collection of atoms in thermal equilibrium at some temperature T for which $\epsilon_i > kT$, where ϵ_i is the internal energy. The distribution of the atoms among the available energy states can be given to a good approximation by the classical Maxwell-Boltzmann distribution. The statistical weight $g(\epsilon_i)$ will depend on the detailed characteristics of the atoms but typically will not differ drastically between adjacent quantum states. Then we have, using Equation 8–1,

$$n(\epsilon_i) \propto f_{\rm MB} \propto e^{-\epsilon_i/kT} \tag{9–3}$$

The relative numbers of atoms in the various possible states are controlled by the system's temperature T according to the Boltzmann factor $e^{-\epsilon/kT}$. The numbers of atoms in progressively higher energy states 1, 2, and 3 are n_1, n_2, and n_3, where $n_1 \propto e^{-E_1/kT}$, $n_2 \propto e^{-E_2/kT}$, and $n_3 \propto e^{-E_3/kT}$. Since $E_1 < E_2 < E_3$, it follows that $n_1 > n_2 > n_3$. The ground state is more heavily populated than the first excited state, and the number of atoms occupying higher states is still less.

Consider, for simplicity, a collection of atoms that have only two energy states and that are in thermal equilibrium. Such a collection would be atoms with free or nearly free electrons, whose spin direction is aligned or antialigned with an external magnetic field. The n_1 atoms in the lower energy state exceed the number n_2 of atoms in the upper energy state; see Figure 9–18*a*. Suppose further that a beam of photons, each of energy $h\nu = E_2 - E_1$, illuminates these atoms. Ignoring for the moment spontaneous emission and the relaxation processes within the system (or assuming these processes characterized by low probability or long mean life), we concentrate on stimulated absorption and stimulated emission only; these both have the same probability. Stimulated absorption depopulates the lower energy state and reduces the number of photons. Stimulated emission depopulates the upper energy state and increases the number of photons. What is the net effect on the total number of photons?

The number of photons disappearing by stimulated absorption is proportional to $P_a n_1$, and the number of additional photons created by virtue of stimulated emission is proportional to $P_e n_2 = P_a n_2$, from Equation 9–2. For thermal equilibrium, however, we have $n_1 > n_2$. So there is net absorption. Absorption dominates emission simply because more atoms occupy the lower energy state than the upper one. Moreover, the net absorption is accompanied by a tendency toward equalization of the populations of the two states.

If we were somehow to produce a *population inversion,* a situation in which the number of atoms occupying the upper energy state *exceeds* the number in the lower state, then emission would dominate absorption; see Figure 9–18*b*. With a population inversion, incoming light would be amplified coherently, since the number of additional photons produced through stimulated emission would more than compensate for the number of photons removed through stimulated absorption. Such population inversions have been achieved for lasers in a wide variety of materials by several clever procedures, most of which involve a relatively slow relaxation. What follows is a brief description of the first laser operating with a crystal of ruby (1960).

Ruby consists of aluminum oxide, Al_2O_3, with a small fraction of Cr replacing Al; the chromium atoms are the ones responsible for the laser behavior. Figure 9–19 shows the important energy levels of chromium (level 3 actually consists of several closely spaced levels). The excited state E_2 is metastable; its lifetime for spontaneous decay to the ground state is

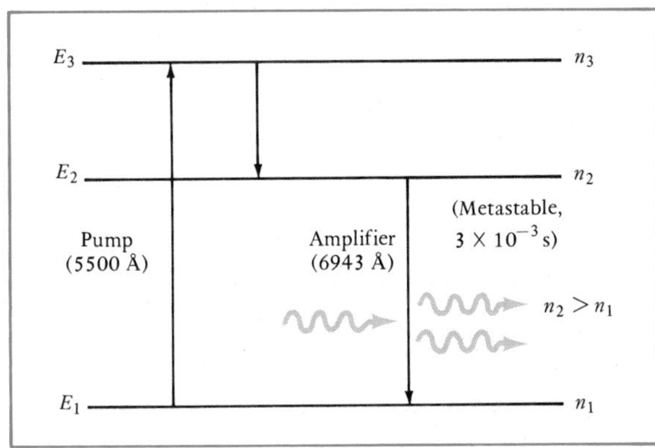

FIGURE 9–19. *Energy levels of Cr in a ruby laser. The pumping radiation between states 1 and 3 depletes the population of state 1 and increases state 2. The radiation between states 2 and 1 is amplified because of the population inversion, with* $n_2 \geq n_1$.

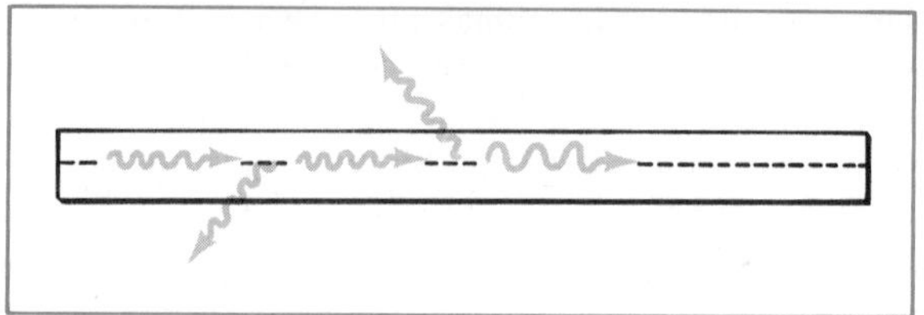

FIGURE 9–20. *Laser cell with flat, parallel reflecting ends. Only those photons traveling perpendicular to the end plates undergo appreciable photon multiplication; oblique photons escape through the sides before they have a chance to stimulate many emissions.*

unusually long, about 3×10^{-3} s. The atoms are initially in thermal equilibrium, with $n_1 > n_2 > n_3$. Then the system is "optically pumped" by sending light of wavelength 5500 Å (yellow green) at the atoms. Atoms make transitions from state 1 to state 3 by absorbing photons of this wavelength. Almost immediately, atoms brought to state 3 are deexcited by transitions to state 2. But atoms arriving in excited state 2 remain in this state for relatively long times. The optical pumping depletes the population of state 1 and enhances the population of state 2. Indeed, n_1 is decreased and n_2 increased to such a degree that $n_2 > n_1$; population inversion is achieved. If a few photons of 6943 Å (ruby red light) appear, possibly from the spontaneous decay from state 2 to state 1, they will stimulate transitions in which emission dominates absorption. Thus, the light is amplified; each photon that joins the train is exactly in phase, or coherent, with the initial beam.

In practice the light is sent through the ruby crystal many times by being reflected from optically flat and parallel ends, as in Figure 9–20. At the ends, some light escapes—more at the output end, which has lower reflectivity. The remainder is reflected into the ruby crystal. But only the light that is precisely perpendicular to the reflecting ends will traverse the crystal in many round trips; photons traveling obliquely to the crystal axis escape from the crystal before substantial photon multiplication can take place. The amplified light is highly monochromatic, unidirectional, intense, and coherent.

The ruby laser involves a solid material and it is operated in pulsed fashion. An example of a common continuously operating laser that uses gas as the active material is the helium-neon laser. Here the population inversion, the essential condition for photon multiplication, is produced through inelastic collisions between excited helium atoms and neon atoms in the ground state. The process can be written

$$\text{He}^* + \text{Ne} \rightarrow \text{He} + \text{Ne}^*$$

For this process to take place, however, the energy He loses must match the energy Ne gains; that is, the two types of atoms must have excited states with the same or very nearly the same energy. Fortuitously, this is indeed the case, as shown in the energy-level diagram of Figure 9–21. The metastable 2s state of helium has an energy of 20.61 eV above the ground state; the 5s state of neon has essentially the same energy, 20.66 eV.† The lifetime of a spontaneous transition of neon from the 5s state to the 3p state is relatively long, but the following transition (3p → 3s) is short. This behavior of the neon atom results in a population inversion between the 5s and 3p states,

† The states for both helium and neon atoms are indicated by the state of a single excited electron. All other electrons in an atom remain in the ground state. For example, the 5s state of Ne has the electron configuration $1s^2 2s^2 2p^5 5s^1$.

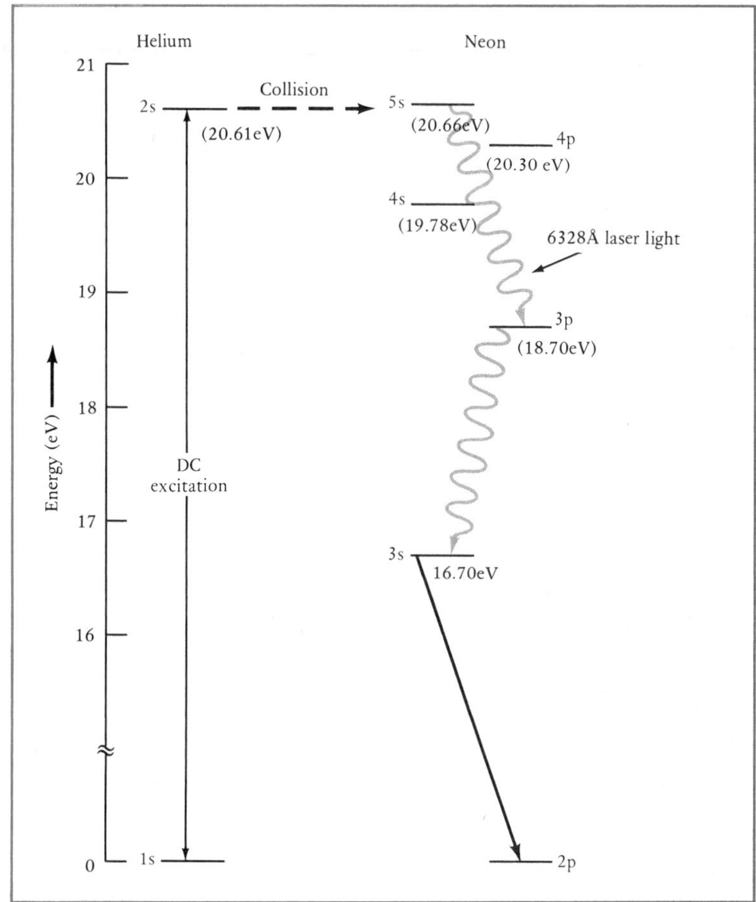

FIGURE 9–21. *Energy-level diagram for He-Ne atoms, illustrating laser action between the 5s and 3p states of neon.*

with the lower state less populated—a necessary requirement for laser action. The 5s → 3p transition corresponds to a photon wavelength of 6328 Å, the orange-red color characteristic of ordinary neon tubes.‡

A typical helium-neon laser has a mixture of He–Ne gases enclosed in a sealed tube with parallel silvered mirrors (~99% reflecting) at its ends. The gas is excited by a dc voltage source, thereby raising some helium atoms to the 2s metastable state. Inelastic collisions then transfer energy to neon atoms and increase the population of the neon atoms in the 5s metastable state. Laser action can then occur, since 6328-Å light reflecting back and forth between the two ends induces more downward transitions than upward transitions because of the population inversion between the 3p and 5s state. The coherent, monoenergetic, directional beam is therefore amplified, and laser light emerges from the tube end with the smaller reflectivity.

The energy "lost" through the emission of photons in the many downward transitions of the excited atoms (Figure 9–21) is restored through the continuous excitation of atoms by the dc power supply. Only a small percentage of the energy supplied to the excitation of the atoms is converted to the

‡ Laser action can also take place between other energy states of the neon atom in which population inversion occurs.

energy of the output coherent laser beam. A typical working efficiency (output laser power to input excitation power) of a He–Ne gas laser is 1×10^{-3} percent.

Besides the two lasers just described, which use solid and gaseous materials, there are lasers that involve liquids and semiconducting materials. Moreover, lasers operating from the far infrared to the ultraviolet region of the electromagnetic spectrum have been constructed. In every instance, the condition for laser operation is the existence of a pair of quantized energy levels for which a population inversion has been achieved.

Lasers have many technological applications. They are all possible because, with lasers, one can produce in the visible and nearby regions electromagnetic radiation that has the coherence properties heretofore available only in radio waves.

9-5 Spontaneous and Stimulated Emission Probabilities

Our detailed treatment of spontaneous emission and stimulated absorption and emission follows the analysis Albert Einstein first made in 1917. We consider a system consisting of identical atoms that interact weakly. The atoms of this system not only are in thermal equilibrium with one another at temperature T but also are immersed in and in thermal equilibrium with electromagnetic radiation at the same temperature T. Therefore, the distribution of the electromagnetic radiation, as a function of both frequency and temperature, is given by the blackbody relation (Equation 8–27).

For simplicity, we suppose that each atom has but two quantized energies, E_1 and E_2 (or equivalently, that only the two lowest energy levels are appreciably populated). The atoms emit and absorb photons with an energy

$$h\nu = E_2 - E_1$$

The number of atoms per unit volume n_1 and n_2, which have the energies E_1 and E_2, are, through the Boltzmann distribution relation

$$n_1 \propto e^{-E_1/kT} \qquad \text{and} \qquad n_2 \propto e^{-E_2/kT}$$

Therefore

$$\frac{n_1}{n_2} = \frac{e^{-E_1/kT}}{e^{-E_2/kT}} = e^{(E_2 - E_1)/kT} = e^{h\nu/kT} \tag{9–4}$$

Atoms in state 2 may spontaneously decay to state 1 and emit a photon of energy $h\nu$. The number of spontaneous decays per unit time must be proportional to the number n_2 of those atoms that can decay spontaneously, and we can write

$$\frac{\text{Spontaneous decays } (2 \rightarrow 1)}{\text{Time}} = n_2 A_{21}$$

where the coefficient A_{21} is the transition probability for spontaneous decay from state 2 to state 1.

Atoms in state 2 can also be stimulated to make a downward transition to state 1 by a photon of energy $h\nu$ of the radiation field. The number of stimulated downward transitions per unit time is proportional to (1) the number of atoms n_2 that can be stimulated to emit radiation, and (2) the number of photons (each of energy $h\nu$) per unit volume in the radiation field. Since the photon density is, in turn, proportional to the energy density $E(\nu)$ of the electromagnetic radiation at frequency ν, we can then write

$$\frac{\text{Stimulated emissions } (2 \rightarrow 1)}{\text{Time}} = n_2 E(\nu) B_{21}$$

where the proportionality constant B_{21} is the coefficient for stimulated emission at frequency ν.† From this definition, we see that B_{21} is proportional to the probability that an atom in state 2 will be stimulated by the radiation field to make a downward transition to state 1 (P_e in Equation 9–2).

In just the same way, the stimulated absorption process, which causes transitions from state 1 to 2, is proportional to the number n_1 of atoms that can be stimulated to absorb a photon of energy $h\nu$ and also proportional to the energy density $E(\nu)$ of the radiation field:

$$\frac{\text{Stimulated absorption } (1 \rightarrow 2)}{\text{Time}} = n_1 E(\nu) B_{12}$$

where the proportionality constant is the stimulated absorption coefficient at frequency ν. The parameters A_{21}, B_{21}, and B_{12} are usually referred to as the *Einstein coefficients,* each of which is proportional to the probability for making the indicated transition.

If the collection of atoms is in thermal equilibrium, the number of atoms n_1 in the lower state and also the number n_2 in the upper state each remains constant. In other words, downward transitions $(2 \rightarrow 1)$ through both spontaneous and stimulated emission over some time interval are matched exactly by upward transitions $(1 \rightarrow 2)$ through stimulated absorption over the same time interval:

$$\frac{\text{Downward transitions } (2 \rightarrow 1)}{\text{Time}} = \frac{\text{Upward transitions } (1 \rightarrow 2)}{\text{Time}}$$

or by the definitions given above,

$$n_2 A_{21} + n_2 B_{21} E(\nu) = n_1 B_{12} E(\nu)$$

Since $E(\nu)$ is given by the blackbody radiation formula, we can use this relation to interpret the Einstein coefficients A_{21}, B_{12}, and B_{21}. Solving for $E(\nu)$, we arrive at

$$\begin{aligned} E(\nu) &= \frac{n_2 A_{21}}{n_1 B_{12} - n_2 B_{21}} \\ &= \frac{A_{21}}{(n_1/n_2) B_{12} - B_{21}} \\ &= \frac{A_{21}}{B_{21}} \frac{1}{(n_1/n_2)(B_{12}/B_{21}) - 1} \end{aligned}$$

† Strictly, $E(\nu)$ is so defined that the electromagnetic energy per unit volume in the small frequency range from ν to $\nu + d\nu$ is given by $E(\nu)\, d\nu$.

Using Equation 9–4 in the above relation for n_1/n_2, we have

$$E(\nu) = \frac{A_{21}}{B_{21}} \frac{1}{(B_{12}/B_{21})e^{h\nu/kT} - 1} \qquad (9\text{–}5)$$

As shown in Example 8–2 and Equation 8–27, the Planck blackbody radiation relation is

$$E(\nu) = \frac{8\pi h\nu^3}{c^3} \frac{1}{e^{h\nu/kT} - 1} \qquad (9\text{–}6)$$

In comparing Equations 9–5 and 9–6, it then follows that

$$\frac{B_{12}}{B_{21}} = 1$$

and

$$\frac{A_{21}}{B_{21}} = \frac{8\pi h\nu^3}{c^3} \qquad (9\text{–}7)$$

Equation 9–7 shows that $B_{12} = B_{21}$; the probabilities for transitions between two states by either stimulated absorption or stimulated emission *are* the *same,* a result stated (Equation 9–2) but not proved in Section 9–4.

Using the first equation of Equation 9–7 in Equation 9–5 yields

$$E(\nu) = \frac{A_{21}}{B_{21}} \frac{1}{e^{h\nu/kT} - 1}$$

We can use this relation to arrive at the ratio of stimulated emission to spontaneous emission. Since $n_2B_{21}E(\nu)$ gives the number of stimulated transitions from the upper state per unit time and n_2A_{21} is the number of spontaneous transitions between the same levels per unit time, the equation above implies that

$$\frac{\text{Stimulated emission}}{\text{Spontaneous emission}} = \frac{n_2B_{21}E(\nu)}{n_2A_{21}} = \frac{1}{e^{h\nu/kT} - 1} \qquad (9\text{–}8)$$

Example 9–3. (*a*) What is the ratio of stimulated to spontaneous emission for downward transitions between two states for a system of atoms that is emitting visible light (2.0-eV photons) and is in thermal equilibrium with this radiation at 300 K and 30,000 K? (*b*) What is this ratio for radio waves of 21 cm at 300 K and 3.0 K?

(*a*) For 300 K,

$$\frac{h\nu}{kT} = \frac{(2.0 \text{ eV})(1.6 \times 10^{-19} \text{ J/eV})}{(1.38 \times 10^{-23} \text{ J/K})(300 \text{ K})} = 77.3$$

Therefore, from Equation 9–8,

$$\frac{\text{Stimulated emission}}{\text{Spontaneous emission}} = \frac{1}{e^{h\nu/kT} - 1} = \frac{1}{e^{77.3} - 1}$$

$$= 3 \times 10^{-34}$$

At room temperature, stimulated emission for this transition in the visible region is negligible compared with spontaneous emission.

At $T = 30{,}000$ K, $h\nu/kT = 0.773$, and the ratio of stimulated to spontaneous emission is

$$\frac{1}{e^{0.773}-1} = 0.86$$

At 30,000 K, stimulated and spontaneous emission are comparable for visible light.

(*b*) What is ordinarily regarded as a single ground state for the hydrogen atom comprises two very closely spaced distinct states that come from the interaction between the electron spin and the proton spin. The upper state is the state in which the electron and proton spins are aligned; the lower state is the one in which they are antialigned. The energy difference between the two states (hyperfine structure) corresponds to a radio photon with wavelength 21 cm. Since hydrogen is the most abundant element in the universe, 21-cm radiation from interstellar atomic hydrogen is the most prominent "line" observed by radio astronomers.

For $\lambda = 21\text{ cm} = 2.1 \times 10^9$ Å, the photon energy is

$$E = \frac{12{,}400\text{ eV}\cdot\text{Å}}{\lambda} = \frac{12{,}400\text{ eV}\cdot\text{Å}}{2.1\times 10^9\text{ Å}} = 5.90\times 10^{-6}\text{ eV}$$

and for room temperature,

$$\frac{h\nu}{kT} = \frac{(5.90\times 10^{-6}\text{ eV})(1.6\times 10^{-19}\text{ J/eV})}{(1.38\times 10^{-23}\text{ J/K})(300\text{ K})} = 2.28\times 10^{-4}$$

Therefore, from Equation 9–8,

$$\frac{\text{Stimulated emission}}{\text{Spontaneous emission}} = \frac{1}{e^{h\nu/kT}-1} = \frac{1}{e^{2.28\times 10^{-4}}-1} = 4.37\times 10^3$$

For these very low energy photons at room temperature, stimulated emission is much greater than spontaneous emission.

On the other hand, at 3 K, near the absolute zero of temperature, $h\nu/kT = 2.29\times 10^{-2}$, and the ratio of stimulated to spontaneous emission is 43.7. Even at this very low temperature, stimulated emission still exceeds spontaneous emission for the 21-cm hydrogen line.

9–6 Holography

An important application of lasers, one that critically depends on the *coherent* light produced by a laser, is in *holography,* a procedure for producing three-dimensional images. The holographic phenomenon is based on the reconstruction of electromagnetic fields reflected from a three-dimensional object. We first consider the rays, and then the waves, that are emitted by such an object and reach the eye of an observer.

Imagine that each point on the surface of a solid, the opaque cube of Figure 9–22, emits light of a single wavelength in all directions and that the cube can be viewed by an observer from several locations. We shall concentrate on three point sources on the cube surfaces: S_1 on the top face, S_2 on the front face, and S_3 on the bottom face. When an observer's eye is at the location L_1, light from S_1 and S_2 but not S_3, reaches L_1; an eye at L_1 sees S_1 and S_2 but not S_3. With an eye at location L_2, light reaches it from S_2 but not S_1 and S_3. And with an eye at location L_3, light from sources S_2 and S_3, but

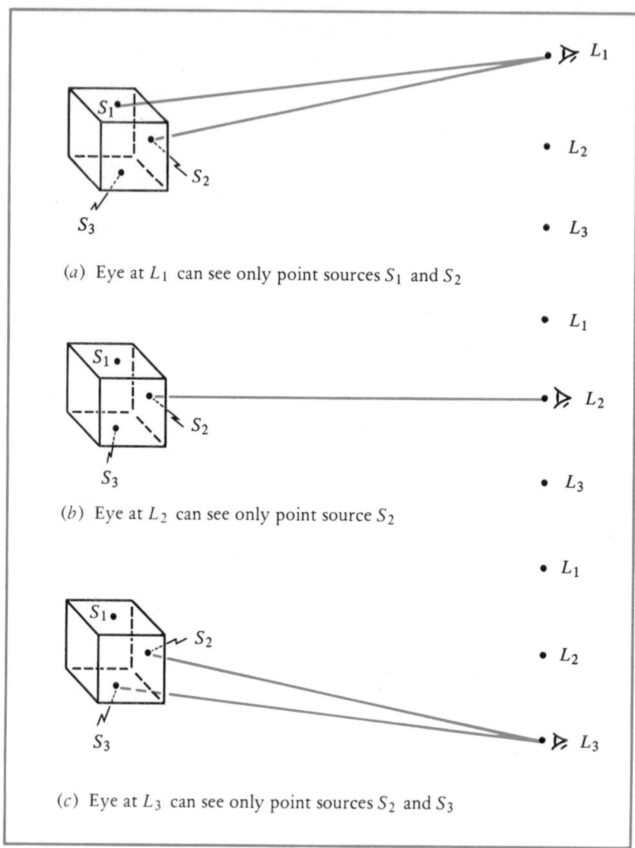

FIGURE 9–22. *View of three point sources on an opaque cube by an observer at three different locations L_1, L_2, and L_3.*

not S_1, reaches L_3. In short, whether light reaches an eye at any location depends on whether an uninterrupted straight line, or path, can be drawn between the point source and the eye. Clearly, the views of the cube—which surfaces can be seen and which are hidden—depend on the location of the observer. Only by changing one's perspective, by placing one's eye at different locations, can an observer truly study the cube as a three-dimensional object.

Each of the point sources S_1, S_2, and S_3 of Figure 9–22 produces an outgoing electromagnetic wave. (The entire cube could, of course, be made visible by shining light on it, so that S_1, S_2, and S_3 become point sources of reflected light.) What an eye sees at locations L_1, L_2, and L_3 depends on the resultant electromagnetic wave reaching each of these locations. We now concentrate on the plane P in Figure 9–23, which lies between the cube and the three eye locations. Certainly any of the electromagnetic radiation that reaches the eye at any of the three locations from the cube must pass through P. At any one instant, the electromagnetic field in plane P can be characterized by the magnitude and direction of the resultant electric field at each point in plane P.† At a particular instant, the resultant electric field will differ in magnitude and direction from point to point over the plane P.

† The magnetic field of the electromagnetic wave need not be specified in addition to the electric field because its direction is always perpendicular to the electric field and also perpendicular to the direction of wave propagation, and the magnitude of the magnetic field is proportional to the electric field.

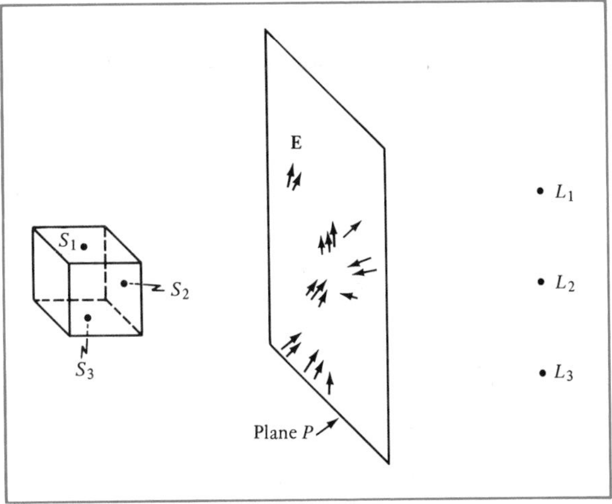

FIGURE 9–23. *The resultant electric field* **E** *at one instant at a few points on plane P. In the upper portion of the plane, the resultant field is the vector sum of the fields from sources* S_1 *and* S_2 *only; the middle portion, the field from* S_2 *only; and the lower portion, the fields from* S_2 *and* S_3 *only.*

Now we assume that the sources S_1, S_2, and S_3 radiate not only at the same frequency but also *coherently*. Therefore, after a time interval of one period, the electric field at each location in plane P will be identical in magnitude and direction to the initial conditions. After one half-period, the electric field at every point in the plane has the same magnitude and is merely reversed in direction. Indeed, the electric field at any one location in plane P undergoes sinusoidal variations at the frequency of the source. Although the phase of oscillation in the electric field at some one point in plane P will differ from the phase of oscillation at another point, the *phase differences* among the oscillations for all points in the plane will remain *constant*. A *complete* knowledge of the wave field consists then of the magnitude, direction, and phase of the electric field at each point in the plane P. Moreover, for coherent sources, knowing the wave field at some one instant allows the wave field at an earlier or later time to be known.

Since light that reaches any of the three eye locations passes through plane P, *all* the information carried by the electromagnetic radiation from the sources to the several eye locations—what the scene looks like from the vantage point of location L_1 (the lower face cannot be seen), how it appears from location L_3 (where the upper face cannot be seen)—is contained in the wave field at any instant in plane P. This implies that if one can somehow record and reconstruct the wave field in plane P, one can reconstruct the entire scene as it is viewed from numerous locations and therefore in full three-dimensional perspective.

The basic idea for reconstructing the wave field was originated by Dennis Gabor in 1947; Gabor was awarded the 1971 Nobel Prize in physics for his development of holography (*holo,* means "whole, entire"; *graph,* "written record"). The development of practical holography had to await the introduction of continuous lasers in 1960.

To see in more specific terms how a hologram can be produced and how it is used to reconstruct a wave field, consider the simple situation shown in Figure 9–24. Here a coherent plane wave W_1, a wave that can be considered to have originated from an infinitely distant point source, is traveling horizontally to the right; for reasons soon to be evident, we denote this the *reference wave*. A second plane, coherent wave W_2, is progressing obliquely

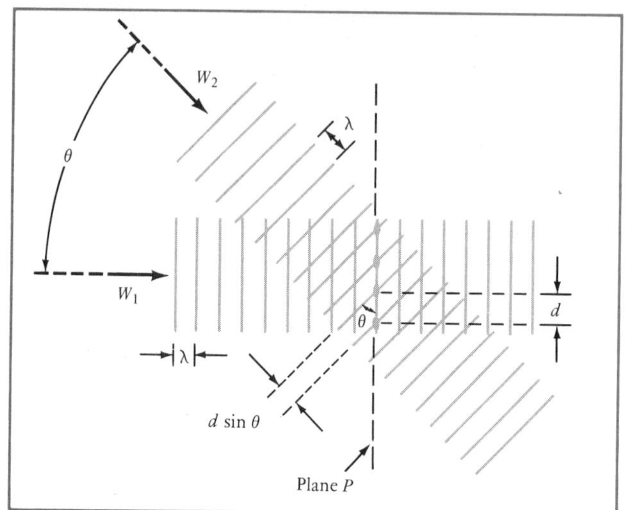

FIGURE 9–24. *Interference pattern in plane P of two-plane coherent waves W_1 and W_2. The propagation directions of the two waves differ by the angle θ. The four broadened bands of constructive interference in plane P are shown as darkened regions and remain unchanged.*

downward and to the right. We concentrate our attention on the resultant wave field in plane P, where the two plane waves overlap and interfere with one another. Imagine that the parallel, equally spaced wave fronts, at right angles to the wave propagation directions, are separated by one wavelength. The interference is constructive where crest meets crest (or where wave fronts intersect) and destructive where trough meets crest (or where wave fronts of W_2 midway between those shown in Figure 9–24 intersect wave fronts of W_1). It is clear that for the two plane waves shown, the regions of maximum constructive interference in plane P consist of equally spaced parallel straight lines (separation distance $= d$) perpendicular to the plane of the paper. The geometry of Figure 9–24 shows that for wave fronts separated by λ and for plane waves differing in propagation direction by the angle θ, the lines of maximum constructive interference are separated by a distance $d = \lambda/\sin\theta$. As time goes on and the two plane waves progress forward, the regions of constructive interference are unchanged.

Suppose now that plane P in Figure 9–24 is not merely a geometrical plane but one in which a sheet of photographic film is placed. When the film is exposed and developed, the image consists of parallel, equally spaced straight lines of maximum density, corresponding to the lines of constructive interference, with lines of zero density (destructive interference) midway between them. The photographic negative is a simple kind of *hologram.* A hologram is not like an *ordinary photographic negative,* which is a record of the *time average* of $E_1^2 + E_2^2$, where $\mathbf{E}_1$ is the instantaneous electric field in the plane P from wave W_1, and similarly for the electric field $\mathbf{E}_2$ from W_2. On the other hand, a hologram is, in effect, a record of the time average of $(\mathbf{E}_1 + \mathbf{E}_2)^2$, which when it is expanded involves an interference term $(2\mathbf{E}\cdot\mathbf{E}_2)$ as well as $E_1^2 + E_2^2$.† For the specific circumstances discussed here, the hologram is a record of the wave field of plane wave W_2, since by its interference with the reference wave W_1, an interference pattern is produced in the hologram.

† When looking at the negative of an ordinary photograph, one sees a picture; on the other hand, when holding up a hologram to ordinary light, one sees no picture since the hologram is a record of the interference pattern.

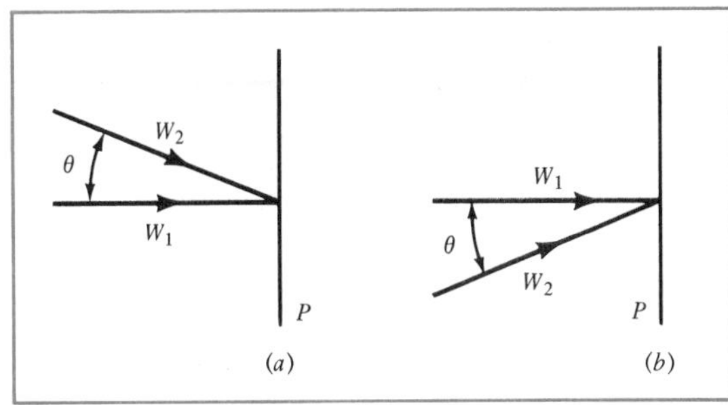

FIGURE 9–25. *Arrangements (a) and (b) produce the same hologram.*

The hologram recorded by the waves of Figure 9–24 could also be produced by another arrangement related to the original by symmetry. Thus, the configuration shown in Figure 9–25*b* produces the same hologram as the one in Figure 9–25*a*.

Suppose that the hologram is illuminated by a beam of light (the *reconstructing light beam*) that is identical to the reference wave W_1 in Figure 9–24. The situation is exactly like what happens when a beam of light is passed through the equally spaced parallel openings of a diffraction grating. Indeed, the hologram here is just a diffraction grating. On emerging from the hologram, the incident beam becomes an undeviated forward beam together with two beams, one deviated upward by an angle θ and the other downward by the same angle θ. See Figure 9–26. (The spacing d of the lines on the hologram is related to the deviation in direction θ by the basic equation for the diffraction grating, arrived at earlier: $d \sin \theta = \lambda$.) The three beams correspond precisely to the undeviated zeroth order and the two first-order

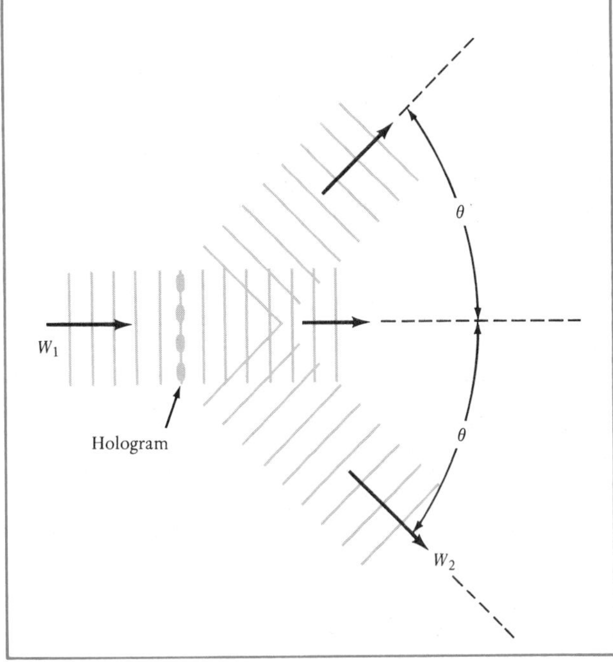

FIGURE 9–26. *Plane wave W_1 diffracted by hologram of Figure 9–24 into undeviated beam and two first-order diffraction beams at angle θ.*

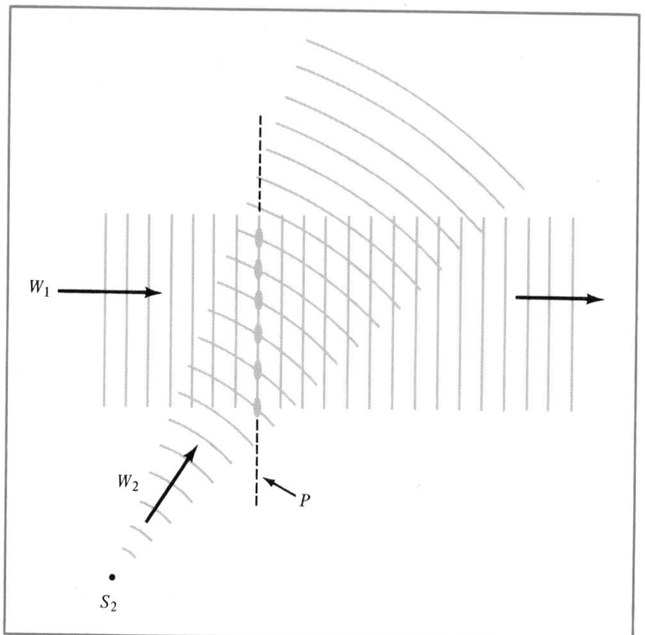

FIGURE 9–27. *Interference pattern in plane P of a coherent plane wave W_1 and a coherent spherical wave W_2.*

diffraction maxima produced by a diffraction grating.† Note especially that in illuminating the hologram with the reconstructing light beam, we have reproduced the light beam W_2, traveling obliquely downward at angle θ, that was used with the reference beam to create the hologram.

There is another arrangement for producing a simple kind of hologram. Once again there is a reference beam, a plane wave W_1 traveling horizontally to the right. A second beam is produced by a relatively nearby point source S_2, as shown in Figure 9–27. The beam W_1 has plane wave fronts; the point

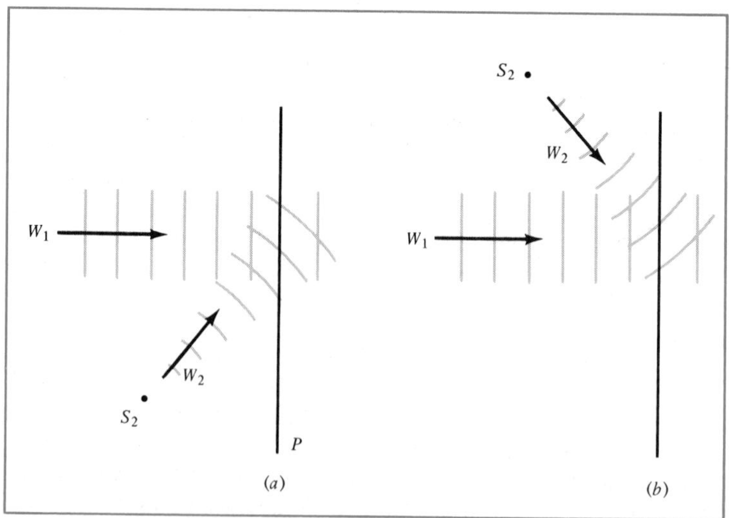

FIGURE 9–28. *Arrangements (a) and (b) produce the same hologram.*

† Second-order and still higher diffraction maxima cannot occur for θ larger than 30°, the case for typical hologram arrangements.

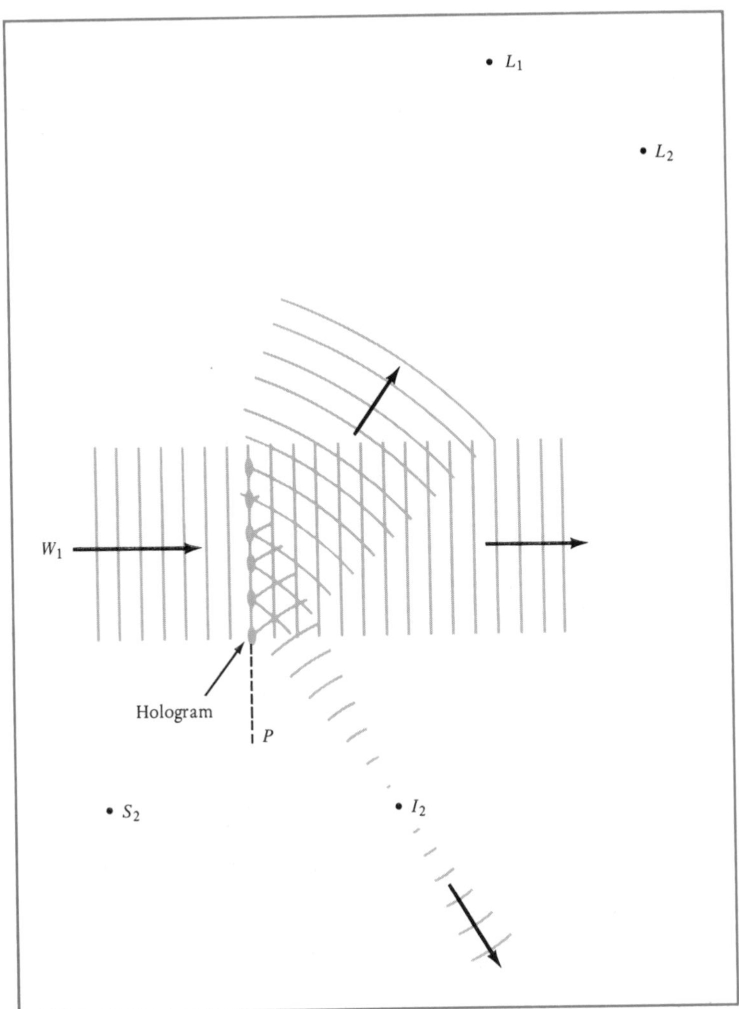

FIGURE 9–29. *Plane wave diffracted by hologram of Figure 9–27 into undeviated plane wave and two spherical wave fronts—one diverging from the virtual source S_2, the other converging onto the real image I_2, then diverging.*

source S_2 produces spherical wave fronts W_2. The interference of the two beams in a plane P is constructive where crest meets crest and destructive where crest meets trough. The lines of maximum constructive interference are now circular arcs in the plane P, all centered about the normal from S_2 to the plane; any point on a given circle is the same distance from S_2. The difference in distance from S_2 to a given circle and to the next circle adjoining it is one wavelength. The circles are more closely spaced as their radii increase. For a photographic film placed at plane P, the developed hologram consists of a group of concentric circular arcs of maximum density; the circles are a record of the interference pattern produced by the reference beam combined with the waves from the point source. (This interference pattern is also referred to as a *zone plate*.) The same hologram would be produced if the reference beam and point source were situated as shown in Figure 9–28.

As before, we consider what happens when the hologram is illuminated by a reconstructing light beam, one identical to the initial reference beam W_1. The result, analogous to that with the plane wave source, is shown in Figure 9–29. Once again there is an emerging plane wave corresponding to the un-

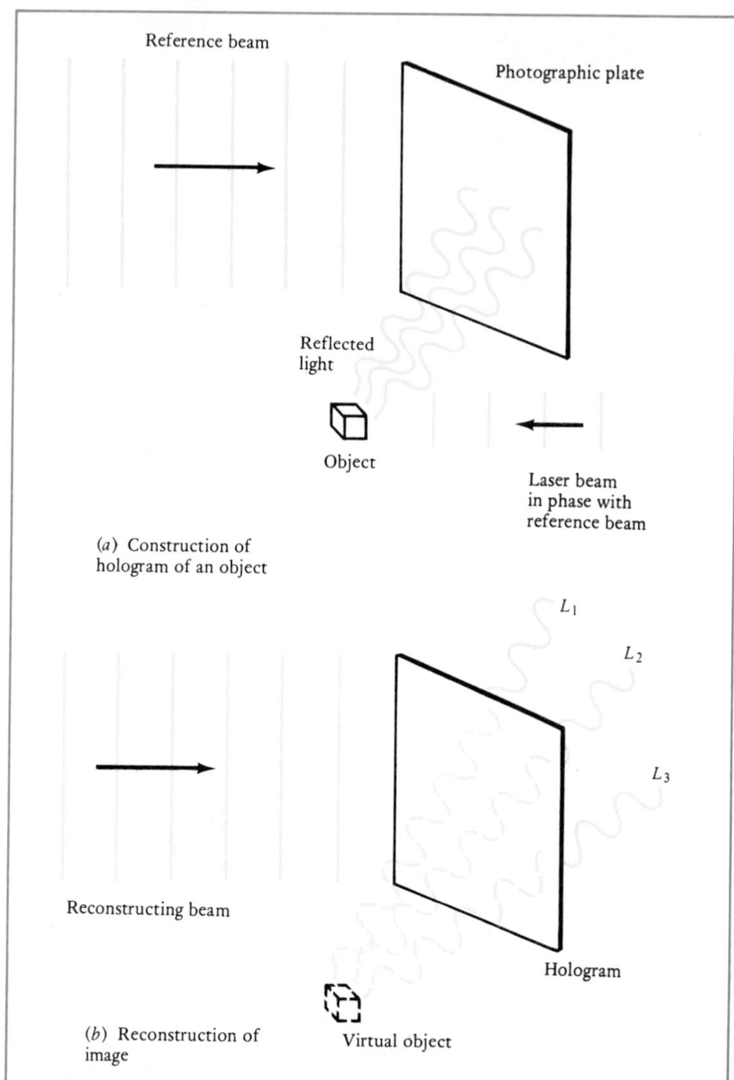

FIGURE 9–30. *Arrangement for (a) construction of the hologram of an object and (b) reconstruction of the object's image.*

deviated reference beam. We have also, corresponding to the two first-order diffraction beams of Figure 9–26, two additional sets of waves: (1) There is a diverging beam of expanding spherical waves fronts. These appear to originate from a virtual source S_2. (This is the same location relative to the viewer as the actual point source in Figure 9–27 used in producing the hologram.) (2) There is also a converging beam of contracting spherical wave fronts. These are symmetrically located relative to the first set; and as they collapse into a point and then expand outward, they constitute a real image at point I_2. Note especially that an observer who shifts his observation location from one place (L_1 in Figure 9–29) to another (L_2) always finds the waves from the virtual source to come from the very same location in space (S_2).

It is simple now to generalize the results as they have been given. Any three-dimensional source of light waves can be regarded as merely a collection of point sources; thus if one can reconstruct a single point source through a hologram, then one can also, in taking the superposition of the wave fields from the collection of all point sources of the object, reconstruct the image of *any* object through a hologram. The procedure is shown in Figure 9–30. Two light beams, both from the *same* laser, interfere at the plane of the photographic film that becomes the hologram: (1) a reference beam of plane waves, and (2) light reflected from the object to be photographed. To reconstruct the image of this object, one illuminates the hologram with a reconstructing plane beam identical to the original reference beam and views through the developed hologram to find a virtual object at precisely the same location as the original object. When the viewer shifts position (from L_1 to L_2 to L_3), he views the image from a new perspective—parts hidden in one location may now become visible, and the contrary. In short, the holographic image is such that the light reaching the viewer through the hologram is identical in all respects to the light that would reach the viewer's eye directly from the object.

PROBLEMS

Conductivity and superconductivity, §9–1

● **9–1.** The temperature of a simple loop of material is reduced until the material becomes superconducting. (*a*) How can a current then be produced in the superconducting loop without interfering appreciably with the loop? (*b*) How can the persistence of the superconducting current be confirmed?

● **9–2.** The microscopic form of Ohm's law is $j = \sigma E$, where j is the current density, E is the electric field, and σ the conductivity. The current density j is related to the density n of charge carriers, the charge e of a charge carrier, and their drift speed v_d by $j = nev_d$. By definition, the mobility of a charge carrier is given by $\mu = v_d/E$. Show that the mobility is given by σ/ne.

● **9–3.** Figures 9–1 and 9–2 show that as the temperature of a superconducting material is lowered, the electronic specific heat rises abruptly at the transition temperature. Show that the increase in the specific heat is consistent with the formation of *bound* Cooper pairs of superconducting electrons as the temperature crosses the transition temperature.

9–4. Verify that a typical speed for a conduction electron in a conductor is of the order of 10^6 m/s; proceed by computing the speed of a conduction electron in copper. Assume that the conduction electron's kinetic energy is equal to the Fermi energy (7.1 eV) for copper.

9–5. Verify that typical *drift* speed for a conduction electron in a conductor is of the order of 10^{-4} m/s; proceed by computing the drift speed of a conduction electron in copper. (The drift speed v_d is given in general by $v_d = i/nqA$, where i is the current through a conductor of cross section A, and n is the number of charge carriers.†) Consider a copper wire of 1.0 mm² cross-sectional area through which a current of 1.0 ampere (A) is passing. The atomic weight of copper is 64; there is approximately one conduction electron per copper atom. The density of copper is 9.0×10^3 kg/m³.

9–6. The mean drift speed for conduction electrons in an ordinary conductor is far lower than the mean speed of conduction electrons (10^{-4} m/s versus 10^6 m/s). Show that this statement is equivalent to a statement of Ohm's law if we assume that the conduction electron speeds are completely randomized after each collision with the lattice. (*Hint:* Assume that after each collision, each conduction electron begins with zero drift speed and is accelerated by a constant electric field. Is the mean time between collisions appreciably affected by the magnitude of the electric field and therefore by the mean drift speed?)

9–7. The wave function of a Cooper pair of electrons in a superconductor must be coherent over the region of space in which a persistent superconducting current exists. Suppose that a superconducting current exists in a loop having the dimensions of the order of 10 cm. What is the maximum fractional uncertainty in the momentum of a superconducting electron having a kinetic energy of a few electron volts?

9–8. What energy is needed to break up a Cooper pair of electrons in the superconductor (an alloy of niobium, aluminum, and germanium) having the very high transition temperature, 21 K?

9–9. One unrealized goal in the technological development of superconducting materials is finding a superconducting alloy or compound whose transition temperature is at or above 77 K; this is the temperature at which liquid nitrogen boils under standard atmospheric conditions. If such a ma-

† See Weidner and Sells, *Elementary Classical Physics, 2,* p. 536.

terial could be found, cooling it to the temperature at which it becomes superconducting would be simple, and the technological applications of superconductivity would be vastly expanded. For a hypothetical superconductor that has a transition temperature of 77 K and is maintained at this temperature, what wavelength of electromagnetic radiation illuminating the material would break up the Cooper pairs and render it a normal conductor?

Band theory of solids, §9–2

● **9–10.** What elements can produce (*a*) donor and (*b*) acceptor impurity atoms for the semiconductor germanium?

● **9–11.** (*a*) Impurity atoms of indium ($Z = 49$, 5p[1]) are added to the semiconductor tin ($Z = 50$, 5p[2]). Is the semiconductor of the *p* or the *n* type? (*b*) Impurity atoms of what element will produce a semiconductor of the reverse type when they are added to pure tin?

9–12. When an impurity atom from the fifth column of the periodic table, such as arsenic, replaces a silicon atom in a silicon crystal, the unbonded electron "sees" a charge of *e* located at the arsenic atom. The dielectric constant of silicon is 12. Assume for simplicity that the unbonded electron moves in a Bohr orbit about the positive charge in silicon. (*a*) Compute the radius of the first Bohr orbit and compare it with the interatomic distance of 2.35 Å of silicon. (*b*) Compute the corresponding electron orbital radius for arsenic impurity atoms in germanium, whose dielectric constant is 16 and interatomic distance, 2.44 Å.

9–13. The conductivity of a semiconductor is directly proportional to the number of charge carriers. It can be shown that the number of conduction electrons in a semiconductor is proportional to $e^{-E_g/2kT}$, where E_g is the energy gap. (*a*) Show that the ratio of the conductivity of a semiconductor at temperature T_2 to the conductivity at T_1 is given by $\exp(E_g/2k)(T_1^{-1} - T_2^{-1})$. (*b*) By what factor does the conductivity of silicon ($E_g = 1.1$ eV) increase as its temperature is increased from 20°C to 100°C? (*c*) What would be the corresponding conductivity ratio for the same temperatures for an ordinary metallic conductor (a conductivity *decrease*), such as copper? The temperature variation of resistance with a temperature increase Δt for an ordinary conductor is given by $R = R_0(1 + \alpha\Delta t)$, where α, the temperature coefficient of resistivity, is 3.93×10^{-3}/C° for copper.

9–14. Compute the energy gap between the valence and conduction bands for germanium from the temperature variation of its conductivity. The conductivity of pure germanium increases by 54 percent when the temperature is increased from 20°C to 30°C. (See Problem 9–13).

Semiconducting devices, §9–3

● **9–15.** A light-emitting diode uses a *p-n* junction of silicon (gap energy = 1.1 eV). (*a*) What minimum potential difference must be applied to the junction to cause it to emit light? (*b*) To which side of the junction, the *p* or the *n* side, must the positive terminal of the battery be connected?

9–16. A *p-n* junction is in thermal equilibrium with no applied electric potential. Show that the Fermi levels of the two types of material must be identical, with other energy levels shifted upward or downward accordingly. (Hint: See Equation 8–35.)

9–17. The total energy consumption in the United States is approximately 10^{20} J/yr. Suppose that this energy is derived solely from solar cells located above the earth's atmosphere and exposed continuously to direct sunlight (intensity = 140 mW/cm²). (*a*) What is the equivalent constant power required, corresponding to the projected U.S. power consumption? (*b*) What would be the required edge length of a square solar-cell array that could produce this power? (Use the data appearing in Example 9–2.)

9–18. The range of wavelengths over which a solar cell using a *p-n* junction will operate is determined by the energy of the gap between the valence and the conduction bands (1.1 eV in silicon). (*a*) What is the theoretical maximum wavelength of electromagnetic radiation (infrared) that can be converted to an electric current in a silicon solar cell? (*b*) Radiation from the sun can be considered blackbody radiation from an object at temperature 6000 K. (See Equation 8–27 or Example 8–2.) By what factor does the energy density of radiation from the sun at the infrared wavelength of part (*a*) exceed the energy density at the visible limit of the electromagnetic spectrum (4000 Å)?

9–19. At room temperature, the number of electron-hole pairs produced by thermal excitation per second in silicon is 10^{14}/cm³. The created free electrons and free holes quickly recombine and become annihilated. The number N of surviving pairs is given by $N = N_0e^{-t/\tau}$, where N_0 is the initial number of pairs and τ is the lifetime of a pair (100 μs for silicon). (*a*) What is the equilibrium number of electron-hole pairs per unit volume in silicon? (*b*) What is the ratio of electron-hole pairs to silicon atoms (at. wt 28.1; density $\rho = 2.33$ g/cm³)?

9–20. A germanium gamma-detector can resolve the energy of a 1.33-MeV photon to within 2 keV. (*a*) How many electron-hole pairs would be produced in germanium (energy gap = 0.7 eV) by the absorption of a 1.33-MeV photon? (*b*) To how many electron-hole pairs does the resolution precision of 2 keV correspond?

The laser, §9–4

9–21. The particles of a certain system have three possible energies—E_1, E_2, and E_3, where $E_1 < E_2 < E_3$. The corresponding number of particles in the three states are n_1, n_2, and n_3, where for thermal equilbrium $n_1 > n_2 > n_3$. The system is irradiated by pumping radiation; each photon has an energy equal to $E_3 - E_1$. Under high-intensity pumping radiation, the populations of the two participating states become essentially equal. Show that a population inversion must exist between the states of one other pair, with either $n_3 > n_2$ or $n_2 > n_1$.

9–22. A carbon dioxide laser produces pulses, each of 1-nanosecond (ns) duration and a power of 2.0×10^{11} W. The transverse cross section of the laser beam is 0.50 mm^2. For each pulse, what is (*a*) the energy? (*b*) the energy density? (*c*) How much water at 20°C would be vaporized by a single pulse, assuming that the energy is completely absorbed by the water? The specific heat and latent heat of water are 4.19×10^3 J/kg·C° and 2.26×10^6 J/kg, respectively.

9–23. A helium-neon laser emits light of 6328 Å in a transition between two states of neon. What is the ratio of the population of the upper state to the population of the lower state for neon atoms in equilibrium at 300 K?

9–24. A helium-neon laser is most commonly operated to produce light of 6328-Å wavelength; but lasing action is possible at other wavelengths. Use Figure 9–21 to identify the transition for producing coherent light of wavelengths (*a*) 3.4×10^3 Å and (*b*) 1.15×10^4 Å.

9–25. The number N of atoms in an excited state with a mean life τ is given as a function of time t by $N = N_0 e^{-t/\tau}$, where N_0 is the initial number of atoms (at $t = 0$). (*a*) Show that if the population of the excited state is changed only through spontaneous emission to a state of lower energy, the number of photons emitted per unit time equals N/τ. (*b*) An atomic excited state with mean life 1×10^{-8} s is populated initially with 1×10^6 atoms. If the excited state is depopulated solely through spontaneous emission, how many atoms must be brought per unit time to the excited state so that the population of the excited state remains constant in time?

■ **9–26.** A population inversion exists between two atomic states in a system of identical atoms; the initial population of the upper state is twice the initial population of the lower state, which is also the atomic ground state. The populations of the two states are affected only by spontaneous emission from the upper state, which has a mean life of 1.0×10^{-6} s, to the lower state. Over what interval of time will the population inversion be maintained? (See Problem 9–25 for the relation governing the decay from an excited state.)

9–27. A collection of atoms has only the ground state (energy E_1) and the first excited state (energy E_2) occupied. The numbers of atoms in the two states are n_1 and n_2 in thermal equilibrium. When the atoms interact, there are upward and downward transitions between the two states *without* emission or absorption of photons. Show that the ratio of the probability for a downward transition to the probability for an upward transition is given by $e^{(E_2-E_1)/kT}$.

9–28. The temperature T of a system of particles can be defined in terms of the relative occupancy of the allowed states. Suppose that the particles of a system have energies E_1 and E_2, with $E_2 > E_1$. The corresponding numbers of particles in the two states are n_1 and n_2. If $n_2 > n_1$, a population inversion has been achieved. (*a*) Show that under these circumstances the system may be said to have a *negative* absolute temperature equal to $(E_2 - E_1)/[k \ln (n_1/n_2)]$ if the number of particles with energy E is taken to be proportional to the Boltzmann factor $e^{-E/kT}$. (*b*) Suppose that such a system, consisting of a collection of protons with spins aligned with or against an external magnetic field, initially has a population inversion and therefore a negative absolute temperature. Although the system is otherwise isolated from external influence, internal relaxation processes may change the relative populations of the two allowed states until the system finally achieves thermal equilibrium with its surroundings. Show that the system's temperature, as defined above, first *rises* to an infinite negative temperature, then becomes infinitely positive, and finally decreases to a finite positive value.

9–29. A 1.0-milliwatt (mW) helium-neon gas laser operates on a 115 V–2.0 A power supply. What is the working efficiency of this laser (the conversion of electrical energy to coherent light energy)?

Emission and absorption probabilities, §9–5

9–30. A system of identical atoms, each with two states differing in energy by $h\nu$, is in thermal equilibrium at temperature T with the radiation in which the system is immersed. Show (*a*) that if $h\nu/kT \ll 1$, the number of stimulated absorptions per unit time equals the number of spontaneous emissions per unit time, and (*b*) that the populations in the two states are essentially equal. Show (*c*) that if $h\nu/kT \gg 1$, the number of stimulated absorptions per unit time equals the number of spontaneous emissions per unit time, and (*d*) that the population in the lower energy state far exceeds the population in the upper energy state.

9–31. Photons from a radio station operating at a frequency of 710 kHz are in thermal equilibrium with their surroundings at a temperature 300 K. What is the ratio of stimulated to spontaneous emission?

9–32. For what frequency will the number of spontaneous emissions per unit time equal the number of stimulated emissions per unit time in a system of atoms in thermal equilibrium at 300 K?

9–33. A carbon dioxide laser produces high-efficiency radiation in the infrared region of the electromagnetic spectrum, centered at a wavelength of 9.4 micrometers (μm). (*a*) What is the ratio of the population of the upper energy state to the population of the lower energy state for the two levels involved in emission of 9.4-μm radiation when the CO_2 is thermal equilibrium at 300 K? (*b*) What is the ratio of stimulated emission to spontaneous emission between the two levels involved in the emission of 9.4-μm radiation, again for the CO_2 in thermal equilibrium with blackbody radiation?

Holography, §9–6

● **9–34.** A hologram is made using two plane waves, as in Figure 9–24. Suppose that a photographic negative is made from the original hologram with the dark and light regions reversed. How does the image of the reversed hologram compare with the image of the original?

● **9–35.** A hologram is broken into two smaller pieces. How does the image from a piece compare with the image from the unbroken hologram.

● **9–36.** Suppose that the reference wave W_1 used in producing a hologram is coherent and that the source wave W_2 is incoherent. What is the appearance of the exposed and developed photographic sheet (the "hologram")?

● **9–37.** What are the advantages of so choosing the relative angles of reference and source waves used in producing a hologram that only the first-order diffraction peaks appear?

9–38. A hologram is produced by a plane reference beam of 4000-Å wavelength together with a coherent source plane beam also of 4000 Å at an angle of 30° relative to the reference beam, as in Figure 9–24. (*a*) Compute the distance on the developed hologram between adjacent parallel lines of maximum intensity. (*b*) Could such alternate lines of darkness and brightness appearing on the hologram be seen by the unassisted eye?

9–39. A hologram is produced by a coherent plane reference beam that also falls on a single point scatterer before reaching the photographic plate. The developed plate is a hologram consisting of concentric circles (a zone plate). Show that when this zone plate is illuminated by a reconstructing wave, a portion of the wave transmitted through the hologram is focused to a point, so that the zone plate behaves (for the particular wavelength) as a converging lens.

9–40. Typically it takes 10^{-8} s for an atom to radiate one photon. Therefore, the spatial extent of the photon is $(3 \times 10^8 \text{ m/s})(10^{-8} \text{ s}) = 3$ m. This distance also represents the *coherence length* of a single photon, or of a collection of photons of the same wavelength, in an *incoherent* beam of light. The coherence length then represents that distance over which a beam of light can be considered coherent.

Show that if a holographic photographic plate is illuminated with *incoherent* monochromatic light instead of coherent light, the developed plate is a hologram and yields a three-dimensional image so long as the difference in distances from the object to various locations on the plate is much less than 3 m.

9–41. Visible light from a laser is so highly monochromatic that the linewidth $\Delta\nu$ is as small as 10^3 Hz. Show that the coherence length (see Problem 9–40) for laser light is $c/\Delta\nu \approx 50$ km.

10
Nuclear Structure

Ernest, Lord Rutherford (1871, New Zealand–1937). Student of J. J. Thomson at Cambridge Cavendish Laboratory. The father of radioactivity and nuclear physics. 1903, properties of alpha particles. 1907, measured Avogadro's number. 1908, Nobel Prize in chemistry, disintegration of the elements. 1911, nuclear model of atom. 1914, knighthood conferred. 1919, first artificial disintegration of nucleus. 1920, speculated on existence of neutron. (Photo courtesy of the American Institute of Physics, Niels Bohr Library.)

So far as atomic structure is concerned, the atomic nucleus can be regarded as a point mass and a point charge. The nucleus contains all the positive charge and nearly all the mass of the atom; it therefore provides the center about which electron motion takes place, and it influences atomic structure primarily through its Coulomb force of attraction with electrons.

The α-particle scattering experiments of Rutherford established that for distances greater than 10^{-14} m, the nucleus interacts with other charged particles by the inverse-square Coulomb force. It was found, however, that when the α particles approached the nuclear center closer than 10^{-14} m, the distribution of the scattered particles could not be accounted for simply by Coulomb's law. These experiments showed then that a totally new type of force, the nuclear force, acts at distances smaller than 10^{-14} m.

The nucleus and the atom show a tremendous difference in their relative sizes—10^{-10} m for atoms compared with 10^{-14} m for nuclei. Along with this size difference are several structural differences. Whereas atoms can be excited to emit their optical or x-ray spectra through gaining an energy of the order of a few electron volts to a few kilo-electron-volts, nuclei generally remain inert until they gain energies of the order of a mega-electron-volt. Some aspects of atomic structure can be understood on the basis of the Bohr model; no such simple model exists for nuclei. The primary force between the particles constituting the atom is the well-understood Coulomb force; additional forces act between the constituents of nuclei and these forces are only partly understood. An atom typically loses energy of excitation by emitting photons; an excited nucleus can lose its energy of excitation by emitting particles as well as photons. Despite these differences, several fundamental laws apply equally well to atoms and to nuclei; these are the rules of quantum and relativity theory, as well as the conservation laws of mass-energy, linear momentum, angular momentum, and electric charge.

10-1 The Nuclear Constituents

The particles of which all nuclei are composed are the proton and the neutron. Their fundamental properties include charge, mass, spin, and nuclear magnetic moment.

CHARGE. The proton is the nucleus of the light isotope of hydrogen; it carries a single positive charge, equal in size to the charge of the electron.

The neutron is so named because it is electrically neutral. Because it carries no charge, it shows only a feeble interaction with electrons, it produces no direct ionization effects, and it can therefore be detected and identified only indirectly. The existence of the neutron was not clearly established until 1932, when J. Chadwick demonstrated its properties in a series of classic experiments.

MASS. The masses of the proton (the bare nucleus of the $^1_1$H atom) and the neutron, in unified atomic mass units, together with the rest energies of these particles in mega-electron-volts, are:

$$\begin{aligned}\text{Proton rest mass} &= 1.007\,276\,63 \pm 0.000\,000\,08 \text{ u}\\ \text{Proton rest energy} &= 938.256 \pm 0.005 \text{ MeV}\\ \text{Neutron rest mass} &= 1.008\,665\,4 \pm 0.000\,000\,4 \text{ u}\end{aligned}$$

Neutron rest energy $= 939.550 \pm 0.005$ MeV

The proton and the neutron have nearly the same mass; the neutron exceeds the proton by slightly less than 0.1 percent. Both particles have rest energies of about 1 GeV. Because the proton carries an electric charge, its mass can be measured directly with high precision by mass spectrometry. The mass of the neutron must be inferred indirectly from experiments, since electric and magnetic fields have virtually no effect on the neutron.

SPIN. Both the proton and the neutron have intrinsic angular momentum, or so-called nuclear spin. Since spin angular momentum is independent of orbital motion, the nuclear spin can be visualized, like electron spin, as the particle's spinning as a whole about some internal rotation axis. The nuclear spin angular momentum L_I corresponds to the nuclear spin quantum number I; its magnitude is given by

$$L_I = \sqrt{I(I+1)}\,\hbar \qquad (10\text{–}1)$$

This is analogous to Equation 7–16, which gives the spin angular momentum of the electron.

The nuclear spin quantum numbers of both the proton and the neutron are $\frac{1}{2}$.

The nuclear-spin angular momentum is space-quantized by an external magnetic field; the permitted components along the direction of the magnetic field are $+\frac{1}{2}\hbar$ and $-\frac{1}{2}\hbar$, as shown in Figure 10–1. The magnitude of the spin angular momentum of a proton or a neutron, as well as the components of the angular momentum along the space-quantization direction, are precisely the same as for an electron.

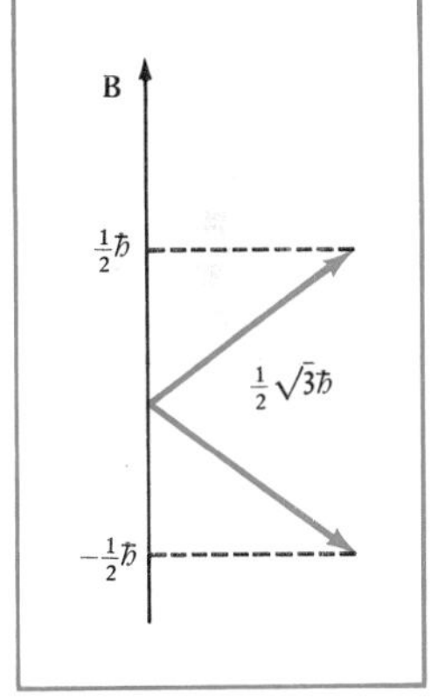

FIGURE 10–1. *Space quantization of a proton spin or a neutron spin.*

NUCLEAR MAGNETIC MOMENT. The component of the magnetic moment associated with electron spin along the direction of an external magnetic field is one Bohr magneton (Section 7–5), $\beta = e\hbar/2m = 0.927\,32 \times 10^{-23}$ J/T. Because the electron has a negative electric charge, its magnetic moment points in the direction opposite to its angular momentum.

Now consider the magnetic moment associated with proton spin. Nuclear magnetic moments are measured in units of the *nuclear magneton* β_I, which is defined as

$$\beta_I = \frac{e\hbar}{2M_p} = (5.050\,50 \pm 0.000\,13) \times 10^{-27}\ \text{J/(Wb/m}^2) \qquad (10\text{–}2)$$

where M_p, the proton mass, replaces the electron mass m in the Bohr magneton. Since the proton mass is 1836.10 times the electron, the nuclear magneton is smaller than the Bohr magneton by this factor. The nuclear magnetic moment of the proton is found by experiment to be

Proton magnetic moment $= +(2.792\,76 \pm 0.000\,02)\beta_I$

The plus sign indicates that the proton's magnetic moment points in the same direction as its nuclear spin; the magnitude of the nuclear moment gives the *component* of the proton's magnetic moment along the space-quantization direction in units of the nuclear magneton. Note that the size of the proton magnetic moment is *not* one nuclear magneton; it is nearly three times larger than what one might expect simply from the proton mass.

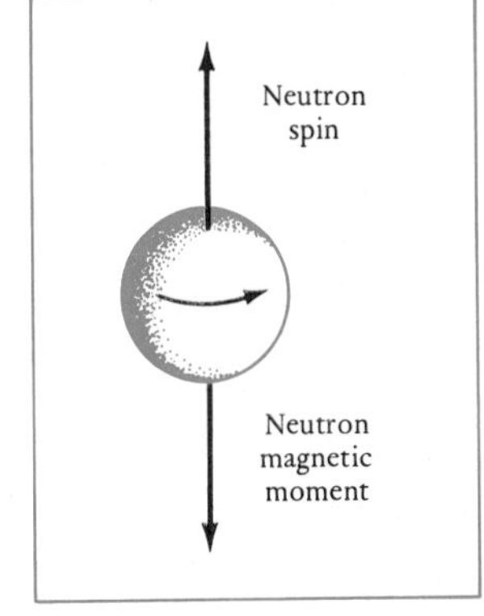

FIGURE 10–2. *Representation of the relative orientations of the neutron spin and magnetic moment.*

Although the neutron as a whole carries no net electric charge, it does have a magnetic moment, whose value is written:

$$\text{Neutron magnetic moment} = -1.913\,15\beta_I$$

The negative sign indicates that the neutron magnetic moment is *opposite* to the direction of the neutron angular momentum; this is shown in Figure 10–2, p. 317.

Because the proton moment is *not* one β_I and the neutron moment is *not* zero, the proton and neutron are more complicated entities than the electron. The nonzero magnetic moment of the neutron implies that although the neutron has zero total charge, there is a nonuniform charge distribution within it.

10–2 Forces Between Nucleons

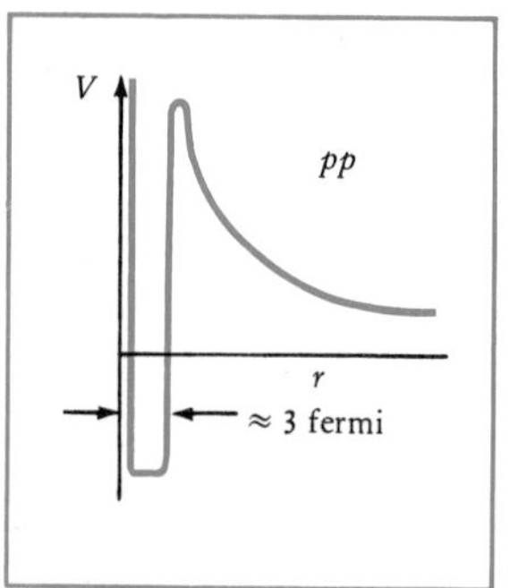

FIGURE 10–3. *Proton-proton potential.*

All nuclei consist of protons and neutrons, bound together to form more or less stable systems; therefore, it is important to know something about the forces that act between these fundamental nuclear constituents. First, there is the force between two protons. The most direct way of examining this force is by a proton-proton scattering experiment, in which monoenergetic protons from a particle accelerator strike a target containing mostly hydrogen atoms, and therefore, protons. From the angular distribution of the scattered particles, one can infer, as in α-particle scattering, the force acting between the incident particles and the target particles; in this instance, they are both protons. Proton-proton scattering experiments show that the force can be represented approximately by the potential curve shown in Figure 10–3. At large distances of separation, the protons repel one another by the Coulomb inverse-square force. At a separation distance of approximately 3×10^{-15} m, a fairly sharp break exists in the potential curve. It indicates the onset of the *nuclear force* between a pair of protons. The force is strongly *attractive* at small distances (although there is a repulsive "core" at very small distances). The "size" of the proton can be taken as the *range,* 3×10^{-15} m, of the nuclear proton-proton force. The customary unit for measuring nuclear dimensions is the fermi, where 1 fermi = 1 fm = 10^{-15} m.

The force between a neutron and a proton can be investigated by neutron-proton scattering experiments. In these experiments a monoenergetic neutron beam bombards a target containing protons. The distribution of the scattered neutrons is analyzed for the force acting between the neutron and the proton. The interaction between a neutron and a proton can be represented approximately by a potential curve, as shown in Figure 10–4. At large distances, there is *no* force between the two particles; but at a distance of about 2 fm the neutron and proton attract one another by a strong nuclear force of well-defined range; a repulsive inner core also exists here. Clearly, the nuclear attraction in no way depends on electric charge, since the neutron is a neutral particle.

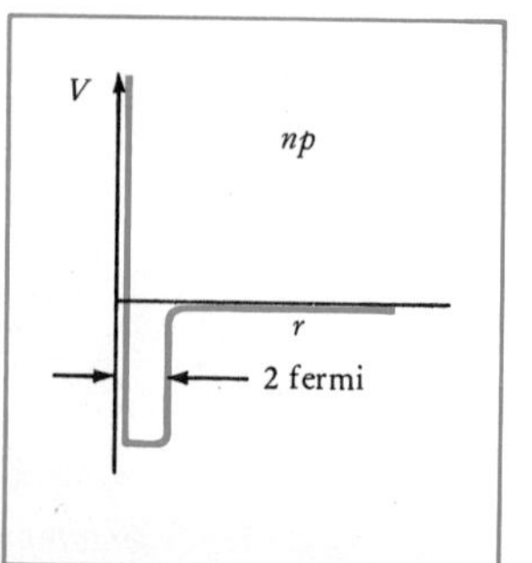

FIGURE 10–4. *Neutron-proton potential.*

The nuclear force between two neutrons cannot be investigated directly by a neutron-neutron scattering experiment; it is impossible to prepare a target consisting of free neutrons. But a variety of indirect evidence indicates that the force between two neutrons approximately equals the force between a neutron and a proton and also the nuclear force between the members of a pair of protons. Because a neutron and a proton are nearly equivalent in their interactions (apart from the Coulomb force between protons), it is customary to refer to a neutron *or* a proton as a *nucleon.* The term designates either a proton or a neutron when the distinction between them is of little importance. The independence of the nuclear force from the charge of the particular participating nucleons is known as the *charge independence* of the nuclear force. More sophisticated treatments of the proton-neutron interactions show that it is possible to consider the proton and the neutron as two different charge states of the *same* particle. (The nuclear force between nucleons, described in terms of the exchange of field particles (pions), is given in Section 12.2.)

10-3 The Deuteron

The simplest nucleus containing more than one particle is the *deuteron,* the nucleus of the deuterium atom. The deuteron consists of a proton and a neutron bound by the attractive nuclear force to form a stable system. The deuteron has a single positive charge, $+e$. Its mass is approximately twice the proton or the neutron; more precisely,

$$\text{Deuteron rest mass} = 2.013\,553 \text{ u}$$

It must be emphasized that the mass given here for the deuteron is that of the bare deuterium nucleus; the mass of the neutral deuterium atom exceeds the mass of the deuteron by the mass of an electron, 0.000 549 u, and is therefore 2.014 102 u.

It is interesting to compare the mass of the deuteron M_d with the sum of the masses of the deuteron's constituents, the proton and the neutron, M_p and M_n:

$$\begin{aligned} M_p &= 1.007\,277 \text{ u} \\ M_n &= \underline{1.008\,665 \text{ u}} \\ M_p + M_n &= 2.015\,942 \text{ u} \\ M_d &= \underline{\underline{2.013\,553 \text{ u}}} \\ \text{Mass difference} = M_p + M_n - M_d &= 0.002\,389 \text{ u} \end{aligned}$$

The total mass of the proton and the neutron when they are separated *exceeds* the mass of the two particles when they are bound, forming a deuteron. This difference is easily interpreted on the basis of the relativistic conservation of mass-energy (see Section 3–3). When *any* two particles attract one another, the sum of their separate masses exceeds the value of the mass of the bound system, since energy (or mass) must be added to the system to separate it into its component particles. The value of this energy, called the binding energy, can be computed from the mass difference by using the mass-energy conversion factor, 1 u = 931.5 MeV/c^2. Thus, the binding energy E_b of the neutron-proton forming a deuteron is given by

$$\begin{aligned} E_b + M_d c^2 &= (M_p + M_n)c^2 \\ E_b &= (M_p + M_n - M_d)c^2 \qquad (10\text{–}3) \\ &= (0.002\,389 \text{ u})(931.5 \text{ MeV/u}\cdot c^2)c^2 = 2.225 \text{ MeV} \end{aligned}$$

If 2.225 MeV is added to a deuteron, the neutron and proton can be separated from one another, beyond the range of the nuclear force; both particles are then left at rest and thus have no kinetic energy.

Such a mass difference arises in *any* system of bound particles. For an atomic system like the hydrogen atom, the difference in mass between the atom and its separated parts is so small, 1 part in 100 million, that it cannot be measured directly. Binding energy is manifest as a measurable mass difference in nuclear systems because the nuclear force is very strong and the binding energy very great. In fact, the nuclear binding energy, 2.225 MeV, of two nucleons forming a deuteron is roughly a *million* times the electrostatic binding energy, 13.58 eV, of a proton and an electron forming a hydrogen atom.

Recall that the binding energy of a hydrogen atom in its ground state (the ionization energy of the hydrogen atom) can be determined from the energy of the photon whose absorption in the photoelectric effect frees the bound electron from the hydrogen nucleus. A completely analogous measurement can be made of the deuteron binding energy. Deuterium gas is irradiated with a beam of high-energy monoenergetic γ-ray photons. If the energy of a photon equals the deuteron's binding energy, photon absorption will produce a free neutron and a proton. If the photon's energy exceeds the binding energy, the deuteron will be dissociated into a neutron and a proton; each particle will also have kinetic energy. This *nuclear reaction* can be written as follows:

$$\gamma + \mathrm{d} \rightarrow \mathrm{p} + \mathrm{n} \tag{10–4}$$

Mass-energy conservation requires that

$$h\nu + M_{\mathrm{d}}c^2 = M_{\mathrm{p}}c^2 + M_{\mathrm{n}}c^2 + K_{\mathrm{p}} + K_{\mathrm{n}} \tag{10–5}$$

where K_{p} and K_{n} are the kinetic energies of the freed proton and neutron. This process, in which the proton and neutron are detached from one another by the absorption of a photon, is a nuclear photoelectric effect, or a nuclear

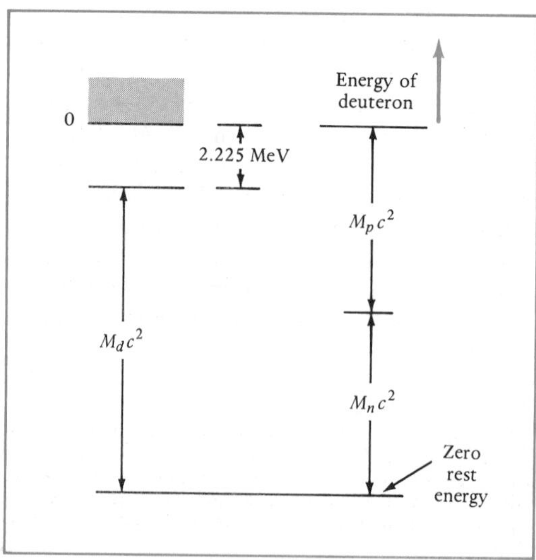

FIGURE 10–5. *Energy-level diagram of the neutron-proton system, the deuteron.*

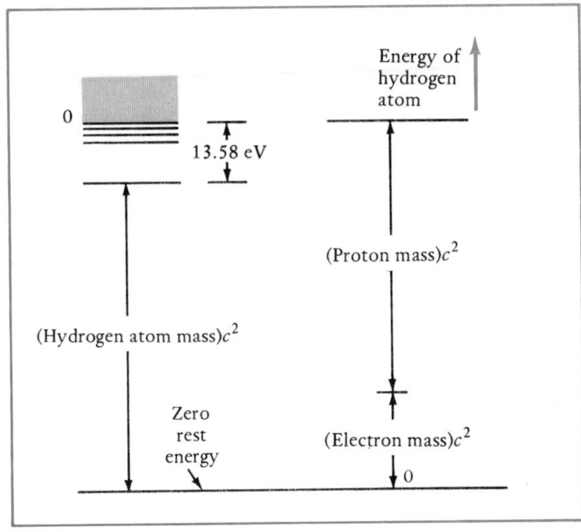

FIGURE 10–6. *Energy-level diagram of the electron-proton system, the* $^1_1$H *atom.*

photodisintegration. The threshold for the reaction corresponds to $K_p = 0$ and $K_n = 0$; then,

$$h\nu_0 = (M_p + M_n - M_d)c^2 = E_b \tag{10–6}$$

That is, the energy of the photon is equal to the binding energy of the deuteron.† If the threshold photon energy $h\nu_0$ is measured and the values of M_p and M_d are known, the neutron mass can be computed from Equation 10–6. This is one of several ways in which the neutron mass can be measured by applying energy conservation to nuclear reactions.

The inverse reaction to deuteron photodisintegration is as follows. A neutron and a proton at rest combine to form a deuteron in an *excited* state; it decays to the ground state with the emission of a photon, of 2.225 MeV,

$$\text{p} + \text{n} \rightarrow \text{d} + \gamma \tag{10–7}$$

This nuclear reaction is merely Equation 10–4 with the arrow reversed.

A *nuclear energy-level diagram* of the deuteron is shown in Figure 10–5. Unlike all atoms and all other nuclei, the deuteron has only *one* bound state; it can exist as a bound system only in this, the ground state. In the continuum of unbound states, the proton and the neutron are free. The rest masses and rest energies of the deuteron, proton, and neutron are also shown (but not to scale) in the diagram. It is useful to compare this diagram of the simplest of all bound nuclear systems with the energy-level diagram of the simplest two-particle atomic system, the hydrogen atom. The simplified but exaggerated energy-level diagram of the bound proton-electron system is shown in Figure 10–6, in which the masses of the electron and proton are also displayed.

The hydrogen atom has a whole series of possible excited states; this

† Strictly, the photon threshold energy for the photodisintegration of a deuteron intially at rest slightly *exceeds* the deuteron binding energy because of the requirement of momentum conservation. The total momentum of proton and neutron out of the collision must equal the photon's momentum $h\nu/c$ entering into the collision. A detailed analysis shows that the photon threshold energy is $h\nu = E_b/(1 - E_b/2M_dc^2)$, greater than the deuteron binding energy by about 1 part in 10^3.

means that the rest mass of the bound electron-proton system can assume any one of many possible quantized values (compare Figure 10–6 with Figure 10–5). Because the binding energy for any one of the possible energy states of the hydrogen atom is small (less than the 13.58 eV), it is not possible in practice to determine the quantized energies of the hydrogen atom by measuring its mass. But the large binding energies of nucleons in nuclear systems permit the binding energy of nuclei to be determined directly from the difference between the rest masses of the constituent particles and the mass of the bound system.

10-4 Stable Nuclei

Consider stable nuclei containing more than two nucleons. The number of protons in a nucleus is represented by the *atomic number Z,* the total number of nucleons by the *mass number A,* and the number of neutrons by $N = A - Z$. The term *nuclide* designates a particular species of nucleus in which all members of the species have the same Z and the same N. Species having the same proton number Z are nuclides known as *isotopes,* those having the same neutron number N are nuclides known as *isotones,* and those having the same nuclear number A are nuclides known as *isobars.* For example, ${}^{37}_{17}\mathrm{Cl}$, which has 17 protons, 20 neutrons, and 37 nucleons, is an isotope of ${}^{35}_{17}\mathrm{Cl}$, an isotone of ${}^{39}_{19}\mathrm{K}$, and an isobar of ${}^{37}_{18}\mathrm{Ar}$. (Here we use the convention that the presubscript to the chemical symbol denotes Z and the presuperscript A.)

The *stable* nuclides found in nature are plotted in Figure 10–7, where neutron number N is plotted against proton number Z. Each point represents a combination of protons and neutrons that forms a stable bound system. If we concentrate on the general features in this diagram, we can see several interesting and significant regularities.

Only those combinations of protons and neutrons that appear as points in the figure are found in nature as stable nuclides in their ground states; all other possible combinations of nucleons are somewhat unstable in that they decay, or disintegrate, into other nuclei. For example, the nuclides ${}^{16}_{8}\mathrm{O}$, ${}^{17}_{8}\mathrm{O}$, and ${}^{18}_{8}\mathrm{O}$, all isotopes of oxygen, exist as stable nuclear systems, but the isotopes ${}^{15}_{8}\mathrm{O}$ and ${}^{19}_{8}\mathrm{O}$ are unstable.

The stable nuclides in Figure 10–7 constitute a *stability line,* the general region in which the most stable nuclides fall. At small values of N and Z, the stable nuclides lie close to the $N = Z$ line at 45°. For example, ${}^{16}_{8}\mathrm{O}$ has $N = Z = 8$. The most stable light nuclides for a given mass number A are those whose number of protons nearly equals the number of neutrons. We might say that light nuclei prefer to have equal numbers of protons and neutrons because such aggregates are more stable than those in which there is a decided excess of protons or neutrons. For the heavier nuclides, the stability line bends increasingly away from the 45° line; that is, $N > Z$ at large A. The stable nuclide ${}^{208}_{82}\mathrm{Pb}$, for example, has $Z = 82$ and $N = 126$. Thus heavy nuclides show a decided preference for neutrons over protons.

The neutron excess can be accounted for through the repulsive Coulomb force that acts between the protons of every pair. If we start with a moderately heavy nucleus and try to construct from it a heavier nucleus by adding

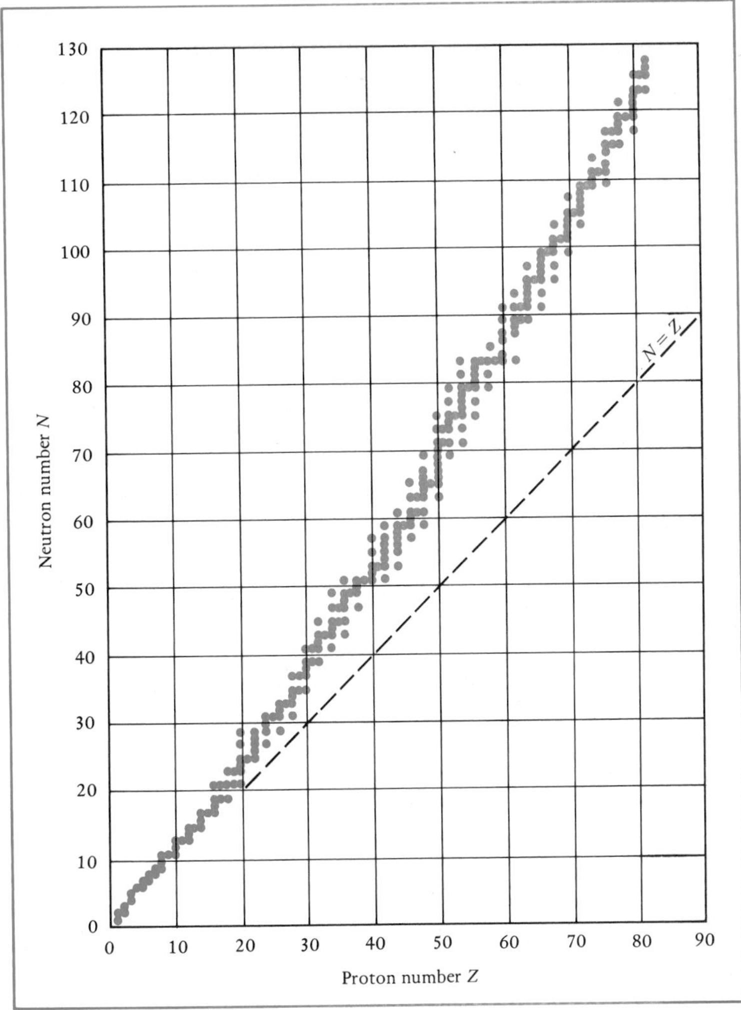

FIGURE 10–7. *Neutron number versus proton number for the stable nuclides.*

one nucleon, the binding of the additional neutron will usually be stronger than the binding of an additional proton. The neutron is attracted by the short-range nuclear force of neighboring nucleons, whereas the proton is attracted by the short-range nuclear force but repelled by the Coulomb repulsive forces of all other protons already in the heavy nucleus. The repulsive Coulomb effect competes noticeably with the strong nuclear attractive force only in heavy nuclides. One could say that if protons had no electric charge but were otherwise distinguishable from neutrons, then the heavy, stable nuclides would have approximately equal numbers of protons and neutrons.

We can understand the near equality of Z and N at small A and the greater N than Z at large A by considering how the Pauli exclusion principle (Section 7–9) operates in the building of stable nuclides. Both the proton and the neutron separately follow this principle; no two identical particles can be placed in the same quantum state in a nucleus. We need not concern

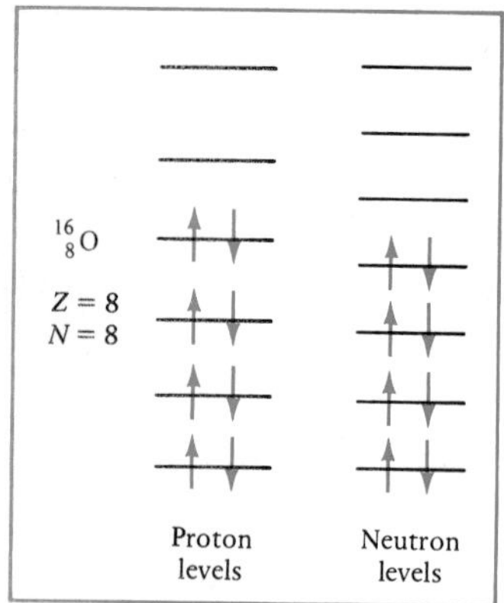

FIGURE 10–8. *Schematic representation of the proton and neutron states of* $^{16}_{8}O$.

ourselves here with details of the quantum theory of nuclear structure; we shall simply recognize that if two protons are in a state having the same three spatial quantum numbers (not necessarily the quantum numbers n, l, and m_l appearing in atomic structure), then the protons must differ in their magnetic spin quantum numbers. This implies that two protons can occupy the same quantum state only if their nuclear spins are antialigned. The same rule applies to neutrons; only two neutrons, one with spin up and one with spin down, can occupy a quantum state that is identical in the three spatial quantum numbers. Apart from the Coulomb interaction between protons, the states available to a proton or a neutron are very nearly the same, since the proton and neutron are essentially equivalent in their nuclear interactions.

Figure 10–8 shows in an oversimplified and schematic fashion the states available to protons and neutrons as they combine to form stable nuclei. For simplicity, the states are shown nearly equally spaced and with one set of levels for protons and a second set for neutrons. The spacing between proton levels increases as the levels get higher, to correspond to the Coulomb repulsion between protons. Two protons, one with spin up and one with spin down, can be accommodated in each proton level and two neutrons in each neutron level.

The first proton level and the first neutron level are filled when a nucleus is formed of two protons and two neutrons. This corresponds to the very stable nuclide $^{4}_{2}He$. With further nucleons added to form stable nuclides of larger A, we should expect the proton and neutron levels to be nearly equally populated. Thus, at small A the stable nuclides will have $Z \approx N$, in accord with observation; the stability is particularly great when both Z and N are even numbers. As the number of nucleons increases, the most stable nuclides are formed when the lowest energy levels available to the protons and neutrons are filled first. This requires a neutron excess, or $N > Z$, for large A. We see, then, that the general features of the stability line can be accounted for by applying the Pauli exclusion principle to the building up of stable nuclides.

10–5 Nuclear Radii

The radius of a nucleus can be defined and evaluated in several ways. A nuclear radius is given approximately by the results of α-particle scattering experiments. (See Section 6–1.) Although the distribution of the scattered particles is accounted for by the Coulomb interaction alone for distances greater than 10^{-14} m, deviations from Coulomb's law occur when the α particles come within approximately that distance from the nuclear center. A nuclear radius, then, is the distance from the nuclear center at which the nuclear force becomes important.

The nuclear radius can, however, be inferred more directly and with higher precision from scattering experiments in which high-energy neutrons bombard target nuclei. Neutrons are not repelled by a Coulomb force; they are deviated from their incident directions or absorbed by the target nucleus only when they come within the range of the nuclear force of the bombarded nuclei. We can define the *nuclear-force* radius as the distance from the center of the nucleus at which a neutron first feels the nuclear attractive force.

The range of the nuclear force is definite, since the nuclear interaction is effectively zero for greater separation distance. Therefore, neutron scattering experiments do not have the complicating effect of the Coulomb force, which must be subtracted in an analysis of the scattering data.

Like all particles, the neutron has a wavelength associated with it. Unless the neutron's wavelength is small enough for the neutron's position to be precisely specified relative to nuclear dimensions, neutron scattering experiments cannot be interpreted simply. Neutrons of 100 MeV have a wavelength of about 1 fermi (fm). Experiments have been designed to measure the absorption of high-energy neutrons in a variety of targets. The results can be summarized by the following relation, in which the nuclear radius R is given as a function of the nuclear mass number A:

$$R = r_0 A^{1/3} \tag{10–8}$$

with $r_0 = 1.4$ fm. The radius of any nucleus, defined in terms of its nuclear interaction with a neutron, can be computed from this equation. For even the most massive nuclei, R is no larger than about 10 fm.

The scattering of very high energy electrons provides another means of assigning a nuclear radius. A 10-GeV electron has a wavelength of only 0.1 fm, *less* than the size of even the smallest nucleus. Such a well-localized particle is a suitable probe for studying nuclear radii. Indeed, experiments with high-energy electrons have led to information about the distribution of electric charge within a nucleus and even within a single proton. Whereas neutrons are scattered only by the nuclear force in nuclear interactions, electrons are scattered only by the charges within nuclei in electric interactions, not by the nuclear force. The results of high-energy electron scattering experiments can be summarized by a similar relation:

$$R = r_0 A^{1/3} \tag{10–9}$$

with $r_0 = 1.1$ fm. The charge, or electromagnetic, radius r_0 is somewhat smaller than the nuclear-force radius. Figure 10–9 shows the distribution of electric charge within nuclei as measured by the scattering of high-energy electrons.

The relation for the nuclear radius leads to an important conclusion

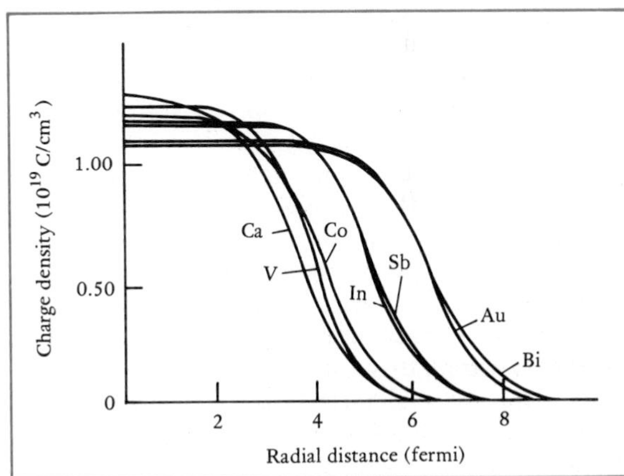

FIGURE 10–9. *Nuclear electric-charge density as a function of radial distance for several elements, determined from scattering experiments with 183-MeV electrons.*

about the density of nuclear material. Cubing Equation 10–8 and multiplying by $4\pi/3$, we have

$$\frac{4\pi R^3}{3} = \frac{4\pi r_0^{\,3}}{3} A \tag{10–10}$$

The quantity $4\pi R^3/3$ is the volume of the nucleus, assumed to be a sphere, or a near sphere, and we can take $4\pi r_0^{\,3}/3$ to be the volume of a single nucleon. Therefore, Equation 10–10 becomes

Volume of the nucleus = (volume of a nucleon) × (number of nucleons)

The total volume of a nucleus is merely the sum of the volumes of the several nucleons composing it. One can imagine the nucleons to be marblelike spheres touching one another to form a spherical cluster. The relation holds for all nuclei. Moreover, all nucleons have nearly the same mass. Therefore, we conclude that *all* nuclei have the same density of nuclear matter. The density of nuclear material is 2×10^{17} kg/m^3. (This extraordinarily high density is consistent with the very large size of the radius of the atom compared with the nucleus—approximately a hundred thousand times greater.)

We are now able to consider a question that has not been raised so far, Why can't electrons be constituents of nuclei? If an electron were confined and localized within a nuclear dimension, say 1 fm, then the electron wavelength could be no greater than this distance. Now, an electron of 1 GeV has a wavelength of approximately 1 fm. Therefore, if the electron were contained within a nucleus, it would have a kinetic energy of at least 1 GeV; but an electron of that energy can be bound and have a negative total energy only if the potential energy is less than -1 GeV. Thus, an attractive potential of at least that much must exist for electrons if they are to be bound within a nucleus. There is no evidence whatever of so strong an attractive force on an electron. We must therefore conclude that electrons cannot exist within a nucleus. A nucleon can, however, be localized within a nuclear dimension when its kinetic energy is only a few MeV; for example, a nucleon of 10 MeV has a wavelength of about 10 fm.

10–6 The Binding Energy of Stable Nuclei

The nuclear force is so strong that the mass of a bound nuclear system is measurably smaller than the sum of the masses of its components. Thus, information on the binding energy of nuclear systems can be arrived at directly by comparing masses.

The masses of atoms can be measured with considerable precision (more than 1 part in 10^5) by the methods of mass spectrometry. Appendix II gives the masses of the *neutral* atoms in unified atomic mass units (u). All measured atomic masses are very close to the integral mass number A. An atom of $^{12}_{6}$C has a mass of precisely 12 u by definition. The mass of a *nucleus* is the *atomic* mass minus Z electron masses (since the number of electrons equals the number of protons). Because the energy binding atomic electrons to a nucleus is small compared with the atom's rest energy, we can take the neutral atom's mass to be the mass of the nucleus plus the mass of the electrons.

The nucleus of $^{12}_{6}$C has 6 protons and 6 neutrons. Suppose we want to calculate the *total* nuclear binding energy. This, in other words, would be the energy required to separate a $^{12}_{6}C$ nucleus into its 12 component nucleons, each nucleon at rest and effectively out of the range of the forces of the other nucleons. In the computation, we shall use the mass of a *neutral hydrogen atom,* 1.007 825 u, and the mass of an electron, 0.000 549 u.

$$\begin{aligned}
6 \text{ protons} &= 6(1.007\,825 - 0.000\,549)\text{ u} \\
6 \text{ neutrons} &= \underline{6(1.008\,665)\text{ u}\qquad\qquad\quad} \\
\text{Total nucleon masses} &= 12.098\,940 - 6(0.000\,549)\text{ u} \\
^{12}_{6}\text{C nuclear mass} &= \underline{\underline{12.000\,000 - 6(0.000\,549)\text{ u}}} \\
\text{Mass difference} &= 0.098\,940\text{ u} \\
\text{Total binding energy} &= 0.098\,940\text{ u} \times 931.5\text{ MeV/u} = 92.16\text{ MeV}
\end{aligned}$$

Since the electron masses cancel, we can use the masses of the *neutral* atoms of hydrogen and carbon-12 rather than the masses of the proton and the carbon-12 nucleus.

We see that 92.16 MeV must be added to a carbon nucleus to separate it completely into its constituent particles; therefore, the 12 nucleons of the carbon atom are bound together with a total binding energy E_b = 92.16 MeV to form the nucleus in its lowest energy state. The *average* binding energy per nucleon, E_b/A, is 92.16/12, or 7.68 MeV.

Example 10–1. What is the energy needed to remove *just one proton* from $^{12}_{6}C$, leaving a nucleus with 5 protons and 6 neutrons, namely, the nucleus of $^{11}_{5}B$?

The energy binding the last proton to the remaining 11 nucleons, *the separation energy,* is computed by using the rest masses of the particles (again we use the masses of *neutral* atoms):

$$\begin{aligned}
\text{Mass of } ^{1}_{1}\text{H} &= 1.007\,825\text{ u} \\
\text{Mass of } ^{11}_{5}\text{B} &= \underline{11.009\,305\text{ u}} \\
\text{Mass of } ^{1}_{1}\text{H} + {}^{11}_{5}\text{B} &= 12.017\,130\text{ u} \\
\text{Mass of } ^{12}_{6}\text{C} &= \underline{\underline{12.000\,000\text{ u}}} \\
\text{Mass difference} &= 0.017\,130\text{ u} \\
\text{Separation energy} &= 0.017\,130\text{ u} \times 931.5\text{ MeV/u} = 15.96\text{ MeV}
\end{aligned}$$

We see that the binding energy, 15.96 MeV, of one particular nucleon in $^{12}_{6}C$ (the least tightly bound proton) is *not* the same as the *average* binding energy of a nucleon, 7.68 MeV. The separation energy exceeds the average binding energy because, in removing just one proton from $^{12}_{6}C$, one breaks one proton from a proton pair.

Removing this proton from a nucleus corresponds, in atomic structure, to removing the least tightly bound valence electron from an atom by ionization. We know that the ionization energy is, for an outer valence electron in an atom, usually several orders of magnitude less than that for an inner, tightly bound electron (visible light for the former; x-rays for the latter). Unlike electrons, bound by electric interaction, the particles of a stable nucleus are all bound with at least approximately the same energy.

The average binding energy per nucleon E_b/A can likewise be computed for any stable nucleus. When the computed values of E_b/A are plotted against the corresponding values of the atomic mass A, we obtain the results

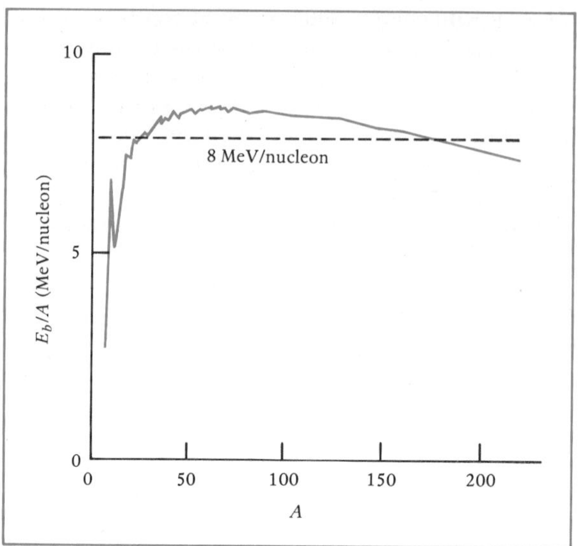

FIGURE 10–10. *Average binding energy per nucleon as a function of mass number, for the stable nuclides.*

shown in Figure 10–10. The characteristics of the curve can be summarized as follows. Apart from sharp peaks for the especially stable nuclides with groups of 2 protons and 2 neutrons—$^4_2$He, $^{12}_6$C, $^{16}_8$O—the curve rises sharply from the lightest stable nuclides to values of $A \approx 20$. At $A > 20$, the curve rises slowly, reaches a maximum near the element $^{56}_{26}$Fe, and then drops slowly toward the heaviest nuclides. It is approximately horizontal from $A = 20$ onward, E_b/A being roughly constant and equal to 8 MeV/nucleon.

Iron and nuclides close to it represent the most stable configurations of nucleons found in nature; in all elements lighter or heavier than iron, the typical nucleon is bound with less energy.

10-7 The Radioactive Decay Law

Any unstable system, including an unstable nucleus, will decay into parts provided that the fundamental conservation laws—mass-energy, momentum, electric charge, and the like—do not preclude such a decay. The only additional conservation law that will concern us at this time is the following:

> *Law of Conservation of Nucleons: The total number of protons and neutrons entering a reaction must equal the total number of nucleons leaving the reaction.*

Therefore, the mass energy, linear momentum, angular momentum, electric charge, and number of nucleons before a reaction or decay all must equal the respective quantities after the reaction or decay.

If unstable nuclei *can* decay, the general question exists about how rapidly such decays take place, how the number of surviving unstable nuclei decreases as time goes on. All unstable nuclei follow the radioactive decay law, whatever the differences in the particles emitted in the decay.

We call the initial unstable nucleus the parent, and the nucleus into which the parent decays, the daughter. The death of the parent gives birth to the daughter. The fundamental assumption of the radioactive decay law is that for any unstable nucleus of a particular species, one instant is like any other. Then the probability that an unstable nucleus will decay spontaneously into one or more particles of lower rest energy is the same at *any* instant. It is independent of the past history of the parent nucleus, it is the same for all nuclei of the same type, and it is very nearly independent of external influences (temperature, pressure, and so on).

There is no way of predicting the precise time that any one unstable nucleus will decay; its survival is subject to the laws of chance. But during an infinitesimally small time interval dt, the probability that an unstable nucleus will decay is directly proportional to this time interval. Thus,

$$\text{Probability that nucleus } \textit{decays} \text{ in time } dt = \lambda\, dt \tag{10–11}$$

where the proportionality constant λ is called the *decay constant,* or the *disintegration constant.* Since the total probability that a nucleus will either survive or decay in the time dt is 1 (100 percent),

$$\text{Probability that nucleus } \textit{survives} \text{ time } dt = 1 - \lambda\, dt \tag{10–12}$$

Then, if we denote the probability that a nucleus will survive to the time t by $p(t)$, and the probability that it will survive to the time $t + dt$ by $p(t + dt)$, if follows from Equation 10–12 that

$$p(t + dt) = (1 - \lambda\, dt)p(t) \tag{10–13}$$

Rearranging terms yields

$$p(t + dt) - p(t) = -\lambda p(t)\, dt$$

or

$$\frac{dp}{p} = -\lambda\, dt \tag{10–14}$$

Integrating Equation 10–14 from $t = 0$ to $t = t$, and p from p_0 to p_t, we have

$$\int_{p_0}^{p_t} \frac{dp}{p} = -\lambda \int_0^t dt$$

$$\ln\left(\frac{p_t}{p_0}\right) = -\lambda t$$

Taking the inverse natural logarithm, we have

$$p_t = p_0 e^{-\lambda t} = e^{-\lambda t}$$

since p_0 is by definition equal to 1 (at $t = 0$, the nucleus has not yet decayed).

$$\text{Probability that nucleus survives time } t = e^{-\lambda t} \tag{10–15}$$

This equation is the necessary consequence of the assumption that the decay of nuclei in unstable states is independent of the present condition and past history of the nucleus. Although we cannot say precisely when a *single* nucleus will decay, we can predict the statistical decay of a *large* number of identical unstable nuclei. If there are initially N_0 unstable nuclei undergoing a decay characterized by the decay constant λ, then the number N surviving a period t is merely N_0 times the probability that any one nucleus will have survived. Therefore, from Equation 10–15, we have

$$N = N_0 e^{-\lambda t} \tag{10–16}$$

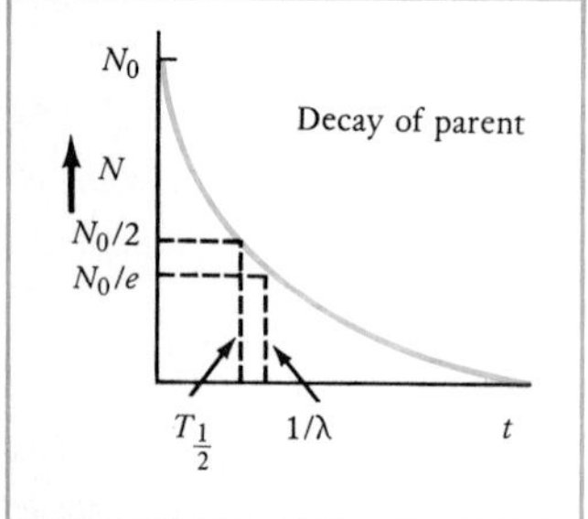

FIGURE 10–11. *Decay in the number of radioactive parent atoms as a function of time.*

The number of unstable nuclei decreases exponentially with time at a rate controlled by the magnitude of λ. Equation 10–16 holds not only for unstable nuclei but also for such unstable systems as atoms in an excited state.

It is customary to measure the rapidity of decay by the *half-life* $T_{1/2}$, which is defined as the time in which half the original unstable nuclei have decayed and half still survive. Therefore, $t = T_{1/2}$ when $N = \frac{1}{2}N_0$, and Equation 10–16 gives

$$\tfrac{1}{2}N_0 = N_0 e^{-\lambda T_{1/2}}$$

$$T_{1/2} = \frac{\ln_e 2}{\lambda} = \frac{0.693}{\lambda} \qquad (10\text{–}17)$$

Thus, if a radioactive material decays with a half-life of 3 s, after 3 s half of the initial nuclei remain; after 6 s, a quarter of the initial nuclei remain; and after 9 s, one-eighth of the initial nuclei remain. The decay constant λ has the units of reciprocal time (for example, s^{-1}), which also follows from its definition as the probability per unit time for decay.

The half-life is *not* the same as the *average lifetime,* or mean life T_{av}, of an unstable nucleus; a straightforward calculation (see Problem 10–24) shows that

$$T_{av} = \frac{1}{\lambda} = \frac{T_{1/2}}{\ln_e 2} \qquad (10\text{–}18)$$

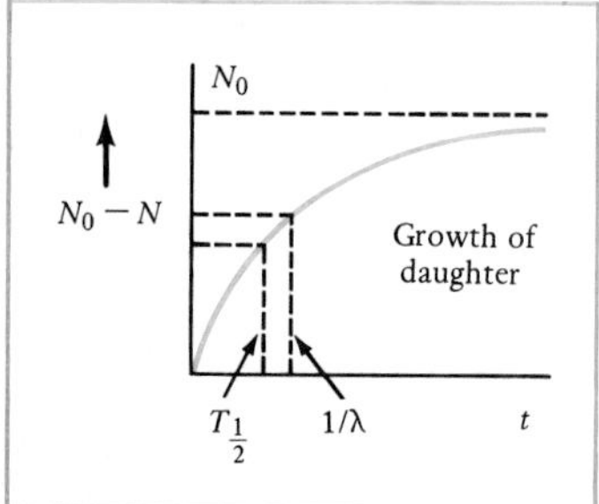

FIGURE 10–12. *Growth in the number of daughter atoms (stable) as a function of time.*

The decay of the parent nuclei and the concomitant growth in the number of daughter nuclei with time are shown in Figures 10–11 and 10–12. The number of daughter nuclei produced after a time t is $N_0 - N = N_0(1 - e^{-\lambda t})$, where N_0 is again the initial number of parent nuclei.

A useful quantity describing radioactive decay is the *activity,* defined as the number of decays per second. It follows from Equation 10–16 that

$$\frac{dN}{dt} = -\lambda N_0 e^{-\lambda t} = -\lambda N$$

$$\text{Activity} = -\frac{dN}{dt} = \lambda N = (\lambda N_0)e^{-\lambda t} \qquad (10\text{–}19)$$

A minus sign appears in this equation because the number of unstable nuclei *decreases* with time. The activity λN, originally λN_0, falls off as $e^{-\lambda t}$. The activity of unstable nuclei that radiate particles or photons, the *radioactivity,* can be measured with nuclear-radiation detectors over periods that are short compared with the half-life of the decay. This provides, then, a simple and direct method of measuring λ or $T_{1/2}$.

The common unit for measuring activity is the *curie* (Ci), defined as exactly 3.7×10^{10} disintegrations per second; a related unit, the millicurie, is $3.7 \times 10^{7}\ s^{-1}$. Another unit sometimes used for activity is the *rutherford,* defined as 10^6 disintegrations per second.

Example 10–2. A 1.0-mg sample of uranium-238 emits 738 α particles a minute. This sample shows the same activity, 738 α particles per minute, even after many years have elapsed. What is the half-life of ^{238}U?

Here the half-life is so very long that there is no appreciable decrease in the activity, following Equation 10–19, over a period that is much shorter than the half-life. But the decay constant, and therefore the half-life, can be computed from the known activity and the number of unstable nuclei by writing Equation 10–19 in the form

$$\lambda = \frac{\text{activity}}{N}$$

Taking the mass of a nucleon to be 1.67×10^{-24} g, we can determine by computation that in a 1.0-mg sample of ^{238}U there are

$$\frac{10^{-3}\text{ g}}{238(1.67 \times 10^{-24}\text{ g})} = 2.52 \times 10^{18}\text{ atoms}$$

The relation above yields

$$\lambda = \frac{\frac{738}{60}\text{ decays/s}}{2.52 \times 10^{18}\text{ atoms}} = 4.88 \times 10^{-18}\text{ s}^{-1}$$

or

$$T_{1/2} = \frac{0.693}{\lambda} = 1.42 \times 10^{17}\text{ s} = 4.50 \times 10^{9}\text{ yr}$$

The half-life of an unstable element, such as ^{238}U, found in nature is extraordinarily long. It is comparable to the age of the universe. The naturally radioactive elements were created when an aging star exploded; their very long half-lives ensure that they still survive in appreciable amounts since the supernova.

10–8 Gamma Decay

All stable nuclides are ordinarily in their lowest, or ground, states. If such nuclei are excited and gain energy by photon or particle bombardment, they may exist in any one of several excited, quantized energy states. Indeed, all radioactive nuclides are initially in energy states from which they may decay with the emission of photons or particles. We are concerned here with the decay of a nucleus from an excited state by the emission of a photon. A photon emitted from a nucleus in an excited state is called a gamma (γ) ray.

Some nuclear energy levels of the radioactive element thallium, $^{208}_{81}$Tl, are shown in Figure 10–13. The energy of the nucleus in the ground state is

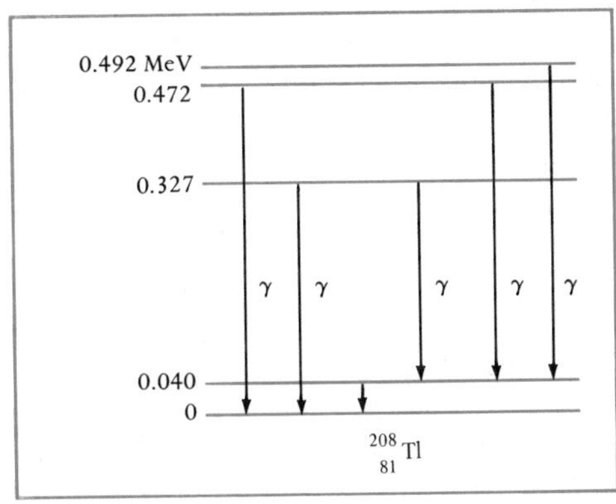

FIGURE 10–13. *Nuclear energy-level diagram and γ-ray transitions of* $^{208}_{81}$Tl.

chosen as zero. The figure also shows transitions giving rise to γ rays. The spacings of nuclear energy levels range from a few keV to a few MeV, in contrast to the much smaller separations associated with atomic energy levels. The nuclear energy levels can be inferred from the γ-ray spectrum emitted when excited nuclei make downward quantum jumps to lower states.

The only nuclear transitions that occur are those in which the conservation laws are satisfied. In the downward transition from an upper nuclear energy state E_u to a lower state E_l through the emission of a γ-ray photon of energy $h\nu$, the conservation of energy requires that

$$h\nu = E_u - E_l \tag{10–20}$$

Linear momentum conservation requires that the total linear momentum following the γ decay equal the linear momentum before the decay. If the decaying nucleus is originally at rest, it must recoil, when the photon is emitted, with a momentum equal to that of the photon, $h\nu/c$. Thus,

$$\frac{h\nu}{c} = mv \tag{10–21}$$

where mv is the momentum of the recoiling nucleus.

Example 10–3. What is the exact relation for the energy of a photon in γ decay?

Equation 10–20 is approximate; it excludes the kinetic energy of the recoiling nucleus. If the parent is free and initially at rest, and the daughter nucleus recoils with a momentum mv, given by Equation 10–21, and a kinetic energy $(mv)^2/2m$, the difference between the upper energy state and the lower is

$$E_u - E_l = h\nu + \frac{(mv)^2}{2m}$$

which can be written, from Equation 10–21,

$$E_u - E_l = h\nu + \frac{(h\nu)^2}{2mc^2}$$

$$= h\nu\left(1 + \frac{h\nu}{2mc^2}\right)$$

Thus, the energy difference between the two states differs from the photon energy by $(h\nu)^2/2mc^2$, typically negligible, because the photon energy $h\nu$ is much less than the rest energy mc^2 of the nucleus. For example, $^{57}Fe^*$ decays from an excited state with the emission of a 14.4-keV photon. For the excited nucleus free and initially at rest, the daughter nucleus recoils with a kinetic energy of

$$\frac{(h\nu)^2}{2mc^2} = \frac{(14.4 \text{ keV})^2}{2 \times 57 \text{ u} \times 0.93 \text{ GeV/u}} = 2.0 \times 10^{-3} \text{ eV}$$

and the difference in energy states differs from the energy 14.4 keV of the photon by $(2.0 \times 10^{-3} \text{ eV})/14.4 \text{ keV} = 1.4 \times 10^{-7}$.

In γ decay, the photon is created when the nucleus in the excited state $^AZ^*$ decays to a lower state, the ground state ^AZ. We use here the conventional notation, in which a nucleus in an excited state is labeled with an asterisk. We can write symbolically

$$^{A}Z^* \rightarrow {}^{A}Z + \gamma$$

The decay is consistent with charge conservation (Ze before $=$ Ze after), and the conservation of nucleons (A before $= A$ after).†

It is useful to have a criterion for judging the relative rapidity with which nuclear decays take place. For this purpose, we may speak of a *nuclear time* t_n as the time required for a typical nucleon, having an energy of several million electron volts and traveling at a speed $\approx 0.1c$, to travel a nuclear distance, ≈ 3 fm. It follows that $t_n = (3 \times 10^{-15}\text{ m})/(3 \times 10^{7}\text{ m/s}) \approx 10^{-22}$ s. It is expected, then, that any rapid nuclear decay will have a half-life that is of the order of 10^{-22} s, an immeasurably short time.

The half-life in a typical γ decay is predicted by theory to be much longer, of the order of 10^{-14} s; even this decay is too fast to be followed in time. Some γ decays are, however, so strongly forbidden that the half-life is greater than 10^{-6} s, which can readily be measured. Such nuclides, having a measurably long half-life in γ decay, are called *isomers*. An isomer is not chemically distinguishable from the lower-energy nucleus into which it slowly decays. An extreme example of isomerism is niobium, $^{91}_{41}$Nb, which undergoes γ decay with a half-life of 60 days.

The γ decay of excited nuclei is a direct way of signaling the instability of the nuclei. An analysis of the γ-ray energies allows nuclear energy-level diagrams, such as Figure 10–13, to be constructed. Although no complete theory of nuclear structure yet exists, the energy-level diagrams of many nuclides are at least partly explained by the quantum theory.

10–9 Alpha Decay

Gamma decay clearly demonstrates that excited nuclear energy states are discrete. Another nuclear decay mode, also verifying the discreteness of nuclear energy states, is alpha (α) decay. Certain radioactive nuclei, those for which $Z > 82$, spontaneously decay into a daughter nucleus and a helium nucleus. Since the α particle has a very stable configuration of nucleons, it is perhaps not surprising that such a group of particles might exist within the parent nucleus before α decay.

The laws of conservation of charge and of nucleons require that for α decay:

$$^{A}_{Z}\text{P} \rightarrow {}^{A-4}_{Z-2}\text{D} + {}^{4}_{2}\text{He}$$

where P and D refer to the parent and daughter nuclei. The subscripts and superscripts give the electric charge in units of e and nucleon numbers, respectively; the conservation laws require that the sums on both sides of the reaction equation be equal. For example, bismuth-212 decays by α emission to thallium 208:

$$^{212}_{83}\text{Bi} \rightarrow {}^{208}_{81}\text{Tl} + {}^{4}_{2}\text{He}$$

†Another process that competes with γ decay is *internal conversion*. In this process, a nucleus in an excited state converts its excitation energy internally (within the atom) to one of the inner atomic electrons, bound with an energy E_b; therefore, $E_u - E_l = E_b + K_e$, where K_e is the kinetic energy of the freed electron.

If the radioactive parent is taken to be initially at rest, the conservation laws of energy and of linear momentum yield

$$M_P c^2 = (M_D + M_\alpha)c^2 + K_D + K_\alpha \tag{10–22}$$

$$M_D v_D = M_\alpha v_\alpha \tag{10–23}$$

where the M's are the atomic rest masses of the parent, daughter, and helium atom, and the K's and v's are the kinetic energies and velocities of the daughter and α particle. In Equation 10–22, *atomic* rest masses may be used rather than nuclear masses because the same number of electron rest masses must be subtracted from atomic masses to yield nuclear masses on each side of the equation. (Nonrelativistic expressions for kinetic energy and momentum may be used in these equations because the energy released in α decay is never greater than 10 MeV, whereas the α-particle rest energy is about 4 GeV.)

Obviously, the kinetic energies K_D and K_α can never be negative; therefore, from Equation 10–22, α decay is energetically possible only if

$$M_P > M_D + M_\alpha$$

If this inequality is not satisfied, α decay simply cannot occur.

The energy released in the decay, $K_D + K_\alpha$, is called the *disintegration energy* and is represented by Q. Using Equation 10–22, we can write

$$Q = K_D + K_\alpha = (M_P - M_D - M_\alpha)c^2 \tag{10–24}$$

Decay is energetically possible only for $Q > 0$.

In an α decay, it is the energy K_α of the α particle that is usually measured, for example, by measuring its radius of curvature in a magnetic field. Let us see how this measured energy K_α is related to the total energy Q released in the decay. Squaring 10–23 and multiplying by $\frac{1}{2}$ gives

$$M_D(\tfrac{1}{2}M_D v_D^2) = M_\alpha(\tfrac{1}{2}M_\alpha v_\alpha^2)$$

$$M_D K_D = M_\alpha K_\alpha$$

In atomic mass units, the daughter and the α masses are approximately $A - 4$ and 4 u, respectively. Then the equation above becomes

$$(A - 4)K_D = 4K_\alpha$$

But

$$Q = K_\alpha + K_D = K_\alpha\left(1 + \frac{4}{A - 4}\right)$$

and therefore

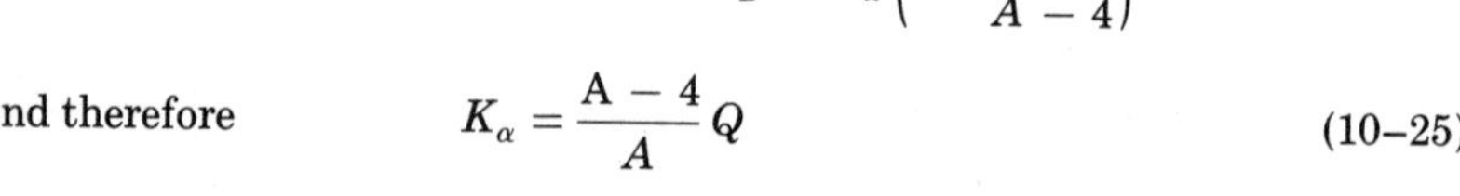

$$K_\alpha = \frac{A - 4}{A} Q \tag{10–25}$$

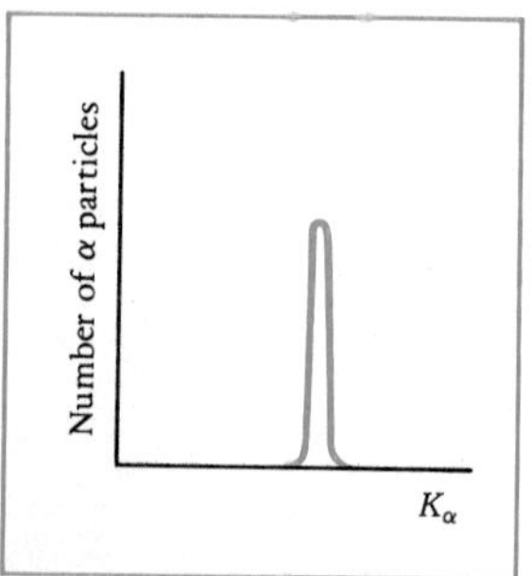

FIGURE 10–14. *Energy spectrum of α particles from a radioactive substance.*

This equation shows that in *two-particle* emission from an initially unstable nucleus at rest, the α particle emerges with a *precisely defined energy;* since Q has a precise value, so does K_α. The energy spectrum of the α particles emitted from a radioactive substance in a simple α decay is shown in Figure 10–14. The α particles are *monoenergetic.*

Radioactive materials unstable to α decay are heavy elements with $A \gg 4$. Equation 10–25 shows that K_α is only slightly less than Q; nearly all the energy released in the decay is carried away as kinetic energy by the light particle.

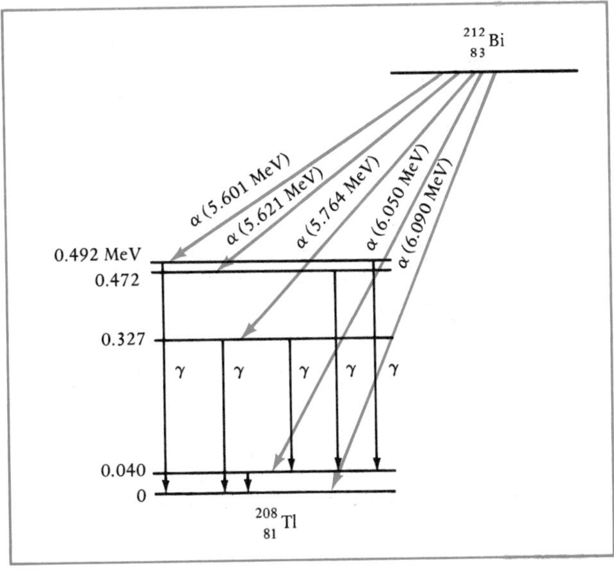

FIGURE 10–15. *Nuclear energy-level diagram showing the decay of bismuth-212 by α emission to the ground and excited states of thallium-208.*

Most α emitters show a group of discrete α-particle energies instead of a single energy. This is easily understood from a nuclear energy-level diagram, such as Figure 10–15 for the decay of bismuth-212. The parent nucleus can decay by α emission to several energy states, the ground state and excited states. The most energetic α particles correspond to those transitions involving decay to the ground state of the daughter.

A decay to an excited state of a daughter is followed by one or more γ emissions leading to the ground state. Because the half-life for γ decay is usually extremely short compared with α-decay half-lives, the γ rays appear to be coincident in time with the α decays. The energies of the γ rays are found to be completely consistent with the differences in the energies of the emitted α particles.

Over 300 α emitters have been identified. The emitted α particles have discrete energies from about 4 to 10 MeV, a factor of 2, but half-lives ranging from 10^{-6} s to 10^{17} s, a factor of 10^{23}. Short-lived α emitters have the highest energies, and the contrary, as indicated by the examples in Table 10–1.

Table 10–1

α EMITTER	K_α (MeV)	$T_{1/2}$ (s)	λ (s^{-1})
$^{238}_{92}U$	4.19	1.42×10^{17}	4.9×10^{-18}
$^{212}_{83}Bi$	6.09	3.64×10^{3}	1.90×10^{-4}
$^{215}_{85}At$	8.00	10^{-4}	10^{4}

BARRIER PENETRATION IN ALPHA DECAY. For an example of alpha decay, let us look at the decay of uranium-238 to thorium-234 by α emission. Figure 10–16 shows the potential energy of the *daughter* as seen by an α particle. When the α particle is at a greater distance from the center of the nucleus than R, the range of the nuclear force (about 10^{-14} m), then the force between the particles is given by Coulomb's law. This is established by α-particle scattering experiments, in which α particles having energies

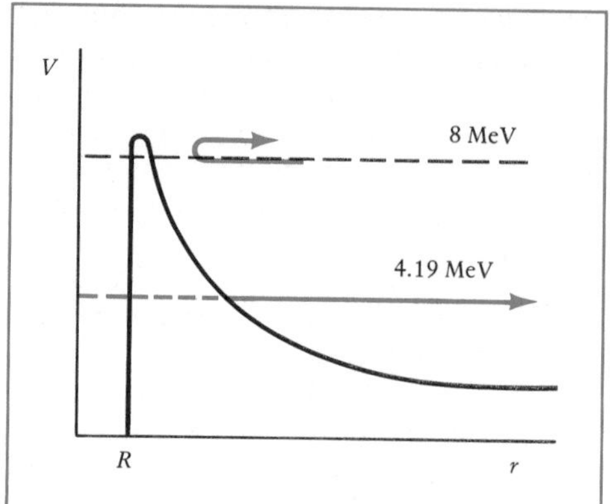

FIGURE 10–16. *Potential energy of the nucleus of thorium-234 as seen by an α particle.*

as great as 8 MeV are scattered from the thorium nuclei by the Coulomb repulsive force. At distances less than R, the α particle is subject to a strong attractive force that holds it to the thorium nucleus. But this bound system, composed of the daughter nucleus $^{234}_{90}\text{Th}$ and an α particle, is just the parent nucleus, $^{238}_{92}\text{U}$. The two protons and two neutrons making up the α particle form a particle within the parent, which exists for a time that is long compared with the nuclear time of 10^{-22} s.

It is known from experiment that uranium-238 emits α particles with a kinetic energy of 4.19 MeV; see Figure 10–16. Since the potential energy is zero when the α particle is very far from the daughter nucleus, 4.19 MeV also represents the *total* energy of the particle at any distance from the nucleus. Within the nucleus, the total energy of the α particle is again 4.19 MeV, the algebraic sum of the potential energy (negative) and the kinetic energy (positive). Classically, if an α particle is confined within nuclear "walls," it moves back and forth between them indefinitely, striking them roughly 10^{21} times per second. But the particle cannot penetrate the walls and escape, for it cannot have a negative kinetic energy. Classically, α decay cannot take place!

Alpha decay is readily understood in terms of the wave-mechanical phenomenon known as the *tunnel effect* (Section 5–9), first proposed by G. Gamow (1928) and R. W. Gurney and E. U. Condon (1928).

In wave mechanics, the probability of locating an α particle is related to its wave function $\psi(r)$. The wave function for the potential of Figure 10–16

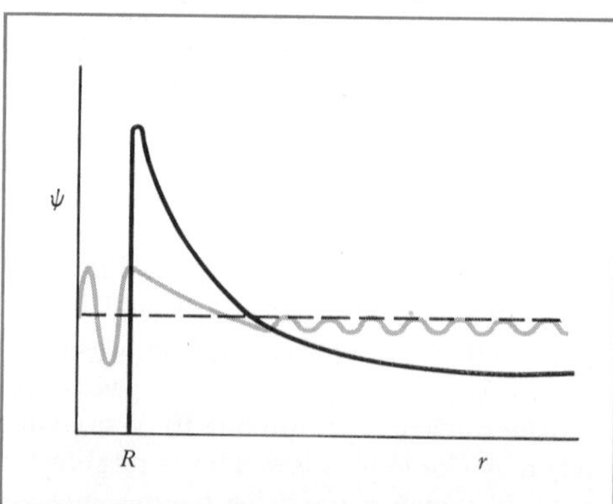

FIGURE 10–17. *Wave function corresponding to the penetration of an α particle through a nuclear barrier.*

is that shown in Figure 10–17: oscillatory within the attractive potential-energy well, drastically attenuated through the potential barrier, and again oscillatory outside the nucleus, with a small but finite amplitude. This means that there is a very small but finite probability that an α particle originally within the nucleus will be at some time outside the nucleus. The probability of tunneling through the barrier strongly depends on the height and thickness of the barrier; it is greater with greater particle energy.

One way of visualizing the decay is to imagine the particle bouncing between the nuclear walls until it finally escapes by penetrating the potential-energy barrier. Let us compute the number of tries the particle must make before it breaks through the potential barrier. The half-life of uranium-238 is about 10^{17} s; on the average then, an α particle must make 10^{21} tries per second for 10^{17} s, or 10^{38} tries altogether, before it escapes.

10–10 Beta Decay

Beta (β) decay can be defined as the radioactive decay process in which the charge of a nucleus is changed without any change in the number of nucleons. Examples of β instability are furnished by two of the three nuclides boron-12, carbon-12, and nitrogen-12, whose proton and neutron occupation levels are shown schematically in Figure 10–18. All three nuclides have 12 nucleons but differ in proton and neutron numbers. Only the carbon nucleus, with 6 protons and 6 neutrons, is stable. Evidently the boron nucleus has too many neutrons, and the nitrogen nucleus too many protons, to be stable. The unstable boron nucleus decays to a lower energy state by changing one of its nucleons from a neutron to a proton, the last neutron jumping, as it were, to the lowest available proton level. In this process, the ${}^{12}_{5}\text{B}$ nucleus has been transformed into the stable ${}^{12}_{6}\text{C}$ nucleus; and to conserve electric charge, one unit of negative charge must be created. We know that an electron cannot exist *within* the nucleus; therefore, the created electron, or β particle, must be emitted from the decaying nucleus, according to the transformation

$${}^{12}_{5}\text{B} \rightarrow {}^{12}_{6}\text{C} + \beta^- + \bar{\nu}$$

where the minus sign indicates the negative charge. The decay relation as written includes the chargeless, massless particle known as the antineutrino $\bar{\nu}$, whose role in β^- decay will be treated shortly.

The decay of nitrogen-12 is analogous. This isotope of nitrogen has too many protons and too few neutrons to be stable. Therefore, it decays to a

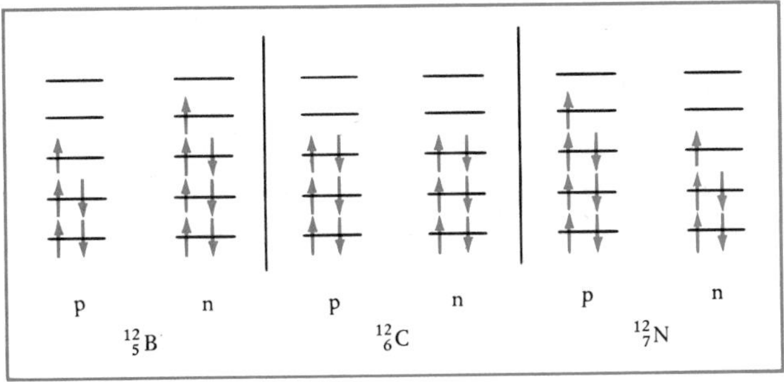

FIGURE 10–18. *Proton and neutron occupation levels of boron-12, carbon-12, and nitrogen-12.*

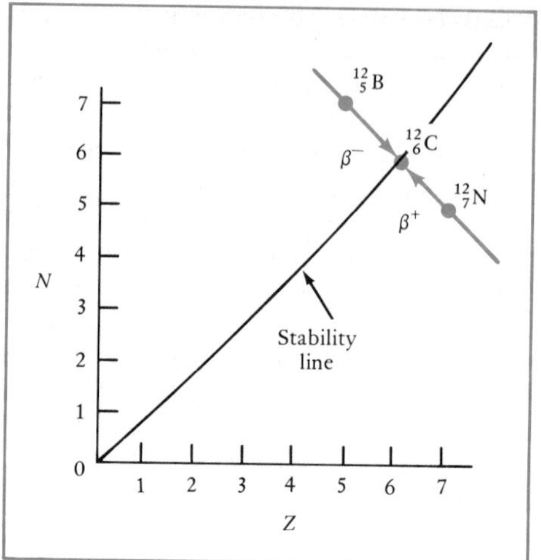

FIGURE 10–19. *The β^+ decay of nitrogen-12 and the β^- decay of boron-12 to carbon-12.*

lower energy state by converting one of its protons into a neutron; the last proton jumps to the lowest available neutron level. In this decay, the unstable $^{12}_{7}N$ nucleus is transformed into the stable $^{12}_{6}C$ nucleus, and charge is conserved by the creation of a positive beta particle, the positron. Because a positron cannot exist within a nucleus, it must be emitted. The decay may be shown as

$$^{12}_{7}\mathrm{N} \rightarrow {}^{12}_{6}\mathrm{C} + \beta^+ + \nu$$

The chargeless, massless neutrino ν is also emitted in the decay.

The β-decay processes are also shown in Figure 10–19 on a plot of N versus Z. The carbon nucleus lies on the stability line; the boron lies above it and the nitrogen below it. The β-decay transformations occur along an isobaric ($-45°$) line in such a way as to bring the unstable nuclides closer to the stability line. The β^+ decay can be readily identified, because an emitted positron will undergo annihilation with an electron and produce two annihilation photons, each with an energy of 0.51 MeV, the rest energy of an electron (or positron). Thus, β^+ decay is always characterized by the appearance of annihilation quanta of 0.51 MeV.

The third type of β decay is *electron capture*. In electron capture, an atomic orbital electron combines with a proton of the nucleus to change it into a neutron. The number of nucleons is unchanged here too, but a proton is converted into a neutron, as in β^+ decay. The electrons of the atom have a finite probability of being at the nucleus (see Figure 7–7), and one of the innermost, or K, electrons has the highest probability of being captured within the nucleus. Beta decay resulting from the nuclear capture of a K-shell electron is often referred to as *K capture.*

No charged particle is emitted in the decay by electron capture. The absorption and annihilation of a particle is equivalent to the creation and emission of its antiparticle; in K capture an electron is absorbed, but in β^+ decay an electron antiparticle, or positron, is emitted. Both processes change a proton into a neutron. An example of electron capture is the decay of unstable beryllium-7 to lithium-7:

$$e_K^- + {}^{7}_{4}\mathrm{Be} \rightarrow {}^{7}_{3}\mathrm{Li} + \nu$$

Again ν denotes the neutrino.

Electron capture cannot be identified by an emitted charged particle. The process can be inferred from the change in the chemical identity of the element undergoing the decay, or it can be detected by observing the *x-ray photons* emitted when the decay takes place. When a K electron is absorbed into the nucleus, there is a hole, or a vacancy, in the K shell; this vacancy is filled as electrons in outer shells make quantum jumps to inner vacancies, thereby emitting characteristic x-ray spectra. Because the x-ray emission must take place *after* the K vacancy is created, that is, after the nuclear decay, the x-rays are characteristic of the *daughter* element, not the parent.

Many hundreds of nuclides are known to decay by emitting an electron or a positron or by capturing an orbital electron. Nearly all unstable nuclides with Z less than 82 decay by at least one of the three processes. Beta decay differs from α and γ decay in several respects:

- The parent and daughter have the same number of nucleons.
- Unlike α emission, the electron or the positron (and antineutrino or neutrino) is created at the time it is emitted.
- Whereas γ-ray photons and α particles are emitted with a discrete spectrum of energies, β particles have a *continuous* energy spectrum.
- The half-lives in β decay are never less than about 10^{-2} s, in contrast to γ decay (as small as 10^{-17} s) and α decay (as small as 10^{-7} s).

β^- DECAY. Let us consider β^- decay in more detail. By the conservation of electric charge and the conservation of nucleons, the decay of a parent nucleus P into the daughter D can be represented by

$$^{A}_{Z}\mathrm{P} \rightarrow {}_{Z+1}^{\;\;A}\mathrm{D} + {}^{\;0}_{-1}e + {}^{0}_{0}\bar{\nu} \qquad (10\text{–}26)$$

For example, boron-12 decays into carbon-12 and an electron with a half-life of 2.0×10^{-2} s:

$$^{12}_{5}\mathrm{B} \rightarrow {}^{12}_{6}\mathrm{C} + {}^{\;0}_{-1}e + \bar{\nu}$$

Mass-energy conservation requires that the rest mass (energy) of the parent *nucleus*, $M_\mathrm{P} - Zm_\mathrm{e}$, exceed the rest masses (energies) of the daughter nucleus, $M_\mathrm{D} - (Z + 1)m_\mathrm{e}$, and the electron m_e, where M_P and M_D are the *neutral atomic* masses of the parent and daughter. (The antineutrino has *zero* rest mass.) Any excess energy Q, that is, energy released in the decay, appears as kinetic energy of the particles emerging from the decay. Therefore,

$$M_\mathrm{P} - Zm_\mathrm{e} = [M_\mathrm{D} - (Z + 1)m_\mathrm{e}] + m_\mathrm{e} + \frac{Q}{c^2}$$

or for β^- decay,

$$M_\mathrm{P} = M_\mathrm{D} + \frac{Q}{c^2} \qquad (10\text{–}27)$$

Equation 10–27 shows that β^- decay is energetically possible whenever $M_\mathrm{P} > M_\mathrm{D}$, that is, whenever the mass of the parent atom exceeds the mass of the daughter atom. Moreover, when β^- decay is energetically possible, it occurs, although the probability may be small and the half-life extremely long.

We can compute Q for the β^- decay of boron-12 to carbon-12 directly from atomic masses by using Equation 10–27:

$$\begin{aligned} \text{Mass } {}^{12}_{5}\mathrm{B} &= 12.014354\ \mathrm{u} \\ \text{Mass } {}^{12}_{6}\mathrm{C} &= \underline{12.000000\ \mathrm{u}} \\ M_\mathrm{P} - M_\mathrm{D} &= 0.014354\ \mathrm{u} \end{aligned}$$

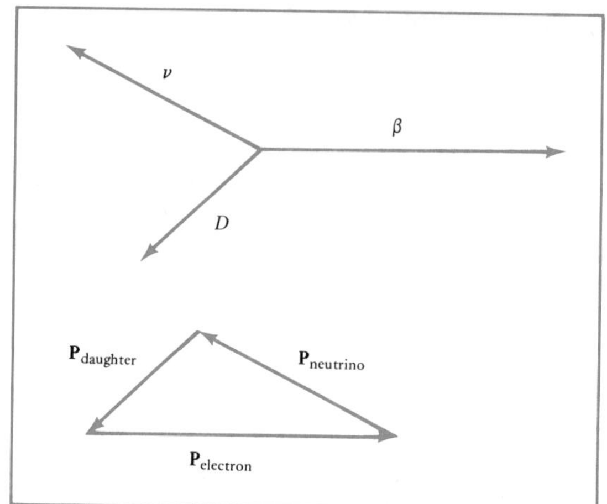

FIGURE 10–20. *Linear-momentum vectors of the daughter nucleus, electron, and neutrino in β⁻ decay.*

$$Q = 0.014\ 354\ \text{u} \times 931.5\ \text{MeV/u} = 13.37\ \text{MeV}$$

The 13.37 MeV of energy released in this decay appears as kinetic energy of (1) the daughter nucleus, (2) the electron, and (3) the antineutrino. The total momentum of the parent nucleus at rest is zero; therefore, the momenta of the three outgoing particles also add to zero, as shown in Figure 10–20. Since the momentum vectors can have varying magnitudes and directions to yield zero total momentum, while the total kinetic energy of the three particles always totals Q, the kinetic energy of any one of the three particles is not discrete but *continuous*. For a decay into *two* particles (as in γ or α decay), each of the emerging particles has a single energy; but for a decay into three particles, there is a distribution of kinetic energies.

Since the rest mass of the daughter nucleus far exceeds the rest mass of the electron or of the antineutrino, the kinetic energy of the daughter nucleus compared with the two light particles is negligible. How are the electron kinetic energies distributed? Consider two extreme cases: (1) The antineutrino carries all the energy released in the decay (13.37 MeV for the decay of ${}^{12}_{5}\text{B}$). Then the electron energy is zero. (2) The antineutrino has zero kinetic energy (and zero rest energy). Then the electron carries all the energy released in the decay (13.37 MeV for the decay of ${}^{12}_{5}\text{B}$). In short, the electron energy ranges from zero to the maximum amount Q released in the decay, as shown in Figure 10–21.

THE NEUTRINO AND THE ANTINEUTRINO. We are now prepared to examine the special and peculiar properties of the neutrino and the antineutrino (the antiparticle of the neutrino), and the unique function of these particles in β decay. The neutrino (and the antineutrino) has electric charge zero and rest mass also *zero*. Therefore, the particle exists only when it is moving at speed c. For a zero-rest-mass particle, the total relativistic energy E is given by $E = pc$, where p is the magnitude of the particle's momentum.

Since the neutrino and the antineutrino have zero charge, they can produce no ionization. Each interacts very weakly with nuclei. These particles are extremely difficult to detect.

A neutrino or an antineutrino has intrinsic, or spin, angular momentum

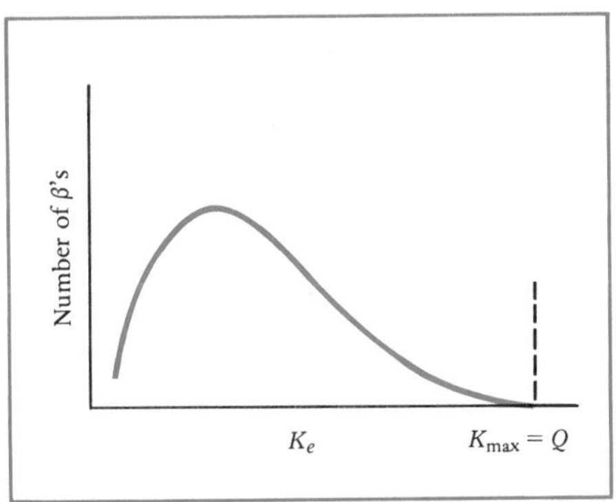

FIGURE 10–21. *Distribution in energy of emitted β^- particles in boron-12 decay.*

$\frac{1}{2}\hbar$, like that of an electron. The distinction between the neutrino and the antineutrino is related to the relative directions of the particle's linear momentum p and its spin angular momentum. The spin-angular-momentum vector must be either aligned with the particle's momentum or antialigned. The neutrino has the linear and spin-angular-momentum vectors in *opposite* directions, whereas the antineutrino has its linear and spin-angular-momentum vectors in the *same* directions. If one takes the curled fingers of a hand to indicate the spin sense and the thumb of the same hand to indicate the direction of the particle's motion, then the *neutrino* can be said to be *left*-handed and the antineutrino *right*-handed. (The experimental basis for determining the handedness of these particles is given in Section 12–6.)

Historically, the neutrino was postulated before the particles had been observed directly. This was done by W. Pauli in 1931 to make sense out of the observations in β decay. If there were no neutrino created and emitted in the decay, all outgoing light particles (electrons or positrons) would have a *single* kinetic energy, equal to Q. The neutrino avoided what would otherwise be a violation of the energy conservation principle. Moreover, without the neutrino, the daughter nucleus and light particle (electron or positron) would also move in opposite directions, contrary to observation. Without the neutrino, there would be a violation of the conservation of momentum principle. Further, the $\frac{1}{2}\hbar$ spin angular momentum of the neutrino avoided what would otherwise have been a violation of the angular-momentum principle.

Example 10–4. As shown in Equation 10–27, the total energy released in β^- decay is equal to the rest energy of the parent *atom* minus the rest energy of the daughter *atom*. How is the total energy Q that is released in (*a*) β^+ decay and (*b*) electron capture related to the atomic masses of the parent and daughter?

(*a*) β^+ *Decay.* The general relation giving the β^+ decay is

$$^{A}_{Z}\text{P} \rightarrow {}_{Z-1}^{\ \ A}\text{D} + {}^{\ 0}_{+1}e + \nu$$

Positron decay occurs only if mass-energy is conserved. This means that the rest mass of the parent *nucleus* must exceed the sum of the rest masses of the daughter nucleus and the positron (the neutrino rest mass is zero). Any excess energy appears as the kinetic energy of the three particles emerging from the decay. Therefore,

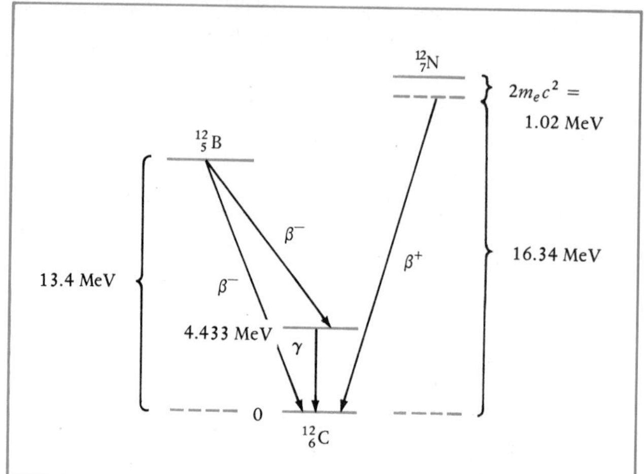

FIGURE 10–22. *Energy-level diagram showing the decay of boron-12 and nitrogen-12 to carbon-12.*

$$M_P - Zm_e = [M_D - (Z - 1)m_e] + m_e + \frac{Q}{c^2}$$

or for β^+ decay,

$$M_P = M_D + 2m_e + \frac{Q}{c^2} \qquad (10\text{–}28)$$

where M_P and M_D are the *neutral atomic* masses of the parent and daughter, m_e is the rest mass of the positron (or electron), and Q is the energy released in the decay and shared by the positron, daughter nucleus, and neutrino. We see from this equation that β^+ decay is energetically possible (Q is greater than zero) only if $M_P > M_D + 2m_e$.

Positron decay is possible, then, only if the mass of the parent atom *exceeds* the mass of the daughter atom *by at least two electron masses,* 2(0.000 549) u or its energy equivalent, 1.02 MeV. (There is nothing especially significant in the appearance here of two electron masses; it merely reflects that neutral *atomic* masses are used instead of *nuclear* masses.)

(*b*) *Electron Capture.* The general relation for electron capture is written

$$^{0}_{-1}e + {}^{A}_{Z}P \rightarrow {}_{Z-1}^{A}D + \nu$$

which shows that an orbital electron is captured by the parent nucleus $^{A}_{Z}P$, and the products of decay are the daughter nucleus $_{Z-1}^{A}D$ and a neutrino.

Mass energy is conserved when the energy Q released in the decay equals the sum of the rest masses entering the reaction minus the sum of the rest masses leaving the reaction. Therefore,

$$m_e + (M_P - Zm_e) = [M_D - (Z - 1)m_e] + \frac{Q}{c^2}$$

or for electron capture,

$$M_P = M_D + \frac{Q}{c^2} \qquad (10\text{–}29)$$

where M_P and M_D are the neutral atomic masses of the parent and daughter. This equation shows that electron capture is energetically possible if the atomic mass of the parent compared with the daughter is larger.

In an electron-capture decay, energy is released. Where does it go? Unlike β^+ and β^- decay, electron capture produces only *two* particles. By momentum conservation, the neutrino and the daughter nucleus must move in opposite directions with the same momentum magnitude; the sum of their kinetic energies is the disintegration energy Q. Because only two particles appear in electron capture, each has a precisely defined energy. The antineutrino's rest mass is zero; therefore, almost all the energy is carried by a virtually unobservable antineutrino, and the nucleus recoils with an energy of only a few electron volts. Nevertheless, some very delicate experiments have confirmed that the recoiling nuclei are monoenergetic and their energy is precisely the amount required to satisfy the laws of momentum and energy conservation. Without the accompanying neutrino in electron capture, this decay process is completely inexplicable.

Figure 10–22 is an energy-level diagram of the decay of boron-12 and nitrogen-12 to the stable nuclide carbon-12. By convention, a nuclide undergoing β^- decay is shown to the left of the daughter, and the nuclide undergoing β^+ decay or electron capture is shown to the right (Z increases toward the right). We see that the decay of the boron consists of electron emission to two states of the carbon, the ground state and an excited state 4.433 MeV above the ground state. Decay from the excited state to the ground state by the emission of a γ-ray photon is essentially simultaneous with the corresponding β^- decay. This near coincidence of the electron and photon emission can be verified experimentally by using two detectors, one for electrons and one for photons, and by noting that the pulses in the two detecting systems are coincident in time (within the resolving time of the detecting instruments).

An energy-level diagram of the radioactive element copper-64, which decays by β^- emission to zinc-64 and also by β^+ emission and electron capture to nickel-64, is shown in Figure 10–23.

A general rule applies to any two isobaric nuclides that differ in Z by 1. Clearly, the atomic mass of one nuclide must exceed the atomic mass of

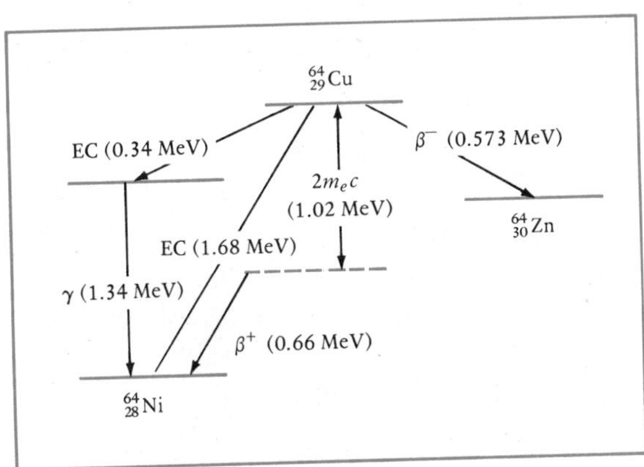

FIGURE 10–23. *Energy-level diagram showing the decay of copper-64 by β^- emission, β^+ emission, and electron capture.*

the other. Therefore, the nucleus of the more massive atom can decay to the nucleus of the lighter atom either by β^- decay or by electron capture. It follows that no two neighboring isobars can both be stable against β decay, and this indeed is in accord with observation of the known nuclides (see Figure 10–7).

The four *basic* reactions associated with β decay are the following:

$$\begin{aligned} \beta^- \text{ decay: } & \mathrm{n} \to \mathrm{p} + \mathrm{e} + \bar{\nu} \\ \beta^+ \text{ decay: } & \mathrm{p} \to \mathrm{n} + \bar{\mathrm{e}} + \nu \\ \text{Electron capture: } & \mathrm{e} + \mathrm{p} \to \mathrm{n} + \nu \\ \text{Neutrino absorption: } & \bar{\nu} + \mathrm{p} \to \mathrm{n} + \bar{\mathrm{e}} \end{aligned} \tag{10–30}$$

The symbol e represents the electron (charge, -1), and $\bar{\mathrm{e}}$ represents the positron (charge, $+1$), the electron's antiparticle; similarly, ν represents a neutrino, and $\bar{\nu}$ an antineutrino.

The basic process of β^- decay, the decay of the neutron into a proton, an electron, and an antineutrino, given by Equation 10–30, occurs in a *free neutron,* not merely a neutron bound within a nucleus. The decay is energetically allowed because the neutron mass exceeds the mass of the hydrogen atom, ^{1}H; the Q is 0.78 MeV. The half-life of this decay is found, in very difficult experiments, to be 11 min. Because a free neutron is typically absorbed in less than 10^{-3} s when it passes through materials, the decay of a neutron is usually unimportant in situations involving free neutrons.

The basic β^+ decay, in which a proton is converted into a neutron, a positron, and a neutrino, is *not* permitted for a free proton, since the mass on the left-hand side of the reaction is less than the mass on the right-hand side. Positron decay is possible only when protons are bound within a nucleus.

Electron capture is closely related to the β^+ decay. Note that in Equation 10–30, the second reaction becomes the third reaction when the antielectron is transferred to the left side and thereby becomes an electron. This follows the general rule that the emission of a particle is equivalent to the absorption of an antiparticle, and the converse. By using this rule together with the permitted reversal of the arrow, it can be seen that all four β reactions are equivalent.

The last reaction in Equation 10–30 is that in which an antineutrino combines with a proton to become a neutron and a positron. Although the relative probability of neutrino capture is extremely small, the capture was observed directly, with the very large neutrino flux from a nuclear reactor, by C. L. Cowan and F. Reines in 1956; the existence of the neutrino (strictly, the antineutrino) was thereby confirmed directly. The antineutrino of the absorption is identified by observing the neutron and the positron produced simultaneously when the antineutrino is captured by a proton; the neutron is detected by observing the photon emitted from an excited nucleus that has absorbed the neutron; and the positron is detected by observing annihilation photons. The difficulty of this experiment can be appreciated by realizing that a neutrino or an antineutrino has only 1 chance in 10^{12} of being captured while traveling completely through the earth. Because neutrinos have such a very small probability of interacting with matter and being absorbed, a large fraction of the energy released in all β-decay processes is effectively lost.

10-11 Natural Radioactivity

We have investigated the three common modes of radioactive decay, α, β, and γ, without concern about how unstable nuclides are produced. It is customary to divide radioactive nuclides into two groups: the unstable nuclides found in nature, which are said to exhibit *natural radioactivity,* and the unstable nuclides made by man (usually by bombarding nuclei with particles), which are said to exhibit *artificial radioactivity.* So far well over a thousand artificially radioactive nuclides have been produced and identified. The number of identified isotopes of a given element varies; hydrogen has 2 stable isotopes and 1 unstable, and xenon has 9 stable and 23 unstable isotopes. Here we examine only the natural radioactive nuclides.

It is believed that a cataclysmic cosmological event occurred about 20 billion years ago, when the universe was formed. All *nuclides,* stable and unstable, are held to have been formed in varying amounts at that time. Those unstable nuclides with half-lives much shorter than 20×10^9 yr have long since decayed into stable nuclides. There are, however, a number of unstable nuclides that have half-lives comparable to the age of the universe or even greater and are therefore still found in measurable amounts in nature. For example, ${}^{40}_{19}K$, which is found in nature, has a half-life of 4.0×10^{16} s, and it decays by both β^- decay and electron capture into a stable daughter. The natural nuclides ${}^{232}_{90}Th$, ${}^{235}_{92}U$, and ${}^{238}_{92}U$, all near the upper end of the periodic table, have half-lives in excess of 10^{16} s, and decay into daughters that are themselves radioactive.

Figure 10–24 shows the complete decay scheme of the uranium (${}^{238}_{92}U$) series, plotted on a neutron-proton diagram. A decrease of 2 in Z and in N

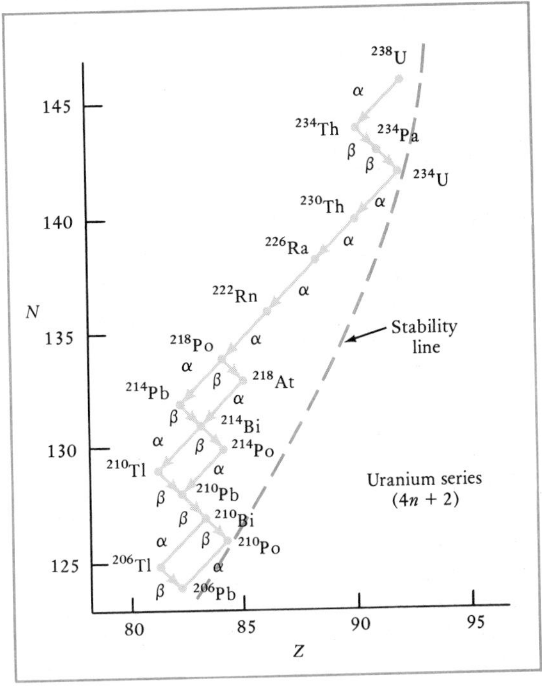

FIGURE 10–24. *Uranium radioactive series.*

represents an α decay, an increase of 1 in Z and decrease of 1 in N represents a β^- decay. The stability line represents the least unstable nuclides for the particular value of A. Both α and β^- decay produce daughter nuclei in excited nuclear states, which leads to subsequent decay by γ emission.

All nuclides with $A > 209$ are unstable. We might say that all such nuclides are too big to be stable and must lose nucleons to become more stable. The only mode of decay in which a heavy, naturally radioactive nucleus loses nucleons is α emission, in which both Z and N are reduced by 2. We see from Figure 10–24 that α decay tends, however, to displace the daughter to the left of the stability line; β^- decay is needed to bring the nucleus back. Some nuclides, such as $^{218}_{84}\text{Po}$, are unstable to both α and β decay, and a *branching* of the series then occurs.

The first nuclide in the uranium series, ^{238}U (Figure 10–24), has a mass number of 238, which divided by 4 yields a remainder 2. Since the only decay that changes the number of nucleons is α decay, in which the mass number is reduced by 4, the A values of any of the members of the uranium series can be written as $4n + 2$, where n is an integer. The thorium series, beginning with ^{232}Th, has members with mass numbers given by $4n$; and the so-called actinium series, beginning with ^{235}U, has members with mass numbers given by $4n + 3$.

There must exist a fourth radioactive series whose members have A values given by $4n + 1$. None of its members, however, has a half-life comparable to the age of the universe. Therefore, nuclides in this series do not exist naturally; they can be produced by nuclear reactions with the very heavy elements of the other series (for example, the capture of a neutron by uranium-236, followed by a β^- decay). This series is named the *neptunium series,* after the longest-lived nuclide in it, $^{237}_{93}\text{Np}$, which decays with a half-life of 2.14×10^6 yr.

A sample of a naturally radioactive substance emits α, β, and γ rays simultaneously, because *all* members of the radioactive series are present and decaying. The β and α emissions result in changes in Z or in both Z and A; the γ emissions result in changes in energy level. The early investigators of radioactivity distinguished the three types of radiation emitted from radioactive substances by noting the deflection of the emitted rays in a magnetic field. The α rays were deflected in the same direction as positively charged particles; the β rays were deflected in the same direction as negatively charged particles; and the γ rays were undeflected. Furthermore, it was observed, the penetration of the radioactive emanations increased in the order α, β, and γ; thus these rays were labeled by the first three letters of the Greek alphabet.

Uranium-238 is the heaviest nuclide found in nature. Still heavier and relatively short-lived nuclides, corresponding to *transuranic elements,* are man-made; they can be produced by bombarding heavy elements with energetic particles. Transuranic elements up to $^{262}_{105}\text{Ha}$ (hahnium) have been produced, at least momentarily, and identified.

The term *natural radioactivity* usually refers to those radioactive materials produced in the very distant past and to their descendants. Radioactive materials are, however, *continuously* being produced in nuclear collisions of high-energy cosmic-ray particles with nuclei in the earth's upper atmosphere. An example of this is the production of carbon-14 by the collision of neutrons with nitrogen nuclei, according to the reaction

$$^{14}_{7}\text{N} + ^{1}_{0}\text{n} \rightarrow ^{14}_{6}\text{C} + ^{1}_{1}\text{p}$$

This radioactive isotope of carbon, *radiocarbon,* decays by β^- emission with a half-life of 5730 yr:

$$^{14}_{6}\text{C} \rightarrow ^{14}_{7}\text{N} + \beta^- + \bar{\nu}$$

A small fraction of the CO_2 molecules in the air thus will contain radioactive carbon-14 atoms in place of stable carbon-12 atoms. Living organisms exchange CO_2 molecules with their surroundings, using both kinds of carbon in their structure. When the organisms die, their intake of carbon-14 ceases, and from that moment on the number of carbon-14 atoms relative to those of carbon-12 atoms decreases by virtue of the ^{14}C decay; only half of the original ^{14}C atoms are present after 5730 yr. This offers a very sensitive method of determining the age of organic archeological objects. One merely determines the relative numbers of the two isotopes. The number of carbon-14 atoms is determined by measuring their activity and using Equation 10–19. This ingenious method of measuring the age of organic relics many thousands of years old was originated by W. F. Libby in 1952; it is known as *radioactive dating.* Libby received the 1960 Nobel Prize in chemistry for this achievement.

A second cosmic-ray nuclear reaction continuously producing a naturally radioactive element is

$$^{14}_{7}\text{N} + ^{1}_{0}\text{n} \rightarrow ^{12}_{6}\text{C} + ^{3}_{1}\text{H}$$

where $^{3}_{1}H$, called *tritium* (with a nucleus known as a *triton*), is a heavy radioactive isotope of hydrogen. Tritium decays into the stable helium isotope $^{3}_{2}He$, with a half-life of 12.3 yr by β^- emission.

$$^{3}_{1}\text{H} \rightarrow ^{3}_{2}\text{He} + \beta^-$$

SUMMARY

Table 10–2 *Properties of the nuclear constituents*

PROPERTY	PROTON	NEUTRON
Mass, u	1.007 277	1.008 665
Charge, e	1	0
Spin, $\hbar$	$\frac{1}{2}$	$\frac{1}{2}$
Magnetic moment, $e\hbar/2M_pc$	+2.79	−1.91

Properties of the nuclear force
- Attractive and much stronger than the Coulomb force
- Short-range, $\approx$ 3 fermi (fm) (3×10^{-15} m)
- Charge-independent; all three nucleon interactions, np, pp, and nn, are approximately equal.

Nomenclature
- Nucleon: proton or neutron
- Atomic number Z: number of protons
- Neutron number N: number of neutrons

- Mass number A: total number of nucleons $(Z + N)$
- Nuclide: nucleus with a particular Z and a particular N
- Isotopes: nuclides with same Z
- Isotones: nuclides with same N
- Isobars: nuclides with same A

Properties of the nuclides

Stable nuclides: $N \approx Z$ at small A, and $N > Z$ at large A.

The nuclear radius is given by $R = r_0 A^{1/3}$, where $r_0 = 1.4$ fm (neutron scattering) or $r_0 = 1.1$ fm (electron scattering). All nuclei have the same nuclear density.

The total binding energy E_b of a nucleus ^AZ is given by

$$\frac{E_b}{c^2} = ZM_H + (A - Z)M_n - M$$

where all masses are those of the neutral atoms. For $A > 20$, the energy is $E_b/A \approx 8$ MeV per nucleon.

In the decay of all unstable nuclei, the laws of conservation of electric charge, nucleons, mass-energy, and momentum are satisfied.

The law of radioactive decay is $N = N_0 e^{-\lambda t}$, where the decay constant λ, the probability per unit time that any one nucleus will decay, is related to the half-life by $T_{1/2} = 0.693/\lambda$.

Table 10–3 *Radioactive decay modes*

	ALPHA (HELIUM NUCLEUS)	BETA (ELECTRON, POSITRON)	GAMMA (PHOTON)
Half-lives	10^{-6} s to 10^{10} yr	$> 10^{-2}$ s	10^{-17} to 10^5 s (isomer)
Energies	4 to 10 MeV	a few MeV	keV to a few MeV
Decay mode	${}^A_ZP \rightarrow {}^{A-4}_{Z-2}D + {}^4_2\alpha$	β^-: ${}^A_ZP \rightarrow {}_{Z+1}^{A}D + {}_{-1}^{0}e + \bar{\nu}$ β^+: ${}^A_ZP \rightarrow {}_{Z-1}^{A}D + {}_{+1}^{0}e + \nu$ EC: ${}_{-1}^{0}e + {}^A_ZP \rightarrow {}_{Z-1}^{A}D + \nu$	${}^AZ^* \rightarrow {}^AZ + \gamma$
Disintegration energy equation (all neutral atom masses)	$M_P = M_D + M_\alpha + Q/c^2$	β^-: $M_P = M_D + Q/c^2$ β^+: $M_P = M_D + 2m_e + Q/c^2$ EC: $M_P = M_D + Q/c^2$	$E_u = E_l + h\nu$
Energy distribution of decay products	Monoenergetic	β^- and β^+: polyenergetic EC: monoenergetic	Monoenergetic

PROBLEMS

Note: See Appendix II for values of the atomic masses.

Nuclear constituents, §10–1

● **10–1.** The neutron is known to have an intrinsic spin-angular-momentum component of $\hbar/2$ and an associated magnetic moment of 1.9 nuclear magnetons. Show that these values preclude the assumption that the neutron can be considered a composite proton and electron.

10–2. High-energy scattering experiments with electrons on nuclei show that although the total electric charge of a neutron is zero, the charge density of the neutron is positive close to its center and negative farther away. Show that this charge distribution is consistent with the direction of the neutron's magnetic moment opposite to its angular momentum. (A neutron can, in fact, from elementary-particle theory, be regarded to consist a fraction of the time as a virtual proton together with a virtual negative pion particle. See Section 12–2 and Problem 12–11.)

10–3. A free proton in an external magnetic field **B** has two possible orientations of its spin and associated magnetic moment relative to the magnetic field lines, because of the space quantization of the proton nuclear spin. A photon whose energy equals the difference in energy between the two proton states can induce the proton to make a transition from one state to the other. The phenomenon is known as *nuclear magnetic resonance* (NMR). Compute the proton resonance frequency for free protons in a magnetic field of 0.500 T.

Forces between nucleons, §10–2

● **10–4.** Two deuterons with equal kinetic energy collide head-on. What is the minimum kinetic energy of each deuteron that would produce the separation of one deuteron into a proton and a neutron?

The deuteron, §10–3

● **10–5.** The deuteron has only one bound state, that in which the spins of the proton and the neutron are parallel. What can you say about the energy of the deuteron for the condition in which the spins of the proton and neutron are antiparallel?

10–6. The total energy to remove all three electrons from a $^{7}_{3}$Li atom is 200 eV. Find the effect this has on the difference between the atomic mass and the nuclear mass of $^{7}_{3}$Li. Express the correction in atomic mass units and state whether it is a significant correction in determining the nuclear mass from the value of the atomic mass of $^{7}_{3}$Li, using Appendix II.

10–7. The nuclear magnetic moment of the deuteron in its ground state is +0.8574 nuclear magneton; its nuclear spin is $I = 1$. Show by comparing the magnetic moment of the deuteron with the separate magnetic moments of the proton and the neutron that the deuteron can be assumed to exist in an $^3S_1$ state (primarily). This means a quantum state of zero orbital angular momentum, with intrinsic angular momenta of proton and neutron so aligned as to produce a net deuteron nuclear spin of one unit.

10–8. Prove that the incident photon energy necessary for the photodisintegration of a deuteron initially at rest in the laboratory system is given by

$$E_\gamma = \frac{E_b}{1 - E_b/2M_d c^2}$$

where M_d and E_b are the mass and binding energy of the deuteron.

Stable nuclei, §10–4

● **10–9** Using Figure 10–7 and Appendix II, find the stable nuclides that (*a*) are isotopes of $^{40}_{20}$Ca; (*b*) are isotones of $^{40}_{20}$Ca; (*c*) are isobars of $^{40}_{20}$Ca. (*d*) What are the total numbers of stable even-even, odd-even, and odd-odd nuclides in parts (*a*), (*b*), and (*c*)? (An even-even nuclide is one with an even number of protons and an even number of neutrons.)

More generally, not only for calcium but for *all* stable nuclides, the number of even-even nuclides is much larger than the number of even-odd, odd-even, or odd-odd nuclides (155 versus 53, 50, and 4). One can infer from this observation that pairs of protons together with pairs of neutrons yield the more stable nuclear configurations.

● **10–10.** Identify the four stable odd-odd nuclides, using Figure 10–7 or Appendix II or both. (An odd-odd nuclide has an odd number of both protons and neutrons.)

Nuclear radii, §10–5

● **10–11.** Verify that the wavelength of 183-MeV electrons is short enough to ascertain the internal charge distribution within the nuclei of Figure 10–9.

● **10–12.** Assume that the radius R of a nucleus is the distance at which the positive electric-charge density within the nucleus is half the density at the center. Estimate the radius of the bismuth nucleus (see Figure 10–9) and use Equation 10–9 to find the number of nucleons in the nucleus. Compare this with the mass number of stable bismuth, 209.

10–13. (*a*) Show that the density of nuclear matter is approximately 3×10^{11} kg/m^3. (*b*) A neutron star, since it is composed of neutrons packed together within a nuclear radius, has the density of nuclear matter. If the sun were to collapse into a neutron star, what would be its radius? (Mass of the sun $= 2 \times 10^{30}$ kg.)

10–14. (*a*) Show that the wavelength of any particle whose kinetic energy is large compared with its rest energy is given by $\lambda = 1.24$ GeV·fermi$/E$, where E is the particle's total energy in GeV. (*b*) At what kinetic energies (GeV) will a photon, an electron, and a proton have a wavelength of 1 fm?

10–15. Assume that the electric charge density arising from Z protons is constant within a nucleus of radius R. (*a*) Find the electric potential energy between an electron and the positive charge of a nucleus as a function of the distance r from the nuclear center to the electron. (*b*) What is the maximum kinetic energy of the electron that will allow the electron to be bound to the nucleus?

10–16. Assume that the electric-charge density within a gold nucleus is constant (an approximation of the experimental charge distribution shown in Figure 10–9). For gold, $Z = 79$ and $R \approx 6.4$ fm. (*a*) Using the uncertainty principle, find the minimum kinetic energy (MeV) that an electron can have if it is confined to a distance of one gold nucleus diameter. (*b*) Compare this with the maximum kinetic energy for which an electron can be electrically bound to the gold nucleus (see Problem 10–15). This is one argument showing that electrons cannot be nuclear constituents.

Binding energy of stable nuclei, §10–6

● **10–17.** (*a*) Starting with the $^4_2$He nuclide, find the energy required to remove one nucleon at a time until there are four separate nucleons. (*b*) Explain why the values of the binding energies in the three steps of part (*a*) are different.

10–18. (*a*) Show that an even-Z nuclide usually has many more stable isotopes than an odd-Z nuclide. (*b*) Between $^{16}_{8}$O and $^{32}_{16}$S there is one stable isotope for each odd-Z nuclide and there are three stable isotopes for each even-Z nuclide. Explain this in terms of the neutron and proton pairs.

10–19. The nuclide $^8_4$Be is highly unstable, with a half-life of 3×10^{16} s. (*a*) What is its mode of decay? (*b*) What is the disintegration energy?

10–20. In Example 10–1, the energy necessary to separate a $^{12}_{6}$C nuclide into a proton and a $^{11}_{5}$B nuclide was found to be 15.96 MeV. (*a*) Find the energy necessary to separate a $^{12}_{6}$C nuclide into a neutron and a $^{11}_{6}$C nuclide. (*b*) Why does it take less energy to separate a proton than to separate a neutron from $^{12}_{6}$C?

10–21. (*a*) Calculate the binding energies of the least tightly bound proton in the three isotones $^{11}_{5}$B, $^{12}_{6}$C, and $^{13}_{7}$N. (*b*) Calculate the binding energies of the least tightly bound neutron in the three isotopes $^{11}_{6}$C, $^{12}_{6}$C, and $^{13}_{6}$C. (*c*) In parts (*a*) and (*b*), why is the binding energy so much lower for the last nuclide than for the first?

■ **10–22.** The charge independence of the strong nuclear forces between any two nucleons in a nucleus can be verified by considering a pair of *mirror nuclides*. Two nuclides form a mirror-nuclide pair if one nuclide becomes the other by the interchange of proton and neutron numbers; for example, $^{11}_{5}$B and $^{11}_{6}$C are mirror nuclides. Since mirror nuclides contain the same number of nucleons A, the charge-independent strong nuclear interaction between nucleons is the same for each. Therefore, the only difference in the total binding energies and masses of a mirror-nuclide pair arises (1) because of the neutron-proton mass difference and (2) because of the increase in the total Coulomb energy for the nuclide of higher Z number. (*a*) What is the increase in the mass of $^{11}_{5}$B over $^{11}_{6}$C because of the additional neutron in $^{11}_{5}$B? (*b*) What is the increase in mass of $^{11}_{6}$C over $^{11}_{5}$B because of the additional proton in $^{11}_{6}$C? Assume that the average Coulomb energy between two protons, each with its charge spread uniformly throughout the nucleus, is given by $\frac{6}{5}(ke^2)/R$, where $R = (1.4 \text{ fm})\,A^{1/3}$. (*c*) Using the results of parts (*a*) and (*b*), compute the mass difference between $^{12}_{6}$C and $^{11}_{5}$B and compare this with the difference in masses given in Appendix II.

Radioactive decay law, §10–7

● **10–23.** Strontium-90 decays with a half-life of 28.1 yr. What fraction of strontium-90 will remain after one century?

10–24. Show that the mean life T_{av} of a radionuclide having a decay constant λ is given by

$$T_{av} = \frac{\int_{N_0}^{0} t\,dN}{\int_{N_0}^{0} dN} = \frac{1}{\lambda} = 1.44T_{1/2}$$

10–25. Show that 1 g of radium-226 ($T_{1/2} = 1620$ yr) has an activity of 1.00 curie (Ci). (This was the basis of the original definition of the curie.)

10–26. What is the probability that a free neutron with a kinetic energy of $\frac{1}{25}$ eV will decay to a proton in traveling a distance of 1.0 km? The neutron's half-life is 693 s. Because the average kinetic energy per particle in a gas is $\frac{3}{2}kT = \frac{1}{25}$eV for room temperature, $T = 300$ K, a neutron with this kinetic energy is said to be a *thermal neutron*.

10–27. The activity of niobium-90 drops by a factor 12.2 in a time interval of 1 h. What is the half-life of ^{90}Nb?

10–28. (*a*) How many grams of cobalt-60 (half-life, 3.65 yr) will have an activity of 100 mCi? (*b*) What would the activity of such a sample be after 10 yr?

10–29. One can assume that, at the time of the supernova explosion that subsequently gave birth to our solar system, the two isotopes ^{235}U and ^{238}U had the same relative abundance. At the present time, the relative abundance of ^{235}U is 0.7 percent and of ^{238}U, 99.3 percent. Uranium-235 has a half-life of 2.25×10^{16} s and uranium-238 has a half-life of 1.42×10^{17} s. From this information, compute the time of the supernova explosion and compare this with the age of our solar system ($\sim 5 \times 10^9$ yr).

10–30. A heart pacemaker is powered by a radioactive source of plutonium-238. The electric current for the pacemaker is provided by the 5.6-MeV α particles emitted by the decaying plutonium. The half-life of $^{238}_{94}$Pu is 86 yr; the half-life of its daughter, $^{234}_{92}$U, is 2.5 × 10^5 yr. The decays of the daughter are followed by still further decays, which lead finally to the stable nuclide $^{206}_{82}$Pb. The plutonium source in a pacemaker is typically replaced every 10 yr. (*a*) Show that the power provided by all decays except α decay from $^{238}_{94}$Pu is negligible. (*b*) Show that if the power output of the plutonium source necessary to run the pacemaker is 500 μW, the mass of the ^{238}Pu source is 860 μg.

Alpha decay, §10–9

● **10–31.** In the α decay of a $^{212}_{83}$Bi source, alpha particles with discrete energies of 6.0901 MeV and 6.0510 MeV are observed. Compute the wavelength of the γ rays that accompany the α decays.

10–32. Polonium-213 decays into actinium-209 with the emission of an alpha particle, where $Q = 8.54$ MeV. (*a*) With what kinetic energy does the actinium-209 nucleus recoil? (*b*) If the polonium-213 atom was initially bound within a molecule, will the actinium-209 atom have enough energy to break free of its chemical bond?

10–33. Polonium-213 decays by α-particle emission either to an excited state or to the ground state of actinium-209. The energy difference between the ground states of ^{213}Po and ^{209}At is 8.54 MeV. (*a*) Show that in transitions to the ground state of ^{209}At, the alpha particles have an energy of 8.38 MeV. (*b*) Alpha particles emitted in a transition to the excited state of ^{209}At have kinetic energy 7.61 MeV. What is the energy of the associated γ ray?

Beta decay, §10–10

10–34. What modes of decay are energetically possible for the following unstable nuclides? (*c*) $^{11}_{6}$C (*b*) $^{15}_{6}$C (*c*) $^{12}_{5}$B (*d*) $^{48}_{24}$Cr and (*e*) $^{64}_{29}$Cu

10–35. Beryllium-7 decays by electron capture. What is (*a*) the energy and (*b*) the momentum of the neutrinos emitted by $^{7}_{4}$Be? (*c*) With what kinetic energy does the $^{7}_{3}$Li nucleus recoil?

10–36. Boron-12 decays into carbon-12. What is the maximum kinetic energy of (*a*) the electrons and (*b*) the antineutrinos emitted?

10–37. The radioisotope $^{24}_{10}$Ne decays by emitting two distinct β^- spectra with maximum electron energies of 1.10 MeV and 1.98 MeV. The total energy difference between $^{24}_{10}$Ne and the ground state of its daughter $^{24}_{11}$Na is 2.47 MeV. The β^- decays are also accompanied by gamma rays with energies 0.47 MeV and 0.88 MeV. Draw a nuclear energy-level diagram (similar to Figure 10–23) that displays all transitions.

Natural radioactivity, §10–11

● **10–38.** The neptunium series begins with the very long lived nuclide $^{237}_{93}$Np and ends with the stable nuclide $^{209}_{83}$Bi. What is (a) the total number of alpha decays in this series? (*b*) The total number of beta decays?

● **10–39.** What is the total energy released when $^{238}_{92}$U nucleus and its unstable daughters decay into the stable nuclide $^{206}_{82}$Pb?

● **10–40.** A neutron at rest decays into a proton, an electron, and an antineutrino. Explain which of the following two circumstances will result in the maximum kinetic energy for the proton: (*a*) the electron at rest or (*b*) the antineutrino with zero energy?

10–41. In a particular sample of mica, the ratio of $^{87}_{38}$Sr atoms to $^{87}_{37}$Rb atoms is 0.020. Assume that the ^{87}Sr has been produced solely by the decay of the naturally radioactive nonheavy element ^{87}Rb, having a half-life of 4.68 × 10^{10} yr. What is the maximum age of the mica?

10–42. The unstable nuclide A_1 decays with half-life $(T_{1/2})_1$ into the unstable nuclide A_2, which decays with a half-life $(T_{1/2})_2$ into the unstable nuclide A_3, and so on, as in a radioactive series. Assume that enough time has elapsed so that the decay series reaches *radioactive equilibrium,* with each nuclide (except A_1) decaying at the rate at which it is formed. Then, the number of nuclei of A_2, A_3, . . . , remains constant. Show that relative numbers of any members of the series are directly proportional to their respective half-lives; that is,

$$\frac{N_2}{(T_{1/2})_2} = \frac{N_3}{(T_{1/2})_3} = \frac{N_4}{(T_{1/2})_4} = \cdots$$

Therefore, long-lived nuclides are relatively abundant and short-lived nuclides scarce.

11
Nuclear Reactions and Devices

J. Robert Oppenheimer (1904, United States—1967). 1925 graduate, Harvard; excelled in Greek, Latin, physics, chemistry, oriental philosophy; 1927, doctorate, Göttingen. Trained a generation of theoretical physicists at the University of California, Berkeley, and California Institute of Technology. 1943, directed Manhattan Project for development of nuclear weapons. 1947, director, Institute of Advanced Study. 1963, recipient Fermi Award, Atomic Energy Commission. (Photo, CERN—courtesy of the American Institute of Physics, Neils Bohr Library.)

11-1 Low-Energy Nuclear Reactions

Although unstable nuclei decay spontaneously, changing their nuclear structure without external influence, one can induce a change in the identity or characteristics of nuclei by bombarding them with energetic particles. The change is known as a *nuclear reaction*. Thousands of nuclear reactions have been produced and identified since Rutherford observed the first one in 1919. The bombarding particles were, until the development of charged-particle accelerators, those emitted from radioactive substances. It is now possible to accelerate charged particles to energies up to 500 GeV. When particles of such great energy strike nuclei, they severely disrupt them and may create new particles. These so-called high-energy reactions and the particles participating in them will be studied subsequently (Chapter 12).

There are also *low-energy* nuclear reactions, reactions in which the incident particles have energies no greater than, say, 100 MeV. All such reactions have several features in common:

- The bombarding particle is typically a lightweight particle, such as an α particle, a γ ray, a proton, a deuteron, a neutron, or an electron.
- The reactions typically involve the emission of *one* other of such particles.

Several types of nuclear reaction can be illustrated with examples that have been important in the history of nuclear physics. In the first observed nuclear reaction (1919), Rutherford used α particles of 7.68 MeV from the naturally radioactive element ${}^{214}_{84}\text{Po}$. When the α particles were sent through a nitrogen gas, most of them were either undeflected by the nitrogen nuclei or elastically scattered in close encounters with them. Rutherford found, however, that in a few collisions (about 1 in 50,000), protons were produced, according to the nuclear reaction

$$ {}^{14}_{7}\text{N} + {}^{4}_{2}\text{He} \rightarrow {}^{1}_{1}\text{H} + {}^{17}_{8}\text{O} $$

In this reaction, an α particle strikes a nitrogen-14 nucleus, producing a proton and an oxygen-17 nucleus. The reaction represents *induced transmutation* of the element nitrogen into a stable isotope of oxygen; α or β radioactive decay represents the *spontaneous transmutation* of one element into another.

The conservation laws for electric charge and nucleons are satisfied in all nuclear reactions; therefore, the presubscripts giving the electric charge of the particles and the presuperscipts giving the number of nucleons in each particle each add to the same amount on both sides of the equation. The reaction can be written in abbreviated form as follows:

$$ {}^{14}_{7}\text{N}(\alpha, \text{p}){}^{17}_{8}\text{O} $$

where the light particles going into and out of the reaction are written in parentheses between the symbols for the target and product nuclei.

Until 1932, all nuclear reactions were produced by the relatively high-energy α particles or γ rays from naturally radioactive materials. In that year, J. D. Cockcroft and E. T. S. Walton, using a 500-keV accelerator, observed the first nuclear reaction produced by artificially accelerated charged particles. They found that α particles were emitted when a lithium target was struck by protons with energies of 500 keV, according to the reaction

$$^{7}_{3}Li + ^{1}_{1}H \rightarrow ^{4}_{2}He + ^{4}_{2}He$$

$$^{7}_{3}Li(p, \alpha)^{4}_{2}He$$

The emitted α particles had a mean energy of 8.9 MeV; thus, an energy of 0.5 MeV had been put into the reaction, and 17.8 MeV was released as kinetic energy of the emerging particles. Here is a striking example of the release of nuclear energy. The total amount of energy released was trifling, of course, since most of the collisions between the incident protons and target nuclei did *not* result in nuclear disintegrations.

In the reactions involving nitrogen and lithium just described, the product nuclei were stable. The first nuclear reaction leading to an unstable product nucleus was observed by I. Joliot-Curie and F. Joliot in 1934. In the reaction, an aluminum target was struck by α particles, leading to

$$^{27}_{13}Al + ^{4}_{2}He \rightarrow ^{1}_{0}n + ^{30}_{15}P$$

$$^{27}_{13}Al(\alpha, n)^{30}_{15}P$$

The product nuclide was not stable but decayed with a half-life of 2.6 min into a stable isotope of silicon by β^{+} emission:

$$^{30}_{15}P \rightarrow ^{30}_{14}Si + \beta^{+} + \nu$$

where ν was a neutrino. The production of unstable nuclides that spontaneously disintegrate by the law of radioactive decay is a feature of many nuclear reactions. The nuclides are said to exhibit *artificial radioactivity.* Indeed, nuclear reactions furnish the only way of getting artificial radioactive isotopes, or *radioisotopes.* The radioisotopes are chemically identical with the element's stable isotopes. If a small amount of radioisotope is added to stable nuclides of the same element, it can act, through its radioactivity, as a *tracer* of the element; that is, the presence and concentration of the element can be determined by measuring the radioisotope's activity.

The discovery of the neutron came as a result of a nuclear reaction that W. Bothe and H. Becker observed in 1930, the bombardment of beryllium by α particles:

$$^{9}_{4}Be + ^{4}_{2}He \rightarrow ^{1}_{0}n + ^{12}_{6}C$$

$$^{9}_{4}Be(\alpha, n)^{12}_{6}C$$

It was thought at first that the products were a γ ray and the stable nucleus $^{13}_{6}C$ instead of a neutron and $^{12}_{6}C$, because an extremely penetrating radiation was found to result. Then Curie and Joliot in 1932 found that when the resulting radiation fell on paraffin (which consists largely of hydrogen), protons with energies of about 6 MeV were emitted. This was interpreted at first in terms of the Compton effect, in which a γ-ray photon makes a Compton collision with a proton and ejects it from the paraffin. The photon energy needed to transfer 6 MeV to protons is easily found, from Equation 4–15, to be nearly 60 MeV. It is easy to show from the conservation of energy that *less* than this amount of energy would be released in a $^{9}_{4}Be(\alpha, \gamma)^{13}_{6}C$ reaction. Therefore, the photon hypothesis was untenable.

J. Chadwick properly interpreted these experiments in 1932; he showed that all experimental results were consistent with the assumption that an uncharged and therefore highly penetrating particle having a mass nearly that of the proton was being emitted. By energy conservation, such a particle would be emitted with an energy of about 6 MeV, and when the neutron should strike a proton head-on, it would come to rest, transferring its momentum and energy to the proton.

Neutrons are emitted in many nuclear reactions and can themselves be used as bombarding particles. In one of the important neutron-induced reactions, a neutron is captured by a target nucleus and a γ-ray photon is emitted. This reaction is known as neutron *radiative capture*. For example,

$$^{27}_{13}\text{Al} + {}^{1}_{0}\text{n} \rightarrow \gamma + {}^{28}_{13}\text{Al}$$

$$^{27}_{13}\text{Al}(\text{n}, \gamma)^{28}_{13}\text{Al}$$

The product nucleus, an unstable isotope of the target nucleus, decays by β^- decay:

$$^{28}_{13}\text{Al} \rightarrow {}^{28}_{14}\text{Si} + \beta^- + \bar{\nu}$$

where $\bar{\nu}$ is an antineutrino.

Since the neutron has no electric charge, the neutron radiative capture can occur when a neutron of almost any energy strikes almost any nucleus; the heavier isotope thus produced frequently is radioactive, and the absorption of neutrons is therefore a common method of producing radioisotopes.

In another important type of reaction, resulting from neutron bombardment, a charged particle, such as a proton or an α particle, is emitted. Such a reaction offers a way of detecting neutrons, since the emitted charged particles produce detectable ionization. One reaction often used in neutron detection is

$$^{10}_{5}\text{B} + {}^{1}_{0}\text{n} \rightarrow {}^{4}_{2}\text{He} + {}^{7}_{3}\text{Li}$$

$$^{10}_{5}\text{B}(\text{n}, \alpha)^{7}_{3}\text{Li}$$

Photodisintegration is the nuclear reaction in which the absorption of a γ-ray photon results in the disintegration of the absorbing nucleus. An example is

$$^{25}_{12}\text{Mg} + \gamma \rightarrow {}^{1}_{1}\text{H} + {}^{24}_{11}\text{Na}$$

$$^{25}_{12}\text{Mg}(\gamma, \text{p})^{24}_{11}\text{Na}$$

followed by $^{24}_{11}\text{Na} \rightarrow {}^{24}_{12}\text{Mg} + \beta^- + \bar{\nu}$.

A special type of low-energy nuclear reaction is *nuclear fission*. In this reaction (which is detailed in Section 11–6), a low-energy neutron is captured by a very heavy nucleus, and the resulting aggregate splits into two moderately heavy nuclei along with a few neutrons.

These are only a few of the many known nuclear reactions. One general statement concerning low-energy nuclear reactions involving the light particles (p, n, d, α, e, and γ), either bombarding or emerging, can be made: Nuclear reactions with essentially all possible combinations of ingoing and outgoing light particles occur.

11–2 The Energetics of Nuclear Reactions

A generalized nuclear reaction is $X(x, y)Y$, where X is the target nucleus, x the bombarding particle, y the emergent light mass particle, and Y the product nucleus. The target nucleus is assumed to be at rest ($K_X = 0$), and the kinetic energies of x, y, and Y are denoted by K_x, K_y, and K_Y.

The disintegration energy, or Q *value*, of a radioactive decay has been defined as the total energy released in the decay (Equation 10–25). The Q value of a nuclear reaction is likewise defined as the total energy released in the reaction; that is, Q is the kinetic energy coming out of the reaction less the kinetic energy going into the reaction:

$$Q = (K_y + K_Y) - K_x \tag{11–1}$$

The total relativistic energy of a particle is the sum of its rest energy and its kinetic energy; mass-energy conservation requires, then, that

$$(m_x c^2 + K_x) + M_X c^2 = (m_y c^2 + K_y) + (M_Y c^2 + K_Y) \tag{11–2}$$

where m_x, M_X, m_y and M_Y are the *rest* masses. Combining the two equations gives

$$\frac{Q}{c^2} = (m_x + M_X) - (m_y + M_Y) \tag{11–3}$$

This equation shows that Q/c^2, the mass equivalent of the energy released in the reaction, is simply the total rest mass going into the reaction less the total rest mass coming out of the reaction. Thus, the nuclear energy released in a reaction can be computed directly from the masses of the participating particles; or if one of the masses (most often the product heavy nucleus) is not known precisely, it can be computed if the Q value is determined from measurement of particle kinetic energies.

Nuclear energy is released in a reaction when $Q > 0$; such a reaction, with mass converted to kinetic energy, is known as an *exothermic,* or *exoergic,* reaction. A reaction in which nuclear energy is absorbed, or consumed, with $Q < 0$, is called *endothermic,* or *endoergic.* An endothermic reaction can be thought of as an inelastic collision in which the identity of the colliding particles changes and kinetic energy is at least partially converted to mass.

A special sort of reaction is that in which the incoming and outgoing particles are identical, $x = y$ and $X = Y$. If no kinetic energy is lost ($Q = 0$), the reaction is an elastic collision; if energy is lost ($Q < 0$), the reaction is an inelastic collision.

Let us compute Q for the reaction ${}^{7}_{3}\text{Li}\ (p, \alpha){}^{4}_{2}\text{He}$, in which $y = Y$, an atypical reaction. We can use the *neutral atomic* masses of the four particles, since in the change from nuclear masses to atomic masses, an equal number of electron masses is added to both sides of Equation 11–2.

Mass ${}^{1}_{1}\text{H}$ = 1.007 825	Mass ${}^{4}_{2}\text{He}$ = 4.002 603
Mass ${}^{7}_{3}\text{Li}$ = 7.016 004	Mass ${}^{4}_{2}\text{He}$ = 4.002 603
$m_x + M_X$ = 8.023 829 u	$m_y + M_Y$ = 8.005 206 u

$$\frac{Q}{c^2} = (m_x + M_X) - (m_y + M_Y)$$

$$= 8.023\,829 - 8.005\,206 = 0.018\,623 \text{ u}$$

$$Q = 0.018\,623 \text{ u} \times 931.5 \text{ MeV/u}$$

$$= 17.35 \text{ MeV}$$

This reaction is exothermic, since 17.35 MeV is released; thus, the total kinetic energy of the two outgoing α particles exceeds the kinetic energy of the incoming proton by that amount. In the original Cockcroft-Walton experiment, the incident protons had an energy of 0.50 MeV; therefore, the total energy carried by the two α particles was expected to be $17.35 + 0.50 = 17.85$ MeV, or about $\frac{1}{2}(17.85) = 8.93$ MeV for each α particle. The measured energy of the α particles agreed well with the expectation. This and all other reactions between particles whose masses and kinetic energies are known gives striking confirmation of the relativistic mass-energy equivalence.

Because $y = Y$ in the $^7_3\text{Li}(\text{p}, \alpha)^4_2\text{He}$ reaction, the emerging particles have nearly equal kinetic energies and momenta (magnitude). When a reaction involves masses $M_Y \gg m_y$, then the energies are $K_Y \ll K_y$; most of the kinetic energy is carried by the light particle.

Energy is released in an exothermic reaction; therefore, it is energetically possible for an exothermic reaction to occur even when the energy of the bombarding particle is nearly zero, although the probability may be small. On the other hand, an endothermic reaction cannot occur unless the incident particle carries kinetic energy. At first thought, it might seem that an endothermic reaction with a Q of, say, -5 MeV would be energetically possible if 5-MeV kinetic energy were carried into the collision by the bombarding particle, but this is *not* so. The value of K_x must exceed the magnitude of Q for the reaction to go. (The specific relation is derived in Section 11–3.)

11-3 Conservation of Momentum and Nuclear Reaction Threshold

An endothermic reaction, one with a negative Q, is a reaction for which the total rest energy of the particles $y + Y$ out of the reaction exceeds the total rest energy of the particles $x + X$ entering the reaction (Equation 11–3). Since the total energy that goes in or comes out of the reaction $X(x, y)Y$ is constant, this means that with $Q < 0$, the kinetic energy of particles entering the reaction must exceed the kinetic energy of particles leaving (see Equation 11–1). Moreover, with the target particle X at rest, all kinetic energy entering the reaction is carried by the incident particle x alone. The *threshold energy* is defined as the minimum kinetic energy of the incident particle that will just allow the reaction to take place.

That the threshold energy must, in fact, *exceed* the magnitude of Q (negative) for an endothermic reaction follows immediately from the momentum conservation principle. The incident particle carries momentum as well as kinetic energy, so the total momentum of the emerging particles $\mathbf{P}_y + \mathbf{P}_Y$ must equal the momentum $\mathbf{p}_x$ of the incident particle. See Figure 11–1. Emerging particles y and Y must then be in motion and carry at least some kinetic energy. Consequently, only a part of the incident kinetic energy K_x can be transformed into rest energy.

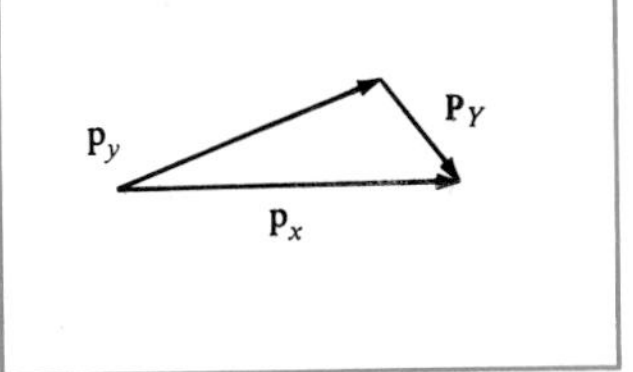

FIGURE 11–1. *Linear momenta of the incident particle x and the emerging particles y and Y in a nuclear reaction.*

All these observations on energy apply to the laboratory reference frame, in which the target particle X is initially at rest. The reaction can also be examined, and it would be simpler so to analyze, in a reference frame in which the system's center of mass is at rest and remains so. In the center-of-mass reference frame, the system's total momentum is by definition *zero;* and x and X are approaching one another in opposite directions with equal momentum magnitude before the reaction, and y and Y are departing in

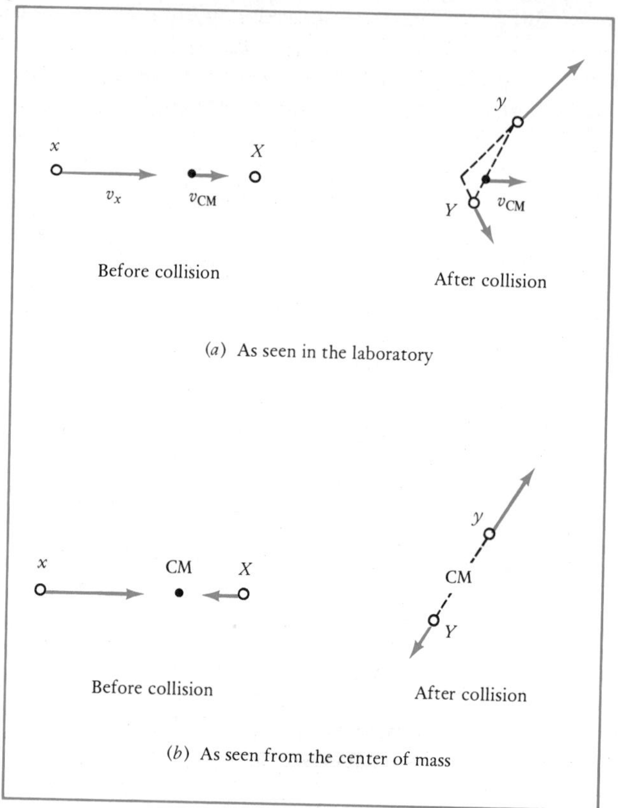

FIGURE 11–2. *Nuclear collision or reaction as seen (a) in the laboratory and (b) from the center-of-mass reference frame.*

opposite directions with equal momentum magnitude after the reaction. See Figure 11–2.

To compute the threshold kinetic energy K_{th} in the laboratory reference frame in terms of the Q of an endothermic reaction and the rest masses m_x, M_X, m_y, and M_Y of the participating particles, we exploit the zero value of the total system momentum in the center-of-mass reference frame. To arrive at a general relation, one applicable for incident particles of the highest energies and possibly involving the creation of additional particles, we shall use the relativistic relations for energy and momentum.

The rest energy E_0 of any one particle is the same in all reference frames, whereas the particle's relativistic momentum **p** and energy E depend on the particular reference frame in which these quantities are measured. The invariant quantity E_0^2 is given by Equation 3–15:

$$E_0^2 = E^2 - (pc)^2 \tag{11–4}$$

The very same relation applies to an isolated *system* of particles, such as those involved in a nuclear reaction, for which E_0 then represents the system's total rest energy, and E and **p** are the total relativistic energy and the relativistic momentum (vector) of all particles in the system, either before or after the reaction.

Writing Equation 11–4 as it applies both in the laboratory and in the center-of-mass reference frames, we have

$$E_0{}^2 = E_{\text{lab}}^2 - (p_{\text{lab}}c)^2 \tag{11–5}$$

and

$$E_0{}^2 = E_{\text{cm}}^2 - (p_{\text{cm}}c)^2 \tag{11–6}$$

We first apply Equation 11–5 in the laboratory frame and at the threshold of the reaction. Before the reaction, only particle x carries momentum. Therefore,

$$p_{\text{lab}} = p_x \tag{11–7}$$

The total energy in the laboratory before the reaction is then given by

$$E_{\text{lab}} = E_x + E_X = (E_{0_x} + K_{\text{th}}) + E_{0_X} \tag{11–8}$$

where K_{th} is the kinetic energy of x and E_{0_X} is its rest energy. The kinetic energy of X is zero, and its rest energy is E_{0_X}.

In the center-of-mass reference frame, the system's total momentum is zero; therefore,

$$p_{\text{cm}} = 0 \tag{11–9}$$

Furthermore, at the reaction threshold, the kinetic energies of y and Y in this reference frame are zero. (The system's center of mass and the particles y and Y are, of course, in motion relative to the laboratory frame.) Therefore, the total energy E_{cm} in the center-of-mass frame after the reaction can be written

$$E_{\text{cm}} = E_{0_y} + E_{0_Y} \tag{11–10}$$

Substituting Equations 11–7 and 11–8 in Equation 11–5, and Equations 11–9 and 11–10 in Equation 11–6, we have

$$[(E_{0_X} + K_{\text{th}}) + E_{0_X}]^2 - (p_xc)^2 = (E_{0_y} + E_{0_Y})^2 \tag{11–11}$$

Applying Equation 11–4 to particle x alone in the laboratory, we have

$$E_{0_x}^2 = E_x{}^2 - (p_xc)^2 = (E_{0_x} + K_{\text{th}})^2 - (p_xc)^2 \tag{11–12}$$

Combining Equation 11–12 and 11–11 to eliminate p_x, we have

$$(E_{0_x} + K_{\text{th}} + E_{0_x})^2 + E_{0_x}^2 - (E_{0_x} + K_{\text{th}})^2 = (E_{0_y} + E_{0_Y})^2$$

Solving then for K_{th} in the above equation yields

$$K_{\text{th}} = \frac{(E_{0_y} + E_{0_Y})^2 - (E_{0_x} + E_{0_X})^2}{2\,E_{0_X}}$$

$$= \frac{[(E_{0_y} + E_{0_Y}) - (E_{0_x} + E_{0_X})][(E_{0_y} + E_{0_Y}) + (E_{0_x} + E_{0_X})]}{2E_{0_X}}$$

and using the relation 11–3 for Q defined in terms of rest masses,

$$\frac{Q}{c^2} = (m_x + M_X) - (m_y + M_Y)$$

$$K_{\text{th}} = -\frac{Q(m_x + M_X + m_y + M_Y)}{2M_X} \tag{11–13}$$

Recall that Q itself is intrinsically negative for an endothermic reaction.

Note that the quantity within parentheses in Equation 11–13 is simply the total rest mass of *all* particles going into and out of the reaction. Although we have focused on two emerging particles (y and Y), the derivation allows for any number of emerging particles; so a more general form of Equation 11–13 is

$$K_{\text{th}} = -\frac{Q\begin{pmatrix}\text{rest mass of } all \text{ particles entering} \\ \text{and leaving reaction}\end{pmatrix}}{2(\text{rest mass target particle})} \tag{11–14}$$

For a low-energy nuclear reaction, one with kinetic energy and Q of the order of a few MeV and therefore much lower than the rest energies of any of the participating particles, Equation 11–3 becomes

$$m_X + M_X \approx m_y + M_Y$$

If this approximation is used in Equation 11–13, that reduces to

$$K_{\text{th}} = -Q\left(\frac{m_X + M_X}{M_X}\right) \tag{11–15}$$

which is the classical threshold relation, applicable to low-energy nuclear reactions.

For example, the first observed nuclear reaction ${}^{14}_{7}\text{N}(\alpha, \text{p}){}^{17}_{8}\text{O}$ has a Q of -1.18 MeV, as computed from the rest masses of the four particles. The rest masses of an α particle and a ${}^{14}_{7}\text{N}$ nucleus can be taken as 4 u and 14 u, respectively. Then Equation 11–15 yields for the threshold α-particle kinetic energy $K_{\text{th}} = -(-1.18 \text{ MeV})(\frac{18}{14}) = +1.52$ MeV. The reaction takes place only if the α-particle energy equals or exceeds 1.52 MeV, a condition that was satisfied (7.68 MeV α particles) in Rutherford's experiment. If the incident particle has a much smaller mass than the target does, the threshold energy equals the magnitude of the Q (if $m_x \ll M_X$, and $K_{\text{th}} = -Q$ from Equation 11–15).

Example 11–1. With what minimum kinetic energy must a proton collide with a second, free proton at rest to create a proton-antiproton pair according to the reaction

$$\text{p}^+ + \text{p}^+ \rightarrow \text{p}^+ + \text{p}^+ + (\text{p}^+ + \overline{\text{p}^-})$$

where p^+ denotes a proton and $\overline{\text{p}^-}$ an antiproton?

The rest energy of a proton or an antiproton is designated E_0. Since two additional particles, each of rest energy E_0, are created in the reaction, we have $Q = -2E_0$. The total rest energy of all particles entering the reaction (2 protons) and leaving the reaction (3 protons and 1 antiproton) is $6E_0$. Then with the use of Equation 11–14, we find

$$K_{\text{th}} = -\frac{(-2E_0)(6E_0)}{2E_0} = 6E_0$$

The incident proton's kinetic energy must be at least *six* times its rest energy, or 6(0.94 GeV) = 5.64 GeV, to create an antiproton (together with a proton). On the other hand, when two protons, each with a kinetic energy E_0 of only 0.94 GeV collide head-on in the laboratory, an additional proton and an antiproton may be created in the collision; in this instance, the laboratory reference frame *is* the center-of-mass reference frame, and none of the kinetic energy of the colliding particles is "wasted" to conserve momentum. Since only a fraction of the incident particle's kinetic energy is available for creating particles when the target particle is at rest in the laboratory, causing beams of particles that are moving in opposite directions to collide is an efficient way, in terms of particle energy, of creating particles (Section 11–14).

Example 11–2. What is the minimum energy of a photon needed to create an electron-positron pair through interaction with a free *electron* initially at rest?

We know that when a photon interacts with a massive particle, such as a nucleus, and produces an electron-positron pair, its threshold energy $h\nu_{min}$ is $2E_0 = 2(0.51 \text{ MeV}) = 1.02 \text{ MeV}$, where E_0 is the rest energy of an electron or a positron (Section 4–5). The massive particle carries away most of the incident photon's momentum but hardly any of its energy, so that essentially all the photon's energy $h\nu$ is available for creating an electron-positron. For the following reaction, however,

$$h\nu + e^- \rightarrow e^- + (e^- + e^+)$$

we must invoke the relativistic threshold relation.

The total reaction energy is $Q = -2E_0$. The total rest energy of all particles entering and leaving the reaction is $4E_0$ (1 electron enters the reaction, the incident photon has zero rest energy, and 2 electrons and 1 positron leave). Thus, we have, through Equation 11–14,

$$h\nu_{min} = K_{th} = -\frac{(-2E_0)(4E_0)}{2E_0} = 4E_0 = 2.04 \text{ MeV}$$

which is *twice* the minimum photon energy for pair production by interaction with a massive particle.

11–4 Cross Section

The decay of unstable nuclei is characterized not only by the energy released in the decay products but also by the half-life, or decay constant, of the disintegration. For an unstable nucleus, the decay constant λ gives a measure of the probability in *time* that the decay will occur. We wish to introduce a quantity, called the reaction *cross section,* that measures the probability *in space* that a nuclear reaction will occur. The conservation laws of energy and momentum tell us whether the reaction is possible; the cross section gives us a quantitative measure of how probable the reaction is.

Consider Figure 11–3, which shows several target nuclei X exposed to an incident beam of particles x. Each X nucleus has associated with it an area σ, called the cross section, which is imagined to be oriented at right angles to the incident particles (regarded as point masses). Each cross-sectional area is assumed to be so small that in a reasonably thin target material no one nucleus is "hidden" from the incident particles by any other nucleus. The area of the cross section is so chosen that if an incident particle strikes the area σ, the reaction $X(x, y)Y$ takes place, and if it misses, the

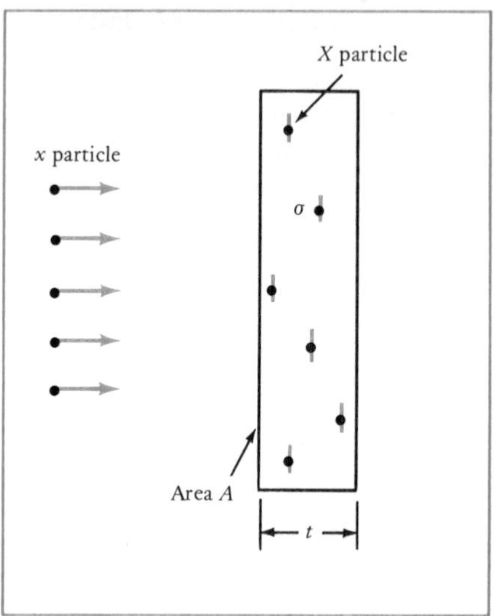

FIGURE 11–3. *Target nuclei X in a target of thickness t, area A, and cross section σ, being struck by x particles.*

reaction does not take place. The intrinsic probability of a nuclear reaction is, therefore, directly proportional to its cross section σ.

We take the number of incident particles on a thin foil of thickness t and area A to be n_i. The number of these particles undergoing the nuclear reaction $X(x, y)Y$ is n, which also represents the number of y, or Y, particles produced. The number of target nuclei per unit volume is N, each with nuclear-reaction cross section σ. Since the total number of nuclei in the target foil is $N(At)$, the total exposed area resulting in reactions is σNAt. The ratio n/n_i of the x particles undergoing reactions to the total number incident on the foil must equal the ratio of the total target area σNAt to the total foil area A; then $n/n_i = \sigma NAt/A$, or

$$\frac{n}{n_i} = \sigma N t \qquad (11\text{–}16)$$

This derivation of the reaction cross section is analogous to the derivation of the scattering cross section, described in Section 6–1. It shows that the probability n/n_i that an incident particle will undergo a nuclear reaction is proportional to the reaction cross section σ, the number N of target nuclei per unit volume, and the thickness t of the target foil. The common unit for measuring nuclear cross sections is the *barn*, which is 10^{-24} cm^2 = 10^{-28} m^2 = 100 fm^2. As nuclear cross sections go, one of 10^{-28} m^2 is relatively large; for an incident particle to hit a nuclear target having a cross section of 1 barn is as easy as hitting the side of a barn. Cross sections vary greatly from one reaction to another; furthermore, they may strongly depend on the energy of the bombarding particle.

Example 11–3. Consider the radiative capture of 500-keV neutrons by aluminum in the reaction $^{27}_{13}\text{Al}(\text{n}, \gamma)^{28}_{13}\text{Al}$. The neutron-capture cross section in aluminum is 2.0 millibarns = 2.0×10^{-31} m^2. Suppose that a neutron flux of 1.0×10^{10} neutrons/cm$^2\cdot$s is incident on an aluminum foil 0.20 mm thick. What is the number of neutrons captured per second in a 1-cm^2 area of the foil?

We can compute the density N of aluminum nuclei from the ordinary density of aluminum, 2.7 g/cm³, Avogadro's number, and the atomic mass of aluminum, 27. Then

$$N = (2.7\ \text{g/cm}^3)(6.0 \times 10^{23}\ \text{atoms/g}\cdot\text{mol})/(27\ \text{g/g}\cdot\text{mol})$$
$$= 6.0 \times 10^{22}\ \text{nuclei/cm}^3$$

From Equation 11–16 we have

$$n = n_i \sigma N t$$
$$= (10^{10}\ \text{neutrons/cm}^2\cdot\text{s})(2 \times 10^{-27}\ \text{cm}^2) \times (6.0 \times 10^{22}\ \text{nuclei/cm}^3)(2 \times 10^{-2}\ \text{cm})$$
$$= 2.4 \times 10^4\ \text{neutrons/cm}^2\cdot\text{s}$$

Since there were 10^{10} incident particles per square centimeter per second, only 2.4 out of every 10^6 neutrons striking the foil are captured in the reaction $^{27}_{13}\text{Al}(\text{n}, \gamma)^{28}_{13}\text{Al}$.

Because the cross section gives a measure of the probability that the nuclear reaction will occur, its measurement and interpretation in the light of nuclear structure have been important activities in nuclear physics. In reaction cross-section experiments, monoenergetic x particles strike a target; the cross section is measured by determining the number of either y particles or Y particles produced by a known number of x particles. The y particles emerging from a target can be counted by particle detectors, and the Y particles (often unstable) can be counted by measuring the radioactivity resulting from their decay. Chemical quantitative analysis of the Y atoms is difficult because their concentration is typically very small.

A few general remarks about reaction cross sections follow. In an endothermic reaction, one that cannot proceed unless energy is added to the combining particles, the reaction cross section is necessarily zero until the threshold energy is exceeded.

Reactions in which the incident particles are neutrons, particularly the radiative-capture reactions (n, γ) may show large cross sections even when the energy of the bombarding particle is extremely small. Unlike a charged particle, a neutron is undeflected by the electric charge of the nucleus, and it can quite easily come within the range of the nuclear force and react with the nucleus at slow speeds. A typical (n, γ) cross section is shown in Figure 11–4 as a function of the neutron energy. It is seen that apart from the pronounced peaks, the cross section increases as the energy or speed of the neutrons decreases. In fact, σ is found to be closely proportional to $1/v$, where v is the neutron speed. This $1/v$ law can be stated as follows. The probability that a neutron will be captured is directly proportional to the time it spends in the vicinity of any one bombarded nucleus, and inversely proportional to its speed. The peaks in the cross-section curve are referred to as *resonances;* their interpretation gives information on nuclear energy levels (Section 11–5).

When *charged* particles strike a target nucleus, the size of the reaction cross section is influenced since the particles are repelled by the Coulomb force. If it were not for the phenomenon of the tunnel effect, or *barrier penetration* (Section 10–9), low-energy charged particles could not come within the range of the nuclear force of the target nucleus, the nuclear reac-

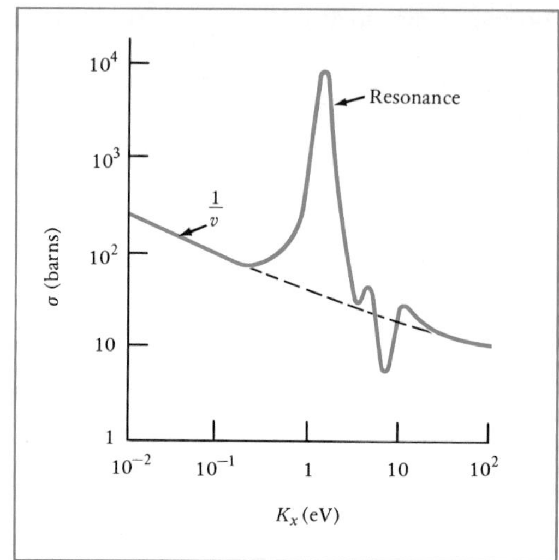

FIGURE 11–4. *Neutron capture cross section for indium as a function of neutron energy.*

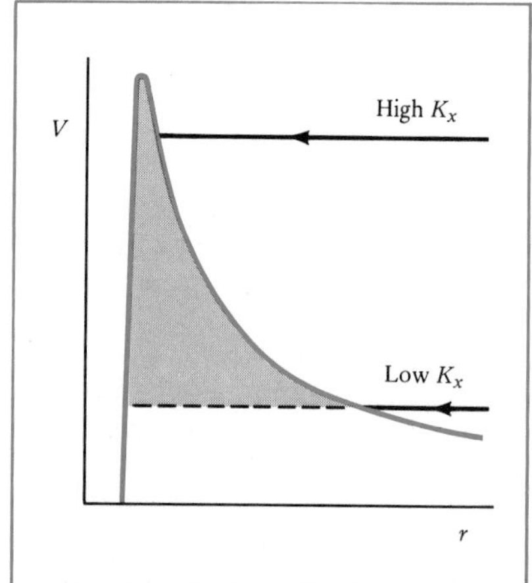

FIGURE 11–5. *Representation of the relative nuclear barriers to be penetrated by low-energy and high-energy incident charged particles.*

tion could not occur, and the reaction cross-section would be zero. Incident charged particles with energies even less than 1 MeV (much less than the height of the Coulomb potential barrier), do undergo nuclear reactions, indicating that the Coulomb barrier has been penetrated. The probability of barrier penetration depends very much on the barrier height and thickness. The more energetic the incident charged particle, the easier its penetration of the barrier; see Figure 11–5. It follows that the reaction cross section will generally increase with the energy K_x, as shown in Figure 11–6.

11–5 The Compound Nucleus and Nuclear Energy Levels

Let us, for an introduction to the compound-nucleus concept, compute the energy with which the "last" (least tightly bound) neutron in the stable cadmium nuclide $^{114}_{48}$Cd is bound to the other 113 nucleons. This separation energy is just the amount of energy that must be added to the nuclide to separate the last neutron, leaving the stable isotope $^{113}_{48}$Cd. The separation

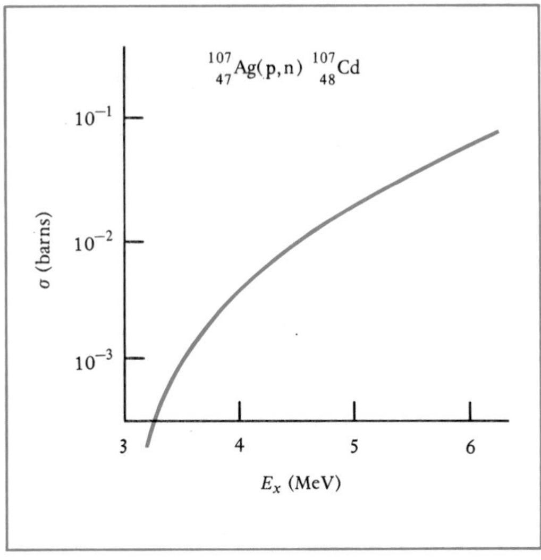

FIGURE 11–6. *Increase in the cross section of a nuclear reaction with proton energy.*

energy E_s is found directly by comparing the masses of $^1_0n + {}^{113}_{48}Cd$ and $^{114}_{48}Cd$:

$$E_s = [(1.008\,665 + 112.904\,401) - (113.903\,361)]\,u \times 931.5\ MeV/u$$
$$= 9.040\ MeV$$

Therefore, if 9.04 MeV of energy is absorbed by the $^{114}_{48}Cd$ nucleus, a free neutron and a $^{113}_{48}Cd$ nucleus are formed, both particles being at rest. Symbolically,

$$^{114}_{48}Cd + 9.04\ MeV \rightarrow {}^{113}_{48}Cd + {}^1_0n$$

Now imagine the process to be reversed, so that we bring together a neutron and a $^{113}_{48}Cd$ nucleus, both with zero kinetic energy, to form a $^{114}_{48}Cd$ nucleus. No energy has to be added to the particles to make them amalgamate, since the neutron will be attracted by the nuclear force of the nuclide when it is close enough to it. The cadmium-114 nucleus then formed will *not*, however, be in its ground state; it will be in an excited state, with an excitation energy of 9.04 MeV. The nucleus $^{114}_{48}Cd^*$ (the asterisk denotes an excited state) is unstable and will quickly decay to its ground state by emitting a γ-ray photon of 9.04 MeV. The overall process can be written

$$^{113}_{48}Cd + {}^1_0n \rightarrow {}^{114}_{48}Cd^* \rightarrow \gamma(9.04\ MeV) + {}^{114}_{48}Cd$$

We have just described the particular neutron radiative-capture reaction $^{113}_{48}Cd(n, \gamma)^{114}_{48}Cd$. This reaction takes place in *two* stages: the amalgamation of the two original particles to form a single nucleus in an excited state and the decay from this intermediate state to the products of the reaction. The energetics of the process are shown in Figure 11–7, where the total energies of the particles going into and coming out of the reaction are displayed.

The neutron radiative-capture reaction illustrates a feature that is common to most low-energy nuclear reactions, the formation and decay of a compound nucleus. The assumptions are these.

1. For the reaction $X(x, y)Y$, the particles x and X combine to form the compound nucleus C, invariably in an excited state: $X + x \rightarrow C^*$.

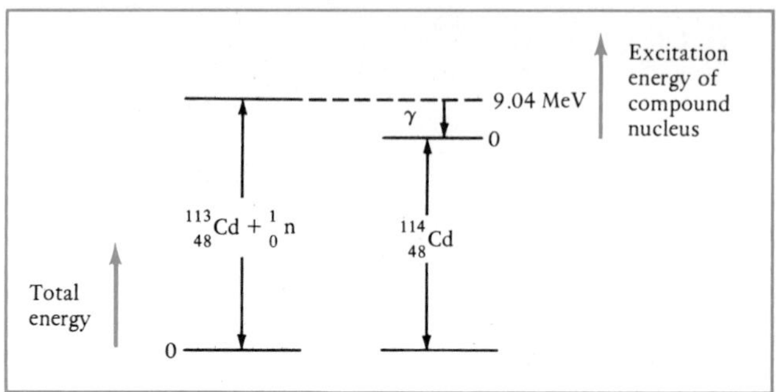

FIGURE 11–7. *Energy-level diagram of a neutron radiative-capture reaction.*

The energy carried into the reaction by x is quickly shared among all the nucleons in the compound nucleus.

2. The compound nucleus C^* exists for a long time compared with the nuclear time ($\approx 10^{-22}$ s), the time for a nucleon with a few MeV of energy to traverse a nuclear dimension. The average lifetime of a typical compound nucleus is nevertheless so short that C* is not directly observable. Various x and X particles can form the same nucleus C* in the same excited state, as shown in Table 11–1 and Figure 11–8. The resultant compound nucleus lives so long that it has no "memory" of how it was formed.
3. The compound nucleus decays into the products of the reaction, $C^* \rightarrow y + Y$, as follows. After a fairly long time has elapsed (on a nuclear scale), the excitation energy of the compound nucleus, which was earlier distributed among the several nucleons, is finally concentrated on some one particle y; this particle is ejected, leaving the nucleus Y. A compound nucleus in some particular excited state may decay, then, through the formation of any one of several y and Y combinations, as shown in Table 11–1 and Figure 11–8. For a particular excited state of C*, one particular decay mode typically dominates all others.
4. Since the nuclear reaction is to be regarded as taking place in two distinct stages (the formation of C* and the decay of C*), the reaction cross section, which gives a measure of the probability that the *complete* reaction will take place, is proportional to *two* probabilities: the probability that x and X will amalgamate to form C* and the probability that C* will decay into some particular y and Y particles.

Table 11–1 shows that there are several ways in which the compound nucleus $^{14}_{7}N^*$ can be formed and several ways in which this nucleus can decay from an excited state. For example, when a proton of 1 MeV combines with $^{13}_{6}C$, the most probable reaction is $^{13}_{6}C(p, \gamma)^{14}_{7}N$; but when a proton of 6 MeV strikes the same target and forms the same compound nucleus, the reaction $^{13}_{6}C(p, n)^{13}_{7}N$ is the most likely. In the first instance, the excitation energy

Table 11–1

$$X + x \rightarrow C^* \rightarrow y + Y$$

$$\left.\begin{matrix} ^{13}_{6}C + p \\ ^{12}_{6}C + d \\ ^{10}_{5}B + \alpha \end{matrix}\right\} {}^{14}_{7}N^* \left\{\begin{matrix} p + {}^{13}_{6}C \\ d + {}^{12}_{6}C \\ \alpha + {}^{10}_{5}B \\ n + {}^{13}_{7}N \\ \gamma + {}^{14}_{7}N \end{matrix}\right.$$

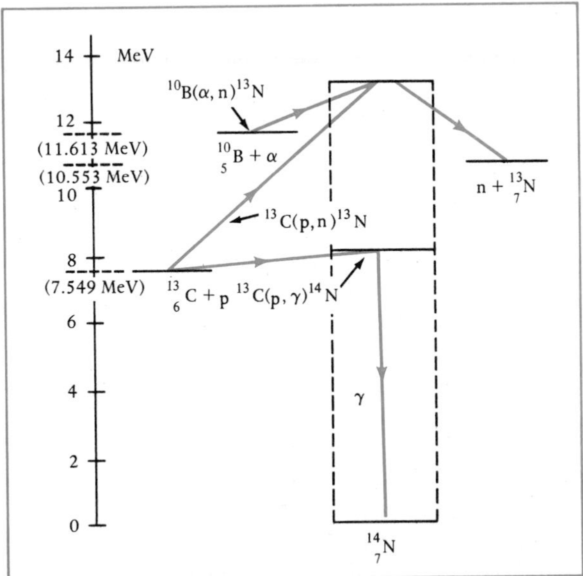

FIGURE 11–8. *Energetics of several nuclear reactions for which nitrogen-14 is the compound nucleus. The scale gives energies relative to the ground state of nitrogen-14 as zero.*

of $^{14}_{7}N^*$ is about 8 MeV, and in the second it is about 13 MeV, as shown in Figure 11–8. Furthermore, when an α particle of 2 MeV combines with $^{10}_{5}B$, again forming $^{14}_{7}N^*$ with an excitation of about 13 MeV, the observed reaction is $^{10}_{5}B(\alpha, n)^{13}_{7}N$. The *decay* mode of the compound nucleus depends only on its excitation energy and *not* on the particles that form it. Note that Q values of the reactions can be read directly from the energies given in Figure 11–8.

Let us return to the $^{113}_{48}Cd(n, \gamma)^{114}_{48}Cd$ reaction, now noting how the cross

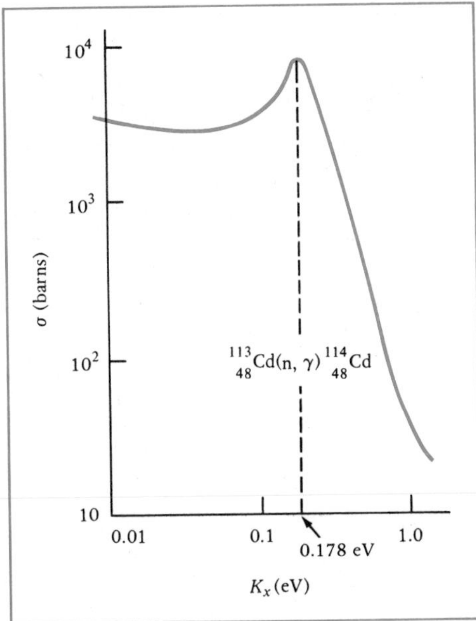

FIGURE 11–9. *Resonance in the cross section of a nuclear reaction.*

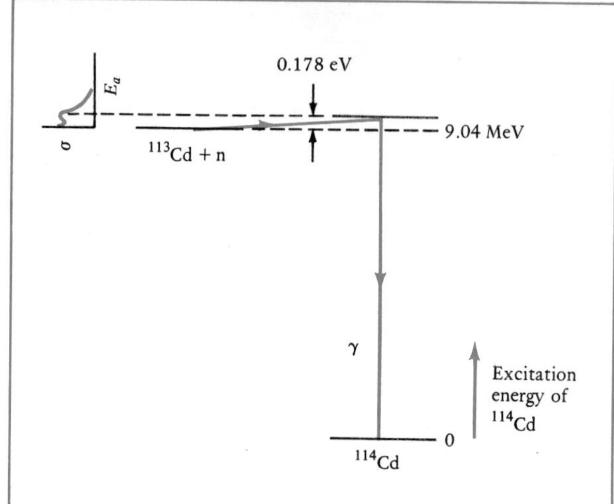

FIGURE 11–10. *Resonance capture at an excited state of a compound nucleus in the reaction* $^{113}_{48}Cd(n, \gamma)^{114}_{48}Cd$. *The energy differences are not to scale.*

section for this reaction varies with the energy of the incident neutrons. The capture cross section for very low neutron energies is shown in Figure 11–9, p. 367. It is well over 10 barns for all neutron energies, but there is a well-defined maximum, or *resonance,* at $K_x = 0.178$ eV. This means that a particularly high probability exists that the compound nucleus $^{114}_{48}Cd^*$ will be formed when it has an excitation energy approximately 0.18 eV higher than what it has (9.04 MeV) when it is combined with the neutrons of zero kinetic energy.† It indicates further that the nucleus $^{114}_{48}Cd$ has a well-defined quan-

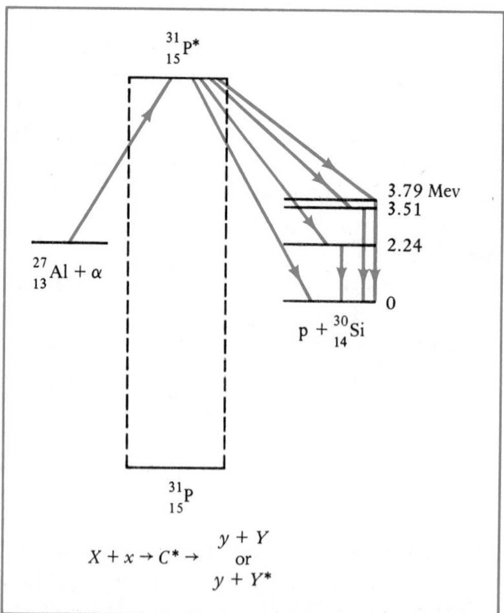

FIGURE 11–11. *Energetics of the* $^{27}_{13}Al(\alpha, p)^{30}_{14}Si$ *reaction, showing the excited states of the product nucleus.*

† Strictly, the additional energy given to the compound nucleus by a neutron with $K_x = 0.178$ eV is $K'_x = K_x[M_X/(M_X + m_x)] = (0.178)(113/114)\text{eV} \approx 0.177$ eV, where K'_x is the kinetic energy of x relative to the system's center of mass.

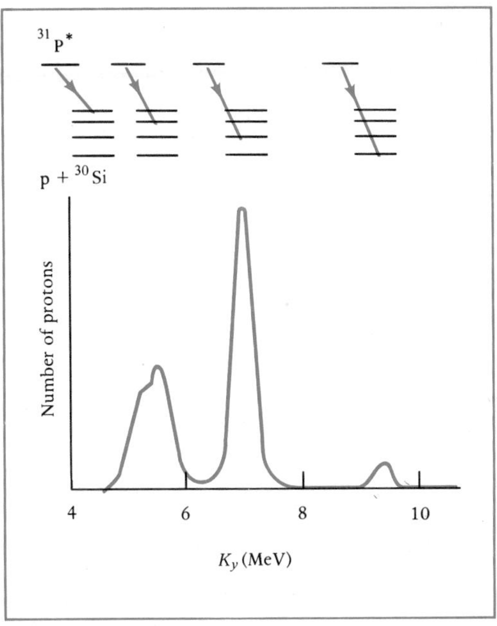

FIGURE 11–12. *Energy spectrum of the proton groups from the reaction* $^{27}_{13}Al(\alpha, p)^{30}_{14}Si$.

tized energy level when its excitation energy is just 0.178 eV higher than 9.04-MeV, as shown in Figure 11–10. This is just one of several discrete excited states of this nucleus. The excited states of a *compound nucleus* can generally be evaluated by observing the well-defined resonances in the reaction cross section; and the existence of these resonances is strong evidence that the compound-nucleus concept is correct.

Nuclear-reaction data can also be used for deducing the excited energy levels of the *product nucleus.* Consider, for example, the reaction $^{27}_{13}Al(\alpha, p)^{30}_{14}Si$, having the compound nucleus $^{31}_{15}P^*$. The compound nucleus decays not only to the ground state of the product nucleus $^{30}_{14}Si$ but also to several of its excited states; these decays are followed by γ decay to the ground state. The energy-level diagram is shown in Figure 11–11.

The existence of these excited states is indicated experimentally, not only by the γ rays emitted when $^{27}_{13}Al$ is bombarded by α particles, but also by the spectrum of energies from the protons observed (at some particular angle to the incident beam), as shown in Figure 11–12. The several *proton groups* correspond to the several possible decay modes of the compound nucleus, and the excitation energies of $^{30}_{14}Si^*$ can be evaluated by measuring the proton energies K_y.

11–6 Nuclear Fission

A special type of low-energy nuclear reaction occurs in very heavy nuclides. Unlike most nuclear reactions, in which a light particle and a heavy particle appear as products, this reaction results in the splitting, or fissioning, of the heavy nucleus into two parts of comparable masses; it is appropriately called *nuclear fission.* O. Hahn and F. Strassman first identified, in 1939, the nuclear-fission reaction.

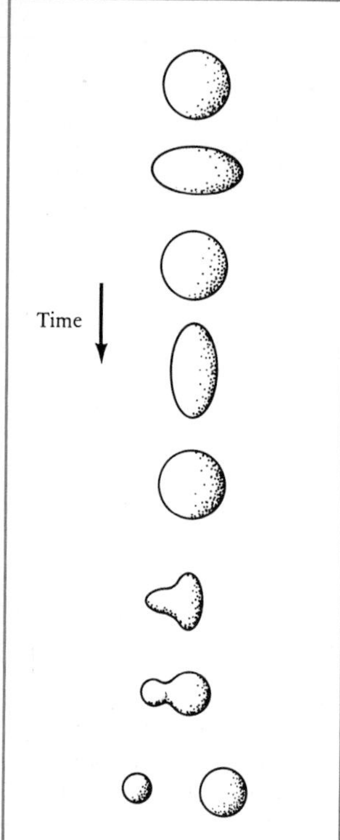

FIGURE 11–13. *Stages in the deformation of an oscillating nucleus, leading to nuclear fission.*

Consider the capture of a neutron of very low energy, such as a thermal neutron, by the very heavy nucleus, uranium-235. The compound nucleus $^{236}_{92}U$ formed in this reaction is in an excited state, with an excitation energy of 6.4 MeV. Almost all lighter excited compound nuclei formed from neutron capture decay emit γ-ray photons; the resulting heavier nuclei usually decay by β^- emission. But an excited uranium-236 nucleus can decay also by nuclear fission, splitting into two, or less frequently, three or more moderately heavy nuclei.

The behavior of a very heavy, excited compound nucleus can be understood by regarding the nucleus in its gross features to be like a liquid drop. The *liquid-drop model* was proposed by Bohr in 1936. Each nucleon in the interior of a nucleus is surrounded by and attracted by the strong short-range forces of immediately adjacent nucleons; more distant nucleons exert no appreciable attractive force. If this were the only effect, the total binding energy of a nucleus would be directly proportional to the total number of nucleons composing it. But a nucleon on the nuclear outer surface, not totally surrounded by other nucleons, is not so tightly bound as a nucleon in the interior. This is like the behavior of molecules in a drop of liquid; the molecules interact by short-range attractive forces, and the unsaturated force on molecules at the surface gives rise to surface-tension effects, including the tendency of a liquid drop to be spherical. For a nucleus, we have, besides the attractive nuclear force between all nucleons, the repulsive Coulomb force between all pairs of protons.

Suppose that owing to a nuclear collision a very heavy nucleus gains energy of excitation. The nucleus as a whole will oscillate and change its shape. One probable mode of deformation is shown in Figure 11–13, in which the nucleus assumes, in turn, the shapes of sphere, prolate ellipsoid (cigar), sphere, oblate ellipsoid (pancake), and so on. During the ensuing oscillations, the nuclear volume is constant. The surface area changes, however, and is greatest in the prolate and oblate deformations. The surface tension is manifested as a tendency of the nucleus to resume its spherical shape. On the other hand, the Coulomb repulsion increases when the nucleus assumes, for example, the prolate ellipsoidal shape, and the positive charges at the two ends of the ellipsoid tend to increase the deformation even further. Thus, two competing influences are at work: the nuclear surface tension, which tends to keep the nucleus spherical, and the Coulomb repulsion, which tends to deform it. If the excitation is great enough, the Coulomb force will succeed in shaping the nucleus into a dumbbell. With so great a distortion, the surface tension is not strong enough to restore the nucleus to sphericity, and the Coulomb force increases the separation between the ends, until they split into two distinct nuclei, or *fission fragments,* usually of unequal size. Then the fission fragments repel one another by the Coulomb force, and they move apart, each gaining kinetic energy as the system loses potential energy.

The transformations of a fission are shown in Figure 11–14 on the plot of N versus Z of Figure 10–7. Two fission fragments, (Z_1, N_1) and (Z_2, N_2), of a heavy compound nucleus share the protons and neutrons of the original compound nucleus (Z, N); that is, $Z = Z_1 + Z_2$ and $N = N_1 + N_2$. Both fragments fall to the *left* of the stability line; that is, both nuclei have too many neutrons to be stable. The neutron excess is so great that it is relieved almost instantaneously by the release of two or three neutrons from the fission fragments. The nuclei still have too many neutrons, and they finally reach

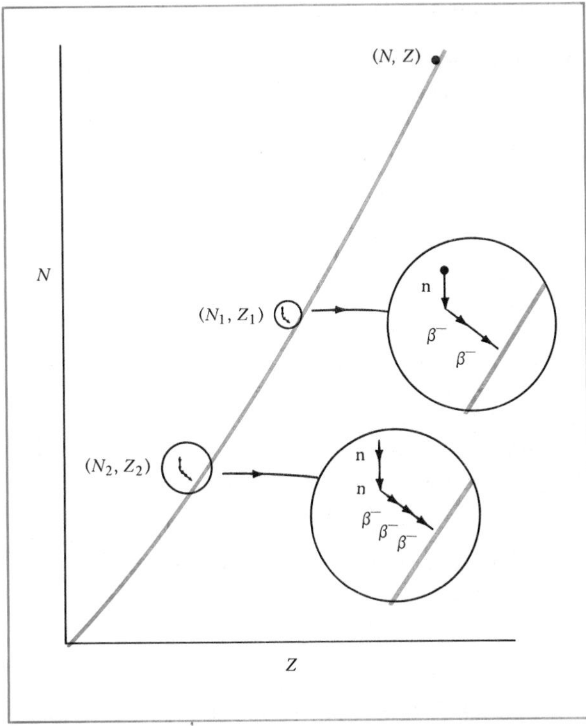

FIGURE 11–14. *Transformations in nuclear fission as they appear on a neutron-proton diagram.*

stability by changing neutrons into protons, by the much slower β^- decay. The β^- decay is accompanied by γ decay from excited nuclear states.

Two of the many known fission reactions resulting from neutron capture in uranium-235 are shown below, with the subsequent β^- decays of the fission fragments.

$$ {}^{1}_{0}\mathrm{n} + {}^{235}_{92}\mathrm{U} \rightarrow {}^{236}_{92}\mathrm{U}^* \rightarrow {}^{144}_{56}\mathrm{Ba} + {}^{89}_{36}\mathrm{Kr} + 3\,{}^{1}_{0}\mathrm{n} $$

$$ {}^{144}_{56}\mathrm{Ba} \xrightarrow{\beta^-} {}^{144}_{57}\mathrm{La} \xrightarrow{\beta^-} {}^{144}_{58}\mathrm{Ce} \xrightarrow{\beta^-} {}^{144}_{59}\mathrm{Pr} \xrightarrow{\beta^-} {}^{144}_{60}\mathrm{Nd} $$

$$ {}^{89}_{36}\mathrm{Kr} \xrightarrow{\beta^-} {}^{89}_{37}\mathrm{Rb} \xrightarrow{\beta^-} {}^{89}_{38}\mathrm{Sr} \xrightarrow{\beta^-} {}^{89}_{39}\mathrm{Y} $$

$$ {}^{1}_{0}\mathrm{n} + {}^{235}_{92}\mathrm{U} \rightarrow {}^{236}_{92}\mathrm{U}^* \rightarrow {}^{140}_{54}\mathrm{Xe} + {}^{94}_{38}\mathrm{Sr} + 2\,{}^{1}_{0}\mathrm{n} $$

$$ {}^{140}_{54}\mathrm{Xe} \xrightarrow{\beta^-} {}^{140}_{55}\mathrm{Cs} \xrightarrow{\beta^-} {}^{140}_{56}\mathrm{Ba} \xrightarrow{\beta^-} {}^{140}_{57}\mathrm{La} \xrightarrow{\beta^-} {}^{140}_{58}\mathrm{Ce} $$

$$ {}^{94}_{38}\mathrm{Sr} \xrightarrow{\beta^-} {}^{94}_{39}\mathrm{Y} \xrightarrow{\beta^-} {}^{94}_{40}\mathrm{Br} $$

The basic requirement for fission in the very heaviest nuclides is that the compound nucleus formed should have enough excitation energy to split. Neutron capture is just one of the several ways in which nuclear fission can be induced. Fission can also result from the bombardment of heavy nuclei by protons, deuterons, α particles, and γ rays (*photofission*).

Let us compute the total energy released in a typical fission. We see from Figure 10–10 that in the very heavy elements, in which $A \approx 240$, the average binding energy per nucleon, E_b/A, is approximately 7.6 MeV and in moderately heavy elements, $A \approx 120$, it is approximately 8.5 MeV. Thus, the total energy released in the fission is approximately

$$ 240 \text{ nucleons} \times (8.5 - 7.6)\ \text{MeV/nucleon} \approx 200\ \text{MeV} $$

The total energy released in a fission reaction is very large indeed compared with the few MeV of energy released in a typical low-energy exothermic nuclear reaction.

Nuclear fission is characterized by the decay of the compound nucleus into two moderately heavy nuclei, the emission of a few neutrons, and the β^- decay of the radioactive fission fragments. In an average fission reaction, about 200 MeV is released and distributed approximately as follows:

- Kinetic energy of fission fragments, 170 MeV
- Kinetic energy of fission neutrons, 5 MeV
- Energy of β^- and γ rays, 15 MeV
- Energy of antineutrinos associated with β^- decay, 10 MeV

The light isotope of uranium, $^{235}_{92}U$, undergoes fission with thermal neutrons; the excitation energy gained by the compound nucleus $^{236}_{92}U^*$ in capturing a slow neutron is great enough to cause the fission. (Uranium-235 is the only *natural* nuclide that undergoes fission with slow neutrons.) The much more abundant (99.3 percent) heavy isotope of uranium, $^{238}_{92}U$, will undergo fission, but only if bombarded by fast neutrons, neutrons having a kinetic energy of at least 1 MeV. Low-energy neutrons are captured by $^{238}_{92}U$, but the excited compound nucleus $^{239}_{92}U^*$ has too little excitation energy to decay by fission; it decays instead by γ emission.

Uranium-235 is fissile with both low- and high-energy neutrons. The (n, γ) reaction is less probable than fission at any energy. On the other hand, uranium-238 is fissile with high-energy neutrons only; low-energy neutrons are captured without fission. Clearly, the compound nucleus $^{239}_{92}U^*$ does not gain enough excitation energy from the capture of low-energy neutrons to decay by fission, whereas the compound nucleus $^{236}_{92}U^*$ is sufficiently excited by them to undergo fission. This difference can be explained by the following reasoning. The nuclide ^{236}U is an even-even nuclide, meaning that it has Z even and N even, and therefore its last neutron is tightly bound (6.4 MeV). The nuclide $^{239}_{92}U$ is even-odd (Z even and N odd) and has its last, odd neutron weakly bound (4.9 MeV). Therefore, $^{236}_{92}U$ gains more excitation energy in capturing a zero-energy neutron than does $^{239}_{92}U$.

Since nuclear fission with uranium-235 can be initiated by low-energy neutrons, with an average 2.5 neutrons released in the fission, it is possible to extract useful energy from uranium. The energy released in exothermic nuclear reactions produced by particle bombardment from accelerators cannot be used in a practical way, since the number of reactions is typically very small. The total energy released in such reactions is much less than the total energy supplied to the many accelerated particles, of which only a small fraction causes reactions. The fission can, on the other hand, be made efficient since a *self-sustaining chain reaction* can be induced.

In essence, the neutrons from one fission reaction may initiate other fission reactions, with a further release of fission energy, which ideally continue until all the nuclear fuel, or fissionable material, is consumed. For the fission reactions to continue, once initiated, several conditions must be fulfilled. These conditions are achieved in a nuclear reactor. The engineering problems connected with nuclear reactors lie within nuclear technology, and we shall merely outline the physical principles on which reactor operation is based.

11–7 Neutrons and Nuclear Reactors

When neutrons pass through a material, two kinds of nuclear interaction usually dominate all others: neutron radiative capture (n, γ) and *elastic collisions* between the neutrons and the nuclei of the material. In certain materials, the capture cross section is so small that the neutrons interact with the nuclei of the material primarily in elastic collisions. In passing through such materials, called *moderators,* the incident, originally energetic, neutrons are slowed down, or *moderated.*

In any elastic collision between two particles, the greatest amount of kinetic energy is transferred from the incident particle to the target particle initially at rest when the masses of the two particles are equal. Thus, a neutron can lose all its kinetic energy when it collides head-on with a proton. Less energy is transferred in oblique collisions. Therefore, hydrogenous materials, such as paraffin and water, are effective in moderating neutrons. When neutrons collide with nuclei more massive than protons, they lose a smaller fraction of their kinetic energy, even in a head-on collision, and many such encounters are needed to slow them down.

Although neutrons are slowed down in a moderator, they are never brought completely to rest. The nuclei in a moderating material are in thermal motion at any finite temperature. It can be said that a collection of neutrons is in thermal equilibrium with the moderating material when a typical neutron is just as likely to gain kinetic energy as to lose it on colliding with a nucleus within the moderator. Such neutrons, which have a distribution of speeds like that of molecules in a gas, can be assigned a temperature equal to the temperature of the moderator. The average kinetic energy of neutrons in equilibrium with a moderator at a temperature T is given by $\frac{1}{2}mv^2 = \frac{3}{2}kT$.

Neutrons in thermal equilibrium with a moderator at room temperature, 300 K, are said to be *thermal neutrons.* Their average kinetic energy is 0.04 eV, their average speed is 2200 m/s, and their average wavelength is 1.80 Å. High-energy neutrons (several MeV) incident on a typical moderator such as graphite (carbon) or heavy water are *thermalized* in less than 1 ms. The most probable fate of moderated neutrons is capture by the nuclei of the moderator, since the capture cross section increases rapidly as the neutron energy falls. Recall that a free neutron is radioactive and decays into a proton and an electron (and an antineutrino) with a half-life of 11 min. The decay of a free neutron, although possible, occurs very infrequently in a material, because it must compete with the much faster process of neutron moderation and then capture.

The first self-sustaining nuclear-fission chain reaction was achieved by E. Fermi in 1942. The reactor used natural uranium (0.7 percent ^{235}U and 99.3 percent ^{238}U) as fuel and graphite as a neutron moderator. Although there are many different kinds of nuclear reactors, we show their basic features with a simple reactor that uses natural uranium and a graphite moderator and is based on the fission of uranium-235 by slow neutrons.

For a fission chain reaction to be self-sustaining, there must be, for each uranium atom split, at least one neutron that will split one more uranium atom. In the fission of uranium, each decay produces about 2.5 neutrons. Therefore, if the chain reaction is not to stop, no more than 1.5 neutrons can be lost. The important ways in which neutrons become unavailable for uranium-235 fission are capture without fission by uranium-238 (and to a lesser extent, uranium-235), capture by the moderator, and leakage from the interior of the reactor to the outside.

First consider the problem of neutron leakage. If the reactor (the fuel elements and the moderator) is a very small one, many of the neutrons produced in some initial fission reactions will leak out of the reactor (through the walls) before inducing further fission reactions. In a bigger reactor, the neutron losses are fewer and the fission reactions more numerous. The fission-reaction production rate is roughly proportional to the volume of the reactor, and the leakage rate is roughly proportional to the surface area of the reactor.

The fission cross section in uranium-235 increases as the neutron energy decreases (reaching 550 barns with thermal neutrons); on the other hand, the neutron-capture cross section of uranium-238 increases as the neutron energy also increases. Therefore, the problem in operating a reactor with natural uranium is to slow the high-energy neutrons (of a few MeV) emitted in the uranium-235 fission to thermal energies at which further fission reactions in uranium-235 are more likely, without losing the neutrons by capture in uranium-238 on the way down. These conditions are met by using a moderator to slow down (but not capture) the neutrons and by properly arranging uranium fuel blocks within the moderator.

The function of the moderator, then, is to slow down neutrons without capturing them. Although hydrogen atoms, whose mass essentially equals the mass of the neutron, cause the greatest fractional loss in the kinetic energy of neutrons, they are not always used as moderators because of the relatively high probability of the neutron-capture reaction ${}^{1}_{1}\text{H}(\text{n}, \gamma){}^{2}_{1}\text{H}$. Other moderator materials are heavy water (D_2O), beryllium (Be), and graphite (${}^{12}\text{C}$); most other light materials are not usable, because of their large neutron-capture cross sections. The fuel elements are distributed throughout the medium of the moderator. Under ideal conditions, a fast neutron from a fuel element escapes into the moderator and is slowed down, thereby avoiding capture (without fission) in uranium-238. Then the thermal neutron enters another fuel element, causing another fission in uranium-235. The whole process takes place in less than a millisecond.

When sources of neutron loss have been minimized, it is possible for the reactor to "go critical"; each fission reaction then leads to at least one more fission reaction. The power level of the reactor, or the rate at which fission reactions occur in it, can be controlled by inserting materials like cadmium, which has a very high neutron-capture cross section and therefore readily absorbs neutrons. Such materials are usually in the form of rods, called *control rods.* A reactor is said to be *subcritical* if on the average, each fission reaction produces *fewer* than one further fission; if that happens, the fission reaction is not self-sustaining. On the other hand, if each fission reaction produces more than one further fission, the reactor is said to be *supercritical;* an extreme example of a supercritical fission reaction is an atom bomb.

The control of reactors by mechanically actuated control rods would be virtually impossible if the only neutrons available were the *prompt neutrons,* those released at the instant of fission; there are also, however, *delayed neutrons* (0.7 percent), which are emitted by a few of the fission fragments, usually *after* one or more β^- decays have occurred. A delayed neutron can cause a further fission about 10 s after the fission that released it. This is in

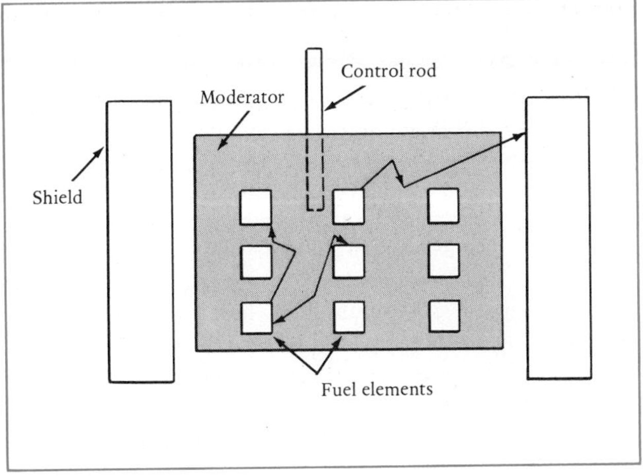

FIGURE 11–15. *Simple elements of a nuclear reactor.*

contrast with a prompt neutron, which causes a further fission in less than 10^{-12} s.

One example of a fission-fragment decay leading to a delayed neutron is

$$\underset{(56\text{ s})}{{}^{87}_{35}\text{Br}} \xrightarrow{\beta^-} {}^{87}_{36}\text{Kr}^* \rightarrow {}^{86}_{36}\text{Kr} + {}^{1}_{0}\text{n} \xrightarrow{\beta^-} {}^{87}_{36}\text{Kr} \xrightarrow{\beta^-} {}^{87}_{37}\text{Rb} \xrightarrow{\beta^-} {}^{87}_{38}\text{Sr}$$

Nuclear reactors are of many designs. They may differ in the following respects: the fuel (natural uranium, uranium enriched with uranium-235, other artificially produced fissionable materials), the moderator (water, graphite, beryllium), the distribution of fuel within the moderator (homogeneous, heterogeneous), the energy of neutrons producing fission (fast, intermediate, slow), and the heat exchanger (gas, water, liquid metals). Figure 11–15 is a schematic of a nuclear reactor.

Nuclear reactors can be classified according to their use: (1) for power generation, (2) as neutron sources, (3) for the production of radioisotopes, and (4) for the production of fissionable material:

1. The large kinetic energy of fission fragments in a nuclear reactor is a source of thermal energy, which can be extracted through a heat exchanger to do useful work, such as generating electric energy.
2. The interior of a reactor is a region in which the neutron flux can be as high as 10^{19} neutron/m$^2\cdot$s.
3. Such a flux can be used in experiments in physics or for irradiating materials so as to produce radioisotopes through (n, γ) reactions.
4. Materials such as uranium-238 and thorium-232, which do *not* undergo fission with low-energy neutrons, can be converted in a nuclear reactor into nuclides that undergo fission with thermal neutrons. Two such reactions are:

$${}^{1}_{0}\text{n} + {}^{238}_{92}\text{U} \rightarrow {}^{239}_{92}\text{U}^* \xrightarrow[(23\text{ min})]{\beta^-} {}^{239}_{93}\text{Np} \xrightarrow[(2.3\text{ days})]{\beta^-} \underset{(24{,}000\text{ yr})}{{}^{239}_{94}\text{Pu}}$$

$${}^{1}_{0}\text{n} + {}^{232}_{90}\text{Th} \rightarrow {}^{233}_{90}\text{Th}^* \xrightarrow[(23\text{ min})]{\beta^-} {}^{233}_{91}\text{Pa} \xrightarrow[(27\text{ days})]{\beta^-} \underset{(1.6\times10^5\text{ yr})}{{}^{233}_{92}\text{U}}$$

Uranium-238 and thorium-232 *cannot* be fissioned by thermal neutrons; but when they capture neutrons, the reactions lead to plutonium-239 and uranium-233, which can be fissioned by thermal neutrons. These two reactions lead to the possibility of a *breeder reactor.* In a breeder reactor, there are two fuel materials, one of which is fissionable (such as plutonium-239) and the other *fertile* (such as uranium-238), in that it can be converted in the reactor into fissionable material. In the fission of plutonium-239, three neutrons are released, on the average; of these, one must sustain the reaction producing the fission of a plutonium-239 nucleus. Of the remaining two neutrons, at least one must be captured by uranium-238, leading to plutonium-239, to maintain the same amount of fissionable fuel in the reactor. When more than one of these two neutrons is captured by uranium-238, the reactor can breed fissionable plutonium-239; that is, more fissionable material is produced than consumed.

11-8 Nuclear Fusion

The origin of energy radiated from the sun and other stars is a series of exoergic nuclear reactions. The atoms participating in such reactions in the interior of the star are completely ionized; all electrons have been removed from them. Such a collection of electrically charged particles—electrons and bare nuclei—is called a *plasma.* The particles are at a very high temperature (up to 10^8 K), move at high speeds, and collide frequently with one another. The average kinetic energy per particle, $\frac{3}{2}kT$, is of the order of 1 keV for $T = 10^7$ K. Despite the Coulomb repulsion between positively charged nuclei, some of the faster-moving nuclei may approach one another close enough to interact through nuclear forces, and nuclear reactions take place with high probability. A nuclear reaction that occurs by virtue of the increased thermal motion of the interacting particles at a high temperature is called a *thermonuclear reaction.*

One cycle of thermonuclear reactions releasing energy in stars is the proton-proton cycle:

$$^{1}\text{H} + {}^{1}\text{H} \rightarrow {}^{2}\text{H} + \beta^{+} + \nu$$

$$^{1}\text{H} + {}^{2}\text{H} \rightarrow {}^{3}\text{He} + \gamma$$

$$^{3}\text{He} + {}^{3}\text{He} \rightarrow {}^{4}\text{He} + 2\,{}^{1}\text{H}$$

This cycle, involving three distinct nuclear reactions, fuses four protons into an α particle, two positrons, and two neutrinos. The first reaction in this cycle, in which a positron is created in the collision of two protons, has a very small cross section. It occurs in the sun's interior, because the temperature there is about 2×10^7 K. The overall Q of the cycle is about 25 MeV, or approximately 6 MeV released for each nucleon participating in the reaction. (Since 200 MeV is released in a typical nuclear fission reaction, the energy per nucleon in a fission reaction is about 1 MeV.)

A second cycle of thermonuclear reactions, operating in some stars, is the *carbon cycle:*

$$^{1}\text{H} + {}^{12}\text{C} \rightarrow {}^{13}\text{N} + \gamma$$

$$^{13}\text{N} \rightarrow {}^{13}\text{C} + \beta^{+} + \nu$$

$$^{1}\text{H} + {}^{13}\text{C} \rightarrow {}^{14}\text{N} + \gamma$$

$$^{1}H + {}^{14}N \rightarrow {}^{15}O + \gamma$$

$$^{15}O \rightarrow {}^{15}N + \beta^{+} + \nu$$

$$^{1}H + {}^{15}N \rightarrow {}^{12}C + {}^{4}He$$

In this process, the carbon-12 nucleus acts merely as a catalyst; the process begins with one carbon-12 nucleus and ends with one carbon-12 nucleus. However, four protons are, in effect, fused into one α particle, two positrons, and two neutrinos. Since the particles entering and leaving the carbon cycle are the same as in the proton-proton cycle, the energy released is again about 25 MeV.

Why do the fission of the heaviest nuclides and the fusion of the lightest both result in highly exoergic nuclear reactions? How can both the splitting and the amalgamation lead to the release of nuclear energy? The answer can be found in Figure 10–10, which gives the average binding energy per nucleon as a function of the number of nucleons in a stable nuclide; both fission and fusion reactions lead to more tightly bound nuclear configurations. The fractional conversion of rest mass into nuclear energy is greater in a fusion reaction (0.66 percent) than in a fission reaction (0.09 percent).

Much attention has been given to the possibility of producing *controlled* thermonuclear-fusion reactions with the resultant very large energy release. A nuclear-fusion energy source has significant advantages over a nuclear-fission energy source. There is a virtually unlimited supply of fuel, the reactions do not result in radioactive wastes, and there is the possibility of generating electric energy more directly than through conventional heat exchangers and turbines.

Formidable technical difficulties, however, must be surmounted in achieving a practicable power source based on controlled nuclear fusion. Chief among these are the extraordinarily high temperatures that are required to overcome the Coulomb repulsion between the interacting nuclei; a very hot plasma must be confined for long periods, so that many collisions can take place between the plasma particles. Ordinary containers are unsuitable, not primarily because they would be melted by the very hot plasma, but because they would chill the plasma below the temperature of spontaneous nuclear fusion.

The high-temperature plasma in a thermonuclear-fusion reactor can be prevented from striking the walls of its container by means of magnetic fields in the same way that charged particles are trapped within the earth's Van Allen belts by the earth's magnetic field.† The design of such containers, called magnetic bottles, is well advanced. A plasma at a temperature of millions of degrees will exert an uncontainable pressure unless its density is very low indeed; therefore, the pressure of the unheated plasma must be no greater than about 10^{-4} atm.

In addition to magnetic confinement of thermonuclear fuel, the other method of "igniting" a thermonuclear fusion reaction is through inertial confinement. In the inertial confinement method, a fuel pellet is imploded by an external energy source; the small fuel pellets may, for example, be bombarded from all sides by very high energy electrons, ions, or laser pulses.‡

† See Weidner and Sells, *Elementary Classical Physics*, 2d ed., Section 29–3.

‡ See for example Yonas, G., "Fusion Power with Particle Beams." *Scientific American* 239 (November 1978): 50–61.

Among the reactions possible with isotopes of hydrogen (^2H, deuterium; ^3H, tritium) are the following:

$$^2H + {}^3H \rightarrow {}^4He + n \qquad Q = 17.6 \text{ MeV}$$

$$^2H + {}^2H \rightarrow {}^3He + n \qquad Q = 3.2 \text{ MeV}$$

$$^2H + {}^2H \rightarrow {}^3H + {}^1H \qquad Q = 4.0 \text{ MeV}$$

Because the Coulomb barrier between charged particles increases with increasing nuclear charge, a thermonuclear reaction with isotopes of hydrogen requires a lower temperature than such a reaction with other elements. Deuterium is particularly attractive as a nuclear-fission fuel because it is readily available in almost unlimited quantity; it is found, for example, in seawater, in which there is one DHO molecule for every 6000 H_2O molecules.

11-9 Ionization and Absorption of Nuclear Radiation

The rest of our consideration of nuclear reactions and devices will concern the physical principles underlying the instruments used in nuclear physics—devices for detecting particles, recording their tracks, measuring their momentum, velocity, and mass, and accelerating them to higher energies. We first consider the detection and control of nuclear particles, processes that pose difficult problems simply because the particles are so minute. If a particle is identified and known to be at a certain place at a certain time, we can say that we have detected it. The detecting instrument must be of such sensitivity that a minute change produced by the particle's presence will influence a large-scale readily discernible characteristic of the instrument.

Detectors of such charged particles as electrons and protons depend basically on the electric charge these particles have. The photon has no electric charge itself, but in interacting with materials, it produces electrons through the photoelectric effect, Compton effect, or pair-production process, and these electrons produce, through their charge, effects ascribable to the initiating photon. Similarly, the neutron, with no electric charge, is detected by the charged particles that it produces in interacting with materials—through a nuclear reaction, for instance, in which a proton is released, or through a collision with a proton. In short, detection of particles depends fundamentally on the effects of electric charge, directly or indirectly.

A charged particle traversing a medium and coming to rest—that is, a charged particle being absorbed—loses energy as it excites atomic electrons to higher-lying states, and most important, as it ionizes atomic electrons. The ionizing particle loses 3 to 40 eV, depending on the absorbing medium, in producing one ion pair consisting of the detached electron and the positively charged atom. If the ionizing particle is massive compared with an electron, the path of the ionizing particle is essentially straight, with a trail of ions along its wake. The probability that such a particle will collide with a nucleus is minute; and the particle is effectively undeflected in collisions with electrons. Any given massive ionizing particle has, for a particular absorbing medium, a well-defined range, or total path length, that depends on the particle's initial kinetic energy. For example, a 5-MeV proton has a range of 34 cm in air and 0.019 cm in aluminum. The path of an ionizing electron, on the other hand, which is deflected strongly by atomic electrons it meets, is tortuous.

One does not associate a specific range with absorption of photons. The reduction in the intensity I of a beam of many photons through the photoelectric, Compton, and pair-production effects is governed by the exponential absorption relation (Section 4–7), $I = I_0 e^{-\mu x}$, where the reciprocal of the absorption coefficient, $1/\mu$, gives the distance over which the number of photons drops to $(1/e)$th of the initial number.

Note the fundamental difference between the absorption of a photon beam and the absorption of charged particles. A photon loses *all* its energy in a *single* collision, it *always* moves at speed c, and its absorption *cannot* be characterized by a sharply defined range.† A charged particle, on the other hand, loses its energy little by little in many collisions, its speed gradually decreases, and its absorption is characterized by a definite range. Photon absorption is accompanied by ionization effects, which can be detected electrically. The electrons (and positrons in pair production), to which energy is imparted in photon collisions, can themselves produce ions in the absorbing material. Photons are far more penetrating than electrons or heavy charged particles of the same energy.

11-10 Detectors

Every detector of nuclear radiation, whether of charged or uncharged particles, ultimately yields an electric signal, or voltage pulse, to be fed into a counting circuit, which then registers the arrival of the particle in the detecting device. The medium in which the incident particles produce effects that are converted finally to electric signals can be of various forms. Gas-filled detectors and scintillation detectors are two of the kinds. The commonly used solid state detector depends on the special properties of semiconducting materials; this was described in Section 9–3.

GAS-FILLED DETECTORS. The simplest detector sensitive to the ionization effects of nuclear radiation through a gas is an electroscope; an initially charged electroscope loses its charge when ions produced by nuclear radiation migrate to the charges on the electroscope and neutralize them.

In the gas-filled detector shown in Figure 11–16, the chamber, filled with a gas, has an outer negatively charged cylindrical electrode and a central positively charged wire electrode. The electric field between the two electrodes is highly inhomogeneous, very strong near the wire and far weaker at the cylinder. Particles enter the detecting chamber through a thin window. The ionization effects of particles entering the chamber are registered as an electric current, which arises from the arrival of electrons and ions at the two electrodes. Then the detector acts as an *ionization chamber*. The combined ionization effects of many nuclear particles, including photons, entering the chamber produces a nearly constant current, whose magnitude is a direct measure of the intensity of the ionizing radiation.

If the potential difference between the electrodes is raised to a few

† In a Compton collision, the incident photon is annihilated and the scattered photon is created.

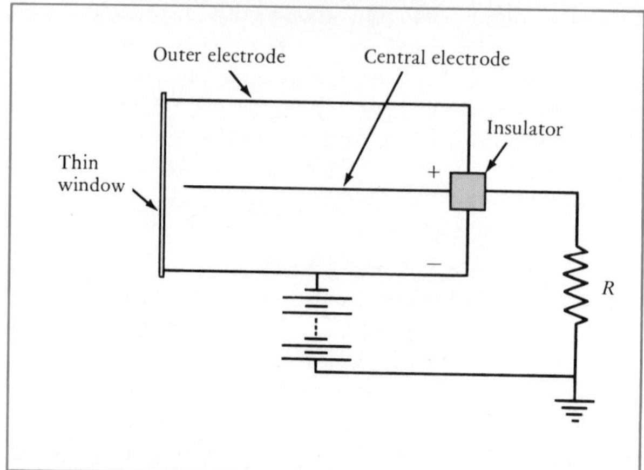

FIGURE 11–16. *Elements of a gas-filled detector.*

hundred volts, the detector may act as a *proportional counter.* Here ion multiplication within the gas takes place. When an initial ion pair—typically an electron and a positive ion—is produced by nuclear radiation, the electrons are strongly attracted to the central wire electrode and the positive ions toward the outer electrode. An electron can gain enough kinetic energy, as it is accelerated toward the wire, to produce still more ions in its collisions with gas molecules. Through ion multiplication, a single ion pair can produce up to 10^6 secondary ion pairs, thereby producing a current pulse whose magnitude is directly proportional to the energy of the ionizing particle.

For still higher potential differences between the electrodes, the detector acts as a G-M (Geiger-Müller) counter. *Any* ionizing particle producing an initial ion pair triggers an avalanche of ion pairs through ion multiplication, so the current pulses are *independent* of the energy of the initiating particle.

SCINTILLATION DETECTORS. A scintillation detector depends on the circumstance that certain materials, called *phosphors,* emit visible light when struck by particles or irradiated by ultraviolet light or x-rays. When a particle collides with a phosphor, it excites an electron into a higher energy state. The deexcitation of the phosphor and the return to the ground state is accompanied by the emission of photons lying in the visible region of the spectrum. A familiar example is the cathode-ray oscilloscope, or television picture tube, in which high-speed electrons strike a phosphor and cause the emission of visible radiation.

The schematic diagram of Figure 11–17 is a scintillation detector with the associated photomultiplier. The phosphor, a transparent material, may be sodium iodide with a trace of thallium (for γ rays), an organic substance, such as anthracene (for electrons), or zinc sulfide with a trace of silver (for massive charged particles, such as α particles). The phosphor produces light flashes after its atoms are excited by collisions with particles or photons. The phosphor is sealed in a light-tight envelope, and photons reach the photocathode of the photomultiplier tube, possibly after reflections within the phosphor. A photon striking the cathode undergoes a photoelectric collision, and one electron is dislodged from the cathode surface. This photoelectron is accelerated by a potential difference of about 100 V to the first *dynode* of the photomultiplier tube. When it collides with the surface of the dynode,

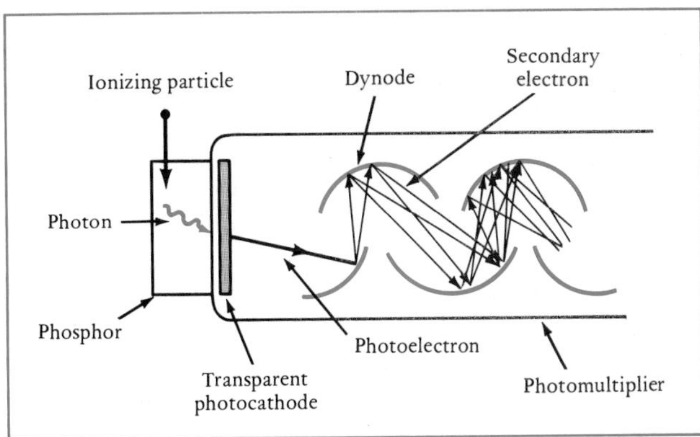

FIGURE 11–17. *Schematic diagram of a scintillation detector with photomultiplier tube.*

now with kinetic energy of at least 100 eV, secondary electron emission occurs, and two or more electrons are released from the surface by the kinetic energy they gain from the initial electron. The secondary electrons are then accelerated to the second dynode, through another 100-V potential difference, and multiplication of the electrons by secondary emission again occurs. A typical photomultiplier tube has 10 dynodes, or 10 stages of electron amplification. The original photoelectron produces at the final dynode a readily measured pulse of current owing to the arrival of millions of electrons.

An important feature of the scintillation detector is that the output voltage pulse from the photomultiplier is very nearly proportional to the energy of the particle or photon that initiates scintillation in the phosphor; not only can particles be detected with a scintillation detector, but their energies can also be measured. Scintillation detectors have other advantages; they are capable of handling very high counting rates, with pulse durations as short as 1 ns (10^{-9} s), and their efficiency in counting γ rays approaches 100 percent.

A scintillation detector together with a pulse-height analyzer, an electronic device that sorts the output pulses from a photomultiplier according to their size, constitutes a *scintillation spectrometer*, by which the energies of monochromatic γ rays in particular can be measured straightforwardly and very precisely. The magnitude of the output pulse is directly proportional to the kinetic energy of the *electrons* produced by the photoelectric effect, the Compton effect, and electron-positron pair production (Chapter 4), three processes in which γ rays interact with the scintillation material.

Example 11–4. Consider the pulse-height distribution, or spectrum, from a scintillation detector and pulse-height analyzer shown in Figure 11–18. The number of detected pulses for a given pulse height is displayed as a function of the pulse height, which is measured in volts but shown here as kinetic energy of the electrons in MeV. The monochromatic γ-ray source, ^{40}K, emits 1.48 MeV-photons. Interpret the peaks in Figure 11–18.

The peak of highest energy originates from the photoelectric effect. The binding energy of atomic electrons is only a few electron volts, small compared with the photons' energy; consequently, the photoelectrons have essentially the entire photon energy. The energy at the so-called photopeak is almost exactly the same as the γ-ray energy. The second peak arises from the Compton effect. When a 1.48-MeV photon collides head-on

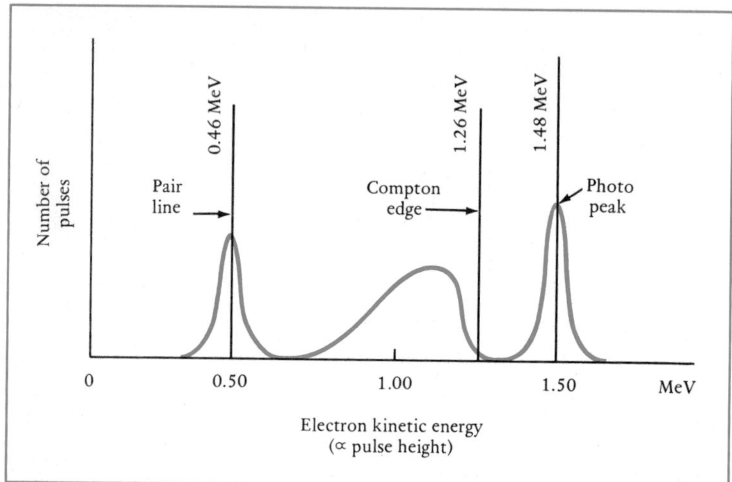

FIGURE 11–18. *Idealized pulse-height spectrum for a scintillator detecting 1.48-MeV photons.*

with an essentially free atomic electron, the Compton electron recoils in the forward direction with a kinetic energy of 1.26 MeV (computed from Equation 4–15), while the scattered photon travels in the reverse direction with the remaining energy, 1.48 − 1.26, or 0.22 MeV. Non-head-on Compton collisions produce less energetic electrons and more energetic scattered photons. Therefore, the Compton peak has a relatively sharply defined high-energy edge—in the figure, corresponding to 1.26 MeV; it trails off gradually on the low-energy side because of Compton collisions producing electrons with less than the maximum kinetic energy. The third peak in the scintillation spectrum originates from electron-positron pair production. Since the rest energy of an electron or a positron is 0.51 MeV, a total of 1.02 MeV is needed to bring a pair into existence. The difference in energy, 1.48 − 1.02 = 0.46 MeV, is the sum of the kinetic energies of the electron and positron (see Section 4–5); this total kinetic energy, after exciting scintillator electrons, produces the pulses at the pair-production peak.

The relative sizes of the three peaks described depend on the energy of the photons and the size, shape, and identity of the scintillation material. Still other peaks may be found. For example, γ rays with an energy above the threshold energy of 1.02 MeV for pair production will produce positrons, and if these positrons become annihilated by electrons before leaving the scintillator, annihilation photons of 0.51 MeV are produced (Section 4–5); these photons also can give rise to photoelectric and Compton peaks.

11-11 Track-Recording Devices

Among the devices by which one can photograph in three dimensions or otherwise record the trail of ions that a charged particle produces as it traverses a medium are the bubble and cloud chambers, the spark chamber, and nuclear photographic emulsions.

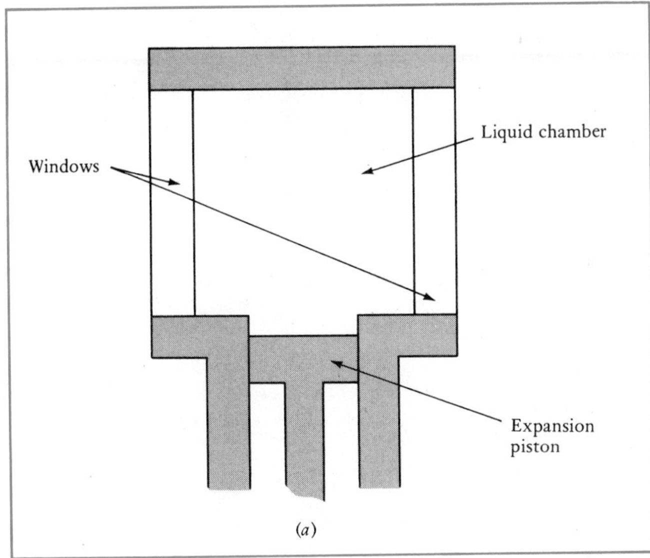

FIGURE 11–19. *(a) Schematic diagram of bubble chamber. (b) The 15-ft bubble chamber at the Fermi National Accelerator Laboratory. High-energy particles from the 500-GeV accelerator enter from the right. Seven cameras for photographing tracks are located in the top of the chamber; the expansion piston is at the bottom. (Courtesy, Fermi National Accelerator Laboratory, Batavia, Illinois)*

BUBBLE AND CLOUD CHAMBERS. The earliest of the particle track-recording devices is the cloud chamber, which C. T. R. Wilson invented in 1907. It has been superseded in all present-day high-energy experiments by the bubble chamber, which D. A. Glaser invented in 1952.

The operation of the bubble chamber depends on the behavior of superheated liquid, a liquid that has been brought by a sudden adiabatic expansion (in which no thermal energy enters or leaves the liquid) to the condition in which its temperature is just above the known boiling point of the liquid. Bubbles will then form readily at points within the liquid that are disturbed. A simple example is the formation of bubbles when the cap of a chilled bottle of beer is removed.

In a bubble chamber, the trail of ions that an energetic charged particle leaves becomes a trail of small bubbles; when this trail is photographed shortly after the formation of the bubbles, a permanent record is made of the track of the charged particle through the superheated liquid. The use of two or more cameras, for stereophotography, permits the particle's path to be analyzed in three dimensions.

A typical bubble chamber uses liquid hydrogen, which boils at 20 K at atmospheric pressure. When the pressure is increased to 5 atm, the temperature of the liquid rises to 27 K. Then, if the pressure is suddenly reduced, the liquid becomes superheated, with its temperature momentarily greater than the boiling point.

The basic elements of a bubble chamber are shown in Figure 11–19a, liquid in a chamber with a transparent cover and with an expansion piston. Figure 11–19b is a photograph of a large bubble chamber.

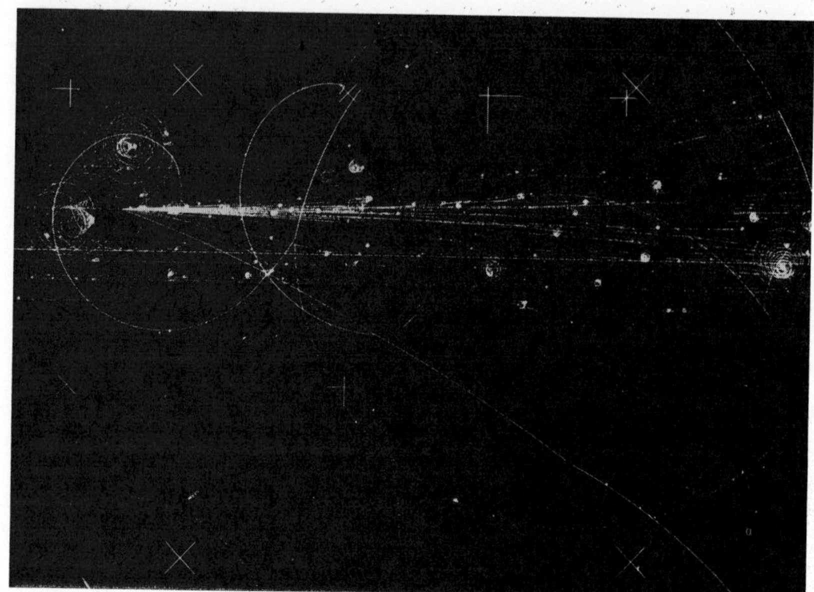

FIGURE 11–20. *A 300-GeV proton from the Fermilab accelerator collides with a nucleon in a 30-inch hydrogen bubble chamber and produces 26 particles. (Courtesy, Fermi National Accelerator Laboratory, Batavia, Illinois)*

The path of a single particle is of limited interest. Much more important are nuclear events, in which an incident particle collides with particles within the chamber, possibly creating new particles, or in which an unstable incident particle decays, or explodes, in flight into other particles. A nuclear event is then typically one in which the tracks of both incident and emerging particles appear. (A photon or an electrically neutral particle leaves no tracks.) By measuring the momenta of the particles (by measuring their curvature r in a magnetic field B through the relation $p = mv = qrB$ and the relative directions of the tracks) and by applying the laws of momentum and energy conservation, one can analyze the event in detail. Figure 11–20 is a bubble-chamber photograph.

Triggering devices, particularly counters in coincidence, can be used with bubble chambers to ensure that photographs are taken only when events are interesting. The dimensions of bubble chambers may be of the order of meters; and they are always operated with external magnetic fields to separate positively and negatively charged particles and to measure their momenta.

The cloud chamber is like the bubble chamber in operation. A supersaturated vapor replaces the superheated liquid; droplets form when the vapor expands, rendering it supersaturated. The principal shortcomings of the cloud chamber are its relatively slow operation and the low density of vapor relative to liquid.

THE SPARK CHAMBER. In the photographic type of spark chamber, the path of a charged particle is registered by a series of sparks, which can be photographed. In its simplest form, a spark detector operates as follows. A high potential difference is applied across a pair of electrodes immersed in an inert gas. A charged particle passes through the region, forming ions, and the ions multiply through intermolecular collisions under the influence of

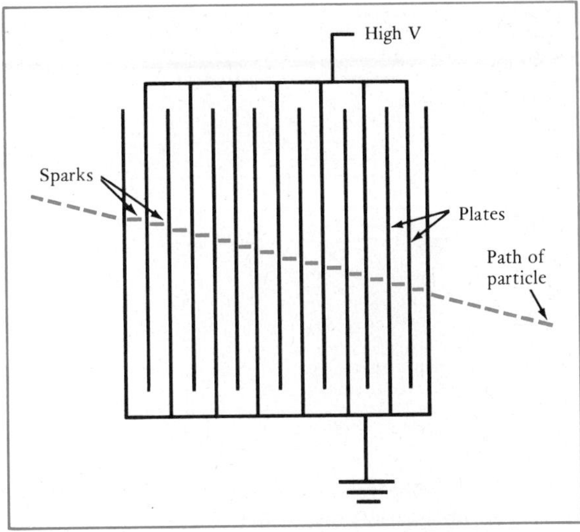

FIGURE 11–21. *Schematic diagram of a spark chamber.*

the accelerating electric field, until a visible spark flashes between the electrodes. Residual ions then are swept away by a clearing electric field, and the spark counter is ready to register again.

A spark chamber is merely a collection of spark counters. Parallel plates separated by several millimeters are immersed in an inert gas such as neon; see Figure 11–21. The region between the plates is viewed edge-on by a camera. *After* the particle whose track is to be photographed has passed transversely through the plates, as signaled by separate counters, an electric potential of tens of kilovolts is applied to alternate electrodes; the ions initially formed by the passing charged particle multiply, sparks form in the gaps between adjacent plates, and a photograph is taken. As Figure 11–22 shows, two or more tracks can be registered simultaneously in a spark chamber. Although the spark chamber lacks the very high spatial resolution

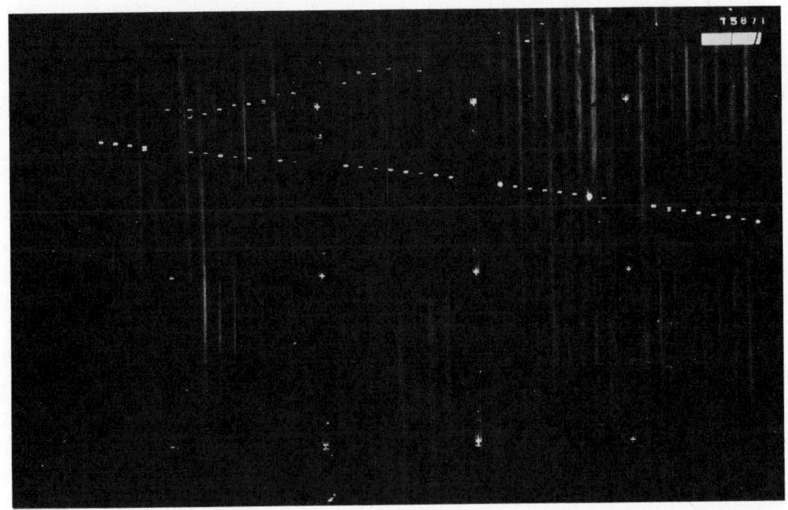

FIGURE 11–22. *Two tracks in a spark-chamber photograph used in experiments to establish the existence of the neutrino associated with the muon. (Courtesy, Brookhaven National Laboratory)*

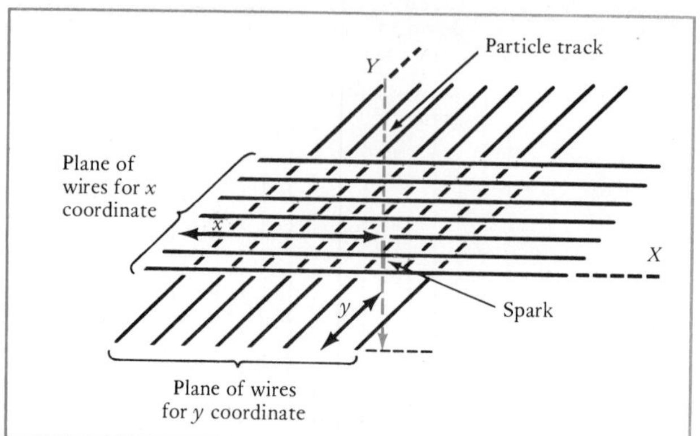

FIGURE 11–23. *Two layers of uniformly spaced, parallel wires in a wire spark chamber.*

of the bubble chamber, its special advantage is the very short insensitive time interval between successive firings, or dead time (as short as 10 μs); consequently, the ratio of interesting events to background events available from high-intensity beams is high.

A more sophisticated device is the *wire spark chamber,* in which uniform electrode plates of the photographic spark chamber are replaced with layers of uniformly spaced parallel wires. Wires in alternate layers are parallel and connected to a high potential, with the second set of wires at right angles to the first, which is connected to ground. When a spark initiated by the passage of a charged particle jumps between two wires in adjacent layers, electric signals travel at constant speed along the two wires; the respective times of arrival at the wire ends of the two signals, typically indicated by a magnetic effect, correspond to the respective distances of the spark from the wire ends (see Figure 11–23). Indeed, a series of sparks resulting from the passage of a particle through the several layers of the spark chamber produce corresponding signals in still other wires. The identity of the wires carrying signals and the times of arrival of the signals can then be fed into a computer, which can reconstruct the path of the particle in three dimensions. Since the wire spark chamber is completely electronic, there is no delay arising from the development of a photographic record, and the system can be triggered to record events as frequently as 100 times a second.

NUCLEAR EMULSIONS. A photographic emulsion used to record the tracks of charged particles is called a *nuclear emulsion;* see Figure 11–24. The emulsion, usually thicker and more sensitive than emulsions used in ordinary photography, renders the trail of ions visible on development, because a latent image can be produced by the track of a charged particle traversing the sensitive volume. A particle's range in a nuclear emulsion depends on its energy, and the measurement of the range can be used to determine the particle's energy. In a typical emulsion, a proton of 10 MeV produces a track 0.5 mm long, whereas one of 20 MeV produces a track 2.0 mm long. The mass of the particle is related to the density of grains of the emulsion along the track; moreover, the number of grains increases as any one particle slows down. Like bubble and cloud chambers, a nuclear emulsion permits collisions and nuclear reactions to be analyzed through the measurement of the energies, masses, and relative directions of the participating

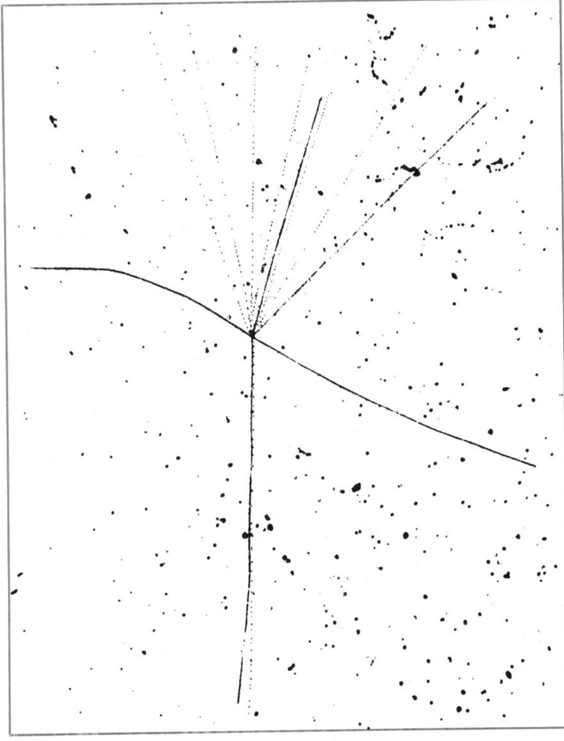

FIGURE 11–24. *Tracks produced in a nuclear emulsion plate by the interaction of high-energy cosmic ray particles with emulsion grains. (Courtesy, Brookhaven National Laboratory)*

particles. Although examining emulsions for particle tracks by microscope is tedious, these track-recording devices have the advantages of small size, light weight, and simplicity. Moreover, a nuclear emulsion is continuously sensitive.

11–12 Devices for Measuring Velocity, Momentum, and Mass

A charged particle is appreciably influenced in its motion through a vacuum only by an electric force $\mathbf{F} = Q\mathbf{E}$ and a magnetic force $\mathbf{F} = Q\mathbf{v} \times \mathbf{B}$ arising from external electric and magnetic fields. All devices for measuring velocity, momentum, and mass involve merely using electric and magnetic fields, singly or in combination, to determine the path of a charged particle. Each device consists of three parts: a source or a beam of charged particles, a region in which electric or magnetic fields act on the particles, and a detector for registering their arrival. What is done in every instrument is to set up, so to speak, an obstacle course for the charged particles in such a way that if the particles succeed in moving from the source to the detector, one can infer, from knowing how the electric or magnetic fields act on the particle, some quantity of interest, like the velocity of the particle.

THE VELOCITY SELECTOR. Consider first a velocity selector; see Figure 11–25. A narrow beam of charged particles is projected into a region of space where a uniform electric field **E** acts to the left, and simultaneously a

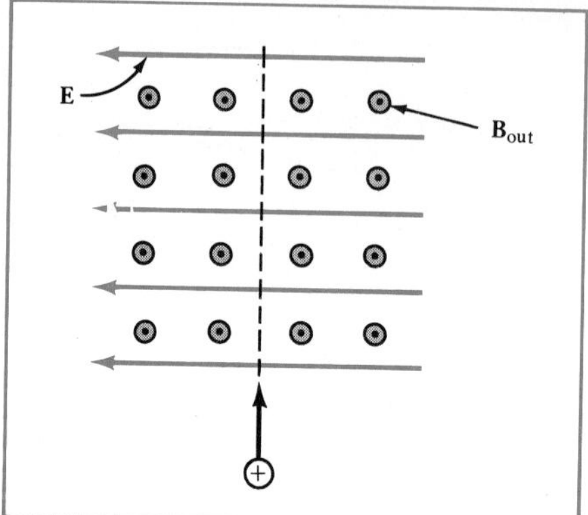

FIGURE 11–25. *A velocity selector, consisting of crossed electric and magnetic fields.*

uniform magnetic field **B** is applied in the direction out of the paper. The incident beam is composed of particles that may have a variety of masses, charges (magnitude and sign), and velocities. A particle of mass m and charge $+Q$ entering the region of the *crossed* electric and magnetic fields **E** and **B** at right angles to them is acted on by an electric force $Q\mathbf{E}$ to the left and a magnetic force QvB to the right, where v is the speed of the particle. If the particle is to travel through the selector undeflected, the net force on it must be zero; that is,

$$QE = QvB$$

$$v = \frac{E}{B} \tag{11–17}$$

Only those particles with speeds equal to the ratio E/B will emerge from the selector undeflected, despite differences in mass or in sign or magnitude of electric charge.

THE MOMENTUM SELECTOR. Only a uniform magnetic field is needed for measuring the momentum of a charged particle. This field is directed into the paper in Figure 11–26, and negatively charged particles move at right angles to the magnetic field lines. The magnetic force, acting at right angles to the velocity, deflects the particle into a circular path of radius r, where

$$QvB = \frac{mv^2}{r}$$

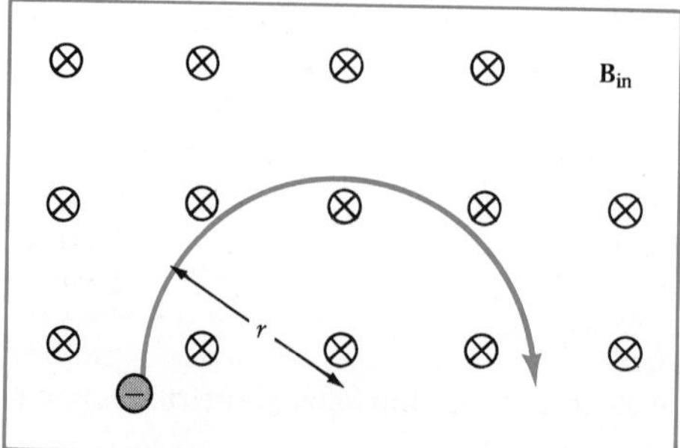

FIGURE 11–26. *A momentum selector, consisting of a uniform magnetic field.*

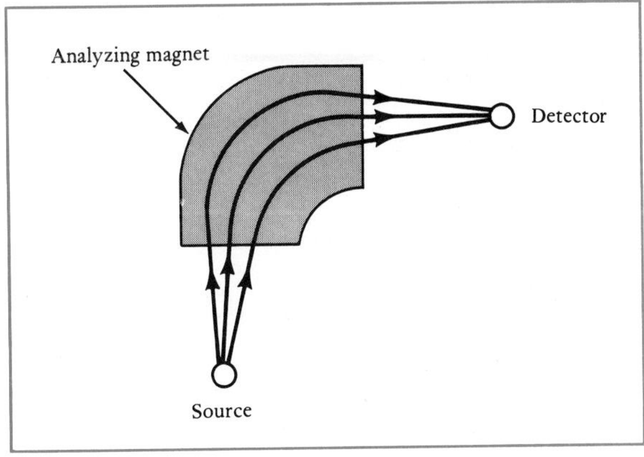

FIGURE 11–27. *An analyzing magnet focusing the particles diverging from the source and converging on the detector.*

$$mv = QBr \tag{11–18}$$

The momentum p is directly proportional to the radius r; all particles having the same charge Q and momentum mv will move in paths having the same radius of curvature. Recall that the mass m appearing in Equation 11–18 is the *relativistic* mass, and mv the relativistic momentum (see Section 3–1). The quantity Br, to which the relativistic momentum is proportional for particles of a given charge Q, is sometimes called the *magnetic rigidity.*

If a charged particle's identity—its charge Q, rest mass m_0 (or rest energy E_0)—is known, then a determination of its momentum $p = QBr$ through a measurement of its curvature r in a known magnetic field B permits the particle's relativistic kinetic energy $E_k = E - E_0$ to be computed directly. One merely applies the general relativistic relation between energy and momentum:

$$E^2 = (pc)^2 + E_0^2 \tag{3–15}$$

Magnetic-momentum analyzers are usually designed to accept particles traveling a variety of paths from source to final detector; see Figure 11–27. Particles diverge from a small source, travel in circular arcs through a uniform magnetic field, and then converge to a small region at the detector. The beam of particles is focused, as well as deflected, by the magnetic field; the source and detector can be thought of as equivalent to an optical object and image. The obvious advantage of such an arrangement is that more particles can be accepted from the source and focused on the detector, thereby improving the sensitivity of the magnetic analyzer.

One form of magnetic spectrograph is the *β-ray spectrometer.* It can be used to measure the momentum of electrons emitted from nuclei with energies up to a few MeV or to measure the momentum of electrons released in photoelectric or Compton collisions of x-rays and γ rays.

The kinetic energy of a high-energy particle is most easily determined by measuring its momentum with a magnetic field. In principle, one could measure the energy of electrons of 1 MeV by finding that they were brought to rest by a retarding potential of 1 million volts. This is difficult in practice, and one must resort to the indirect way—determining the energy of a very-high-speed particle by measuring its momentum.

THE MASS SPECTROMETER. A device for measuring the mass of an ionized atom is a mass spectrometer.

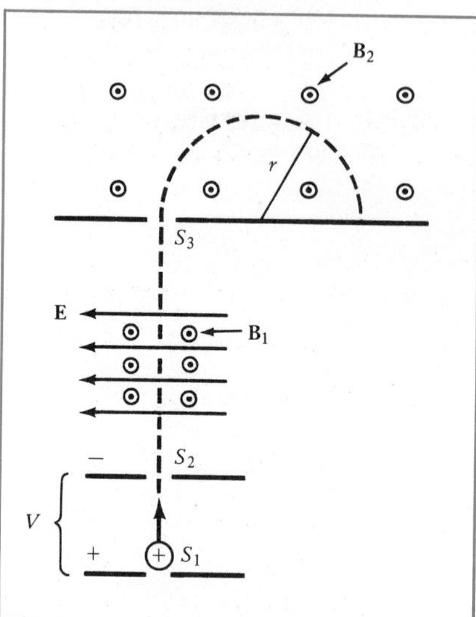

FIGURE 11–28. *Simple form of a mass spectrometer, consisting of a velocity selector and a momentum selector.*

Since a particle's linear momentum p is mv, it is obvious that a mass selector, or mass spectrometer, can be constructed by combining a velocity selector and a momentum selector. A simple type is shown schematically in Figure 11–28. Ions from a source pass through a slit S_1 and are accelerated through a potential difference V. After they pass through slit S_2, they enter a velocity selector. Only those ions moving with a velocity E/B_1 emerge through slit S_3, where E is the uniform electric field between vertical plates and B_1 the uniform magnetic field, directed out of the paper and confined to the region of the velocity selector. The surviving ions leaving S_3 enter a second uniform magnetic field B_2, directed out of the paper, and are deflected so as to move in a circle of radius r. We have, from Equations 11–17 and 11–18,

$$\frac{m}{Q} = \frac{B_2 r}{v} = \frac{B_2 r}{E/B_1}$$

from which the mass-to-charge ratio m/Q can be computed directly. If the charge of the ion is known, the mass itself can be evaluated. The mass m is directly proportional to the radius r. When ions of various masses fall on a photographic plate (*mass spectrograph*), the mass spectrum of the ions is recorded. Alternatively, if the ions are collected in a detector located behind a slit at a fixed distance $2r$ from the entrace slit S_3, a plot of collector current versus variable magnetic field B_2 yields the mass spectrum.

11-13 Nuclear Accelerators

To produce nuclear reactions, particles with initial kinetic energies typically of the order of several MeV are required. Similarly, scattering experiments to explore the forces between nuclei, or even nucleons, require that energetic particles be directed at targets. Apart from cosmic rays and naturally radioactive materials, particles with energies measured in keV, MeV, and even GeV are attainable only to the degree that such particles have been accelerated with machines. The highest-energy particles, those with energies measured in hundreds of GeV, are used to study particle properties, and they can create copious numbers of more exotic elementary particles.

An ideal accelerating machine produces a beam of charged particles with a well-defined high energy and with a high beam intensity, or particle flux. The beam energy must be high; only then can the nuclear structure be appreciably changed by particles colliding with target nuclei or creating new particles. The beam intensity should ideally be high, because the probability of a collision between an incoming particle and a target nucleus is very small owing to the extremely small target area.

All charged-particle accelerators use the principle that a charged particle has its energy changed when it is acted on by an *electric* field. A *constant* magnetic field does *no* work on a moving particle and cannot change its energy; on the other hand, a *changing* magnetic field produces an electric field, which in turn can accelerate a charged particle.

Before describing the basic types of accelerators, we point out two formidable technical problems that must be solved in the design of any of them. They are (1) maintaining a very high *vacuum* in the interior of the machine, and (2) *focusing* the beam of accelerated particles with electric or magnetic fields. A high vacuum reduces the probability of collisions with gas molecules and the consequent loss of useful beam intensity. Focusing ensures that those accelerated particles that deviate slightly from the ideal design path (which may be as long as many miles between the ion source and the target) will be returned to the path and so kept in the useful beam. Although high-vacuum and focusing problems are critical in the design of all accelerators, we shall concern ourselves only with the basic principles underlying the acceleration of particles.

The two general classes of charged-particle accelerators are the *linear accelerators* (referred to colloquially as "linacs"), in which the charged particles move along a straight line, and the *cyclic accelerators,* in which the charged particles move in curved paths and are recycled.

LINEAR ACCELERATORS. The machine most used for accelerating particles along a straight line by applying a *single* large potential difference is the Van de Graaff electrostatic generator, which R. J. Van de Graaff invented in 1931. Such a machine can accelerate singly charged particles to energies of about 30 MeV and particles that are multiply charged to correspondingly higher energies. The chief advantage of a Van de Graaff accelerator is its ability to produce large beam intensities (to a few milliamperes) and precisely controlled particle energies (to within 0.1 percent). Many hundreds of Van de Graaff accelerators are currently in operation for medical, industrial, and research applications throughout the world. The basic principle on which the machine is designed is that electric charge placed within a hollow metal conductor will always move to the outer surface, irrespective of the quantity of charge already existing on that surface.†

The *tandem* Van de Graaff accelerator is a machine in which a *single* high electric potential difference is used *twice* by changing the sign of the

† See Weidner and Sells, *Elementary Classical Physics,* 2d ed., Section 25–7.

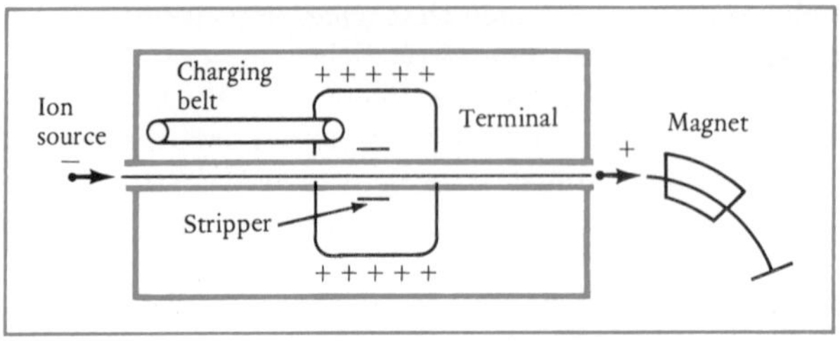

FIGURE 11–29. *Principal parts of a tandem Van de Graaff accelerator.*

particle's charge midway through the acceleration. For example, negative ions of hydrogen, H^-, each consisting of a proton and *two* electrons, are accelerated from ground potential across a potential difference of say 10 MV; each acquires a kinetic energy of 10 MeV midway through the tandem Van de Graaff machine. Then the two electrons are stripped off by passing through a thin foil or a gas (the stripper). The bare, positively charged proton is then accelerated through a 10-MV potential difference back to ground potential, thereby acquiring a final kinetic energy of 20 MeV. See Figure 11–29.

Another form of linear accelerator, one suitable for accelerating particles to kinetic energies of tens of GeV, can be designated a *waveguide linear accelerator,* since the particles are accelerated by the electric field of an electromagnetic wave that is guided through a long, hollow conductor. Particles ride, so to speak, on the traveling electromagnetic waves, or more particularly, are accelerated along the axis of the waveguide tube by the electric field also along the axis. A prime example of such an accelerator is the Stanford linear accelerator (SLAC). It is over 2 miles long, and it accelerates electrons to final kinetic energy of 20 GeV; the electromagnetic energy is provided by 245 klystron microwave oscillators operating at a frequency of 2.9 GHz. See Figure 11–30.

CYCLIC ACCELERATORS. In cyclic accelerators, multiple accelerations are given to charged particles that are restricted to motion along circu-

FIGURE 11–30. *The 20-GeV Stanford electron linear accelerator. Pictured is the target area at the end of the two-mile-long accelerator, where electrons are deflected and magnetically analyzed and directed to such devices as spark and bubble chambers. (Courtesy, Stanford Linear Accelerator Center, Stanford University)*

lar arcs by a magnetic field. Since we shall be discussing particles of very high energy, we shall consider the general relativistic relations.

The relativistic momentum **p** of a particle that has charge Q and relativistic energy $E = mc^2$ and is moving at right angles to a magnetic field **B** in a circular arc of radius r is given by Equations 3–27 and 3–9:

$$p = mv = \frac{Ev}{c^2} = QBr \tag{11–19}$$

Since only the particle's charge Q is fixed, for the particle's energy E to be increased, the magnetic field B or path radius r or both must increase.

The angular velocity ω of the particle in a circular arc is by definition $\omega = v/r$. By Equation 11–19, the angular frequency can be written

$$\omega = \frac{v}{r} = \frac{QBc^2}{E} \tag{11–20}$$

The frequency $f = \omega/2\pi$, known as the *cyclotron frequency,* is

$$f = \frac{QBc^2}{2\pi E} \tag{11–21}$$

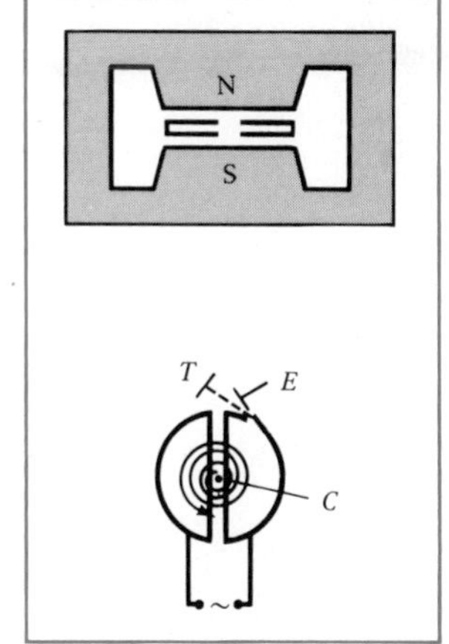

FIGURE 11–31. *Cyclotron accelerator: (a) top view; (b) side view.*

The first cyclic accelerator was the cyclotron that E. O. Lawrence and M. S. Livingston invented in 1932. The cyclotron operates strictly at nonrelativistic energies; it accelerates such particles as the proton, deuteron, and α particle to kinetic energies of tens of MeV, for which the particle's kinetic energy is far less than its rest energy. The general relativistic relation for the cyclotron frequency (Equation 11–21) reduces in the nonrelativistic domain to the following. For $v \ll c$ and $K \ll E_0$, we find

$$f_0 = \frac{QBc^2}{2\pi E_0} = \frac{QB}{2\pi m_0} \tag{11–22}$$

Note that f_0 depends on the charge-to-mass ratio Q/m_0 and the magnitude of the magnetic field **B** but not the particle's speed or the radius of its circular path. Thus, all particles of a given kind circle the magnetic field lines at the same frequency, quite apart from differences in their speeds or energies (nonrelativistic).

In the cyclotron, a charged particle is subjected to *constant* magnetic field while it is accelerated each half-cycle by an electric field.† Ions such as protons, deuterons, α particles, and even negative ions are injected into the central region, point C of Figure 11–31, between two flat, D-shaped, hollow metal conductors (called "dees"). An alternating high-frequency voltage is applied to the dees, producing an alternating electric field in the region between them. During the time that the left dee is positive and the right dee is negative, the ions are accelerated to the right by the electric field between the dees. As the ions enter the interior of the right dee, they are electrically shielded from any electric field and therefore move in a semicircle at constant speed under the influence of the constant magnetic field. When they emerge from the right dee, they are further accelerated across the gap if the left dee is then negative. This means that the frequency of the alternating voltage applied to the dees must equal the orbital, or cyclotron, frequency of the

† See Weidner and Sells, *Elementary Classical Physics,* 2d ed., Section 29–5.

ions, given by Equation 11–22. During each acceleration, the ions gain energy, move at a higher speed, and travel in semicircles of larger radii. As the ions spiral outward in the dees, they remain in resonance with the ac source of constant frequency; this is because the time for an ion to move through 180° is independent of its speed or radius, provided only that $v \ll c$ or $K \ll E_0$. When the accelerated particles reach the perimeters of the dees, they are deflected by the electric field of an ejector plate E and strike the target T. Their final kinetic energy is, if we use Equations 11–19 for v,

$$K = \tfrac{1}{2}m_0 v_{\text{max}}^2 = \frac{Q^2 B^2 r_{\text{max}}^2}{2m_0}$$

The final kinetic energy of the particle varies with the square of the radius of the dees and with the square of the magnetic field B; therefore, to achieve the greatest possible energies, B and r_{max} are made as large as possible. For the largest magnetic fields attainable (with iron, about 2 T), the frequency f, by Equation 11–22, is measured in megahertz (radiofrequencies). The diameter of the dees, also the diameter of the electromagnet's pole faces, can be as large as 8 ft. A typical alternating voltage across the dees is 200 kV. Massive particles—protons, deuterons, and α-particles—can be accelerated in a cyclotron.

For protons up to about 12 MeV and deuterons up to about 25 MeV, the particle's mass does not increase appreciably, and the orbiting particle can remain in synchronism with the alternating voltage. Electrons, on the other hand, can be easily accelerated to relativistic energies (their rest energy is only 0.5 MeV); such light particles cannot be accelerated to high energies by a cyclotron.

A cyclic accelerator that accelerates particles to highest energies (up to hundreds of GeV) is the *synchrotron*. This type of machine accelerates particles in a circular path of *fixed* radius. To attain the largest possible kinetic energies in a cyclic accelerator, one must maximize the particle's relativistic momentum $p = QBr$, which is to say, one must maximize B and r. There is a limit on the magnitude of the magnetic field B over moderately large regions of space; with such magnetic materials as iron, B is limited to about 2 T, although superconducting magnets can raise this somewhat. Therefore, for large p one must have large r.

Example 11–5. What radius must a synchrotron have for protons to have a maximum energy of 500 GeV in a magnetic field of 2 T?

Since $E_0 \approx 1$ GeV for a proton and here $E = (500 + 1)$ GeV, then by Equation 11–19,

$$E = pc = QBrc$$

The radius is then

$$r = \frac{E}{QBc} = \frac{(500 \text{ GeV})(10^9 \text{ eV/GeV})(1.6 \times 10^{-19} \text{ J/eV})}{(1.6 \times 10^{-19} \text{ C})(2 \text{ T})(3 \times 10^8 \text{ m/s})}$$

$$\approx 1 \text{ km}$$

which corresponds to a circumference of 3 miles.

The magnetic field is confined to the very large circumference of the circular path. If the radius were not constant, the magnetic field would have to extend over a correspondingly larger region of space, which is not feasible in terms of cost and materials.

With the orbital radius r fixed in a synchrotron, the particle's energy and momentum $p = QBr$ can increase only if the applied magnetic field B increases from some initial value up to its final maximum value. But if B increases, so must the cyclotron frequency. From Equation 11–20,

$$\omega = \frac{QBc^2}{E} = \frac{QBc^2}{\sqrt{(pc)^2 + E_0{}^2}}$$

By Equation 11–19, the angular cyclotron frequency can be written

$$\omega = \frac{QBc^2}{\sqrt{(QBrc)^2 + E_0{}^2}}$$

Figure 11–32 shows schematically the principal parts of a proton synchrotron. Protons first are accelerated to an energy of several MeV by a linac that acts as an injection accelerator (either a Van de Graaff machine or a resonant linear accelerator). Then they enter a doughnut-shaped evacuated tube contained within an electromagnet that is producing a deflecting magnetic field at the tube. Energy is supplied to the particles once in each revolution by an alternating electric potential difference supplied by a variable radiofrequency source. As the particles acquire speed, momentum, and kinetic energy in successive orbits, the magnitude of the deflecting magnetic field and the frequency of the accelerating electric field both increase with time, so that the particles continue to travel in a path of constant radius and also arrive at the energizing gap at the right time for further acceleration. After the magnetic field has reached its maximum magnitude and the particles their maximum kinetic energy, the particles are deflected and strike an external target. A photograph of the 500-GeV

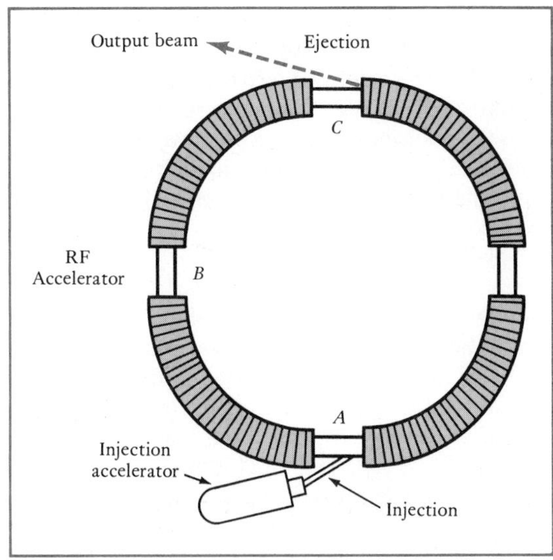

FIGURE 11–32. *Schematic diagram of a synchrotron.*

FIGURE 11–33. *Aerial view of main accelerator at Fermi National Accelerator Laboratory, Batavia, Illinois. The synchrotron has a diameter of 1.24 miles and it accelerates protons to 500 GeV. (Courtesy, Fermi National Accelerator Laboratory, Batavia, Illinois)*

synchrotron constructed at the Fermi National Accelerator Laboratory, Batavia, Illinois, is shown in Figure 11–33.

11-14 Colliding Beams

There is a very important reason for constructing accelerating machines with ever increasing particle energies. When accelerated particles have a high energy, the rest energies of particles that can be created in collisions between the incident and target particles are correspondingly higher. Also, the wavelengths of incident particles that are used as probes of subnuclear structure are then shorter. Only a fraction of the incident particle kinetic energy can, however, be transformed into the rest energy of created particles. This limitation arises from the momentum-conservation principle; particles emerging from the collision between a particle in the accelerator beam and a target at rest must have a net momentum equal to that of the incident particle. Since the emerging particles necessarily have momentum and kinetic energy, only part of the incident particle's kinetic energy can be tranformed into rest energy.

A simple example for particles traveling at very low speeds illustrates the effect. Suppose that a particle of mass m traveling initially at speed v strikes and sticks to an identical particle initially at rest. The composite object of mass $2m$ must, according to momentum conservation, travel after the collision at speed $v/2$: Momentum before $= mv = (2m)(v/2) =$ momentum after. The kinetic energy of the composite object is then $(\frac{1}{2})(2m)(v/2)^2 = \frac{1}{2}(\frac{1}{2}mv^2)$, or half the kinetic energy of the incident particle. Therefore, only half of the incident particle's energy is dissipated in the completely inelastic collision; the remaining half is kinetic energy after the collision.

So that all initial kinetic energy will be "dissipated," with none appearing after the collision, the total momentum of the system must be zero. For particles of equal mass, this implies that the two particles are initially in motion at equal speeds in opposite directions. The particles collide head-on; the system's total momentum is zero, before and after the collision; and all the initial kinetic energy of the two particles may be dissipated.

Although the considerations just given are based on the classical relations for momentum and kinetic energy, the same considerations enter for particles traveling at relativistic speeds. Indeed, the fraction of the incident particle's kinetic energy that can be transformed into rest energy in a collision with a target particle at rest decreases progressively with the incident particle's speed.

The general relativistic relation for the threshold kinetic energy K_{th} of an incident particle striking a target particle at rest has already been derived (Section 11–3):

$$K_{th} = -\frac{Q\left(\begin{array}{c}\text{rest mass of } all \text{ particles}\\ \text{entering and leaving reaction}\end{array}\right)}{2(\text{rest mass of target particle})} \tag{11–14}$$

In Equation 11–14, Q (always intrinsically negative) is the total rest energy into the reaction less the total rest energy that goes out of the reaction. As shown in Example 11–1, a proton-antiproton pair can be created in collision between an incident proton and a target proton at rest only if the incident proton's kinetic energy is at least $6E_0$, where E_0 is the rest energy of the proton (or antiproton). (Note that if the threshold kinetic energy for proton-antiproton production were computed improperly, by applying classical relations, this threshold kinetic energy would be $4E_0$, or *twice* the total rest energy of the created particles, instead of $6E_0$, or *three* times the total rest energy of the created particles.)

Therefore, the motivation for using head-on collisions between high-energy particles is to avoid "wasting" a large part of the incident particle's energy. Colliding-beam arrangements do just this; high-energy particles in oppositely directed beams collide head-on, and the total kinetic energy of the two particles may be transformed into rest energy.

Example 11–6. Two protons, each with an initial kinetic energy K, collide head-on. All the initial kinetic energy goes into creating rest mass, a total of $2K$. Suppose that the same rest energy $2K$ is to be created by a proton colliding with a proton at rest. What is the minimum kinetic energy of the incident proton?

Since the created rest mass is $2K$, the Q for the reaction in Equation 11–14 is given by $Q = -2K$. The total rest energy of all particles that enter and leave the reaction equals $4E_0 + 2K$, where E_0 is the rest energy of the proton. The rest energy of the target particle is also E_0. Substituting these values in Equation 11–14 yields

$$K_{th} = -\frac{(-2K)(4E_0 + 2K)}{2E_0} = 2K\left(2 + \frac{K}{E_0}\right)$$

Suppose that each of the two protons colliding head-on has a kinetic energy of 25 GeV, so that 50 GeV of rest energy can be created in the collision. Then the minimum kinetic energy of a proton colliding with a proton at rest to produce the same rest energy is, from the relation just given,

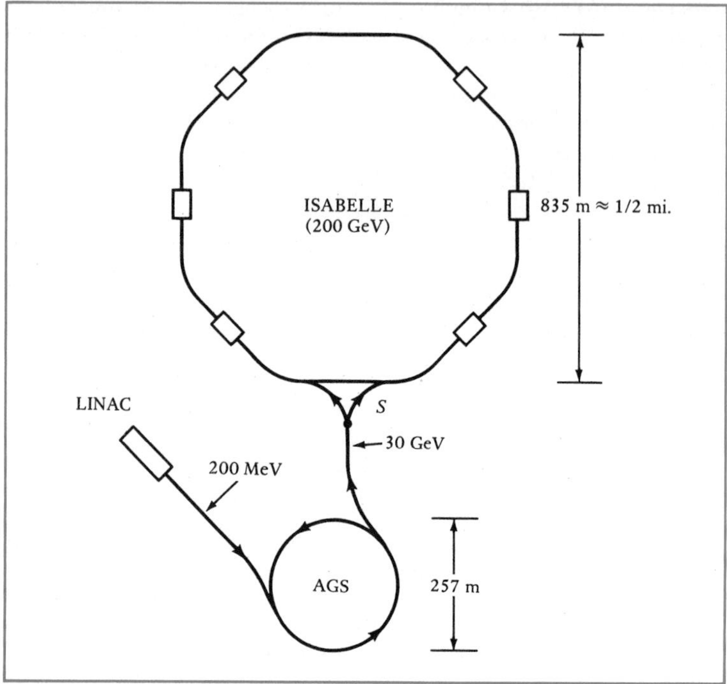

FIGURE 11–34. *Sketch of the design of ISABELLE (Intersecting Storage Accelerator at Brookhaven Laboratory). The machine, which is a combined storage ring for colliding beams and a synchrotron, produces intersecting proton beams, each of energies from 30 to 400 GeV. Protons leave the linac with an energy of 50 MeV; they are accelerated to 30 GeV in the Brookhaven AGS (Alternating Gradient Synchrotron). The output beam from the synchrotron is split into two oppositely circulating proton beams. Each beam travels in a nearly circular path of 2.6-km circumference. The orbit has six straight sections, at the middle of which the two beams intersect and experiments are conducted, and six circular arcs, where the beams are deflected by superconducting magnets producing fields up to 5 T.*

$$K_{\text{th}} = (50\text{ GeV})\left(2 + \frac{25\text{ GeV}}{0.94\text{ GeV}}\right)$$
$$= 1430\text{ GeV}$$

The minimum kinetic energy of a proton colliding with a proton at rest is 57 times greater than for colliding beams of 25-GeV protons.

Although colliding beams are far more efficacious, in converting kinetic energy into rest energy, than a single beam against a fixed target, high-energy particles colliding head-on pose special problems. For one thing, nuclear cross sections are relatively small, simply because the range of the nuclear force is very small. Furthermore, the target for either of the two beams is a moving target, and the number of target particles is limited by the intensity of either of the two beams. Additionally, there are the formidable technical problems or producing two high-intensity beams of high-energy particles and having these oppositely directed.

The essential element in colliding-beam arrangement is a *storage ring,* a circle in which particles that have been accelerated to a high energy are stored at a constant energy under the influence of a constant magnetic field applied perpendicular to the plane of the circle along its circumference. Particles in the storage ring simply coast at constant energy while they are deflected in a circular arc by a magnetic field. A pulse of particles enters the storage ring after leaving the accelerator; indeed, particles are accumulated in a storage ring by successive donations from the accelerator until the number circulating in the ring suffices to yield a beam of maximum intensity, one in which additions from the accelerator just cancel inevitable losses from imperfect focusing.

A colliding-beam arrangement typically involves *two* storage rings with particles, both from the same accelerator, circulating in opposite senses. The two rings intersect at points at which the two oppositely directed beams collide. See Figure 11–34.

SUMMARY

Most low-energy nuclear reactions are of the general form $X(x, y)Y$, where x is the incident particle, X the target nucleus, y the emerging light-mass particle, and Y the product nucleus (often radioactive). The nucleon number, electric charge, momentum, and mass-energy are conserved in a nuclear reaction. The net nuclear energy Q released in the reaction is defined by

$$Q = [(M_X + m_x) - (M_Y + m_y)]c^2 \quad (11\text{–}3)$$

$$= (K_y + K_Y) - K_x \quad (11\text{–}1)$$

where the target nucleus is at rest in the laboratory. The threshold energy of x for an endoergic ($Q < 0$) reaction is given in general by

$$K_{\text{th}} = -\frac{Q\begin{pmatrix}\text{rest mass of } all \text{ particles}\\ \text{entering and leaving reaction}\end{pmatrix}}{2(\text{rest mass of target particle})} \quad (11\text{–}14)$$

The reaction cross section σ gives a measure of the intrinsic probability that a nuclear reaction will occur. The fractional number of x particles undergoing a reaction in a thin foil of thickness t and containing N particles per unit volume is

$$\frac{n}{n_{\text{i}}} = \sigma N t \quad (11\text{–}16)$$

According to the concept of the compound nucleus, nuclear reactions take place in two distinct stages: the formation of the compound nucleus and the decay of the compound nucleus:

$$X + x \rightarrow \text{C}^* \rightarrow y + Y$$

The occurrence of peaks, or resonances, in the reaction cross section is a manifestation of the compound nucleus in quantized excited states.

PROBLEMS

Note: See Appendix II for values of the atomic masses.

Low-energy reactions, §11–1

● **11–1.** By what mode are the unstable products of the following reactions with stable target materials likely to decay: (*a*) (n, γ), (*b*) (p, n), (*c*) (d, p), (*d*) (p, α), and (*e*) (α, n)?

11–2. In the Cockcroft-Walton experiment, protons with a kinetic energy of 0.50 MeV were directed against a target of lithium-7 to produce the nuclear reaction $^7_3\text{Li}(\text{p}, \alpha)^4_2\text{He}$. (*a*) What is the distance of closest approach between a proton and lithium nucleus? (*b*) What is the nuclear radius of lithium-7 according to the relation $R = r_0 A^{1/3}$ with $r_0 = 1.1$ fm? (*c*) Must an incident particle be within the range of the nuclear force for a nuclear reaction to take place?

Energetics of reactions, §11–2

● **11–3.** For a $^{12}_6\text{C}$ target, which of the following reactions have stable product nuclides: (*a*) (n, γ); (*b*) (p, n); (*c*) (d, p), (*d*) (α, p)?

● **11–4.** For a $^{19}_{9}F$ target, find the product nuclides for the following reactions: (*a*) (n, γ); (*b*) (p, n); (*c*) (n, p); (*d*) (α, n).

11–5. Deuterons with energies to 4 MeV are available from an accelerator. What nonradioactive target material should be used to produce radioactive carbon-14 in a nuclear reaction?

Nuclear reaction threshold, §11–3

● **11–6.** The first nuclear reaction producing a radioactive nuclide ($^{30}_{15}P$) was observed by Joliot-Curie and Joliot (Section 11–1). (*a*) Show that if incident α particles produce the reaction $^{27}_{13}Al(\alpha, n)^{30}_{15}P$, they can also produce the reaction $^{27}_{13}Al(\alpha, p)^{30}_{14}Si$. (*b*) How did Curie and Joliot know they were observing the first and not the second of the described reactions?

● **11–7.** Show that the threshold kinetic energy of a low-energy incident particle in an endoergic nuclear reaction is (*a*) equal to the magnitude of reaction Q when the mass of the incident particle is much less than the mass of the target particle initially at rest, and (*b*) much greater than $|Q|$ when the mass of the incident particle is much greater than the mass of the target particle initially at rest.

● **11–8.** A beam of 5.0-GeV protons or a beam of 5.0-GeV photons may be directed against a certain target. Which beam, one of protons or one of photons, is capable of creating the more massive particle?

● **11–9.** Which collision—(*a*) a 10-MeV electron striking a proton at rest or (*b*) a 10-MeV proton striking an electron at rest—is capable of creating particles of larger rest energy?

11–10. Find the threshold kinetic energy (MeV) for the following incident particles to disintegrate the deuteron into a proton and a neutron: (*a*) electrons; (*b*) protons; (*c*) α particles.

11–11. (*a*) With what minimum kinetic energy must protons strike a $^{7}_{3}Li$ target to produce neutrons? (*b*) With what minimum kinetic energy must $^{7}_{3}Li$ ions strike a proton target to produce neutrons?

11–12. What is the minimum kinetic energy (MeV) of an incident α particle on a target $^{27}_{13}Al$ nuclide that will produce a neutron and an unstable $^{30}_{15}P$ nuclide?

■ **11–13.** Find the kinetic energies of the two α particles emitted in the nuclear reaction $^{7}_{3}Li(p, \alpha)^{4}_{2}He$ when a 500-keV proton collides head-on with a ^{7}Li nuclide at rest, and the two α particles are emitted along the line of the incident proton. (The average value of α-particle energies Cockcroft and Walton observed was 8.93 MeV.)

11–14. Use Equation 11–14 to show that, in general, the kinetic energy of particles striking a target at rest and causing an endoergic nuclear reaction to take place must exceed the magnitude of the reaction Q.

■ **11–15.** The Q of a nuclear reaction can be evaluated by measuring the kinetic energies K_x and K_y of the incident and emerging particles x and y at some known angle θ between the directions of x and y. Show that the reaction Q for low energies is given by

$$Q = K_y\left(1 + \frac{m_y}{m_Y}\right) - K_x\left(1 - \frac{m_x}{m_Y}\right) - 2\frac{\sqrt{m_x m_y K_x K_y}}{M_Y}\cos\theta$$

Hint: Construct a momentum-vector diagram and apply the law of cosines.

11–16. Neutrons with a kinetic energy of 10.0 MeV strike a target of ^{3}He and effect the reaction $^{3}He(n, d)^{2}H$. What is the kinetic energy of deuterons observed at 90° relative to the incident beam? (See Problem 11–15.)

Cross section, §11–4

● **11–17.** How does the cross section in high-energy neutron absorption vary with the mass number A of the target material?

11–18. The fraction of incident particles undergoing a nuclear reaction with a cross section σ in a thin foil of thickness t is given by $\sigma N t$, where N is the number of target nuclei per unit volume (Equation 11–16). Show that if the target foil is not thin but has a thickness x, the fraction of incident particles emerging through the target is given by $e^{-\sigma N x}$.

11–19. A thin 8.0-mg foil of cadmium-113 is exposed to a beam of thermal neutrons with a neutron flux of 1.0×10^{13} neutrons/cm²·s for a period of 1.0 h. The capture cross section in cadmium-113 for thermal neutrons is 2.0×10^{4} barns. How many nuclei of cadmium-114 are formed?

Compound nucleus, §11–5

11–20. A 1.375-MeV proton incident on a $^{11}_{5}B$ target nuclide produces a single $^{12}_{6}C$ nuclide in an excited state. (*a*) What is the Q (MeV) for the reaction? (*b*) Show that if the excited carbon nuclide quickly decays to its ground state by γ emission, the emitted photon will have an energy of 17.2 MeV.

11–21. (*a*) Using Figure 11–8, estimate the excitation energy of the second excited state of $^{14}N^*$ shown in this figure. Find the kinetic energy of the incident particles producing this particular excited state of $^{14}N^*$ for (*b*) protons, (*c*) alpha particles, (*d*) neutrons, and (*e*) deuterons.

11–22. One of the excited states of the ^{15}N nuclide has an excitation energy of 17.47 MeV above the ground state. Find the threshold kinetic energy of the incident particle that will produce $^{15}N^*$ in this excited state for the reaction (*a*) ^{13}C + d; (*b*) ^{14}N + n; (*c*) ^{14}C + p; and (*d*) ^{11}B + α.

Nuclear fission, §11–6

11–23. A nucleus of $^{236}_{92}U$ decays into the fission fragments, $^{144}_{56}Ba$ and $^{89}_{36}Kr$. Assume that the two fragments are spherical and just "touching" after their formation. (*a*) What is the Coulomb potential energy (MeV) of this pair of fragments? (*b*) Compare this potential energy with the total energy released in the fission of ^{236}U.

Neutrons and nuclear reactors, §11–7

11–24. Verify that a thermal neutron (at 300 K) has an average kinetic energy of 0.039 eV, a speed of 2.7×10^3 m/s, and a wavelength of 1.5 Å.

11–25. In one example of delayed neutrons in nuclear fission given in Section 11–7, the nuclide ^{87}Br decays by β^- emission into ^{87}Kr* in an excited state, and ^{87}Kr* can in turn decay into stable ^{86}Kr by neutron emission. The maximum energy of the β^- particles is 2.6 MeV. Find the kinetic energy of the delayed neutrons. (The mass of ^{87}Br is 86.922 38 u; the mass of ^{86}Kr is 85.910 61 u.)

11–26. Neutrons are most effectively thermalized by a moderator of hydrogen; only if the incident particle (neutron) and target (moderator atom) have the same mass can an incident neutron striking a free proton intially at rest be brought to rest in a *single* collision. Only with equal masses do two particles moving at the same speed have the same momentum and same kinetic energy. (*a*) Show that in general (for nonrelativistic speeds) the fraction of energy transferred from an incident particle m to a target particle of mass M initially at rest in a head-on collision is given by $4mM/(m + M)^2$. (*b*) What fraction of a neutron's energy is transferred in a head-on collision to an atom of carbon-12?

11–27. Use order-of-magnitude arguments to show that a prompt neutron released in nuclear fission initiates a further nuclear fission in a time of about 10^{-12} s. A thermal neutron travels a distance of the order of 10^2 atomic diameters before initiating a fission.

Nuclear fusion, §11–8

● **11–28.** What is the temperature of a plasma whose particles have an average kinetic energy of 1.0 MeV?

● **11–29.** The proton-proton cycle for the release of thermonuclear energy in stars consists of three distinct nuclear reactions. One reaction is governed by strong nuclear force, another by the weaker electromagnetic force, and the third by the still weaker weak interaction. Identify the reactions according to the dominant basic force.

● **11–30.** In nuclear fission, approximately 5 percent of the energy released is carried away by neutrinos. Is the corresponding fraction in a fusion reaction (p-p or carbon cycle) greater than that for fission?

11–31. Find the energy released in (*a*) the binding of protons with an equal number of electrons to form 1 g of hydrogen atoms; (*b*) the fusion of protons and neutrons to form 1 g of deuterium; (c) the annihilation of 0.5 g of hydrogen with 0.5 g of antihydrogen.

11–32. The temperature in the sun's interior is approximately 2×10^7 K. (*a*) What is the distance of closest approach of two thermal protons at this temperature? (*b*) The result of part (*a*) shows that the closest-approach distance exceeds the range of the nuclear force. Does this preclude a nuclear reaction between protons at this temperature? (*c*) On the basis of the preceding, show that the first reaction in the proton-proton cycle, although not prohibited, has a very low probability of taking place in the interior of the sun.

Ionization, §11–9

● **11–33.** Protons with a kinetic energy of 10 MeV are absorbed in copper having an *areal density* of 0.21 g/cm². (The areal density of a plate of absorber is readily measured through its mass and area.) What is the range (mm) of a 10-MeV proton in copper, whose density is 8.9 g/cm³?

11–34. For air at standard temperature and pressure, one ion pair is produced when an ionizing particle traversing the air loses 35 eV. (*a*) How many ion pairs are produced by a 10-MeV proton absorbed in air? (*b*) What is the magnitude of the total charge of either sign produced by the proton? (*c*) If the two kinds of charge are collected on opposite plates of a 0.2-picofarad (pF) capacitor, by how much would the potential difference between the capacitor plates change?

11–35. Charged particles traversing a gas lose energy little by little in forming ions; energy loss by head-on elastic collisions with free, or nearly free, electrons or with nuclei are relatively unimportant. (*a*) Suppose that a particle with mass M collides head-on with a particle of mass m initially at rest in an elastic collision, where $M \gg m$. Show that the fractional loss in the kinetic energy of M is $4m/M$. (*b*) Suppose that a particle with mass m collides head-on with a particle of mass M initially at rest in an elastic collision, again with $M \gg m$. Show that the fractional loss in the kinetic energy of m is $4m/M$.

Detectors, §11–10

11–36. Monochromatic photons of 2.76 MeV are detected in a scintillation spectrometer. The spectrometer is so adjusted that a pulse height of 100 V corresponds to an electron energy of 1.00 MeV. At what voltages in the pulse-height distribution would one expect to find the (*a*) photopeak? (*b*) Compton edge? (*c*) pair-production peak?

11–37. Radioactive ^{60}Co decays into ^{60}Ni with the emission of 1.18- and 1.33-MeV γ rays. When the γ rays of these two energies are studied with a scintillation spectrometer, what six peak (electron) energies are expected in the electron pulse-height distribution?

Track-recording devices, §11–11

11–38. A cloud-chamber photograph reveals the path of a charged particle before and after it passes through a thin lead plate located within the cloud chamber; see Figure 11–35.

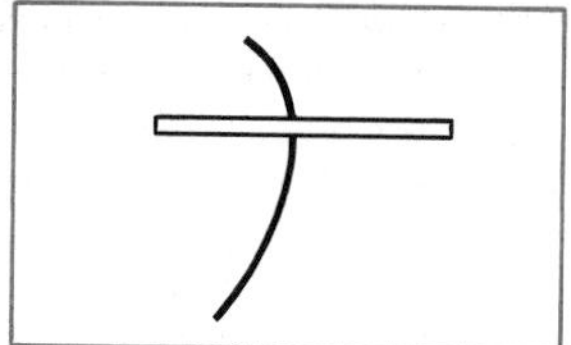

FIGURE 11–35.

From the density of the droplets, it is established that the particle has the same mass as an electron. The radii of curvature of the particle above and below the lead plate are 5.0 and 10.0 cm, respectively. Assume that the particle is traveling perpendicular to the constant magnetic field of 1.0 T directed into the paper. (*a*) In which direction is the particle moving? (*b*) Is the particle a positron or an electron? (*c*) How much energy was lost by the particle in traversing the lead plate?

11–39. Monochromatic photons produce electron-positron pairs in a bubble chamber immersed in a uniform magnetic field. The electron and positron in each pair need not have equal kinetic energies, and the radii of their circular paths are therefore not necessarily the same. Show that if the photon energy exceeds greatly the rest energy of the electron and if all particles travel in a plane perpendicular to the magnetic field, the sum of the electron and positron radii is the same for any pair and this sum is proportional to the photon energy.

Nuclear measuring devices, §11–12

● **11–40.** In a velocity selector, charged particles are sent through crossed electric and magnetic fields. What would be the path of a charged particle sent through (*a*) two crossed uniform electric fields? (b) two crossed uniform magnetic fields?

11–41. What are the radii of curvature in a magnetic field of 10,000 G of (*a*) 10-keV electrons? (*b*) 10-GeV electrons? (*c*) 10-keV protons? (*d*) 10-GeV protons?

11–42. What is the maximum speed for a singly charged particle that can be measured by a velocity selector consisting of crossed electric and magnetic fields if the magnetic field is maintained at a magnitude of 10,000 G while the maximum potential difference that can be applied to parallel plates separated by 4.0 cm is 10,000 V? (As the answer of this example shows, the method of crossed electric and magnetic fields is not suitable for measuring particle speeds in excess of $\sim 10^6$ m/s.)

Nuclear accelerators, §11–13

● **11–43.** Atoms that are multiply ionized may be accelerated in a linear accelerator to produce nuclear reactions of intermediate energy (tens of MeV). (*a*) What is the final energy of a quadruply ionized oxygen-16 atom accelerated in a Van de Graaff machine operating at a potential difference of 10 MV? (*b*) What is the kinetic energy per nucleon of the accelerated O^{4+}?

● **11–44.** Show that for particles accelerated in the most energetic particle accelerators (particles with kinetic energies in GeV), *any* target material can be considered to consist of individual nucleons that are essentially free of binding to other nucleons.

11–45. A 5.00-MeV proton moves at right angles to a uniform magnetic field of 2.00 T. What is the proton's (*a*) cyclotron frequency? (*b*) speed? (*c*) momentum? (*d*) path radius?

11–46. A cyclotron is designed to accelerate low-energy protons, deuterons, and α particles. Its frequency is fixed, but its magnetic field can be reduced from the maximum value. The cyclotron is first adjusted to accelerate α particles to a maximum kinetic energy of 4.0 MeV. (*a*) Show that with the frequency and magnetic field unchanged, the cyclotron can accelerate deuterons to 2.0 MeV. (*b*) By what factor must the magnetic field be reduced to accelerate protons? (*c*) What is the proton's energy then?

11–47. At the Stanford University electron linear accelerator (SLAC), electrons are accelerated from 30 MeV to 20 GeV in an accelerating tube 2 miles long. (*a*) By what fraction does the final electron speed differ from the speed of light? (*b*) What would be the total length of the accelerating tube as measured by an observer traveling with a 20-GeV electron? (*c*) The electron beam at the target has a current of 15 microamperes (μA) and a power of 0.50 megawatts (MW). What is the average number of electrons striking the target per second?

11–48. The synchrotron at Fermilab in Batavia, Illinois, accelerates protons to a final energy of 500 GeV in a circular path of a 1.24-mile diameter. (*a*) What is the magnitude of the magnetic field at the orbit of a 500-GeV proton? (*b*) How long does it take for a 500-GeV proton to complete one loop?

Colliding beams, §11–14

● **11–49.** (*a*) A proton collides with a proton at rest and two photons each of 0.50-MeV energy are produced. What is the minimum energy of the incident proton? (*b*) What is the total kinetic energy of two protons colliding head-on at equal speed that will also produce two photons each of 0.50-MeV energy?

11–50. In the ISABELLE intersecting storage accelerator (Figure 11–34) protons enter the storage ring at 30 GeV and are accelerated to 400 GeV. (*a*) Find the time for a 30-GeV proton to complete one cycle around the 2.623-km circumference of the storage ring and accelerator. (*b*) Find the corresponding time for one cycle for a 400-GeV proton. (*c*) What is the fractional difference in time interval for parts (*a*) and (*b*)?

11–51. At the maximum, the total number of protons in either beam of the ISABELLE storage ring is 5.5×10^{14} (see Problem 11–50). Show that each proton beam corresponds to a current of approximately 10 A. The circumference of the storage ring is 2.6 km.

11–52. Show that when two protons of equal kinetic energy K collide head-on, the total energy that can be dissipated is at least four times greater than the energy for one proton of energy K striking another initially at rest. Assume that the incident and target particles emerge from the collision.

11–53. What is the maximum rest energy of particles that can be created at the Fermilab accelerator with 500-GeV protons striking protons at rest (and with both the incident and target protons existing after their collision)? (*b*) What is the minimum kinetic energy of protons in a colliding-beam arrangement yielding the same rest energy as in part (*a*)?

11–54. The *center-of-mass energy* is by definition the total energy of all particles of a system measured relative to the center of mass of the system. Therefore, in a colliding-beam experiment with identical particles moving with equal energies in opposite directions, the total energy of the particles is the center-of-mass energy. By *accelerator energy* is meant the kinetic energy of a particle striking a fixed target. (*a*) Show that if the center-of-mass energy E_{cm} is large compared with the rest energies of the two colliding particles, the accelerator energy is given by $K_a = E_{cm}{}^2/2E_0$, where E_0 is the rest energy of the target particle. (*b*) Show that the accelerator energy equivalent to two 400-GeV protons in a colliding-beam experiment is 3.4×10^5 GeV, that is, that it requires a 3.4×10^5 GeV proton accelerator (protons against fixed target protons) to produce a center-of-mass energy of 2(400 GeV) = 800 GeV.

12
The Elementary Particles

Murray Gell-Mann (born 1929, United States). B.S. Yale, aged 19; Ph.D., M.I.T., aged 22. 1952, introduced concept of strangeness. 1961, announced "Eight-fold Way", a system for grouping elementary particles taking its name from Buddha's list of eight virtues for attaining harmony in life. 1963, theory of quarks. 1964, predicted existence of the omega-minus elementary particle. 1969, Nobel Prize, discovery of classification of the elementary particles. (Photo courtesy of the American Institute of Physics, Neils Bohr Library.)

The search for the ultimate building blocks of nature goes back to the Greek notion of four elements—earth, water, air, and fire (and possibly an ethereal fifth element, a "quintessence")—that were supposed to be the basic components of all other materials. Twenty-two centuries later came the idea of the chemical elements—molecules and atoms—and finally, of the particles within the atoms, even within the nucleus. Underlying the quest for the elementary particles is the expectation that if one has identified the truly fundamental particles—preferably of only a few distinct types—and learned the rules by which they affect one another, then the remainder of physics will be a straightforward although possibly very difficult exercise.

We have not arrived at the end of the quest and possibly never will. The particles that now are thought to be elementary in some sense are many, and they may be grouped in various ways to form coherent patterns, but the grand pattern still eludes physicists. Indeed, the principal motivation for constructing accelerating machines of higher and higher energies is to produce still more particles, and to study the properties of the subnuclear domain using very short wavelength, high-energy incident particles. Thus, elementary-particle physics is high-energy physics.

The study of elementary-particle physics is partly a study of the four fundamental forces among particles: the strong nuclear interaction, the electromagnetic interaction, the weak interaction, and the still weaker gravitational interaction. It is also a study of conservation laws, not merely the well-known classical laws of mass-energy, momentum, angular momentum, and electric charge, but also of certain others, somewhat more esoteric. Finally, it is concerned with how the fundamental forces, the conservation laws, the intrinsic properties of the particles, and even the properties of space and time can be fitted together to make some sense.

12-1 The Electromagnetic Interaction

We consider first those familiar particles that we have already taken to be elementary, namely the electron (e^-), the proton (p^+), the photon (γ), the positron (the antiparticle to the electron, designated $\overline{e^+}$), and the antiproton (antiparticle of the proton, designated $\overline{p^-}$).† Each has intrinsic properties, such as a definite electric charge, a definite rest mass (or rest energy), a definite intrinsic angular momentum (or spin), and a mean lifetime before decay into other elementary particles. Since all five particles mentioned are found to be stable against spontaneous decay, each, it is assumed, has an infinite lifetime. See Table 12–1.

As we found in Chapter 4, in treating the *interaction* between electromagnetic radiation and a charged particle, the radiation must be regarded as consisting of particlelike photons with each interaction occurring at a single point in space and in time. The several photon-electron interactions are illustrated in Figure 4–19.

It is illuminating to represent these interactions on a space-time diagram, one in which time as ordinate is plotted against position as abscissa.

† Throughout this chapter we will denote an antiparticle with a color overbar to distinguish it from the particle.

Table 12–1 *Some properties of some elementary particles*

PARTICLE	REST MASS (UNITS OF ELECTRON MASS)	REST ENERGY (MeV)	CHARGE (UNITS OF ELECTRON CHARGE)	↑ ANGULAR MOMENTUM COMPONENT ($\times \hbar$)	LIFETIME (s)
Photon γ	0	0 ($<10^{-21}$)	0	1	∞
Electron e^-	1	0.511003	-1	$\frac{1}{2}$	∞ ($>10^{29}$)
Positron $\overline{e^+}$	1	0.511	$+1$	$\frac{1}{2}$	∞
Proton p^+	1836	938.280	$+1$	$\frac{1}{2}$	∞ ($>10^{30}$)
Antiproton $\overline{p^-}$	1836	938.3	-1	$\frac{1}{2}$	∞

For simplicity we show the particle's spatial location in one dimension only; such a two-dimensional plot of time versus a single spatial coordinate reveals all important aspects of the interaction.

The history of a particle in a space-time diagram is shown by a line, known as a *world line.* Since the interaction between particles occurs only at space-time points, each particle is assumed to be free between interactions, traveling at constant velocity; therefore its world line between interactions is straight. A vertical line represents a particle whose coordinate x does not change with time; it is a particle at rest. A line inclined from the vertical represents a particle in motion, the angle between it and the vertical increasing with particle speed. Since a photon or any other particle with zero rest mass travels at the maximum possible speed c, the angle of its world line with the vertical is the maximum.

Figure 12–1 shows in schematic fashion space-time diagrams of the electron-photon interactions corresponding to those in Figure 4–19. Each interaction is observed in the center-of-mass inertial system, and time goes from past to future as the ordinate increases. The heavy-mass particles (p^+, H-atom) are represented by black lines, the light-mass particles (e^-) by color lines, and the photon (γ) by wavy color lines. In the following paragraphs we shall take up each of the parts of this figure in turn.

In the *photoelectric effect* (Figure 12–1*a*), a photon collides with a hydrogen atom, consisting of an electron and a proton bound together. After the interaction, the photon has been annihilated, and the electron and proton move away as separate particles. The space-time event characterizing the interaction corresponds to the vertex in the figure, where the incoming photon and electron lines coalesce into a single outgoing electron line (the proton's motion is virtually unaffected). Since the hydrogen atom consisted initially of an electron and a proton bound together, the net result of the interaction is the absorption of one real photon. Note that the total number of protons before and after the interaction is unchanged.

In *bremsstrahlung* (Figure 12–1*b*), an electron creates a photon in colliding with a proton. Again the interaction is the instantaneous event occurring at the vertex on the space-time diagram, where a photon line joins an electron line. The electron line is bent to indicate that the electron's momentum and energy change. Indeed, we may think of the incoming electron as being annihilated at the vertex while a second outgoing electron of different momentum and energy is created.

In *pair production* (Figure 12–1*c*), a photon is annihilated and an elec-

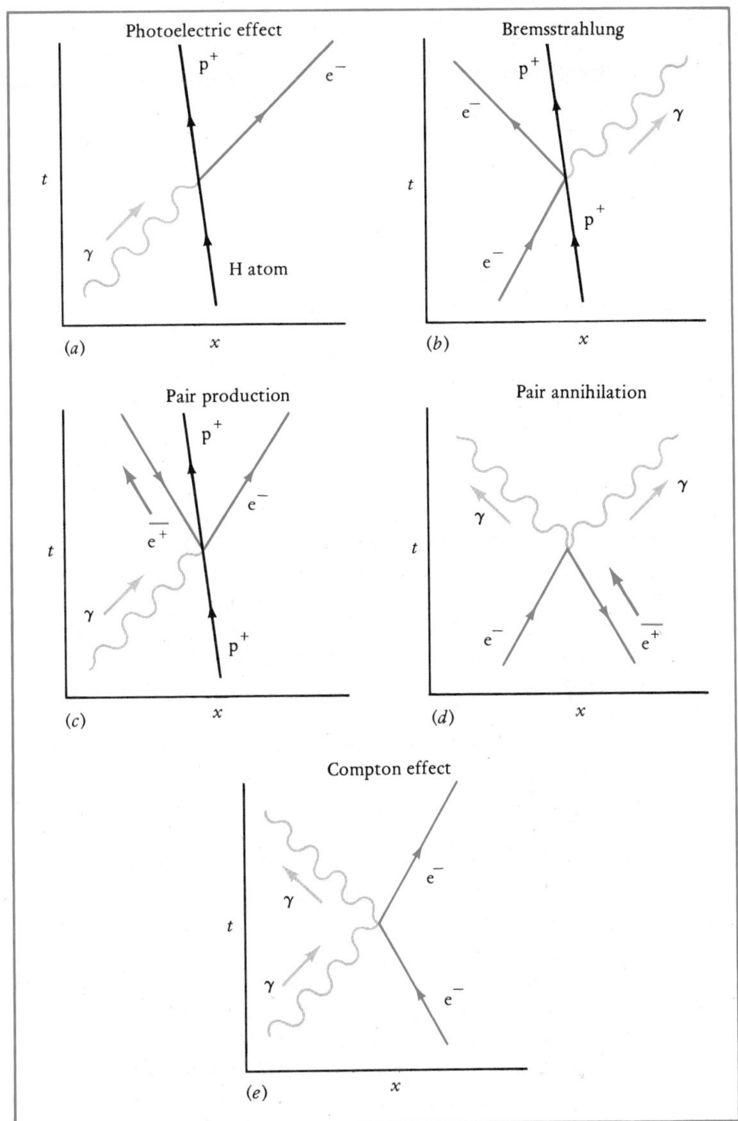

FIGURE 12–1. *Space-time graphs of the basic electron-photon interactions. An electron is represented by a color line, a photon by a wiggly color line, and a proton by a black line. A positron moving forward in time corresponds to an electron moving backward in time.* *(a) photoelectric effect;* *(b)* bremsstrahlung; *(c) pair production; (d) pair annihilation; (e) Compton effect.*

tron and a positron are created. The positron, the electron's antiparticle, is here represented by an electron world line whose arrow is *reversed;* the antiparticle is regarded as an electron moving backward in time. Such a representation—an antiparticle moving forward in time equivalent to a particle moving backward in time—is justified by the considerations of electromagnetic quantum field theory. So is a representation in which the creation of a particle is equivalent to the annihilation of its antiparticle. We see, then, that the electron and photon lines representing the pair production are basically the same as those representing the photoelectric effect and bremsstrahlung, an inclined electron line joined to a photon line at

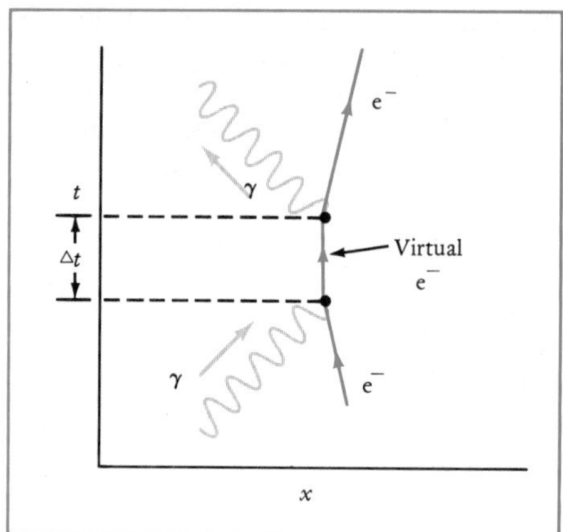

FIGURE 12–2. *Diagram of the Compton effect as two separate events joined by the virtual electron.*

the vertex. These processes, and still others that we shall later treat, differ only in the orientation of the lines on the space-time graph.

In *pair annihilation* (Figure 12–1*d*), an electron and a positron unite to create photons. Typically, two or more photons are created in pair annihilation to conserve momentum.

The *Compton effect* (Figure 12–1*e*) is an electromagnetic interaction in which an incident photon interacts with an electron to produce a scattered photon and a scattered electron.

In the electromagnetic interactions just described, each interaction has been shown as occurring at a single space-time point; total energy and total momentum remain the same before and after the interaction. Actually, quantum mechanics allows more freedom, through the uncertainty principle. Consider, for example, the Compton effect (Figure 12–1*e*). Assume that a photon joins the incident electron to produce an *intermediate electron;* later, this intermediate electron then produces another photon and the final outgoing electron (see Figure 12–2). Charge is conserved at both vertices. At the first vertex, however, where the incident photon joins the incident electron to form the intermediate electron, it is impossible simultaneously to conserve momentum and total energy. In the center-of-mass frame, the total momentum, by definition, is zero. Therefore, after the first vertex, the intermediate electron must be at rest. Since the total energy before the interaction is the sum of the photon's energy E_γ, the incident electron's rest energy E_0, and the electron's kinetic energy E_k, whereas the total energy after collision consists only of the intermediate electron's rest energy (it must be at rest to conserve momentum), there is a loss in total energy of magnitude $\Delta E = E_\gamma + E_k$. This energy "loss" will be regained at the second vertex, where the intermediate electron decays into the final outgoing photon and electron. Therefore, during the lifetime Δt of the intermediate electron, there is a violation of energy conservation by an amount ΔE. If the period of energy violation is less than the time interval determined by Heisenberg's uncertainty principle ($\approx\hbar/\Delta E$), then the intermediate electron is unobservable, and there is no measurable violation of energy (or momentum). Thus, the uncertainty principle allows the system's energy to differ from the initial energy by ΔE during a time interval $\Delta t \leq \hbar/\Delta E$. The intermediate particle, here an electron, is unobservable and therefore differs from a real electron. The world line of such a particle is always between vertices, with the time interval between emission and absorption of the so-called *virtual particle* given by $\Delta t \leq \hbar/\Delta E$.

Similar arguments show that momentum conservation can be violated by an amount Δp over the space interval Δx between emission and absorption of the virtual particle, where $\Delta x \leq \hbar/\Delta p$. For the Compton effect, virtual electrons of all kinetic energies and momenta can be created so long as they later decay into the final outgoing real electron and real photon within the time and space intervals given by the uncertainty principle. (See Example 12–1 for a specific example of the lifetime of a virtual electron in the Compton effect.) Indeed, according to quantum field theory, the basic electromagnetic interaction can be regarded as that instantaneous space-time event in which two world lines of a charged particle (going forward or backward in time) and the world line of a photon unite, provided the following conditions exist. (1) One of the outgoing particles is virtual and terminates its line at a second vertex within the time and space restrictions of the uncertainty principle

$$\Delta t \leq \frac{\hbar}{\Delta E} \quad \text{and} \quad \Delta x \leq \frac{\hbar}{\Delta p_x} \tag{12–1}$$

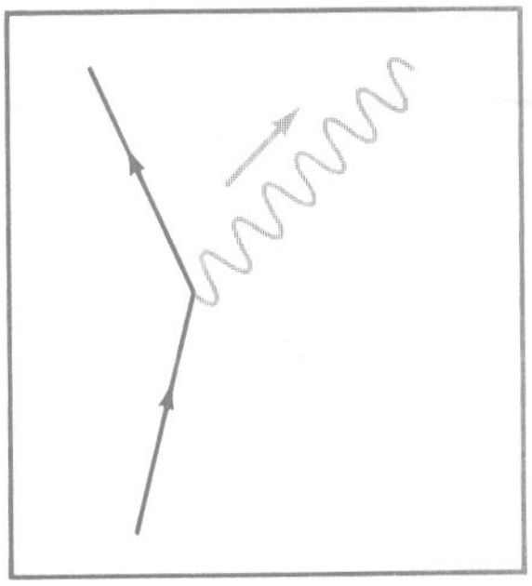

FIGURE 12–3. *Graph of the basic electromagnetic interaction.*

The quantities Δt and Δx here are the time and space intervals between the emission and the absorption of the virtual particle, and ΔE and Δp_x are the amounts by which energy and momentum conservations are violated within the space-time interval. (2) Electric charge conservation holds at all points in time. This basic interaction is shown in Figure 12–3.

Another example of the application of the basic electromagnetic interaction (Figure 12–3) is shown in Figure 12–4, where the bremsstrahlung process (Figure 12–1*b*) is depicted as a combination of three basic electromagnetic interactions of the form of Figure 12–3. Note that the first intermediate particle here is a virtual photon, the second, a virtual proton. Similar

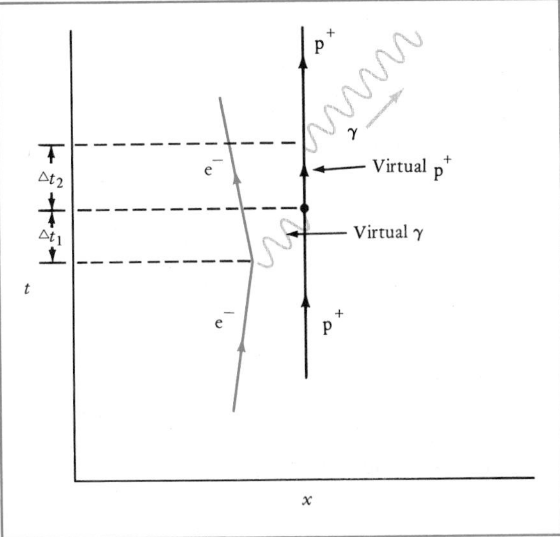

FIGURE 12–4. Bremsstrahlung *according to basic electromagnetic interactions. Both the photon and the proton, having their world lines joined at both ends, are virtual and unobservable.*

schematic graphs can be used to represent the electromagnetic interactions among any number of electrical charges.

All five particles listed in Table 12–1 are stable against spontaneous decay. In any overall interactions among these real particles, there are several fundamental physical properties that remain strictly conserved; we have, for example, the conservation of linear momentum, of relativistic mass-energy, of angular momentum, and of electric charge. Moreover, in every basic interaction (Figure 12–3) in which virtual particles are involved, the number of electrons minus the number of positrons is conserved. In graphical terms, this means that the world line of an electron does not end; for every electron line going into a vertex there is an electron line out of it. (An "electron line" going backward in time signifies a positron going forward in time.) Similarly, in any interaction, the number of protons less the number of antiprotons is constant; graphically, the proton lines are continuous.† Although these conservation laws govern the electron-minus-positron and the proton-minus-antiproton numbers, there is no restriction on the number of photons.

We have described the interaction between electromagnetic radiation and a charged particle in terms of the basic space-time graph. What about the interaction between two electrically charged particles, an interaction that, in classical electromagnetic theory, is familiarly described in terms of the electric and magnetic fields produced by the two particles and in terms of an electric and a magnetic force? This, too, is attributable to the creation and annihilation of photons.

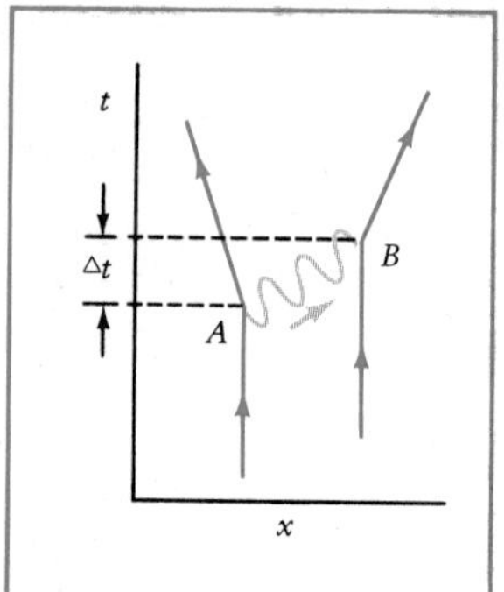

FIGURE 12–5. *Feynman diagram of the interaction between two electrons.*

Consider Figure 12–5, which shows a collision between two electrons. One electron creates a virtual photon spontaneously at vertex A (in the fashion of the bremsstrahlung process of Figure 12–4, and the second electron absorbs the photon at vertex B. Each of the two interacting electrons has its energy and momentum changed by virtue of the exchange of a photon; that is, each charged particle has been acted on by an electromagnetic force. The exchanged particle responsible for the force between the charged particles is a virtual photon (unobservable), which, like real photons (observable), travels at the speed c. Charged particles interact with one another through the exchange of such virtual photons. Graphical representations of interactions in space-time similar to Figure 12–5 are called *Feynman diagrams,* after R. P. Feynman, who used these diagrams to represent in simple fashion and also to compute in detail the electrodynamic interactions between quantum charges.

The electromagnetic force can be described in terms of the continuous interchange of virtual photons between the charged particles. The uncertainty principle limits the borrowed energy ΔE, by which energy conservation is violated, and borrowed momentum, by which momentum conservation is violated, according to Equation 12–1. Virtual photons of all energies, from zero to infinity, can thereby be created; therefore, the time interval and associated space interval ($\Delta x = c\Delta t$ for virtual photons) between the emission and absorption events can have different magnitudes. They can range from very short intervals (associated with interactions at small distances and with virtual photons having very high energies) to very long intervals (associated with large distances and photons of very low energy).

† The conservation laws for electrons and protons illustrate the more general conservation laws for leptons and baryons. See Section 12–4.

In an interaction between two charged particles, a virtual photon is created spontaneously by one particle and then absorbed by the other. Can the virtual photon be absorbed by the same charged particle that created it? It *can,* so long as the limits imposed by the uncertainty principle are satisfied. Figure 12–6*a* shows a single electron (or it may be any other electrically charged particle) emitting and then reabsorbing a virtual photon. Moreover, a photon, whether real or virtual, may spontaneously create a virtual electron-positron pair, as shown in Figure 12–6*b*, even though its energy is less than the threshold energy for pair production. The process is again possible according to, and limited by, the uncertainty principle. The virtual pair may be annihilated and yield the original photon. Still more complicated processes may be constructed, as shown in the Feynman diagram of Figure 12–6; the chain of creation-annihilation processes depicted is merely a collection of space-time graphs, whose basis is Figure 12–3. Thus, every electrically charged particle, even if isolated from other particles, can be considered to emit and reabsorb photons, which can become particle-antiparticle pairs. Although virtual particles cannot be observed directly, the validity of the conception is emphatically proved by the success of theoretical field-theory calculations of subtle electromagnetic effects, based on these ideas. The success of the field theory has caused it to be the model for understanding fundamental forces besides the electromagnetic interaction and has led to the prediction of particles whose existence was later confirmed experimentally.

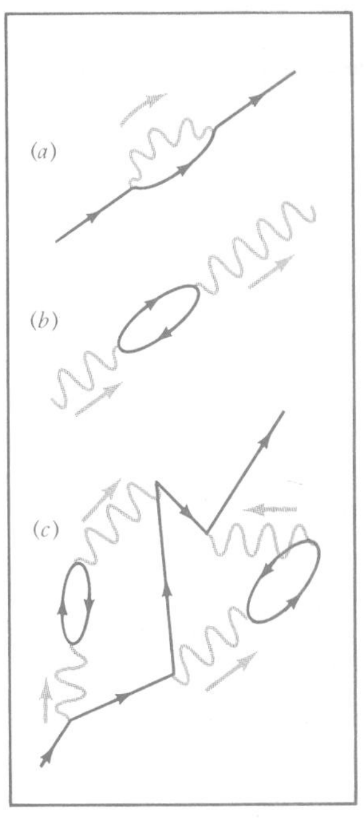

FIGURE 12–6. *Feynman diagrams of the electromagnetic interaction: (a) An electron spontaneously creates and then reabsorbs a photon; (b) a photon spontaneously creates an electron-positron pair, the pair is annihilated, and a photon is created; (c) a complex chain of annihilation-creation processes.*

Example 12–1. Estimate the duration Δt of the virtual electron in Figure 12–2 representing the Compton scattering of a photon by a free electron, assuming the incident photon energy E_γ to be 6200 eV.

If we choose the simple (but unobservable) situation in which the total linear momentum is conserved at the first vertex, then the virtual electron is at rest, since the measurements are made in the center-of-mass reference system. Thus, there must be a violation of total energy at this vertex by the amount

$$\begin{aligned}\Delta E &= \text{energy in} - \text{energy out}\\ &= \mathrm{E}_\gamma + \mathrm{E}_0\,(\text{electron}) + E_\mathrm{k}\,(\text{electron}) - E_0\,(\text{electron})\\ &= \mathrm{E}_\gamma + \mathrm{E}_\mathrm{k}\end{aligned}$$

The magnitudes of the momenta of the incident photon (p_γ) and electron (p_e) must be equal relative to the center of mass. From this we can find the kinetic energy of the incident electron:

$$\begin{aligned}E_\mathrm{k} &= \frac{\mathrm{p_e}^2}{2m_\mathrm{e}} = \frac{\mathrm{p_e}^2c^2}{2E_0} = \frac{(\mathrm{p}_\gamma\, c)^2}{2E_0} = \frac{E_\gamma^{\,2}}{2E_0}\\ &= \frac{(6200\ \mathrm{eV})^2}{2(5.1\times10^5\ \mathrm{eV})} = 38\ \mathrm{eV}\end{aligned}$$

Therefore, $\Delta E = (6200 + 38)\ \mathrm{eV} \approx 6200\ \mathrm{eV}$, and Equation 12–1 gives the maximum duration:

$$\begin{aligned}\Delta t &\le \frac{\hbar}{\Delta E} = \frac{1.1\times10^{-34}\ \mathrm{J\cdot s}}{(6.2\times10^3\ \mathrm{eV})(1.6\times10^{-19}\ \mathrm{J/eV})}\\ &\le 1\times10^{-19}\ \mathrm{s}\end{aligned}$$

Example 12–2. An electron and a proton, both initially at rest, exchange a virtual photon of energy 1 eV. (*a*) Show a Feynman diagram of this electromagnetic interaction and estimate the time duration Δt and spatial extension Δx of the virtual photon. (*b*) Find the increase in kinetic energy of the outgoing electron-proton pair and identify the source of this energy.

(*a*) Figure 12–7 is a schematic representation of the attractive electromagnetic interaction between an electron and a proton, with both charge particles initially at rest. The electron emits a virtual photon at the first vertex and recoils in the *same* direction as that of the photon. At a later time Δt, the virtual photon, after traveling a distance $\Delta x = c\,\Delta t$, is absorbed by the proton, which recoils in a direction *opposite* to that of the photon. Thus, the exchange of this virtual photon results in an attractive force between the two oppositely charged particles. Obviously, the emission of the virtual photon at the first vertex is an event in which neither energy nor momentum is necessarily conserved, and the nonconservation persists until the photon is absorbed by the proton at the second vertex. The time interval Δt and the space interval Δx during which the virtual photon is exchanged are limited by Equation 12–1.

Assume, for simplicity, that momentum is conserved at each vertex (even though the virtual photon is unobservable). This requires that at the first vertex the photon should be created and move to the right; but it must have a negative momentum (that is, $p_\gamma = -E_\gamma/c$). At the second vertex, the photon imparts this negative momentum (and positive kinetic energy) to the proton. We then have the following momenta for the outgoing electron and proton:

$$\mathrm{p_e} = -\mathrm{p_p} = -\mathrm{p}_\gamma = \frac{E_\gamma}{c}$$

where $E_\gamma = 1$ eV.

The time and space intervals for the virtual photon are, by Equation 12–1,

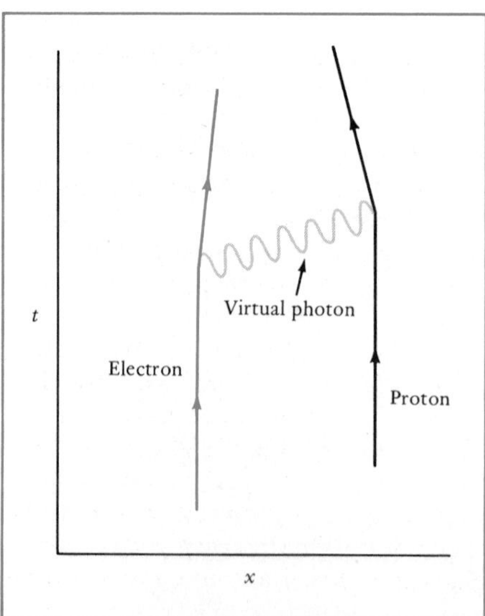

FIGURE 12–7. *Space-time diagram of the interaction between an electron and a proton.*

$$\Delta t = \frac{\hbar}{\Delta E} = \frac{1.1 \times 10^{-34}\ \text{J}\cdot\text{s}}{(1\ \text{eV})(1.6 \times 10^{-19}\ \text{J/eV})} = 7 \times 10^{-16}\ \text{s}$$

and $$\Delta x = c\,\Delta t = c\left(\frac{\hbar}{\Delta E}\right) = \frac{hc}{2\pi\,\Delta E} = \frac{12\,400\ \text{Å}\cdot\text{eV}}{2\pi(1\ \text{eV})} \approx 2000\ \text{Å}$$

(*b*) In magnitude, $p_e = p_p = E_\gamma/c$. Therefore, the final kinetic energies of the outgoing electron and proton are

$$E_{k_e} = \frac{p_e^2}{2m_e} = \frac{(p_e c)^2}{2E_{0_e}} = \frac{E_\gamma^2}{2E_{0_e}} = \frac{(1\ \text{eV})^2}{2(5 \times 10^5\ \text{eV})} = 1 \times 10^{-6}\ \text{eV}$$

$$E_{k_p} = \frac{E_\gamma^2}{2E_{0_p}} = \frac{(1\ \text{eV})^2}{2(1 \times 10^9\ \text{eV})} = 5 \times 10^{-10}\ \text{eV}$$

Thus, the kinetic energy of the outgoing pair has increased by $(1 \times 10^{-6} + 5 \times 10^{-10})\ \text{eV} \approx 1 \times 10^{-6}\ \text{eV}$

The kinetic-energy increase arising from this single virtual photon exchange is but a small part of the total increase in kinetic energy coming from the loss in electrostatic potential energy as the charges move closer together. If we were to add the contributions of all virtual photons being exchanged between the two charged particles as they move closer together, we would get the Coulomb force law, with the kinetic-energy increase attributed to a Coulomb electrostatic potential-energy decrease of magnitude $ke^2(1/r_1 - 1/r_2)$.

12–2 Other Fundamental Interactions

The electromagnetic interaction is one of the four known types of interaction among particles. The other three are the "strong" interaction, the "weak" interaction, and the gravitational interaction. The electromagnetic interaction is responsible for atomic and molecular structure (Sections 6–4 and 8–1). The "strong" interaction is responsible for the large binding energy among protons and neutrons within the nuclei of atoms (Section 10–6). The "weak" interaction is responsible for the radioactive β decay of unstable nuclides (Section 10–10). The gravitational interaction is responsible for the binding energy among the sun and planets in the solar system.

The great success of the quantum field theory in describing the electromagnetic interaction between two charged particles in terms of the exchange of field particles (virtual photons) suggests that each of the other three interactions can also be described by the exchange of virtual particles. We follow this approach and see what successes and also what problems arise from this simple but useful point of view.

First, recall three properties of the photon. (1) There is no conservation law restricting the number of photons. (2) The Pauli exclusion principle does not apply to the photons (Section 8–2). Therefore, photons (real or virtual) must have integral spin. Indeed, the spin of a photon is 1, with a spin angular momentum magnitude of $\sqrt{s(s+1)}\hbar = \sqrt{2}\hbar$. (3) The photon travels at the speed of light c; therefore, by relativity theory, its rest mass must be zero. It follows from the uncertainty principle that the range of the electromagnetic interaction is infinite; by Equation 12–1, a virtual photon of total energy $\Delta E = E_0 + E_k$ can exist for a time

$$\Delta t = \frac{\hbar}{\Delta E} = \frac{\hbar}{E_0 + E_k} = \frac{\hbar}{E_k}$$

since $E_0 = 0$ for a photon. Therefore, the range of a virtual photon is

$$R = c\,\Delta t = \frac{c\hbar}{\Delta E} < \frac{c\hbar}{E_k} = \frac{c\hbar}{h\nu}$$

and $R \to \infty$ as $\nu \to 0$. This is consistent with the $1/r^2$ behavior of the electromagnetic interaction. Field particles representing the other three fundamental interactions have properties like those of the photon. A summary of the properties of the field particles is, then,

1. No conservation law restricting the number of field particles.
2. *Integral* spin angular momentum quantum number (field particles are bosons).
3. Range of interaction limited by the rest mass of the virtual field particle according to uncertainty principle,

$$R = v\,\Delta t < \frac{v\hbar}{E_0 + E_k} < \frac{\hbar c}{E_0} \qquad (12\text{–}2)$$

Therefore, the range in angstroms is

$$R < \frac{12{,}400\ \text{Å}\cdot\text{eV}}{2\pi E_0\ (\text{eV})}$$

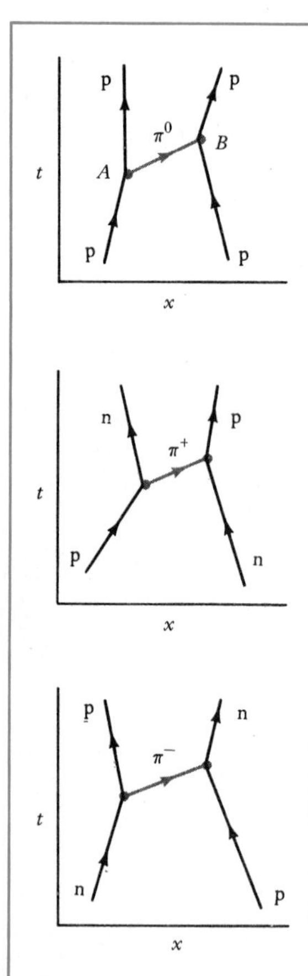

FIGURE 12–8. *Nucleon-nucleon interactions through the exchange of virtual pions π^0, π^+, and π^-.*

Turning first to the strong interaction, with the nuclear force acting between nucleons (it acts also between other particles, to be discussed later), we inquire what properties the field particles would have. As we have seen (Section 10–5), for nucleon separation distances less than 1.4 fm $= 1.4 \times 10^{-15}$ m, the interaction is very large; it is over 100 times the electromagnetic interaction between two protons at this separation. But unlike the electromagnetic interaction, the strong interaction drops abruptly to zero at separation distances greater than 1.4 fm; the strong interaction is a short-range force.

This question of the short-range behavior of the nuclear interaction and the characteristics of the corresponding field particles was first posed (and answered) by the Japanese physicist H. Yukawa in 1935. He hypothesized that the nuclear force between nucleons is mediated through the exchange of virtual particles associated with the nuclear force field. The essential characteristics of these particles—now called *π mesons* (pi mesons) or simply *pions*—which act as agents of the nuclear force, can be deduced by a simple argument based on the uncertainty principle. The strong nuclear interactions between nucleons are described by Feynman diagrams like that depicting the electromagnetic interaction, Figure 12–5. Figure 12–8 shows three possible virtual meson exchanges. The world line of a meson is here represented by a gray line.

In the first diagram of Figure 12–8 a proton creates and emits a virtual neutral π meson, π^0 (pi zero), at the vertex A; a short time Δt later, a second proton absorbs this pion, at the vertex B. During the time Δt of existence of the pion, the energy-conservation principle can be violated, so long as its violation is consistent with the uncertainty principle $\Delta E\,\Delta t \approx \hbar$, where ΔE

now represents the "borrowed" energy during the exchange. When the meson is absorbed by the second proton at the later time, vertex B, the energy of the system is again restored. Similar meson exchanges are shown for the proton-neutron interactions in the other diagrams of Figure 12–8. To describe the three nucleon-nucleon interactions, note that three distinct pions (π^+, π^0, and π^- mesons) with charges of $+1$, 0, and -1 (in units of the electron charge e) are necessary to ensure conservation of electric charge at each vertex.

The strong nucleon-nucleon interaction is characterized by a short-range force of about 1.4 fm (Section 10–2). This limits the travel distance R of the virtual meson before it is again absorbed. If we assume, for simplicity, that the meson travels at a speed near that of light, then it exists only for the time interval Δt of the exchange:

$$\Delta t \approx \frac{R}{c} = \frac{1.4 \times 10^{-15}\ \text{m}}{3 \times 10^{8}\ \text{m/s}} = \tfrac{1}{2} \times 10^{-23}\ \text{s}$$

Using the uncertainty principle, we find the borrowed energy to be

$$\Delta E = \frac{\hbar}{\Delta t} = \frac{10^{-34}\ \text{J}\cdot\text{s}}{\frac{1}{2} \times 10^{-23}\ \text{s}} = 2 \times 10^{-11}\ \text{J} \approx 130\ \text{MeV}$$

Assuming that the "borrowed" energy is primarily the rest energy $E_\pi = m_\pi c^2$ of the virtual pion, we find the pion's approximate rest mass to be

$$m_\pi = \frac{E_\pi}{c^2} \approx \frac{\Delta E}{c^2} = \frac{2 \times 10^{-11}\ \text{J}}{(3 \times 10^{8}\ \text{m/s})^2} \approx 2 \times 10^{-28}\ \text{kg} \approx 200\ m_e$$

Therefore, because of the short-range character, we expect the field particles associated with the strong interaction to have a *finite* rest mass of the order of a few hundred electron masses m_e.

The existence of all three pions Yukawa predicted has been confirmed experimentally: first, by observations of collisions of high-energy nucleons in cosmic rays with nucleons in the earth's atmosphere; then, under controlled laboratory conditions, with accelerators having particle energies of several hundred MeV and more. Typical pion-production reactions are $\text{p} + \text{n} \rightarrow \text{p} + \text{n} + \pi^0$ and $\text{p} + \text{n} \rightarrow \text{p} + \text{n} + \pi^- + \pi^+$. In both reactions, some of the incident kinetic energy is transformed into rest energy of the created pions.

Some important properties of the three charge states of pions are listed in Table 12–2. The pion is found, like the photon, to satisfy the three requirements of field particles (Equation 12–2). In the production and decay of pions, there is no conservation law for the number of pions created or annihilated in an interaction; all pions have an integral spin 0; and the finite pion rest mass ($\sim$140 MeV) is consistent with the short-range nuclear force.

The photon and pion do differ in some ways. Whereas the photon has spin 1, the pion has spin 0; whereas the photon is electrically neutral and has only one charge state (0), the pion exists in three charge states; whereas the photon has zero rest mass corresponding to an infinite range for the electromagnetic interaction, the pion's mass is finite and accounts for the short-range nuclear force; and whereas a free photon is stable, all free pions are unstable.

Table 12–2 *Properties of pions*

	π^+	π^-	π^0
Rest mass, m_e	273.3	273.3	264.3
Rest energy, MeV	139.569	139.569	134.964
Charge, e	+1	−1	0
Spin, $\times\hbar$	0	0	0
Magnetic moment	0	0	0
Mean lifetime, s	2.603×10^{-8}	2.603×10^{-8}	8.3×10^{-17}
Decay modes	$\pi^+ \rightarrow \mu^+ + \nu_\mu$ (99.99%) $\pi^+ \rightarrow e^+ + \nu_e$ (0.0127%)	$\pi^- \rightarrow \mu^- + \bar{\nu}_\mu$ $\pi^- \rightarrow e^- + \bar{\nu}_e$	$\pi^0 \rightarrow \gamma + \gamma$ (98.85%) $\pi^0 \rightarrow e^+ + e^- + \gamma$ (1.15%)

From Table 12–2, we see that free pions decay by one of two fundamental interactions. The two decay modes of the π^0 involve photons; the first decay (98.8 percent of all π^0 decays) produces two photons and the other decay (1.2 percent of all π^0 decays) produces one photon with an electron and a positron. Therefore, the π^0 decay takes place through the relatively fast electromagnetic interaction (lifetime $\sim 10^{-16}$ s). The decay modes of the charged pions, on the other hand, involve electrons; muons, which are new particles like electrons but with a larger rest mass ($\sim 200\ m_e$); and neutrinos, chargeless particles of apparently zero rest mass, and their antiparticles. These decays are reminiscent of the decay of a free neutron into a proton, an electron, and an antineutrino, and they too take place through the relatively slow weak interaction. The strength of this interaction is approximately 10^{-13} times the strength of the strong interaction and is known to have a very short range, less than 10^{-2} fm. The postulated virtual particle for the weak interaction, commonly called the W (for weak) particle, is believed to have three charge states (like that of the pion of the strong interaction) and a spin of 1. The rest mass of the W particle can be estimated from Equation 12–2,

$$E_0 \geq \frac{\hbar c}{R} = \frac{hc}{2\pi R} = \frac{12\,400\ \text{Å}\cdot\text{eV}}{2\pi(10^{-2}\ \text{F})(10^{-5}\ \text{Å/F})} \approx 20\ \text{GeV}$$

Figure 12–9 shows Feynman diagrams for possible weak interactions involving the exchange of each of the three charge states of virtual particles. Note the similarity between Figures 12–9 and 12–8. The virtual mesons and the nuclear world lines on the right-hand side of each diagram in Figure 12–8 are replaced by the corresponding virtual W particles and electron-neutrino world lines.

The fourth fundamental interaction is the gravitational. Although this interaction is the weakest, it is the only one acting between all particles, and it appears to be only attractive in character. Its feeble strength (10^{-43} times the strong interaction) is illustrated by the attraction of an electron and a proton for one another gravitationally by a force that is only 10^{-40} the Coulomb electric force between them at the same separation distance. Gravitational waves have not yet been observed directly, and their quantization into field particles of extremely low energies has also not been observed. But physicists are confident that these gravitational field particles, referred to as *gravitons*, exist. Because the gravitational force is inverse square, and

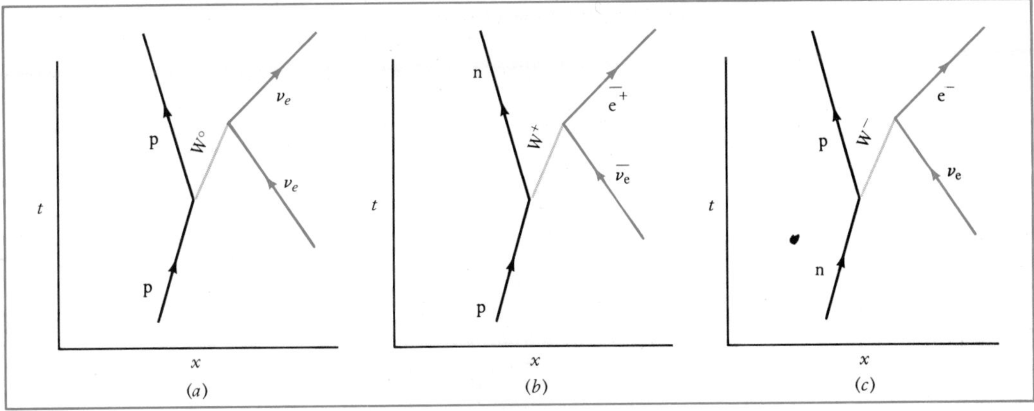

FIGURE 12–9. *Weak interactions through the exchange of virtual* W *particles* W^0, W^+, *and* W^-.

therefore of infinite range, the graviton rest mass is zero. Its spin is expected to be 2.

Table 12–3 lists the four fundamental interactions that govern the forces between elementary particles, together with some of the properties of the field particles mediating the respective interactions. The table shows an inverse relation between the strength and the time of an interaction. This behavior is illustrated in the decay of unstable particles. The stronger the interaction, the more rapid the decay. A particle will always decay by the strongest interaction allowed by the conservation laws. Since the decay time also depends on other factors (such as the decrease in total rest energy and the number of decay products), there is a range of decay lifetimes. For example, particles decaying by the strong interaction have lifetimes ranging from 10^{-25} to 10^{-19} s; those decaying by the electromagnetic interaction, 10^{-19} to 10^{-16} s; and those by the weak interaction, from 10^{-15} to 10^{3} s.

The maximum range of an interaction is also seen to vary inversely with the rest mass of the exchange particle. All four exchange particles, each with an integral spin, are bosons and obey Einstein-Bose statistics (Section 8–2).

Table 12–3. *The four fundamental interactions*

	RELATIVE STRENGTH	INTERACTION TIME (s)	FIELD PARTICLE	MASS	RANGE	CHARGE STATES	SPIN
Strong	1	10^{-23}	Meson	140 MeV (pion)	1 F	$\pi^+\pi^0\pi^-$	0 (pion)
Electro-magnetic	10^{-2}	10^{-17}	Photon	0	∞	γ	1
Weak	10^{-13}	10^{-10}	W particle[a]	~80 GeV	Less than 10^{-2} F	$W^+W^0W^-$	1
Gravita-tional	10^{-40}	10^{17}	Graviton[a]	0	∞	G	2

a Not yet detected.

Are the four "fundamental" interactions really separate? Scientists have found many times in the past that what appeared to be disparate phenomena were in fact unified. A few examples are these:

1. Newton showed that the force between heavenly bodies was the same as that between earth and objects on its surface.
2. Many investigators were involved in the discovery that static electricity had the same origin in electric charge as current electricity.
3. Oersted and others showed that magnetism has its origin in electric currents.
4. Maxwell unified all electric and magnetic effects in his classical electromagnetic theory, which also showed light to be but one special manifestation of electromagnetic waves.
5. Einstein in his general theory of relativity showed that the gravitational interaction was describable in terms of space-time geometry.

Einstein also attempted for many years to produce a single unified field theory amalgamating the gravitational and electromagnetic interactions, but without success. There has, however, been success more recently in combining two of the four interactions—the weak and the electromagnetic—into a unified gauge theory, especially as developed by S. Weinberg and A. Salam.

Still more comprehensive schemes have been postulated which describe and unite properties of fundamental particles and the interactions. For example, so-called supersymmetry relates fermions to bosons, and supergravity unites all four interactions through a combination of general relativity and quantum theory.

12-3 Properties of Observed Fundamental Particles

Except for very rare occasions, what matter we observe on the earth is describable in terms of the particles discussed in Section 12–2. The heavy nucleons (protons and neutrons) are strongly bound together by the virtual pions ($\pi^+\pi^0\pi^-$) to form the nuclei of atoms. The much lighter electrons (e^-) are bound to the positively charged nuclei by virtual photons (γ) to form atoms. The long-lived naturally radioactive nuclides remaining in the earth decay into unstable and stable daughters, resulting in the emission of high-energy ($\sim$MeV)α, β^-, and γ particles.

There is a large abundance of real photons in the earth's environment, coming not only from the sun (and stars), but also from the endless inelastic collisions among the earth's atoms and molecules. Few positrons ($\overline{e^+}$) exist because of the high energy necessary ($\sim$1 MeV) for the creation of an electron-positron pair, and even fewer real pions are detectable because of their large rest mass ($\sim$140 MeV). For particle kinetic energies limited to 100 MeV or less, the fundamental particles detectable at the earth are limited to 15; two nucleons (p, n), three pions ($\pi^+\pi^0\pi^-$), a muon-antimuon pair (μ^-, μ^+), an electron-positron pair (e^-, $\overline{e^+}$), two neutrino pairs ($\nu_e\bar{\nu}_e$ $\nu_\mu\bar{\nu}_\mu$), a photon (γ), and a graviton (G). Of the fifteen particles, six are unstable when free. The neutral pion decays by the *electromagnetic interaction;* the neutron, two charged pions, muon and the antimuon all decay by the weak interaction.

Over the last few decades, experiments involving accelerators to produce particles of higher and higher energy have increased the list and complexity of "particles" to where it now numbers over 200. With so many particles, there is a serious question whether all are really elementary, and scientists have been evolving new theoretical models to explain this large number of observed particles by smaller groups of fundamental particles. Before examining one such model (Section 12–7), we list some of the well-established particles and their properties. Table 12–4 lists those particles that have been observed, directly or indirectly, to exist as distinct particles for a time greater than 10^{-20} s. This therefore excludes the many very short-lived particles (so-called *resonance* particles) that decay by the strong interactions. (See later sections.)

The particles in Table 12–4 are listed in order of increasing rest energy.† They are also grouped according to family. The *photon,* with spin 1, is in a category by itself. The *lepton* (Greek, "light") family, with spin $\frac{1}{2}$, includes the electron and its associated neutrino and the muon and its associated neutrino. The *meson* (Greek, "intermediate") family, with spin 0, includes pions, kaons, an eta particle, and charmed D particles. The *baryon* (Greek, "heavy") family consists of the nucleons (proton and neutron) and also more massive particles called *hyperons* (Greek, "over"), the lambda, sigma, xi, and omega particles. Since all baryons and mesons participate in the strong interaction, they are collectively referred to as *hadrons* (Greek, "strong").

Each of the particles in Table 12–4 has an integral charge (in units of proton charge), either $+1$, 0, or -1. To every particle whose properties are listed in the table there corresponds an antiparticle (indicated by an overbar), which has identical properties except for a *sign reversal* in all properties having positive and negative values: baryon number, lepton number, electric charge, magnetic moment, hypercharge, isospin component, and the charge of each of the decay products. The photon, neutral pions, and eta particle are their own antiparticles and the two charged pion are antiparticles of one another. Note that the sign of the charge indicated for the antiparticle is opposite that for the particle.

Every particle is characterized by a distinctive mean life, the average lifetime from creation to decay of many identical particles. Only nine of the particles and antiparticles are stable against spontaneous decay: the proton, electron, two neutrinos, their associated antiparticles, and the photon. (Experimental measurements show that the lifetime of the proton is greater than 10^{30} yr and of the electron, greater than 10^{21} yr.)

Every elementary particle has a distinctive intrinsic, or *spin,* angular momentum, visualizable classically as the perpetual spinning of the particle about an internal rotation axis. The spin angular momentum is specified by the spin quantum number J, where the magnitude of the spin angular momentum is $\sqrt{J(J+1)}\hbar$. The values of J are either integral or half-integral. As we know, any group of identical particles with half-integral spin values are known as *fermions,* and they obey the Pauli exclusion principle. This

† See Appendix III for the reference source of updated values of the properties of fundamental particles.

Table 12–4. *Fundamental particles with* $\tau > 10^{-21}$ s

	FAMILY NAME	PARTICLE NAME	SYMBOL (AND ELECTRIC CHARGE)	ANTIPARTICLE SYMBOL[a]	REST ENERGY, (MeV)	MEAN-LIFE τ (s)	SPIN ANGULAR MOMENTUM	HYPERCHARGE $2\overline{Q}$ (UNIT, e)	ISOSPIN I	PRINCIPAL DECAY MODES (FRACTION IF > 5%)
	PHOTON	Photon	γ	Same	0	∞	1	0	0 or 1	. . .
	LEPTON	Neutrino	ν_e	$\overline{\nu}_e$	0	∞	$\frac{1}{2}$	. . .	.	. . .
NONHADRON	$L_e = +1$ for		ν_μ	$\overline{\nu}_\mu$	0	∞	$\frac{1}{2}$	. . .	. . .	. . .
	e^- and ν_e	Electron	e^-	e^+	0.511003	∞	$\frac{1}{2}$	. . .	. . .	. . .
	$L_\mu = +1$ for μ^- and ν_μ	Muon	μ^-	$\overline{\mu^+}$	105.6595	2.197×10^{-6}	$\frac{1}{2}$	. . .	. . .	$\mu^- \to e^- + \overline{\nu}_e + \nu_\mu$
	MESON	Pion	π^+,π^-	π^-,π^+	139.59	2.602×10^{-8}	0	0	1	$\pi^+ \to \overline{\mu^+} + \nu_\mu$ (100)
			π^0	Same	134.964	0.83×10^{-16}	0	0	1	$\pi^0 \to \gamma + \gamma$ (98.8)
		Kaon[b]	K^+	$\overline{K^-}$	493.70	1.237×10^{-8}	0	+1	$\frac{1}{2}$	$K^+ \to \overline{\mu^+} + \nu_\mu$ (63.6)
										$K^+ \to \pi^+ + \pi^0$ (21.0)
										$K^+ \to \pi^+ + \pi^+ + \pi^-$ (5.6)
			$K^0 = \frac{1}{2}(K_S{}^0 + K_L{}^0)$	$\overline{K^0}$	497.7		0	+1	$\frac{1}{2}$	. . .
			$(K_S{}^0)$			0.893×10^{-10}				$K_S{}^0 \to \pi^+ + \pi^-$ (68.7)
										$K_S{}^0 \to \pi^0 + \pi^0$ (31.3)
			$(K_L{}^0)$			5.18×10^{-8}				$K_L{}^0 \to \pi^0 + \pi^0 + \pi^0$ (21.4)
										$K_L{}^0 \to \pi^+ + \pi^- + \pi^0$ (12.2)
										$K_L{}^0 \to \pi^\pm + \mu^\mp + \nu_\mu$ (27.1)
										$K_L{}^0 \to \pi^\pm + e^\mp + \nu_e$ (39.0)
		Eta	η^0	Same	548.8	7×10^{-19}	0	0	0	$\eta_0 \to \gamma + \gamma$ (38)
HADRON		Charmed D	D^0	$\overline{D^0}$	1863	$\sim 10^{-13}$	0	0	$\frac{1}{2}$	$D^0 \to K^+ + \pi^-$
			D^+	$\overline{D^-}$	1868	$\sim 10^{-13}$	0	0	$\frac{1}{2}$	$D^+ \to K^- + \pi^+ + \pi^+$
	BARYON	Proton	p^+	$\overline{p^-}$	938.280	∞	$\frac{1}{2}$	1	$\frac{1}{2}$	. . .
	$B = +1$	Neutron	n	$\overline{n}$	939.573	0.92×10^3	$\frac{1}{2}$	1	$\frac{1}{2}$	$n \to p^+ + e^- + \overline{\nu}_e$
		Lambda	Λ^0	$\overline{\Lambda^0}$	1115.60	2.5×10^{-10}	$\frac{1}{2}$	0	0	$\Lambda_0 \to p^+ + \pi^-$ (64)
										$\Lambda_0 \to n + \pi^0$ (36)
		Sigma	Σ^+	$\overline{\Sigma^-}$	1189.4	0.800×10^{-10}	$\frac{1}{2}$	0	1	$\Sigma^+ \to p^+ + \pi^0$ (52)
										$\Sigma^+ \to n + \pi^+$ (48)
			Σ^0	$\overline{\Sigma^0}$	1192.5	$<10^{-14}$	$\frac{1}{2}$	0	1	$\Sigma^0 \to \Lambda^0 + \gamma$
			Σ^-	$\overline{\Sigma^+}$	1197.4	1.49×10^{-10}	$\frac{1}{2}$	0	1	$\Sigma^- \to n + \pi^-$
		Xi	Ξ^0	$\overline{\Xi^0}$	1314.9	3.0×10^{-10}	$\frac{1}{2}$	−1	$\frac{1}{2}$	$\Xi^0 \to \Lambda^0 + \pi^0$
			Ξ^-	$\overline{\Xi^+}$	1321.3	1.65×10^{-10}	$\frac{1}{2}$	−1	$\frac{1}{2}$	$\Xi^- \to \Lambda^0 + \pi^-$
		Omega	Ω^-	$\overline{\Omega^+}$	1672.2	1.3×10^{-10}	$\frac{3}{2}$	−2	0	$\Omega^- \to \Xi^0 + \pi^-$ (?)
										$\Omega^- \to \Xi^- + \pi^0$ (?)
										$\Omega^- \to \Lambda^0 + K^-$ (?)

a Antiparticles are designated by a color overbar. Note that the antiparticle always has a charge opposite that of the particle.
b The neutral kaon K^0 is a superposition of two states that have different decay times.

restricts the number of particles in a given quantum state to one and thereby acts effectively as a repulsive force between identical particles. All the leptons in Table 12–4 have spin $\frac{1}{2}$ and are fermions. Similarly, the baryons have half-integral spin, the omega-minus with spin $\frac{3}{2}$, and all others with spin $\frac{1}{2}$. Therefore, baryons are also fermions. On the other hand, the field particles in Table 12–4, the photon and the mesons, have integral spin and are bosons. They are not constrained by the Pauli principle and can therefore appear without restriction in their number.

The definition and importance of the terms *hypercharge* and *isospin* in Table 12–4 in classifying elementary particles will be given in Section 12–5. The final column lists the principal decay modes of those particles that are unstable. Still other decay modes for some of these particles have been observed but are not shown in the table because the fraction of decays is lower than 5 percent. For every entry of a particle decay in the table, there is a complementary decay in which each particle in the decay is replaced by the corresponding antiparticle. For example, since a neutron decays into a proton, and an electron antineutrino,

$$\mathrm{n} \rightarrow \mathrm{p}^+ + \mathrm{e}^- + \bar{\nu}_\mathrm{e}$$

then we expect to find in replacing each particle by its antiparticle, the decay of the antineutron according to the decay

$$\overline{\mathrm{n}} \rightarrow \overline{\mathrm{p}^-} + \overline{\mathrm{e}^+} + \nu_\mathrm{e}$$

with the same lifetime, 920 s, as the neutron.

The decay times of the unstable particles in Table 12–4 are all much longer than the interaction time 10^{-23} s associated with the strong force (see Table 12–3). Three particles (π°, η°, Σ^0) decay by the relatively fast electromagnetic interaction with mean lives of 10^{-15} to 10^{-19} s; all other unstable particles here decay by the much slower weak interaction, with mean lives ranging from 10^{-10} to 10^3 s.

Thirty-nine particles and antiparticles appear in Table 12–4 (1 photon, 8 leptons and antileptons, 12 mesons and antimesons, and 18 baryons and antibaryons). Adding the postulated graviton and three W particles, we find 43 "elementary" particles, all of which are stable against decay by the strong interaction, and 40 of which are also stable against decay by the electromagnetic interaction. Although this is an uncomfortably large number of particles to be regarded as truly fundamental building blocks, the picture becomes even more disturbing if we include the more than 200 new particles, called *resonance particles,* that also exist. The resonance particles have very short lives, 10^{-24} to 10^{-21} s, before they decay by the strong interaction. The next sections will be devoted to classifying these particles according to symmetry properties and associated conservation laws and discussing one proposed model which describes nature in terms of a smaller set of fundamental building blocks.

12-4 Universally Valid Conservation Laws

Why do the elementary particles that have been observed (those listed in Table 12–4 and also many others) appear in only a severely limited number of charge states? Why do they have the specific masses, mean lives, spins, and other basic properties given in Table 12–4, and not other values? What patterns of coherence can be discerned in their properties and interactions? Why are there so many particles, and are they truly elementary or is it possible that they are composed of fewer more basic constituents?

Although many of the fundamental questions remain unanswered, at least one kind of question can be answered straightforwardly—Why do unstable particles decay in the particular modes, as shown in Table 12–4, and not in other conceivable decay modes? Or if a particle is stable, why does it not decay? The answer lies in conservation laws. Even at the elementary-particle level, there are many conservation principles, and these are very important in understanding the processes that can occur. A useful guiding principle for predicting physically allowed processes has been that there is complete freedom in particle interactions except as limited by all relevant conservation laws for each of the fundamental interactions. This principle still holds. One takes the attitude that *any* conceivable process can occur unless some basic physical principle (conservation law) precludes it, recognizing, however, that not all conceivable processes for a given reaction will occur with the same probability. On the contrary, if a conceivable process is never observed, then some fundamental physical law must exist that prohibits it.

A simple example is given by the behavior of neutrons and protons. In an isolated nuclide, each neutron or proton interacts with the other neutrons and protons through the strong interaction (Figure 12–8). The nuclide is always composed of a fixed number of neutrons and a fixed number of protons.

On the other hand, when a neutron is free of other nucleons, it is unstable and always decays by the weak interaction, whereas a free proton is stable and does not decay by any of the four interactions. First, consider the proton. Why does it not decay into a positron and a photon? Such a process would not violate the four conservation laws of mass-energy, linear momentum, angular momentum, and electric charge. One must conclude that this process is ruled out by one or more additional conservation laws—laws that are formulated simply to summarize the nonexistence of certain otherwise conceivable processes. Similarly, why does the free neutron not decay by the strong interaction into a proton and a negative pion? or by the electromagnetic interaction into a proton and a photon? or by the weak interaction into a neutrino and an antineutrino? The neutron decay by the strong interaction is ruled out by mass-energy conservation, and the decay by the electromagnetic interaction is prohibited by charge conservation. Neutron decay into neutrino and antineutrino by the weak interaction, however, does not violate any of the four conservation laws thus far discussed; this process therefore must be prohibited by still other conservation laws.

Two additional conservation laws, together with the well-established conservation laws of energy, linear momentum, angular momentum, and electric charge, are now believed to be universally valid and thereby applicable to *all* fundamental interactions. These new universally conserved properties are *baryon number* and *lepton number*. (The lepton category is subdivided into separate conservation laws of *electron-type leptons* and *muon-type leptons*.) Members of the lepton family and the more stable members of the baryon family are listed in Table 12–4.

A special form of the baryon number conservation law is the conservation of nucleons (Section 10–7), according to which the total number of nucleons (protons and neutrons) in any nuclear decay or low-energy reac-

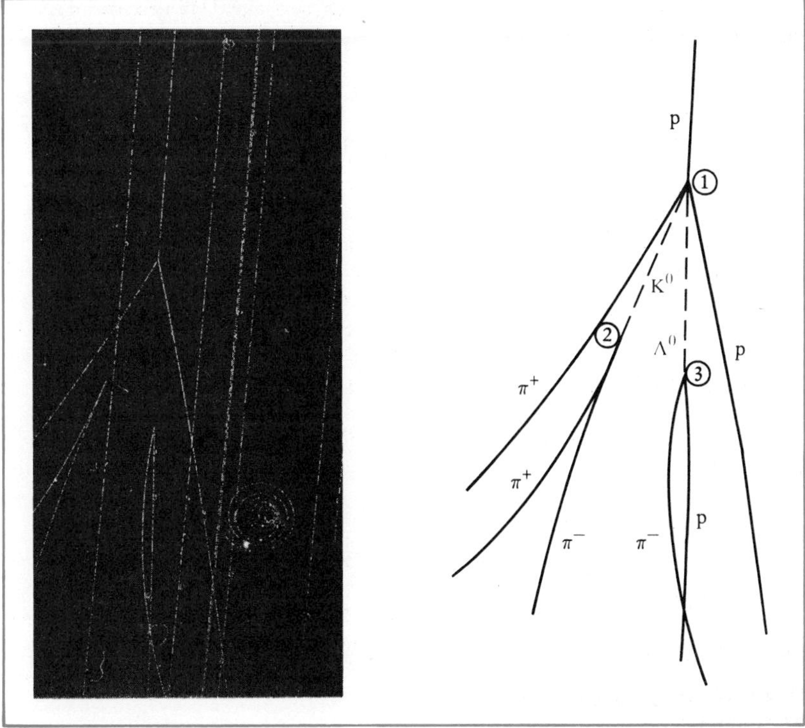

FIGURE 12–10. *A bubble-chamber photograph of a 2.85-GeV proton colliding with a proton and creating a* K^0 *and a* Λ^0. *The* K^0 *decays to two pions.*

tion (particle kinetic energies less than 1 GeV) is constant. In the more general baryon number conservation law, one assigns a baryon number $B = +1$ to particles in the baryon family, a baryon number $B = -1$ to every antiparticle in the baryon family, and a baryon number $B = 0$ to all other particles. Then the conservation law is simply the statement that the *total baryon number* for *any* process is a constant. This conservation law precludes such decays as the decay of the proton ($B = +1$) into a photon ($B = 0$) and a positron ($B = 0$), and the decay of a neutron ($B = +1$) into a neutrino ($B = 0$) and an antineutrino ($B = 0$).

The bubble-chamber photograph in Figure 12–10 illustrates the conservation of baryons. A 2.85-GeV proton collides with a proton at rest at point 1 in the liquid-hydrogen bubble chamber and four particles emerge from the interaction:

$$p^+ + p^+ \rightarrow p^+ + \pi^+ + \Lambda^0 + K^0$$

$$\text{Baryon number:} \quad 1 + 1 = 1 + 0 + 1 + 0$$

The two emerging charged particles (p^+ and π^+) leave tracks in the liquid, and the two neutral particles (Λ^0 and K^0) are identified when they decay later into charged particles at points 2 and 3:

$$\Lambda^0 \rightarrow p^+ + \pi^- \qquad K^0 \rightarrow \pi^+ + \pi^-$$

$$\text{Baryon number:} \quad 1 = 1 + 0 \qquad 0 = 0 + 0$$

Still other illustrations of baryon number conservation are found in the decay schemes in Table 12–4.

The conservation laws for lepton number apply to the lepton family: electron, muon, two varieties of neutrino (ν_e, ν_μ), and the four antiparticles. The necessity of recognizing four distinct types of neutrinos (ν_e, $\bar{\nu}_e$, ν_μ, $\bar{\nu}_\mu$) and two separate conservation laws will be explained shortly. *Electron number* $L_e = +1$ is assigned to the electron e^- and to the electron-type neutrino ν_e, and electron number $L_e = -1$ to the positron e^+ and electron-type antineutrino $\bar{\nu}_e$. All other particles have electron number $L_e = 0$. Similarly, muon number $L_\mu = +1$ is assigned to the muon μ^- and to the muon-type neutrino ν_μ, while the antimuon μ^+ and muon-type antineutrino $\bar{\nu}_\mu$ are assigned muon number $L_\mu = -1$. All other particles have muon number $L_\mu = 0$. With these assignments, all observations are consistent with the *lepton conservation laws:* The *total electron number* L_e is *constant,* and the *total muon number* L_μ is *constant.*

Examples of the conservation laws for leptons are furnished by the principal decay modes of the charged pion and the neutron in Table 12–4.

Charged pion decay:	$\pi^- \to \mu^- + \bar{\nu}_\mu$	$\pi^+ \to \mu^+ + \nu_\mu$
Electron no.:	$0 = 0 + 0$	$0 = 0 + 0$
Muon no.:	$0 = 1 + (-1)$	$0 = (-1) + 1$

The muons then decay as follows (see Figure 12–11).

Muon decay:	$\mu^- \to e^- + \bar{\nu}_e + \nu_\mu$	$\mu^+ \to e^+ + \nu_e + \bar{\nu}_\mu$
Electron no.:	$0 = 1 + (-1) + 0$	$0 = (-1) + 1 + 0$
Muon no.:	$1 = 0 + 0 + 1$	$(-1) = 0 + 0 + (-1)$

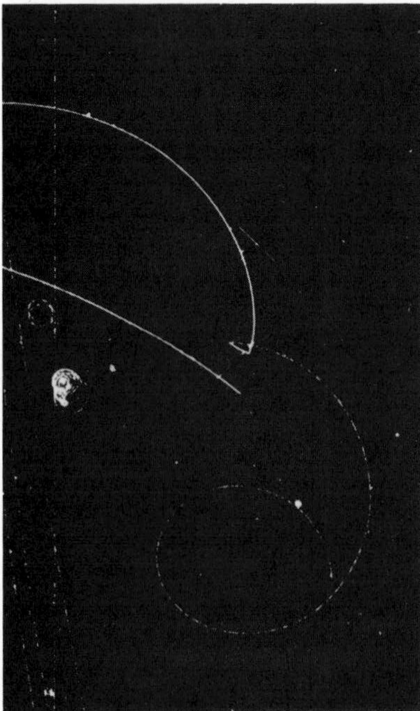

FIGURE 12–11. *A liquid-hydrogen bubble-chamber photograph showing a pion decaying into a muon, which in turn decays into an electron. The pion enters from the upper left in a downwardly curved path. The muon track is short; the electron track is thin. (Courtesy, Fermi National Accelerator Laboratory)*

The decay of the neutron and the related neutrino absorption (see Equation 10–30), in which a proton captures an electron-type neutrino, illustrate the conservation of both baryon and electron number.

Neutron decay:	$n \rightarrow p^+ + e^- + \bar{\nu}_e$	$\bar{\nu}_e + p^+ \rightarrow n + \overline{e^+}$
Baryon no.:	$1 = 1 + 0 + 0$	$0 + 1 = 1 + 0$
Electron no.:	$0 = 0 + 1 + (-1)$	$(-1) + 0 = 0 + (-1)$

The distinction between antineutrinos of the muon variety and of the electron variety was shown first in a high-energy experiment at the Brookhaven National Laboratory in 1962. A beam of 15-GeV protons incident on a target produced, among other particles, charged pions, which decayed into muons and muon-type antineutrinos. After the muons and other particles were stopped with heavy shielding, the remaining antineutrino beam passed through a spark chamber. Each of the high-energy antineutrinos had a small probability (1 chance in 10^{12}) of interacting with a proton in the chamber and producing a neutron and a detectable muon (see Figure 12–12). Since no electron tracks were observed, this experiment established that the electron varieties of neutrino and antineutrino are distinct from the muon neutrino and antineutrino types. We shall subsequently indicate (Section 12–5) what, in addition to the muon and electron number, distinguishes a neutrino from an antineutrino.

The muon has a curious position among the fundamental particles in that it is apparently identical in all properties to an electron, except that it has greater mass and a different lepton number from the electron. There is no known explanation for this mass difference between these otherwise identical particles.

In a sense, the baryon and lepton conservation laws "explain" why the proton, electron, neutrinos, and associated antiparticles are absolutely stable. There exist no lighter particles into which they can decay without violating these conservation laws. By the same token, there are no conservation laws for the number of photons or mesons; such particles can be created

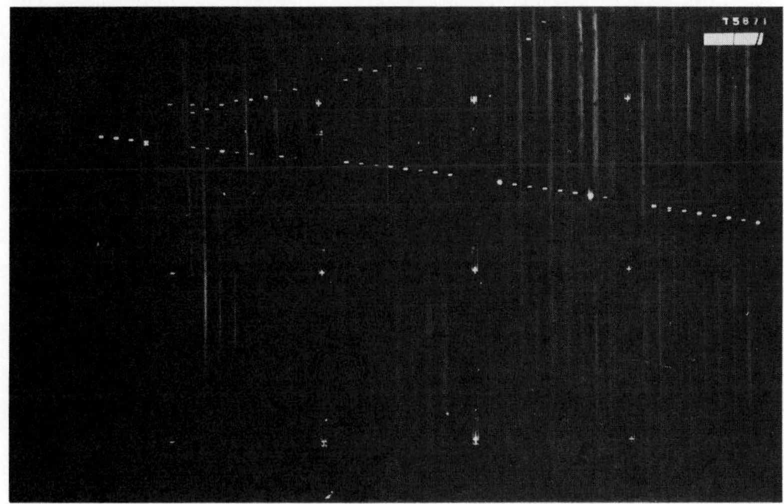

FIGURE 12–12. *Two tracks in a spark-chamber photograph used in experiments to establish the existence of the neutrino associated with the muon.*

(or annihilated) in unlimited number, subject only to limitations imposed by other conservation laws.

In summary, six conservation laws that are universally valid for *all* four fundamental interactions are:

Universally valid conservation laws:

Mass-energy	$\Sigma E_i = \text{constant}$	
Linear momentum	$\Sigma \mathbf{p}_i = \text{constant}$	
Angular momentum	$\Sigma \mathbf{L}_i = \text{constant}$	
Electric charge	$\Sigma q_i = \text{constant}$	(12–13)
Baryon number	$\Sigma B_i = \text{constant}$	
Lepton number	$\Sigma L_e = \text{constant}$	
	$\Sigma L_\mu = \text{constant}$	

12-5 Additional Conservation Laws for Strong and Electromagnetic Interactions

Three new concepts—isospin, hypercharge, and space parity—are now introduced. These concepts are also associated with conservation laws, which, curiously, apply rigorously to the strong interaction but *not universally* to other fundamental interactions.

ISOSPIN. Isospin (sometimes referred to as isotopic spin) is so named because it has quantum behavior that is analogous to ordinary angular momentum. Recall that the total angular momentum of an atom, or a nucleus, is quantized, $L_j = \sqrt{j(j+1)}\hbar$, where j has only integral or half-integral values (Section 7–7). For a given j, there are $2j + 1$ values of the magnetic quantum number m_j. In the absence of an external magnetic field, these $2j + 1$ quantum states are degenerate, having the same energy. The presence of an external magnetic field breaks this symmetry and splits the single energy state into $2j + 1$ closely spaced but distinct energy values (see Figure 7–15). Conversely, by counting the number of distinct energy states observed through the atom's interaction with an external magnetic field, one can determine j.

The energy splittings of the 18 distinct quantum states for the valence electron in the sodium atom in the $n = 3$ state are illustrated in Figure 12–13. Each of the three electromagnetic interactions—"penetration" of valence electron into inner core, spin-orbit interaction, and external magnetic field influence—removes some degeneracy.

Analogous arguments have been successfully used to describe the groupings of the many baryons and mesons in the hadron family. This procedure was first applied to the mass splittings and groupings of the baryons with spin $\frac{1}{2}$ and the mesons with spin 0, as listed in Table 12–4. We note from the table that both the spin-$\frac{1}{2}$ baryons and the spin-0 mesons are clustered into groups of particles with very nearly equal masses (and therefore energies). Let us follow the scheme characterizing the energy differences of the various atomic quantum states denoted by the quantum numbers (n, l, j, m_j) in Figure 12–13

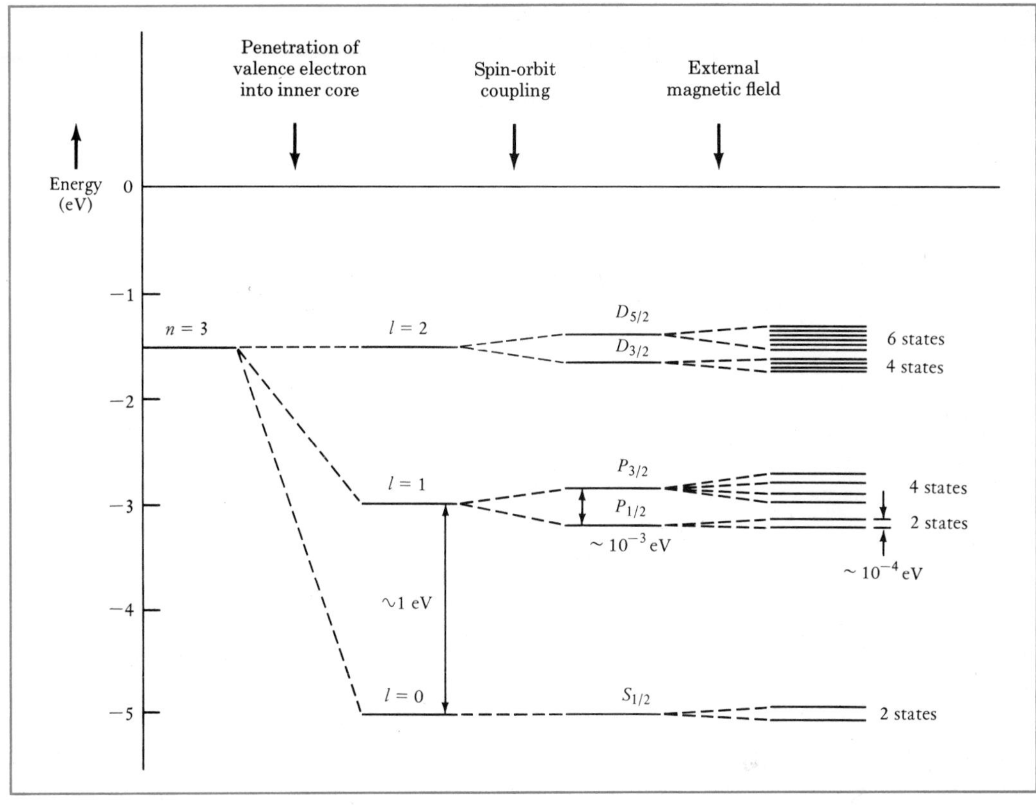

FIGURE 12–13. *Energy splitting of the sodium atom by electromagnetic interactions.*

as we look for interactions that lead to mass-energy differences for the eight baryons of spin angular momentum $\frac{1}{2}$ and for the eight mesons of spin 0.

Figure 12–14 depicts the mass splitting of the baryons. First, there is a *semistrong* interaction, which separates the baryons into four families (see Table 12–4): two nucleons (designated N) with an average mass of 939 MeV; one lambda particle (Λ) with a mass of 1116 MeV; three sigma particles (Σ) with an average mass of 1193 MeV; and two xi particles (Ξ) with an average mass of 1317 MeV. The mass-energy splitting by the semistrong interaction is of the order of 10^2 MeV. In this sense, the eight hadrons can be regarded as different states of a single primordial hadron; if somehow one could turn off the semistrong interaction, all eight hadrons would have identical properties. Next, the smaller mass differences between particles of the same family can be attributed to an electromagnetic interaction that separates the baryons in each family into particles of different electric charge and mass.

The reason for the relatively large splitting of the initial spin-$\frac{1}{2}$ single baryon state by the semistrong force in Figure 12–14 into the four substates can be understood more easily after the smaller mass splittings of the particles of each subfamily are described. If one attributes the relatively small difference in mass ($\sim$ few MeV) within each family (or multiplet) to the

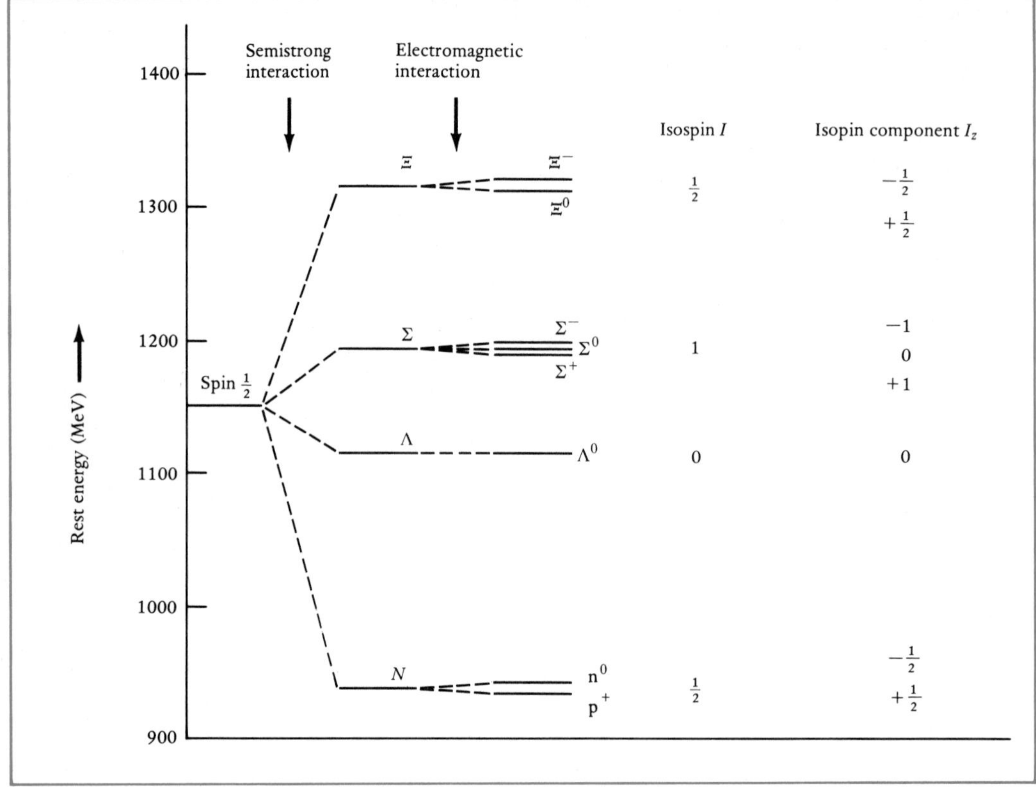

FIGURE 12–14. *Mass splitting of spin-½ baryons by the semistrong and electromagnetic interactions.*

difference in electric charge (and to the relatively weaker electromagnetic interaction associated with electric charge), the particles within each group can be regarded as *different states* of the *same particle*. In analogy with the quantum description of the $2j + 1$ distinct magnetic quantum states of the atom in an external magnetic field (Figure 12–13), one assigns an *isospin quantum number I* to each group in such a way that the multiplicity of particles within the group equals $2I + 1$. Then, to each of the possible "orientations," or components, I_z of the isospin vector **I** in an abstract *isospace* there corresponds a particle. For a given quantum number I, the possible $2I + 1$ values of I_z are $I, I - 1, I - 2, \ldots, -I$. The isospin quantum number I is determined by knowing the number of particles in a group. For example, for the nucleon group (p and n) there are two particles and $2 = 2I + 1$, or $I = \frac{1}{2}$. Similar arguments give the isospin quantum numbers of the other three families in Figure 12–14.

Each of the multiplet states of a group has an integral electric charge, and it is convenient to assign the isospin component of a particle according to the rule

$$I_z = Q - \overline{Q} \tag{12–4}$$

where Q is the electric charge number of the particle and $\overline{Q}$ is the average charge number of its group. Thus, for the nucleon group,

$$Q_p = +1 \qquad Q_n = 0 \qquad \text{and} \qquad \overline{Q} = \frac{Q_p + Q_n}{2} = +\tfrac{1}{2}$$

Therefore, by Equation 12–4,

$$I_z\,(\text{proton}) = Q_p - \overline{Q} = 1 - \tfrac{1}{2} = +\tfrac{1}{2}$$

$$I_z\,(\text{neutron}) = Q_n - \overline{Q} = 0 - \tfrac{1}{2} = -\tfrac{1}{2}$$

The last column of Figure 12–14 lists the values of the isospin components for other spin-$\frac{1}{2}$ baryons.

The isospin properties of the first eight mesons in Table 12–4, determined as we have just done for the nucleons, are shown in Figure 12–15. Only hadrons (baryons and mesons) can have nonzero values of isospin; the photons and all leptons are all assigned an isospin $I = 0$.

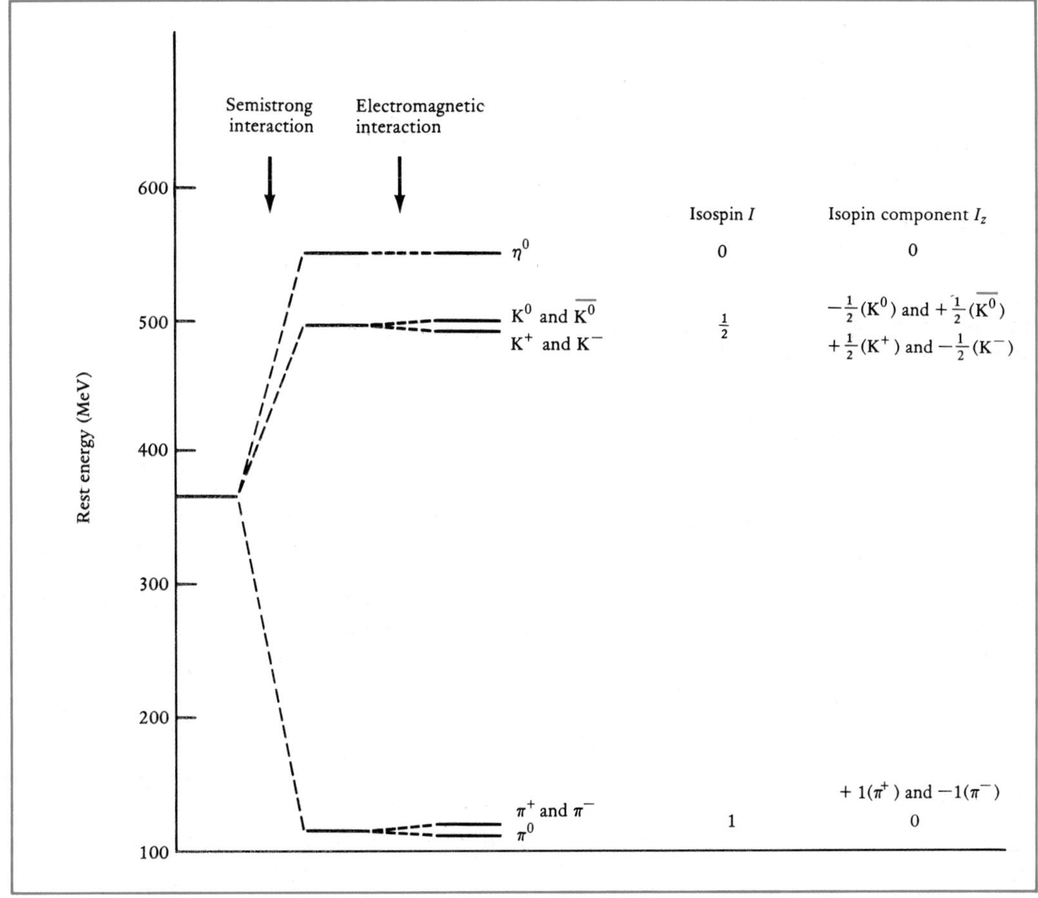

FIGURE 12–15. *Mass splitting of the spin-zero mesons by the semistrong and electromagnetic interactions.*

As is true of all quantum concepts, the isospin quantum number is also associated with a conservation law. The isospin conservation law, unlike the universally valid laws of Equation 12–3, can be stated as follows:

For *any strong interaction,* the total isospin *magnitude* after a reaction is the same as the total isospin magnitude before the reaction. (12–5)

For *any strong or electromagnetic interaction,* the *total z component* of isospin after a reaction is the same as the total isospin component before the reaction. (12–6)

In short, total I is conserved for the strong interaction and total I_z for the strong and electromagnetic interactions. Total I and I_z need not, however, be conserved for the other interactions (see Table 12–6).

The following examples show the isospin conservation principle.

Strong-interaction collision

$$p^+ + p^+ \rightarrow \Lambda^0 + K^0 + p^+ + \pi^+$$

I:	$\vec{\tfrac{1}{2}} + \vec{\tfrac{1}{2}} = \vec{0} + \vec{\tfrac{1}{2}} + \vec{\tfrac{1}{2}} + \vec{1}$	I conserved
I_z:	$\tfrac{1}{2} + \tfrac{1}{2} = 0 - \tfrac{1}{2} + \tfrac{1}{2} + 1$	I_z conserved

Decay of π^0 by electromagnetic interaction

$$\pi^0 \rightarrow \gamma + \gamma$$

I:	$1 \neq 0 + 0$	I not conserved
I_z:	$0 = 0 + 0$	I_z conserved

Decay of Λ^0 by weak interaction

$$\Lambda^0 \rightarrow p + \pi^-$$

I:	$\vec{0} \neq \vec{\tfrac{1}{2}} + \vec{1}$	I not conserved
I_z:	$0 \neq \tfrac{1}{2} + (-1)$	I_z not conserved

Since I is not conserved in the Λ° decay, by Equation 12–5 this decay could not have taken place by the fast strong interaction; and since I_z is also not conserved, it could not have taken place either by the fast strong interaction or by the relatively fast electromagnetic interaction. The decay of Λ^0 into $p + \pi^-$ must then have taken place by the much slower weak interaction, since neither I nor I_z need be conserved for this interaction.

HYPERCHARGE AND STRANGENESS. Two other quantum numbers, *hypercharge* Y and *strangeness* S, are often used in place of the isotopic spin quantum number I_z just described. As we shall see, all three of these quantum numbers express equivalent conservation laws, which hold for both the strong and the electromagnetic interactions but not for the weak interaction.

The hypercharge quantum number Y is simply defined as twice the average charge $\overline{Q}$ (in units of e) of each particle group for hadrons (see Figures 12–14 and 12–15):

$$Y\,(\text{group}) = 2\overline{Q}\,(\text{group}) \tag{12–7}$$

The factor 2 is introduced so that the quantum numbers for the hypercharge will always be *integral* instead of half-integral. For example, for the nucleon group (proton and neutron)

$$\overline{Q} = \frac{+1+0}{2} = +\tfrac{1}{2} \qquad \text{and} \qquad Y = 2\overline{Q} = +1$$

Recalling that the isotopic spin I_z of each particle in a group is given by $I_z = Q - \overline{Q}$ (Equation 12–4), we find that Y can be expressed by the isotopin spin I_z and charge Q of a particle:

$$Y = 2(Q - I_z) \tag{12–8}$$

For *any interaction,* the total charge must be conserved; and for *strong and electromagnetic interactions* (see Equation 12–6) the total z-component of **I** must be conserved. Therefore, it follows from Equation 12–8 that the total hypercharge must be conserved in the strong and electromagnetic interactions.

> For any strong or electromagnetic interaction, the total hypercharge after the reaction is the same as that before the reaction. (12–9)

Values of the electric charge Q, hypercharge Y, and isotopic spin component I_z, along with the strangeness quantum number to be defined shortly, are listed in Table 12–5. For each of the nine baryons listed in the table, there is an antibaryon; the antibaryon, compared with the baryon, has an opposite sign for each property.

An example of the conservation of total hypercharge Y is the strong interaction collision shown earlier for isotopic-spin conservation.

$$\text{p}^+ + \text{p}^+ \rightarrow \Lambda^0 + \text{K}^0 + \text{p}^+ + \pi^+$$

$$Y\!: \quad 1 + 1 = 0 + 1 + 1 + 0$$

Using the values of hypercharge listed in Table 12–5, we find the total hypercharge to be conserved for the collision.

Hypercharge conservation can also be used to explain why some decays of excited particle states must proceed by the slow, weak interaction. The two principal decay modes of the Σ^+ baryon, for example, are $\Sigma^+ \rightarrow \text{p}^+ + \pi^0$ or $\Sigma^+ \rightarrow \text{n}^0 + \pi^+$, with a lifetime of 0.80×10^{-10} s (see Figure 12–4). If it were not for the conservation of hypercharge (or equivalently, isospin component), these decays would take place quickly ($\sim 10^{-23}$ s) by the strong interaction. But since the hypercharge is zero for the Σ^+ and is $1 + 0 = 1$ for the decay products of either decay mode, hypercharge is not conserved, and decay by the strong or electromagnetic interactions is forbidden. Hypercharge need not be conserved, however, for the weak interaction, and processes such as the Σ^+ decay take place in times of the order of 10^{-10} s, characteristic of weak interaction.

Another equivalent form of isotopic spin conservation is the so-called *conservation of strangeness,* with associated *strangeness* quantum numbers S. This concept arose because many of the hadrons, as listed in Table 12–4,

Table 12–5. *Values of Q, Y, I_z, and S for the relatively long-lived hadrons of Table 12–4*

	FAMILY GROUP	PARTICLE	Q	$Y = 2\overline{Q}$	$I_z = Q - \overline{Q}$	S
	Pion (π)	π^+	+1	0	+1	0
		π^0	0	0	0	0
		π^-	−1	0	−1	0
Mesons ($B = 0$)	Kaon (K)	K^+	+1	1	$+\frac{1}{2}$	1
		K^0	0	1	$-\frac{1}{2}$	1
	($\overline{K}$)	$\overline{K^0}$	0	−1	$+\frac{1}{2}$	−1
		$\overline{K^-}$	−1	−1	$-\frac{1}{2}$	−1
	Eta (η)	η^0	0	0	0	0
	Nucleon (N)	p^+	+1	1	$+\frac{1}{2}$	0
		n^0	0	1	$-\frac{1}{2}$	0
	Lambda (Λ)	Λ^0	0	0	0	−1
Baryons ($B = 1$)	Sigma (Σ)	Σ^+	+1	0	+1	−1
		Σ^0	0	0	0	−1
		Σ^-	−1	0	−1	−1
	Xi (Ξ)	Ξ^0	0	−1	$+\frac{1}{2}$	−2
		Ξ^-	−1	−1	$-\frac{1}{2}$	−2
	Omega (Ω)	Ω^-	−1	−2	0	−3

Note: For each baryon above there is an antiparticle having the same magnitude but an opposite sign for each property.

behave "strangely" in interactions. These strange particles are created by the *strong* or *electromagnetic interaction;* but then they decay by the *weak interaction.* To account for this behavior, scientists arbitrarily introduced the strangeness quantum number, assigning a strangeness 0 to the familiar nucleon group (designated by the symbol N in Table 12–5) and to the familiar pion group (designated by the symbol π). Strangeness can then be related to hypercharge by recalling that the hypercharge of the nucleon group (see Table 12–5) is $Y(\mathrm{N}) = 2\overline{Q}(\mathrm{N}) = +1$ and the hypercharge of the pion group is $Y(\pi) = 2\overline{Q}(\pi) = 0$. If one defines strangeness by:

$$S = Y - B \tag{12–10}$$

we have

$$S(\mathrm{N}) = 1 - 1 = 0$$

$$S(\pi) = 0 - 0 = 0$$

This definition for strangeness S ensures that the familiar baryons (the nucleons) and the familiar mesons (the pions) have zero strangeness. The strangeness quantum numbers of the other elementary particles in Tables 12–4 or 12–5 follow from Equation 12–10: $S(\Lambda) = Y(\Lambda) - B(\Lambda) = 0 - 1 = -1$, and so on. In the last column in Table 12–5 is listed the values for the strangeness of baryons; with $B = 1$ for all baryons, the strangeness S of each baryon group will always be one less than the hypercharge Y for that group. Also listed are the values for the strangeness of mesons; with $B = 0$ for all mesons, the strangeness S of each meson group will equal the hypercharge Y for that group.

From Equation 12–10 it follows that in any strong interaction or electromagnetic interaction, the total strangeness must be conserved, since the total baryon number is universally conserved and the total hypercharge number is conserved for both strong and electromagnetic interaction. In summary,

For strong or electromagnetic interactions, three equivalent conservation laws hold: Strangeness, hypercharge, and isotopic spin. (12–11)

To illustrate the conservation of total strangeness, we again consider the collision of a high-energy proton (several GeV) with a target proton producing the following outgoing particles

$$p^+ + p^+ \rightarrow \Lambda^0 + K^0 + p^+ + \pi^+ \qquad (12\text{–}12)$$

$$\text{Strangeness:} \quad 0 + 0 = (-1) + 1 + 0 + 0$$

Total strangeness is conserved. Two strange particles, the lambda particle and the kaon, with opposite strangeness are created, along with an outgoing proton and a positive pion. Strangeness conservation requires that *at least two* strange particles must be produced if *any* strange particles are created at all in any strong or electromagnetic interaction that is proceeding by the collision of nonstrange incoming particles. This is commonly referred to as *associated production.*

Figure 12–16 shows a bubble-chamber photograph of the strong-interaction collision of a 2.85-GeV proton with a target proton at rest in the bubble chamber resulting in the reaction products of Equation 12–12. After production the four outgoing particles have different velocities and travel away from the creation point in different directions. Neither of the two strange particles created in the reaction can then decay by the strong or electromagnetic interaction without violating one of the conservation laws of these interactions. For the K^0 meson with $S = +1$, there is only the K^+ meson ($S = +1$) with slightly smaller mass (see Tables 12–4 and 12–5), and to decay without leptons (weak interaction) in the decay products would require

$$K^0 \rightarrow K^+ + \pi^- \ (?)$$

$$\text{mass (GeV):} \quad 497.7 \rightarrow 493.7 + 140$$

Although this decay does not violate strangeness, it does violate mass-energy conservation. The decay must therefore proceed by the weak interaction.

Similarly, conservation of baryon number and of mass requires that the Λ^0 baryon decay into a lighter baryon (p or n) and mesons. For strangeness to be conserved, one meson must be a K meson (say K^0), and

$$\Lambda^0 \rightarrow n^0 + K^0$$

$$\text{mass (GeV):} \quad 1116 \rightarrow 940 + 498$$

This decay also violates conservation of mass-energy and does not occur.

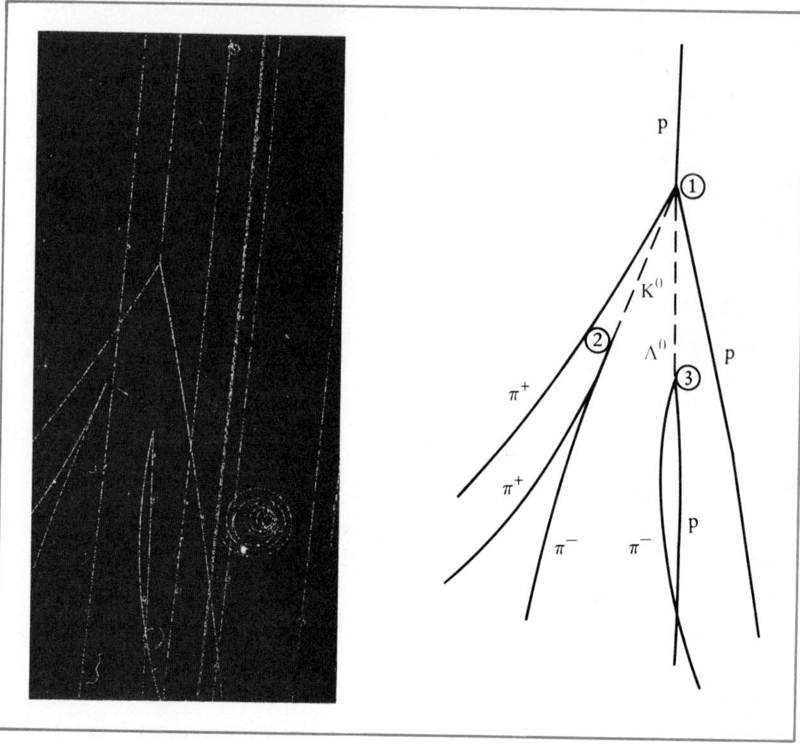

FIGURE 12–16. *A bubble-chamber photograph of a 2.85-GeV proton colliding with a proton at point 1 and creating a K^0 and a Λ^0. The K^0 later decays to two pions at point 2; the Λ^0 decays to a pion and a proton at point 3.*

As observed in Figure 12–16, the K^0 and Λ^0 are unstable and do decay into lighter hadrons:

	$K^0 \to \pi^+ + \pi^-$	$\Lambda^0 \to p^+ + \pi^-$
Strangeness:	$+1 \to 0 + 0$	$-1 \to 0 + 0$
	$\Delta S = -1$	$\Delta S = +1$

Both of these decays violate strangeness conservation. The K^0 decay has a change in total strangeness of $\Delta S = -1$; and the Λ^0 decay, a change $\Delta S = +1$. The observed lifetimes of the two decays (see Figure 12–4) are $\sim 10^{-10}$ s, indicating decay by the weak interaction. This suggests that strangeness need not be conserved in the weak interaction. Inspecting the lifetimes of the unstable hadrons in Table 12–4 shows that all decays do occur by the weak interaction (except the π^0, η^0, and Σ^0 decays, which do not violate strangeness conservation and therefore occur through the faster electromagnetic interaction).

Although weak interactions need not conserve strangeness, there is a selection rule for these processes:

For weak interactions, the change in strangeness may change by one unit or none: $\Delta S = \pm 1 \text{ or } 0$ (12–13)

An example of a weak-interaction process with $\Delta S = 0$ is the familiar decay of the neutron:

$$n^0 \rightarrow p^+ + e^- + \bar{\nu}_e$$

$$\text{Strangeness:} \quad 0 = 0 + 0 + 0$$

$$\Delta S = 0$$

This is obviously a weak interaction, since it involves leptons. On the other hand, some weak interactions do not involve leptons (for example, the K^0 and Λ^0 decays above).

Example 12–3. The existence of the most massive baryon in Table 12–4, the omega-minus (Ω^-), was predicted on theoretical grounds (1961) before it was created and observed experimentally (1964). Figure 12–17 shows the production and subsequent decays of an Ω^- when an incoming K^- meson collides with a proton at rest in the bubble chamber. Discuss the change in strangeness (*a*) at the point of creation, and (*b*) at the subsequent decay of the Ω^-.

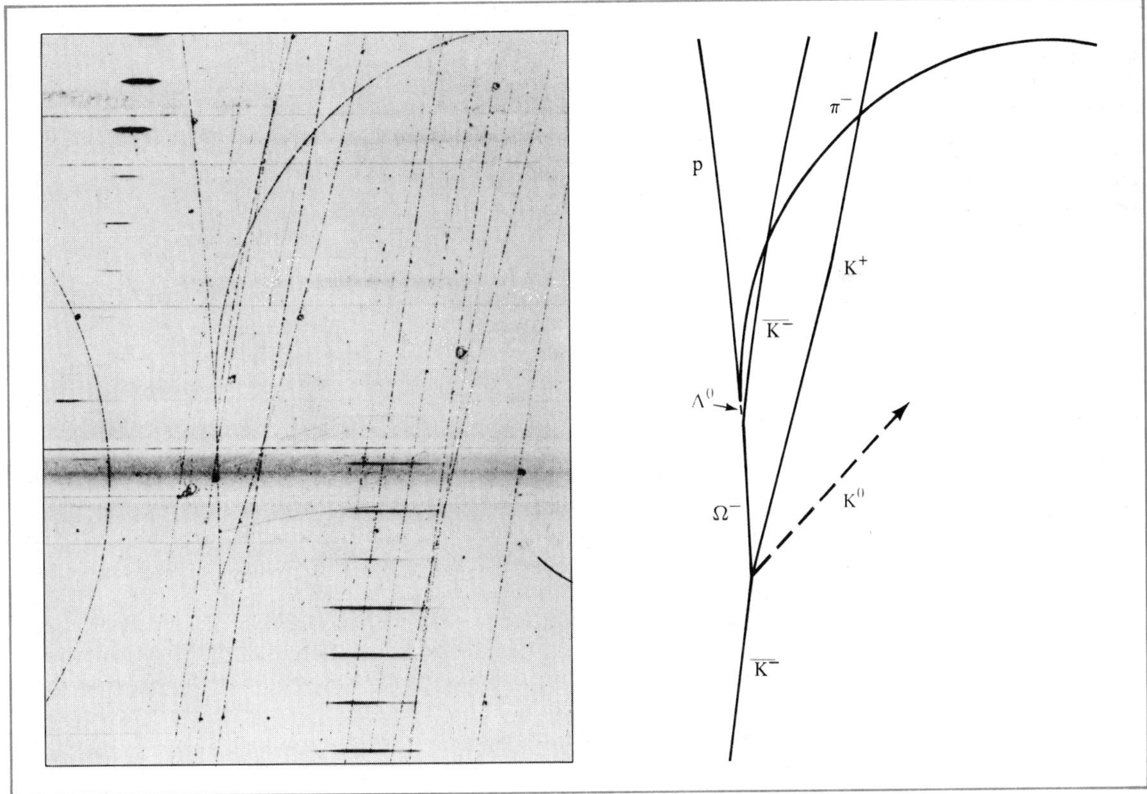

FIGURE 12–17. *A bubble-chamber photograph and sketch of the production of a negatively charged Ω baryon (Ω^-). An incoming K^- meson collides with a stationary proton with the result that a K^0, a K^+, and an Ω^- are produced. The photograph also shows the subsequent decay of the Ω^- into a Λ^0 and a K^-, and the decay of the Λ^0 into a proton and a negative pion.*

(*a*) When the high energy $\overline{K^-}$ meson collides with the proton, as shown in Figure 12–17, the following reaction occurs:

$$\overline{K^-} + p^+ \rightarrow \Omega^- + K^+ + K^0$$

$$\text{Strangeness:} \quad -1 + 0 \rightarrow -3 + 1 + 1$$

$$\Delta S = 0$$

Strangeness is conserved, and the collision takes place by the strong interaction.

(*b*) From Figure 12–17, the Ω^- baryon is observed to decay into a Λ^0 baryon and a $\overline{K^-}$ meson.

$$\Omega^- \rightarrow \Lambda^0 + \overline{K^-}$$

$$\text{Strangeness:} \quad -3 \rightarrow -1 - 1$$

$$\Delta S = +1$$

Strangeness is not conserved in this decay; therefore, the decay cannot take place by the strong or the electromagnetic interaction. By Equation 12–13, the decay can occur by the weak interaction, and it does.

PARITY. Parity is a concept related to the symmetry of physical experiments. Suppose that the position coordinate of each particle is changed from $\mathbf{r}$ to $-\mathbf{r}$ (that is, $x \rightarrow -x$, $y \rightarrow -y$, and $z \rightarrow -z$). This is equivalent to changing a right-handed coordinate system into a left-handed coordinate system, a classical example of a parity change. In quantum mechanics, the parity of a particle is simply defined by the wave function, say $\psi(\mathbf{r})$, that describes the particle. If the probability density $[\psi(\mathbf{r})]^2$ of the particle is to be the same whether we use a right-handed or a left-handed coordinate system, we have just two possibilities for ψ

$$\psi(-\mathbf{r}) = \psi(\mathbf{r}) \qquad \text{even parity, designated } (+1)$$

$$\psi(-\mathbf{r}) = -\psi(\mathbf{r}) \qquad \text{odd parity, designated } (-1)$$

The conservation of parity law for any interaction implies that there is no preference for right or left. Or if a certain process takes place in nature, then that process viewed by reflection in a mirror would also occur with equal probability.

Until 1956 it was believed that parity was a universal conservation law, true for all interactions. We shall see that it is not and that like strangeness conservation, parity conservation is strictly conserved only for the strong and electromagnetic interactions. First, consider an example where parity is conserved, the decay of neutral pions (π^0) by the electromagnetic interaction. Assuming the π^0 to be at rest, then the two created γ particles must leave the decay in opposite directions with equal momentum magnitude P to satisfy linear-momentum conservation. See Figure 12–18. The π^0 has spin zero, whereas each of the outgoing photons has spin 1 (see Table 12–4); therefore, to conserve angular momentum, the spin senses (represented by the angular-momentum vectors $\mathbf{J}$ in Figure 12–18) of the photons must be opposite. There are a priori *two* possible orientations of angular- and linear-

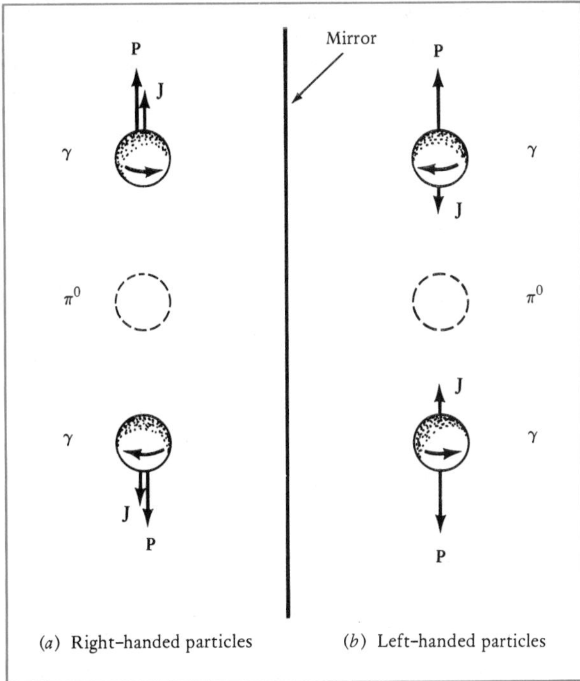

FIGURE 12–18. *Two a priori possibilities for the decay of a π^0 into two γ's. (a) Linear momentum* **P** *and angular momentum* **J** *vectors aligned, corresponding to the* right-handedness *of both γ's. (b)* **P** *and* **J** *antialigned, corresponding to the* left-handedness *of the two γ's. Note that (b) is the mirror image of (a) relative to the vertical line dividing the two decays. Both (a) and (b) decays are observed.*

momentum vectors—**J** aligned with **P** for each photon (right-handed particles, as shown in Figure 12–18*a*) and **J** antialigned with **P** for each photon (left-handed particles, as shown in Figure 12–18*b*). Note that the mirror image of the decay of the π^0 into right-handed particles in Figure 12–18*a* is merely the decay of the π^0 into left-handed particles in Figure 12–18*b*. Therefore parity conservation demands that in the spontaneous decay of many neutral pions, the number of right-handed photons must equal the number of left-handed photons. Experiment confirms this.

Now, consider the observed weak-interaction decay of positive pions (π^+) into positive antimuons ($\overline{\mu^+}$) and muon-type neutrinos ν_μ, according to Table 12–4. Again, linear-momentum conservation requires $\mathbf{P}_{\overline{\mu}} = -\mathbf{P}_{\nu_\mu}$, and angular-momentum conservation requires $\mathbf{J}_{\overline{\mu}} = -\mathbf{J}_{\nu_\mu}$. (See Figure 12–19.) The decay into left-handed particles (Figure 12–19*b*) is again the mirror image of the decay into right-handed particles (Figure 12–19*a*). If parity is to be conserved in this interaction, one expects equal numbers of outgoing right-handed muons and left-handed muons and also equal numbers of right-handed neutrinos and left-handed neutrinos. Since neutrinos undergo virtually no absorption through ordinary matter, one must observe the handedness of the muons and infer that the handedness of the neutrinos is the same. The results of experiment are that positive pions, when observed in an inertial system where they are at rest, decay into *left-handed* antimuons ($\overline{\mu^+}$) and *left-handed* neutrinos (ν_μ) alone, never into right-handed antimuons and right-handed neutrinos! Similar observations of the decay of resting negative pions (π^-) show that only right-handed negative muons (μ^-) and right-handed antineutrinos ($\overline{\nu}_\mu$) are emitted, never left-handed particles.

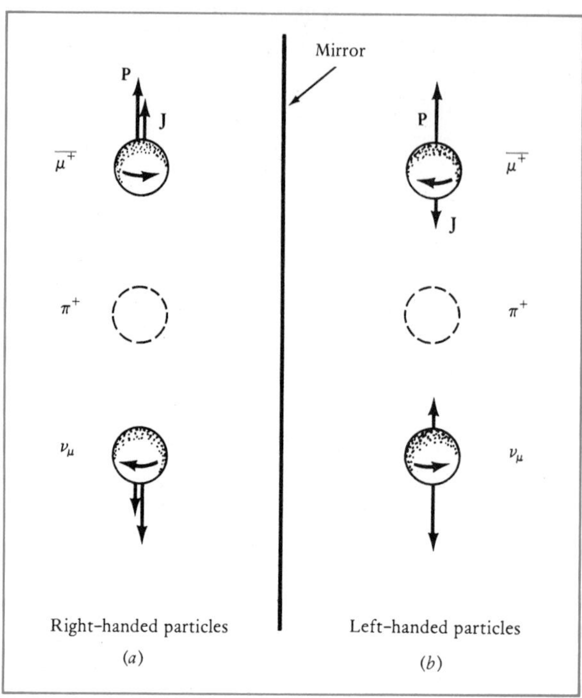

FIGURE 12–19. *Two a priori possibilities for the decay of π^+ into an antimuon $\overline{\mu}^+$ and a neutrino ν_μ. (a) Linear momentum* **P** *and angular momentum* **J** *vectors aligned, corresponding to the* right-handedness *of both particles; (b)* **P** *and* **J** *antialigned, corresponding to the* left-handedness *of both particles. Note that (b) is the mirror image of (a), relative to the vertical line dividing them. Experimentally, only decay according to (b), with both particles exclusively left-handed, is observed.*

A consequence of the observations just noted is that neutrinos and antineutrinos (of either the muon type or the electron type) have different intrinsic handedness—the *antineutrinos* are exclusively *right-handed* and the *neutrinos* are exclusively *left-handed.* Thus, they are distinguishable particles. Although only left-handed antimuons are emitted in the decay of positive pions when viewed in a reference system where the π^+ is at rest, it is easy to see that any particle with a *finite* rest mass, such as a muon, will not have just one intrinsic handedness for all inertial observers. Suppose that an antimuon travels upward with its spin also upward, constituting a right-handed particle, viewed by an observer at rest relative to the meson, as shown in Figure 12–19*a*. If we view the particle from a reference frame traveling upward at a higher speed than the antimuon, then the antimuon moves downward. Thus, the particle's linear-momentum vector **P** has been reversed but not its angular-momentum vector **J**, and the particle is seen as a left-handed particle. Since antimuons may be both right- and left-handed, the appearance of left-handed antimuons only in the decay of positive pions at rest implies that the neutrino must always be left-handed. This is, moreover, in accord with the constant speed c of the neutrinos of zero rest mass; there is no reference frame for which the neutrino's velocity vector can be reversed.

The nonconservation of parity P in the weak interaction has prompted reexamination of two other closely related conservation laws: *Charge conjugation C,* which indicates that interactions and processes are unchanged when every particle is replaced by its antiparticle; and *time invariance T,* according to which any process can proceed either forward or backward in time without change (or more specifically, if a motion picture portrays a

Table 12–6. *Partial conservation laws*

QUANTITY	STRONG INTERACTION	ELECTROMAGNETIC INTERACTION	WEAK INTERACTION
Isospin magnitude I	Yes	No	No
Isospin component I_z (equivalently, S or Y)	Yes	Yes	No
Parity P	Yes	Yes	No
Charge Conjugation C	Yes	Yes	No
Time Reversal T	Yes	Yes	Yes[a]
CP	Yes	Yes	Yes[a]

[a]Except infrequent neutral kaon decay.

possible process, then the motion picture that one sees unfold as the film is run backward also portrays a possible process). Experiments indicate that P, C, and T are each separately conserved in strong and electromagnetic interactions, and T is conserved in weak interactions except for infrequent violation in the decay of the neutral K mesons. Although parity P and charge conjugation C are separately not conserved in the weak interactions, the combined operation CP is conserved for this interaction.

Eight conservation laws holding for all particle interactions are mass-energy, linear momentum, angular momentum, electric charge, baryon numbers, lepton number (electron type), lepton number (muon type), and the combined charge conjugation–parity–time reversal operation (CPT). Those conservation laws not universally true for all interactions are shown in Table 12–6.

12-6 Resonance Particles

Most of the unstable fundamental particles we have studied decay by the weak interaction, with lifetimes long enough ($\sim 10^{-10}$ s) to leave observable tracks in detection devices such as bubble chambers. The others decay much faster by the electromagnetic interaction (lifetimes, $\sim 10^{-16}$ s), and the existence of such particles must be inferred by other means, since their track lengths are too short to observe directly.

Many new particles with lifetimes even shorter than 10^{-16} s have been discovered since the mid 1950s. The existence and properties of these so-called *resonance particles*, with lifetimes comparable to the nuclear-interaction time ($\sim 10^{-23}$ s), must also be deduced indirectly.

To understand one common procedure for identifying an extremely short-lived particle, consider an analogous situation, in which electrons of controllable kinetic energy bombard hydrogen atoms initially in the ground state. Since the hydrogen atom's internal energy is quantized, none of the incident kinetic energy of the electron can be converted to internal energy of the hydrogen atom if the incident electron has kinetic energy less than 10.2 eV; the quantity 10.2 eV is the energy necessary to raise the atom to its first excited state (see Figure 6–12). Therefore, if electrons of less than 10.2 eV are striking hydrogen atoms, the collisions are perfectly elastic and all atoms remain in the ground state. When the energy of the incident electrons reaches 10.2 eV, some inelastic collisions occur. In the inelastic collisions, there is a sudden decrease in the electron's kinetic energy; at the same time, the hydrogen atom with which it collided makes a quantum transition to the first excited state, from which it subsequently decays back to the ground state through the emission of a photon. This phenomenon was first observed in 1914 by J. Frank and G. Hertz (Section 6–7).

The excitation to a quantized state by particle bombardment shows a resonance behavior; the probability for an inelastic excitation process is a maximum when the electron kinetic energy equals the internal excitation energy of the atom and falls off for incident electron energies smaller or larger

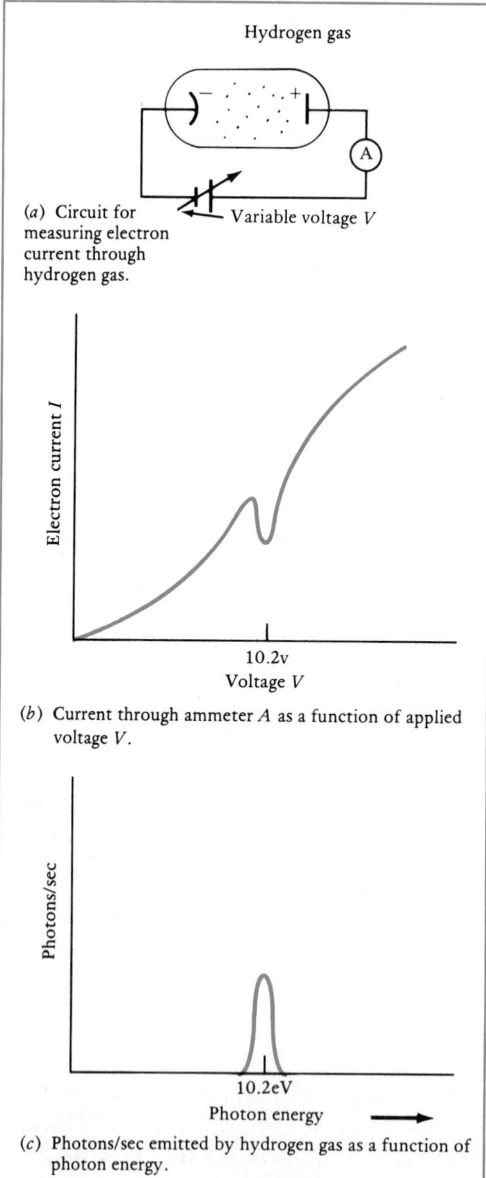

FIGURE 12–20. *(a) Circuit for measuring electron current through hydrogen gas. (b) Current through ammeter* A *as function of applied voltage* V. *(c) Photons per second emitted by hydrogen gas as function of photon energy.*

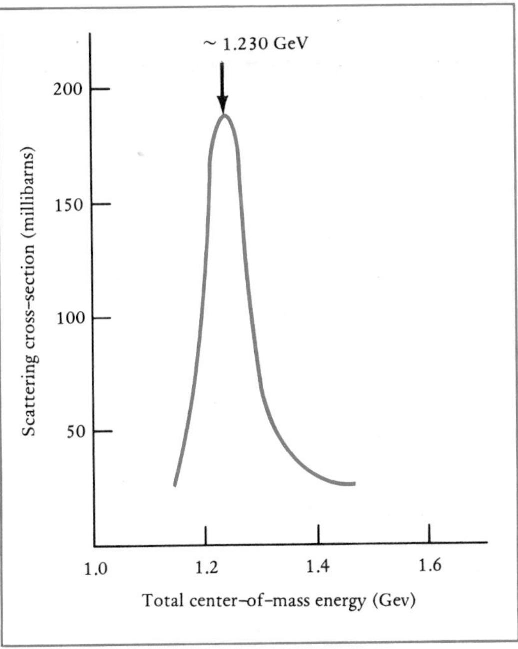

FIGURE 12–21. *Scattering of π^+ mesons by protons at rest in the laboratory system. Resonant peak, at 1.23 GeV, indicates momentary existence of Δ^{++} meson.*

than the atom's excitation energy. See Figure 12–20. Note that when a hydrogen atom exists in, say, the first excited state, it can be regarded as a particle distinct from a hydrogen atom in the ground state, since the two states have different basic properties; for example, their masses differ by exactly 10.2 eV, and they have different angular momenta. The created particle (excited hydrogen atom) is unstable and quickly ($\sim 10^{-8}$ s) decays through the electromagnetic interaction into a photon and a hydrogen atom in the ground state. Its lifetime is much longer than typical electromagnetic interaction times only because the electron and the proton are separated by a large distance (atomic).

Resonance particles can be produced likewise by bombarding nucleons with other nucleons or pions of much higher kinetic energies, in the GeV range instead of the eV range. For example, the lowest-mass resonance particle in the baryon group, now called the Δ particle, was first observed (but not identified as such) in 1952 in the scattering of 40–220-MeV pions by a proton target. (See Figure 12–21.) The resonance, or peak, in the number of incident pions scattered at pion kinetic energies near 200 MeV (total center-of-mass energy, 1.230 GeV) suggests that in addition to elastic scattering, inelastic scattering also is taking place. This can be interpreted in the "particle" model by assuming that the initial baryon, the proton, is momentarily converted to a baryon of a higher energy state, the delta particle, which then quickly decays into a pion and proton. Thus,

$$\pi^+ + p^+ \rightarrow \Delta^{++} \rightarrow \pi^+ + p^+$$

The mean life of the intermediate resonance particle Δ can be estimated from the width of the resonance peak. Because of the uncertainty principle, $\Delta E\, \Delta t \geq \hbar$; and from the experimental curve (Figure 12–21), $\Delta E \approx 100$ MeV. Therefore,

$$\Delta t \approx \frac{\hbar}{\Delta E} = \frac{1 \times 10^{-34}\ \text{J}\cdot\text{s}}{(1 \times 10^{8}\ \text{eV})(1.6 \times 10^{-19}\ \text{J/eV})} \simeq 10^{-23}\ \text{s}$$

The decay time of the Δ particle indicates that it decays through the strong interaction force. The resonance phenomenon and the uncertainty principle provide a way of determining lifetimes of unstable particles over the range 10^{-19} to 10^{-25} s. For times shorter than 10^{-25} s, the energy spread, $\Delta E \approx 6$ GeV, becomes comparable to the rest energy of the intermediate particle, and one cannot ascertain from the energy-distribution curve that the resonance particle exists. For times longer than 10^{-19} s, we find that $\Delta E \approx 6$ MeV; and for incident pion energies in the GeV range it becomes increasingly difficult to resolve energy differences.

Many resonance mesons and baryons have been and are being discovered with rest masses ranging from $\frac{1}{2}$ to over 10 GeV. Table 12–7 lists just a few of the resonance particles in the meson family (hadrons of integral angular momentum $\hbar$), and Table 12–8 gives some in the baryon family (hadrons of half-integral angular momentum $\hbar$). Such fundamental properties as spin angular momentum, rest energy, charge, lifetime, decay schemes, isospin, strangeness, and parity have been determined. All the universally valid conservation laws listed at the end of Section 12–4 and the partial-conservation laws listed in Table 12–6 have been found to hold true for all interactions, including those involving resonance particles.

Table 12–7. *Some established resonance mesons along with important quantum properties.*

MESON	E_0[a] (MeV)	τ ($\times 10^{-23}$ s)	Q ($\times e$)	J^P ($\times \hbar$)	I	S	C	MOST PROBABLE DECAY
ρ	773	43.4	0, ±1	1^-	1	0	0	$\pi\pi$
ω	782	6.6	0	1^-	0	0	0	$\pi^+\pi^-\pi^0$
K*	892	1.3	0, +1	1^-	$\frac{1}{2}$	1	0	$K\pi$
$\overline{K^*}$	892	1.3	0, −1	1^-	$\frac{1}{2}$	−1	0	$\overline{K}\pi$
η'	958	>60.0	0	0^-	0	0	0	$\eta\pi\pi$
φ	1020	16.0	0	1^-	0	0	0	K^+K^-
f	1271	0.37	0	2^+	0	0	0	$\pi\pi$
A_2	1310	0.65	0, ±1	2^+	1	0	0	$\rho\pi$
K*	1420	0.60	0, +1	2^+	$\frac{1}{2}$	1	0	$K\pi$
$\overline{K^*}$	1420	0.60	0, −1	2^+	$\frac{1}{2}$	−1	0	$\overline{K}\pi$
f′	1516	1.6	0	2^+	0	0	0	$K\overline{K}$
ω'	1667	0.45	0	3^-	0	0	0	$\rho^0\pi^0$
g	1690	0.37	0, ±1	3^-	1	0	0	$\pi\pi$
ψ	3098	984.0	0	1^-	0	0	0	$\pi\pi$
P_C	{3267, 3504}	?	0	2^-	1	0	0	$\psi\gamma$
ψ'	3684	280.0	0	1^-	0	0	0	$\psi\pi^+\pi^-$

[a]Definition of column-heading symbols: E_0, rest energy; τ, mean life; Q, charge; J^P, angular momentum and parity; I, isospin magnitude; S, strangeness; C, charm.

Table 12–8. *A partial list of baryon resonance states with important quantum properties. (See Table 12–7 for definition of symbols.)*

PARTICLE, SYMBOL AND REST ENERGY (MeV)	τ ($\times 10^{-23}$ s)	J^P ($\times \hbar$)	Q ($\times e$)	I	S	C	MOST PROBABLE DECAY
Nucleon, N			0, +1	$\frac{1}{2}$	0	0	Nπ
N(1470)	0.3	$\frac{1}{2}^+$	0, +1	$\frac{1}{2}$	0	0	″
N(1520)	0.5	$\frac{3}{2}^-$	0, +1	$\frac{1}{2}$	0	0	″
N(1535)	0.5	$\frac{1}{2}^-$	0, +1	$\frac{1}{2}$	0	0	″
N(1670)	0.4	$\frac{5}{2}^-$	0, +1	$\frac{1}{2}$	0	0	″
N(1688)	0.5	$\frac{5}{2}^+$	0, +1	$\frac{1}{2}$	0	0	″
N(1700)	0.4	$\frac{1}{2}^-$	0, +1	$\frac{1}{2}$	0	0	″
N(2220)	0.2	$\frac{9}{2}^+$	0, +1	$\frac{1}{2}$	0	0	″
N(2650)	0.2	$\frac{13}{2}$	0, +1	$\frac{1}{2}$	0	0	″
N(3030)	0.2	$\frac{17}{2}$	0, +1	$\frac{1}{2}$	0	0	″
Delta, Δ							
Δ(1232)	0.6	$\frac{3}{2}^+$	−1, 0, +1, +2	$\frac{3}{2}$	0	0	Nπ
Δ(1650)	0.4	$\frac{1}{2}^-$	″	$\frac{3}{2}$	0	0	″
Δ(1670)	0.3	$\frac{3}{2}^-$	″	$\frac{3}{2}$	0	0	″
Δ(1950)	0.3	$\frac{7}{2}^+$	″	$\frac{3}{2}$	0	0	″
Δ(2420)	0.2	11/2+	″	$\frac{3}{2}$	0	0	″
Δ(2850)	0.2		″	$\frac{3}{2}$	0	0	″
Δ(3230)	0.1		″	$\frac{3}{2}$	0	0	″
Lambda, Λ			0	0	−1	0	N$\overline{\text{K}}$
Λ(1520)	4.4	$\frac{3}{2}^-$	0	0	−1	0	″
Λ(1815)	0.8	$\frac{5}{2}^+$	0	0	−1	0	″
Λ(2100)	0.3	$\frac{7}{2}^-$	0	0	−1	0	″
Λ(2350)	0.5		0	0	−1	0	″
Λ(2585)	0.3		0	0	−1	0	″
Sigma, Σ			−1, 0, +1	1	−1	0	N$\overline{\text{K}}$
Σ(1385)	1.9	$\frac{3}{2}^+$	−1, 0, +1	1	−1		$\Lambda\pi$
Σ(1670)	1.3	$\frac{3}{2}^-$	−1, 0, +1	1	−1		N$\overline{\text{K}}$
Σ(1765)	0.6	$\frac{5}{2}^-$	−1, 0, +1	1	−1		N$\overline{\text{K}}$
Σ(1915)	0.7	$\frac{5}{2}^+$	−1, 0, +1	1	−1		N$\overline{\text{K}}$
Σ(2030)	0.4	$\frac{7}{2}^+$	−1, 0, +1	1	−1		N$\overline{\text{K}}$
Σ(2250)	0.7			1	−1		N$\overline{\text{K}}$
Cascade, Ξ			−1, 0	$\frac{1}{2}$	−2	0	$\Xi\pi$
Ξ(1530)	7.2	$\frac{3}{2}^+$					
Ξ(1820)	1.3						$\Lambda\overline{\text{K}}$
Ξ(1940)	1.0						$\Xi\pi$

Note: If a property is not known it is left blank in the table.

12–7 Quarks, Subhadronic Particles

The number of "elementary" particles—photon, leptons, hadrons—that have been observed is large, more than twice the number of elements in the periodic table. Although few leptons have been found (four), a multitude of hadrons have been observed. Physicists, trusting that nature can be described by a relatively few fundamental building blocks, have questioned the assumption that hadrons are indeed elementary. Several theoretical models, supported by the results of high-energy collision experiments (well over 1 GeV), have been proposed. These models envision the hadrons as comprising a small number of more fundamental, subhadronic constituents. In this section we give a brief account of one model that has evolved into a successful and appealing description of hadrons. The fundamental subhadronic particles in this model are called *quarks;* the name was originally proposed by M. Gell-Mann in 1963, when he and G. Zweig independently introduced similar models for the internal structure of hadrons.†

First, here is a simple definition of an elementary particle—it has no internal structure, is not decomposable into smaller parts, and has no spatial extent. The four particles (and the four associated antiparticles) in the lepton family appear to satisfy this definition (see Table 12–4). For example, in high-energy scattering experiments, the electron behaves as a point particle (size less than 10^{-2} fm). The quantum properties characterizing each of the eight known members of the lepton-antilepton family are listed in Table 12–9. Each of the eight particles in the table is a distinct elementary particle in that it has at least one quantum number different from that for any other member of the lepton family.

Likewise, it is possible to introduce eight hypothetical quark-antiquark particles that have distinct quantum properties and that combine in small groups to form the many known hadrons. The four quarks are named *up* (symbol u), *down* (d), *charm* (c), and *strange* (s). The four antiquarks are labeled $\overline{\text{u}}$ (antiup), $\overline{\text{d}}$ (antidown), $\overline{\text{c}}$ (anticharm), and $\overline{\text{s}}$ (antistrange). Table

Table 12–9. *Quantum properties of leptons*

	LEPTON SYMBOL	ELECTRIC CHARGE Q	LEPTON NUMBER L_e	LEPTON NUMBER L_μ	SPIN QUANTUM NUMBER s
LEPTONS	e^-	-1	1	0	$\frac{1}{2}$
	ν_e	0	1	0	$\frac{1}{2}$
	μ^-	-1	0	1	$\frac{1}{2}$
	ν_μ	0	0	1	$\frac{1}{2}$
ANTILEPTONS	e^+	1	-1	0	$\frac{1}{2}$
	$\overline{\nu}_e$	0	-1	0	$\frac{1}{2}$
	μ^+	1	0	-1	$\frac{1}{2}$
	$\overline{\nu}_\mu$	0	0	-1	$\frac{1}{2}$

† The name *quark* was chosen by Gell-Mann from the opening words of the last chapter of Part 2 of the esoteric novel *Finnegans Wake,* by James Joyce:

Three quarks for Muster Mark!
Sure he hasn't got much of a bark
And sure any he has it's all beside the mark.

Table 12–10. *Quantum properties of quarks*

	QUARK SYMBOL	ELECTRIC CHARGE Q	BARYON NUMBER B	STRANGENESS S	CHARM C	SPIN s
QUARKS	u	$\frac{2}{3}$	$\frac{1}{3}$	0	0	$\frac{1}{2}$
	d	$-\frac{1}{3}$	$\frac{1}{3}$	0	0	$\frac{1}{2}$
	c	$\frac{2}{3}$	$\frac{1}{3}$	0	1	$\frac{1}{2}$
	s	$-\frac{1}{3}$	$\frac{1}{3}$	-1	0	$\frac{1}{2}$
ANTIQUARKS	$\bar{u}$	$-\frac{2}{3}$	$-\frac{1}{3}$	0	0	$\frac{1}{2}$
	$\bar{d}$	$\frac{1}{3}$	$-\frac{1}{3}$	0	0	$\frac{1}{2}$
	$\bar{c}$	$-\frac{2}{3}$	$-\frac{1}{3}$	0	-1	$\frac{1}{2}$
	$\bar{s}$	$\frac{1}{3}$	$-\frac{1}{3}$	1	0	$\frac{1}{2}$

12–10 lists the quantum properties of the particles in the quark family. All quantum properties except charm are familiar from the discussion of the properties of hadrons (see Tables 12–4, and 12–7).† The fourth quark, c, was postulated in 1964 by J. Bjorken and S. Glashow and was verified ten years later in high-energy collision experiments. Examples of hadrons with charmed-quark constituents will be given later.

The symmetry between the lepton model and the quark model is obvious. There are four leptons and four quarks. The particles in each group are apparently representable as point particles without internal structure. All the particles have the same intrinsic spin quantum number, $\frac{1}{2}$. However, significant differences also exist between these two groups of elementary particles:

1. Each lepton has integral lepton number; each quark, on the other hand, has fractional baryon number.
2. Each lepton has integral charge quantum number, but each quark has fractional charge quantum number ($\pm\frac{1}{3}$e, $\pm\frac{2}{3}$e).
3. Leptons have all been observed as free particles, and they do not combine with one another to form bound systems; however, single quarks have not been observed, and they apparently always combine in small quark groups to form composite hadrons.
4. Leptons do not interact through the strong nuclear force; quarks, on the other hand, do participate in a strong interaction and are responsible for the strong short-range binding of hadrons in nuclei.

The fractional values of charge and baryon numbers of quarks (Table 12–10) and the rules for combining quarks to form hadrons have been so chosen as to be consistent with the known integral values of the quantum properties for the composite hadrons. The rules for quark combinations and the resultant value of bound quark systems are:

1. Quarks combine in only three ways: (*a*) Three bound quarks form a baryon; (*b*) Three bound antiquarks form an antibaryon; and (*c*) A bound quark-antiquark pair forms a meson.

† Besides the quark "flavors"—u, d, c, and s—there is strong evidence for another quantum property, designated "color," with allowed values of "red, green, and blue", which has been introduced to preserve the exclusion principle for quarks.

2. For the quantum property of charge, baryon number, strangeness, or charm, the quantum number of the composite hadron is the algebraic sum of the quantum number of the quark constituents.
3. For the intrinsic angular-momentum property, the total angular momentum of the composite hadron is the vector sum of the intrinsic spins of the quarks and the orbital angular momentum of the quarks about the common center of mass, by the conventional rules of quantum mechanics.

These rules, along with the quantum values of the quarks in Table 12–10, guarantee the proper values of the properties of composite hadrons. For example, since each of the four quarks in Table 12–10 have baryon number $\frac{1}{3}$, any combination of three quarks gives the baryon number $B = 1$, which is characteristic of all baryons. Similarly, any combination of three antiquarks gives $B = -1$, the value for all antibaryons. Further, the combination of any quark with any antiquark gives $B = 0$, which is true for all mesons.

The rules also result in integral charge numbers for baryons, since the total electric charge of any three quarks must be -1, 0, 1, or 2; in integral charge numbers -2, -1, 0, or 1 for antibaryons; and in integral charge numbers -1, 0, or 1 for mesons. The strangeness number of baryons can be -3 (three strange quarks, written (sss)), -2, -1, or 0; for antibaryons, S can be 0, 1, 2, or 3; for mesons, S can be -1, 0, or 1. The values for charm for the hadrons are the same as for strangeness except that there is a reversal in the sign. Finally, rules 1 and 3 for quark combinations result in half-integral spin for baryons and antibaryons, and integral spin for mesons.

The quark model provides a simple description of observed hadrons as bound systems of three quarks, or three antiquarks, or a quark-antiquark pair. There are, however, unresolved questions:

1. Why do quarks only combine in three ways?
2. Can single, isolated quarks exist, and if not, what extraordinarily strong attractive force acts between quarks, confining them to one another in groups of two or three?
3. Are there only four distinct quarks and four distinct leptons, or will still others be found?

Answers to these questions are being pursued, and as particle accelerators achieve even greater energies and intensities, no doubt better theoretical models of the fundamental constituents describing our universe will evolve.

Example 12–4. What combination of bound quarks describes the properties (*a*) of a proton? (*b*) of a neutron? (*c*) of a positive pion π^+?

(*a*) The proton is a baryon ($B_p = 1$); therefore, it must be composed of three quarks from Table 12–10. Since the proton has zero strangeness and charm (see Table 12–4), it can be composed only of up and down quarks. The only combination of three u and d quarks that has the electric charge of the proton, $+1$, comprises two up quarks and one down quark. Thus the proton can be described as a bound (uud) quark trio.

(*b*) The neutron, like the proton, is a baryon with $S_n = C_n = 0$. Therefore, it too must be composed of three up and down quarks. Since the charge of the neutron is zero, the only combination of quarks is one up and two down quarks, or (udd).

(*c*) The positive pion is a meson ($B_\pi = 0$), with $S_\pi = 0$ and $C_\pi = 0$. Therefore, it must be composed of a quark (u or d) and an antiquark ($\bar{\text{u}}$ or $\bar{\text{d}}$) that gives a total charge of 1. From Table 12–10, we see that the only combination leading to the π^+ is an up quark and an antidown quark, (u$\bar{\text{d}}$).

Example 12–5. Find the quark constituents of the following hadrons, listed in Tables 12–4 and 12–7: (*a*) the lambda-zero (Λ^0) baryon; (*b*) the sigma-zero (Σ^0) baryon; (*c*) the charmed D^+ meson with S = 0 and C = +1.

(*a*) Since the Λ^0 is a baryon, it must consist of three quarks. With strangeness number −1, charm number 0, and charge 0, it must have one strange quark, one up quark, and one down quark, or (uds).

(*b*) Like Λ^0, the sigma-zero hadron is a baryon with $S = -1$, $C = 0$ and $Q = 0$. Therefore, it comprises a (uds) combination. The Σ^0 is, however, distinct from the Λ^0 in that its isotopic spin quantum number is different from that of the Λ^0 (1 rather than 0).

(*c*) The charmed D^+ meson must consist of a quark-antiquark pair. With $Q = 1$, $S = 0$, and $C = 1$, it must have one charmed quark (which has charge $\frac{2}{3}$) and one nonstrange, noncharmed antiquark with charge equal to $\frac{1}{3}$. By Table 12–10, this is the antidown quark. Therefore, the charmed D^+ meson is (cd).†

† For an interesting account of the first charmed particles to be discovered, read Roy F. Schwitters, "Fundamental Particles with Charm," *Scientific American*, October 1977, p. 56.

PROBLEMS

Electromagnetic interaction, §12–1

● **12–1.** (*a*) Show that pair production (represented by a single space-time event in Figure 12–1*c*) can be depicted as a combination of three basic electromagnetic interactions of the form shown in Figure 12–3. (*b*) Identify the two intermediate virtual particles.

12–2. Assume that an x-ray photon of energy 1.0×10^5 eV becomes a virtual electron-positron pair during the unobservable time Δt (see Figure 12–6*b*). (*a*) If the virtual pair moves at speed $c/2$, what is the maximum time the pair could exist? (*b*) What is the spatial separation of the two events in part (*a*)?

Other interactions, §12–2

● **12–3.** On a space-time graph, show the decay and indicate the virtual particles for the following: (*a*) $\mathrm{n} \rightarrow \mathrm{p} + \mathrm{e}^- + \bar{\nu}_e$; (*b*) $\pi^- \rightarrow \mu^- + \bar{\nu}_\mu$; (*c*) $\pi^0 \rightarrow \gamma + \gamma$.

● **12–4.** What would be some differences between the structure of our universe and the structure of a universe in which all properties and laws were alike except that the masses of the proton and the neutron were interchanged?

12–5. Any unstable particle existing for a finite time has an uncertainty in its total energy and therefore an uncertainty in its rest mass. In a specific example, find the inherent uncertainty in energy of a charged pion, whose mean life is 3×10^{-8} s, (*a*) when it is at rest with respect to the observer; (*b*) when it has a kinetic energy of 20 GeV with respect to the observer.

12–6. Compare the ranges of a virtual photon, a virtual pion, and a virtual W particle (assumed rest mass, 80 GeV) when they have the same kinetic energy, 140 MeV.

Fundamental particle properties, §12–3

● **12–7.** Identify the smallest rest-mass particle or particles that interact through (*a*) the strong interaction; (*b*) the electromagnetic interaction; (*c*) the weak interaction.

12–8. Show the space-time graph for event 1 in Figure 12–16 and identify the two virtual exchange particles.

12–9. What is the maximum number of antiprotons that can be produced in the collision of a 20-GeV electron with a proton at rest?

12–10. Even though the neutron has no electric charge, it does have a magnetic moment antiparallel to its angular momentum. Show that this is consistent with the neutron being regarded, for short times governed by the uncertainty principle, as a virtual proton and a virtual π^- meson with orbital angular momentum $\hbar$. The spin angular momentum of the neutron and the proton is $\hbar/2$ and of the pion, zero.

Universal conservation laws, §12–4

12–11. Find the threshold kinetic energy for an incident proton striking a target proton at rest to produce the reaction at point 1 in Figure 12–10.

12–12. In the production and subsequent decays of the K^0 and Λ^0 hadrons shown in the bubble-chamber photograph of Figure 12–10, the radius of curvature can be measured for the six outgoing charged pions and protons. Assuming that all tracks lie in the plane of the paper and that the uniform magnetic field is perpendicular to this plane, one can determine the speed and kinetic energy of each emerging charged particle. Applying linear-momentum and energy conservation laws at events 2 and 3 then gives the speeds of the K^0 and Λ^0 particles. Assume that this gives a speed of $0.84c$ for the kaon and $0.64c$ for the Λ^0 particle. (*a*) What are the kinetic energies (GeV) of these two neutral particles? (*b*) What is the sum of the kinetic energies of the π^+ and p following event 1? (*c*) If the actual distances in the bubble chamber are twice the distance on the photographic film, how long (in the laboratory frame of reference) did the K^0 and Λ^0 particles exist? (*d*) Determine the proper lifetime of each of the two neutral particles and compare the values with the mean lives as shown in Table 12–4.

Additional conservation laws, §12–5

● **12–13.** Show that the π^0, η^0, and Σ^0 in Table 12–4 can all decay by the electromagnetic interaction.

● **12–14.** Show, by using the decay values, that Σ^0 and $\overline{\Sigma^0}$ are distinct elementary particles.

12–15. In the creation of the Ω^- shown in Figure 12–17, show that I_z is conserved in the creation of Ω^- but not in the decay of Ω^-.

12–16. (*a*) Show that the $\overline{\mathrm{K}^-}$-meson beam incident on the protons of Example 12–3 also produces Ξ^- baryons and Σ^- baryons. (*b*) Compare the threshold energies of the $\overline{\mathrm{K}^-}$-meson beam in producing Ω^-, Ξ^-, and Σ^- baryons.

12–17. A high-energy proton beam is incident on a solid target; the kinetic energy of the incident protons is much larger than the binding energy among the nucleons in the target nuclides. For each of the following reactions, identify the other reaction particle X.

(*a*) $\mathrm{p} + \mathrm{p} \rightarrow \Sigma^0 + \mathrm{p} + \mathrm{X}$

(*b*) $\mathrm{p} + \mathrm{p} \rightarrow \overline{\Sigma^0} + \mathrm{p} + \mathrm{p} + \mathrm{n} + \mathrm{X}$

(*c*) $\mathrm{p} + \mathrm{n} \rightarrow \Sigma^0 + \mathrm{n} + \mathrm{X}$

(*d*) $\mathrm{p} + \mathrm{n} \rightarrow \overline{\Sigma^0} + \mathrm{p} + \mathrm{n} + \mathrm{n} + \mathrm{X}$

Resonance particles, §12–6

● **12–18.** What conservation laws allow the neutral pion to decay into two photons by the electromagnetic interaction but do not allow the charged pions to decay by this interaction?

● **12–19.** Why are all resonance particles hadrons?

12–20. Why is the track length of the Λ^0 in Figure 12–17 much shorter than the track length of the Ω^-?

12–21. (*a*) Find the threshold kinetic energy of a beam of incident positive pions on a proton target to produce a Δ^{++} particle (mass, 1.23 GeV) and compare this with the range of pion energies (40 to 220 MeV) that Fermi used in the 1952 scattering experiments. (*b*) Show that the distance of closest approach in part (*a*) is small enough to be within the range of the nuclear interaction.

12–22. What charge states of the Δ resonance family can be produced by a 600-MeV beam of positive kaons incident on a target composed of protons and neutrons?

12–23. In the following four strong nuclear reactions, the meson and the nucleon exchange electric charge. Determine which reactions can occur and also the minimum kinetic energy of the incident meson beam necessary for each reaction to take place:

$$\pi^+ + \mathrm{n} \rightarrow \pi^0 + \mathrm{p}$$

$$\pi^- + \mathrm{p} \rightarrow \pi^0 + \mathrm{n}$$

$$\mathrm{K}^+ + \mathrm{n} \rightarrow \mathrm{K}^0 + \mathrm{p}$$

$$\overline{\mathrm{K}^-} + \mathrm{p} \rightarrow \overline{\mathrm{K}^0} + \mathrm{n}$$

12–24. Show that all four charge states of the Δ baryon can be produced by the electromagnetic interaction with high-energy photons incident on a target composed of neutrons and protons. Write down the reaction equation for each and find the threshold energy of the photon beam.

12–25. Show that the decays of the resonance particles in Table 12–8 satisfy all conservation laws associated with the strong interaction.

12–26. Which of the following reactions *cannot* occur? Give the reason or reasons for each case. If the reaction is possible, by what interaction does it take place?

(*a*) $\nu_e + \mathrm{p} \rightarrow \mathrm{e}^- + \Sigma^0 + \mathrm{K}^+$

(*b*) $\overline{\mathrm{e}^+} + \mathrm{e}^- \rightarrow \overline{\Sigma^-} + \mathrm{K}^+ + \mathrm{n}$

(*c*) $\overline{\mathrm{e}^+} + \mathrm{e}^- \rightarrow \Lambda^0 + \overline{\Sigma^0}$

(*d*) $\Xi^- \rightarrow \Sigma^- + \pi^0$

(*e*) $\pi^+ + \mathrm{n} \rightarrow \mathrm{K}^+ + \Sigma^0$

Quarks, §12–7

● **21–27.** What are the quark compositions of the four $\Delta(1232)$ particles listed in Table 12–8?

12–28. What are the quark compositions of the following mesons whose properties are listed in Tables 12–4, 12–5 and 12–7: (*a*) ρ^-; (*b*) K^+; (*c*) $\overline{\mathrm{K}^0}$?

12–29. What are the quark compositions of the following baryons whose properties are listed in Table 12-5: (*a*) Ξ^-; (*b*) Ω^-; (*c*) $\overline{\Sigma^0}$?

12–30. The baryon supermultiplet with the smallest-mass baryons is the group of eight particles, each with $J^P = (\frac{1}{2})^+$; consisting of the nucleon, lambda, sigma, and xi families. (*a*) On a graph of hypercharge Y (ordinate) versus isospin component I_z, show the location of these eight particles. (*b*) Indicate the quark constituents of each particle.

Appendix I
Derivation of Lorentz Transformations

The Lorentz coordinate transformations follow directly from the two postulates of relativity: (1) Physical laws have the same mathematical form in all inertial systems. (2) The speed of light in a vacuum has the same value in all inertial systems.

For simplicity, let us consider motion in the xy-plane only, with the motion of S' chosen to be parallel to the positive direction of the coincident x-x' axes (see Figure 2–1, which is reproduced below as Figure I–1). By choosing an orientation for each of the axes like those used for expressing the Galilean transformations, we shall get the simplest relation between the space-time coordinates of S and S'. Since both the z-axis and the y-axis are perpendicular to the direction of relative motion of the two systems, the behavior will be the same along the one as along the other.

The most general type of transformation equations relating the space and time coordinates (x, y, t) of an event, as observed in inertial system S, to the coordinates (x', y', t') of the same event, as observed in inertial system S', must be of the form

$$
\begin{aligned}
x' &= A_1x + A_2y + A_3t + A_4 \\
y' &= B_1x + B_2y + B_3t + B_4 \qquad \text{(I–1)} \\
t' &= D_1x + D_2y + D_3t + D_4
\end{aligned}
$$

where the 12 quantities $A_1, \ldots, A_4, \ldots, D_4$ are independent of the space and time coordinates but may depend on the relative velocity of one inertial system with respect to the other. We have assumed that Equations I–1 are *linear* equations, involving the variables to the first power only, since only then would some single real event (x, y, t) in S correspond to a *single* real event (x', y', t') in S', and the converse. Note that the time coordinate t' in S' now includes terms involving the spatial coordinates x and y of S, which cannot be precluded a priori. It is our task to find the values of the 12 constant coefficients $(A_1, \ldots, D_4)$.

The components of the velocity of a particle as measured in the two inertial systems can easily be obtained from Equation I–1. From the definition for the components of the velocity observed from system S'

$$
\dot{x}' \equiv \frac{dx'}{dt'} \quad \text{and} \quad \dot{y}' \equiv \frac{dy'}{dt'}
$$

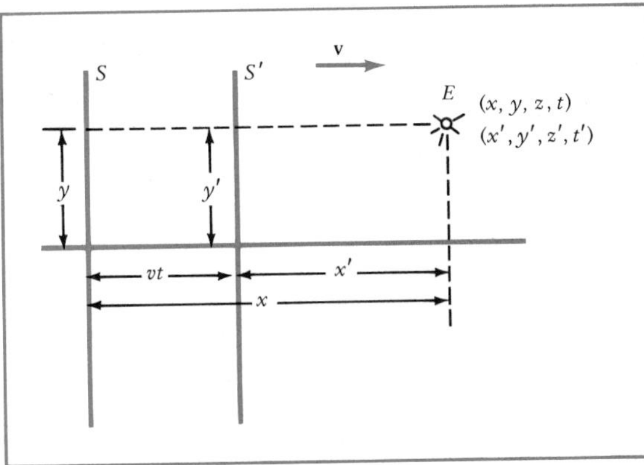

FIGURE I–1. *Coordinates x' and y' of rod as measured by observer S'.*

and for those observed from S,

$$\dot{x} \equiv \frac{dx}{dt} \quad \text{and} \quad \dot{y} = \frac{dy}{dt}$$

we get, using Equation I–1,

$$\dot{x}' = \frac{dx'}{dt'} = \frac{A_1\,dx + A_2\,dy + A_3\,dt}{D_1\,dx + D_2\,dy + D_3\,dt}$$

By dividing both the numerator and the denominator of the right-hand side of this equation by dt, we can express $\dot{x}'$ in terms of $\dot{x}$ and $\dot{y}$:

$$\dot{x}' = \frac{A_1\dot{x} + A_2\dot{y} + A_3}{D_1\dot{x} + D_2\dot{y} + D_3} \tag{I–2}$$

Similarly,

$$\dot{y}' = \frac{dy'}{dt'} = \frac{B_1\dot{x} + B_2\dot{y} + B_3}{D_1\dot{x} + D_2\dot{y} + D_3} \tag{I–3}$$

The denominators of these equations are identical.

We shall first apply some general conditions to the transformations I–1 to I–3; this will reduce considerably the number of nonzero coefficients. Then we shall impose the experimental observation that the speed of light is the same in all inertial systems, to get the proper transformations of space-time events.

TRANSFORMATION OF SPACE-TIME EVENTS. We choose the $+x$-axis parallel to, and in the same direction as, the relative velocity $\mathbf{v}$ of system S' to system S. This means that an observer in S takes the origin of S' to be moving at speed v along the $+x$-axis; conversely, an observer in S' takes the origin of S to be moving at speed v along the $-x'$-axis.

We so set the clocks in s and S' that when the origin of S' passes the origin of S, both clocks read zero. Thus, when $t = 0$, then $t' = 0$; the origins of S and S' coincide at this time; and the space-time coordinates of the origin of S' as seen by observers S and S' are as follows:

Observer S sees (0, 0, 0)

Observer S' sees (0, 0, 0)

Using these values in Equations I–1 gives

$$A_4 = B_4 = D_4 = 0$$

Next, let us consider the motion of system S's origin. As seen by observer S, the origin is always at rest: $\dot{x} = 0$, $\dot{y} = 0$. As seen by observer S', it moves along the $-x$-axis at speed v. Thus, $\dot{x}' = -v$, and $\dot{y}' = 0$. Using these values in Equations I–2 and I–3, we get

$$-v = \frac{A_3}{D_3} \quad \text{and} \quad 0 = \frac{B_3}{D_3}$$

Since D_3 must be finite, we have

$$A_3 = -vD_3 \quad \text{and} \quad B_3 = 0$$

Likewise, if we now consider the motion of system S''s origin, we get $\dot{x}' = \dot{y}' = 0$ and $\dot{x} = +v, \dot{y} = 0$. Substituting these values in Equations I–2 and I–3 gives

$$0 = \frac{A_1 v + A_3}{D_1 v + D_3} \qquad 0 = \frac{B_1 v + B_3}{D_1 v + D_3}$$

Again, with D_1 and D_3 finite, the numerators must be zero, giving

$$A_3 = -A_1 v \qquad \text{and} \qquad B_1 = \frac{-B_3}{v} = 0$$

With B_1, B_3, and B_4 all zero, the equation for y' in Equation I–1 is now reduced to

$$y' = B_2 y \tag{I–4}$$

Before we find the value of B_2, let us show by symmetry arguments that the coefficients A_2 and D_2 must be zero.

First consider a rod of length L_0 lying parallel to the y-axis from 0 to L_0 and at rest in system S. One end is at $y = 0$ and the other end at $y = L_0$. Then from Equation I–2 we have

$$\text{Lower end:} \qquad \dot{x}' = -v = \frac{A_3}{D_3}$$

$$\text{Upper end:} \qquad \dot{x}' = -v = \frac{A_3}{D_3}$$

So relative to observer S' both ends move to the left at constant speed v.

If $A_2 \neq 0$, then at time $t = t' = 0$ a rod along the y-axis will have $x = 0$; and according to Equation I–1 and I–4, observer S' will measure

$$x' = A_2 y$$

$$y' = B_2 y$$

See Figure I–2. But this cannot be so, since we must have symmetry about the axis of the relative velocity **v**. This symmetry requirement will be met only if observer S' sees the rod perpendicular to **v**; therefore,

$$A_2 = 0$$

To show that $D_2 = 0$, consider two events in system S occurring at the same x-coordinate and the same time, say $x = 0$ and $t = 0$. Because they

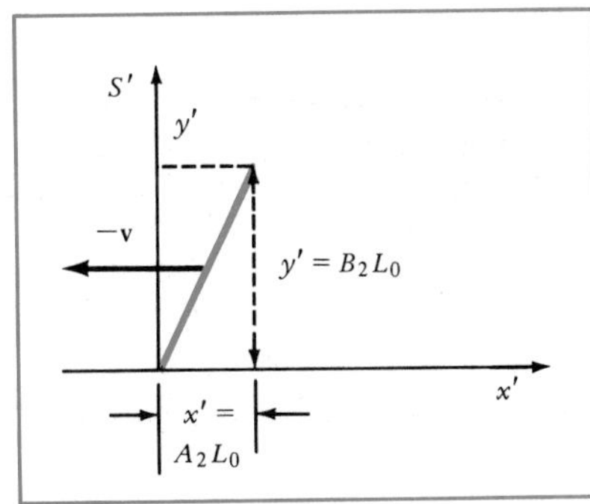

FIGURE I–2. *Light pulse as observed (a) by S and (b) by S'.*

occur at the same time, they are simultaneous events from the point of view of an observer in that system. If one event occurs at the origin and the other at $y = L_0$, then according to the last of Equations I–1, an observer in S' observes the event at L_0 to occur later than that at the origin at $\Delta t' = D_2 L_0$. Thus, events along the y-axis (which are simultaneous in S) will, from the point of view of an observer in S', occur later or earlier than $t = 0$, depending on whether they occur above or below the y-axis. This again violates symmetry about the axis of **v**, unless

$$D_2 = 0$$

We now show that $B_2 = 1$. Consider identical measuring rods, each of length L_0 when measured at rest in either S or S'. We fix one of the rods along the y-axis in S between $Y = 0$ and $y = L_0$. The second rod is fixed along the y-axis between $y' = 0$ and $y' = L_0$. What does observer S' measure as the length of the rod fixed in system S? Since we know that the y' coordinates of the ends of the rod do not depend on time (see Equation I–4), we can determine these coordinates at any time, and their difference will give the length as seen in S'. Using Equation I–4, we have, for the length L' of the moving rod as seen by observer S',

$$L' = B_2 L_0$$

Again, by Equation I–4, we find that observer S measures the length of the rod fixed in system S' as

$$L_0 = B_2 L$$

Now, the ratio of the length of the moving rod to the length of the fixed rod measured by observer S', must be the same as the ratio of the lengths measured by S; otherwise, we could distinguish between the inertial systems! Thus,

$$\frac{L'}{L_0} = \frac{L}{L_0} \quad \text{or} \quad B_2 = \frac{1}{B_2} \quad \text{or} \quad B_2 = +1$$

Equation I–4 then becomes

$$y' = y$$

Notice that we discard the solution $B_2 = -1$, since that implies that the point $y = L_0$ goes into the point $y' = -L_0$, which is not true. Since the z-coordinate is similar to the y-coordinate, the z-components transform as

$$z' = z$$

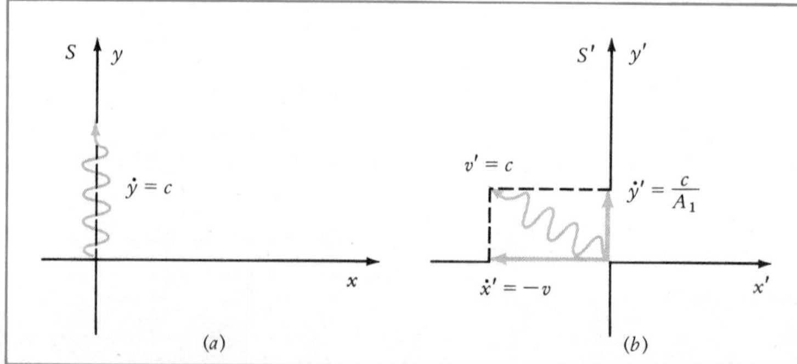

FIGURE I–3. *Light pulse as observed (a) by S and (b) by S'.*

Coordinates perpendicular to the relative velocity **v** of the two inertial systems are the same in both systems. This is nothing new.

Substituting all the determined coefficients in Equations I–1 and I–2, we get the following (the z-coordinates are now included for completeness):

Coordinate transformations		*Velocity transformations*	
$x' = A_1(x_1 - vt_1)$		$\dot{x}' = \dfrac{A_1(\dot{x}_1 - v)}{D_1\dot{x}_1 + A_1}$	
$y' = y_1$	(I–5)	$\dot{y}' = \dfrac{\dot{y}_1}{D_1\dot{x}_1 + A_1}$	(I–6)
$z' = z_1$		$\dot{z}' = \dfrac{\dot{z}_1}{D_1\dot{x}_1 + A_1}$	
$t' = D_1x_1 + A_1t_1$			

We have arrived at Equations I–5 and I–6 by using the homogeneity of space (the fact that a single event in one inertial system corresponds to a single event in any other inertial system) and by arguments from symmetry. There are now only two undetermined coefficients, A_1 and D_1.

Let us now solve for A_1 and D_1, imposing the second postulate of relativity, the requirement that the speed of light must be the same for all observers.

Imagine that observer S shines a beam of light along the $+x$-axis, measuring its speed as c; that is,

$$\dot{x} = c$$

A second observer S' moving to the right with speed v relative to S must also measure the speed of this light beam as c. Substituting c for $\dot{x}'$ in Equation I–6 gives

$$c = \frac{A_1(c - v)}{D_1c + A_1}$$

which gives

$$D_1c^2 = -A_1v$$

$$D_1 = -\frac{v}{c^2}A_1 \qquad \text{(I–7)}$$

This equation reduces the unknown coefficients in Equation I–5 to one, either A_1 or D_1. We choose A_1 as the unknown. To get this coefficient, we again impose the constancy of the speed of light, but this time we suppose that S projects the light beam along the $+y$-axis; see Figure I–3.

Observer S will, of course, measure the speed as c (that is, $\dot{y} = c$ and $\dot{x} = 0$).

Observer S', moving to the right with speed v relative to S, will, by the velocity transformations of Equation I–6, measure the velocity components $\dot{x}'$ and $\dot{y}'$ as

$$\dot{x}' = \frac{A_1(0 - v)}{A_1} = -v$$

$$\dot{y}' = \frac{c}{A_1} \qquad \text{(I–8)}$$

But S' must also observe the light pulse to travel at the speed c. Thus, we have

$$v' = \sqrt{\dot{x}^2 + \dot{y}^2} = c$$

Substituting Equation I–8 in this equation gives

$$\sqrt{v^2 + \frac{c^2}{A_1^2}} = c$$

$$A_1^2 = \frac{1}{1 - (v/c)^2}$$

$$A_1 = \frac{1}{\sqrt{1 - (v/c)^2}} \tag{I–9}$$

Only the positive sign is taken for A_1, because a negative A_1 by Equation I–6 would lead to a reversal in the y- and z-components of the velocity of objects seen from different inertial systems.

Substituting the values of A_1 and D_1 from Equations I–7 and I–9 in Equation I–5 finally gives the Lorentz transformations for any space-time event:

Lorentz coordinate transformations:

$$\begin{aligned} x' &= \frac{x - vt}{\sqrt{1 - (v/c)^2}} \\ y' &= y \\ z' &= z \\ t' &= \frac{t - (v/c^2)x}{\sqrt{1 - (v/c)^2}} \end{aligned} \tag{I–10}$$

Appendix II
The Atomic Masses

Given here are the masses of the neutral atoms of all stable nuclides and some of the unstable nuclides (with their respective half-lives). The terrestrial abundances of the nuclides are also included.

The uncertainties in mass are less than 0.000 001 u for many nuclides of low mass number. For all other listed nuclides except ^{124}Xe, the uncertainties are in the last or next-to-last significant figure. Xenon-124 has an uncertainty of 0.000 15.

The percentage abundance and half-life data are from the *Chart of the Nuclides,* General Electric Company (1977). The symbols used for half-lives of unstable nuclides in this Appendix are: yr (years), d (days), h (hours), min (minutes), s (seconds), and ms (milliseconds).

ELEMENT	A	ATOMIC MASS (u)	NATURAL ABUNDANCE (%)	HALF-LIFE (IF UNSTABLE)
$_0n$	1	1.008 665		10.6 min
$_1H$	1	1.007 825	99.985	
	2	2.014 102	0.015	
	3	3.016 050		12.33 yr
$_2He$	3	3.016 029	0.000 14	
	4	4.002 603	99.999 86	
	6	6.018 891		805 ms
$_3Li$	6	6.015 123	7.5	
	7	7.016 004	92.5	
	8	8.022 487		844 ms
	9	9.026 790		177 ms
$_4Be$	7	7.016 930		53.28 d
	8	8.005 305		$\sim 10^{-16}$ s
	9	9.012 182	100	
	10	10.013 535		1.6×10^6 yr
$_5B$	8	8.024 608		770 ms
	10	10.012 938	20	
	11	11.009 305	80	
	12	12.014 353		20.4 ms
	13	13.017 780		17.3 ms
	14	14.025 397		16 ms
$_6C$	9	9.031 038		127 ms
	10	10.016 858		19.3 s
	11	11.011 433		20.3 min
	12	12.000 000	98.89	
	13	13.003 355	1.11	
	14	14.003 242		5730 yr
	15	15.010 599		2.45 s
	16	16.014 700		0.75 s
$_7N$	12	12.018 613		11.0 ms
	13	13.005 739		9.97 min
	14	14.003 074	99.63	
	15	15.000 109	0.37	
	16	16.006 099		7.10 s
	17	17.008 449		4.17 s
$_8O$	14	14.008 597		70.5 s
	15	15.003 065		122 ms

ELEMENT	*A*	ATOMIC MASS (u)	NATURAL ABUNDANCE (%)	HALF-LIFE (IF UNSTABLE)
	16	15.994 915	99.758	
	17	16.999 131	0.038	
	18	17.999 159	0.204	
	19	19.003 576		26.8 s
$_{9}$F	17	17.002 095		64.5 s
	18	18.000 937		109.8 min
	19	18.998 403	100	
	20	19.999 982		11.0 s
	21	20.999 949		4.33 s
$_{10}$Ne	18	18.005 710		1.67 s
	19	19.001 880		17.2 s
	20	19.992 439	90.51	
	21	20.993 845	0.27	
	22	21.991 384	9.22	
	23	22.994 466		37.5 s
	24	23.993 613		3.38 min
$_{11}$Na	22	21.994 435		2.60 yr
	23	22.989 770	100	
	24	23.990 963		15.0 h
$_{12}$Mg	23	22.994 127		11.3 s
	24	23.985 045	78.99	
	25	24.985 839	10.00	
	26	25.982 595	11.01	
$_{13}$Al	27	26.981 541	100	
$_{14}$Si	28	27.976 928	92.23	
	29	28.976 496	4.67	
	30	29.973 772	3.10	
$_{15}$P	30	29.978 310		2.50 min
	31	30.973 763	100	
$_{16}$S	32	31.972 072	95.02	
	33	32.971 459	0.75	
	34	33.967 868	4.21	
	35	34.969 032		87.2 d
	36	35.967 079	0.017	
$_{17}$Cl	35	34.968 853	75.77	
	36	35.968 307		3.01×10^5 yr
	37	36.965 903	24.23	
$_{18}$Ar	36	35.967 546	0.337	
	37	36.966 776		34.8 d
	38	37.962 732	0.063	
	39	38.964 315		269 yr
	40	39.962 383	99.60	
$_{19}$K	39	38.963 708	93.26	
	40	38.963 999	0.01	1.28×10^9 yr
	41	40.961 825	6.73	
$_{20}$Ca	40	39.962 591	96.94	
	41	40.962 278		1.3×10^5 yr
	42	41.958 622	0.647	
	43	42.958 770	0.135	
	44	43.955 485	2.09	

ELEMENT	A	ATOMIC MASS (u)	NATURAL ABUNDANCE (%)	HALF-LIFE (IF UNSTABLE)
	45	44.956 189		163 d
	46	45.953 689	0.0035	
	47	46.954 543		4.54 d
	48	47.952 532	0.187	
$_{21}$Sc	41	40.969 250		0.60 s
	45	44.955 914	100	
$_{22}$Ti	46	45.952 633	8.25	
	47	46.951 765	7.45	
	48	47.947 947	73.7	
	49	48.947 871	5.4	
	50	49.944 786	5.2	
$_{23}$V	48	47.952 257		15.98 d
	50	49.947 161	0.25	$\sim 10^{17}$ yr
	51	50.943 962	99.75	
$_{24}$Cr	48	47.954 033		21.6 h
	50	49.946 046	4.35	
	52	51.940 510	83.79	
	53	52.940 651	9.50	
	54	53.938 882	2.36	
$_{25}$Mn	54	53.940 360		312.5 d
	55	54.938 046	100	
$_{26}$Fe	54	53.939 612	5.8	
	56	55.934 939	91.8	
	57	56.935 396	2.1	
	58	57.933 278	0.3	
	59	58.934 878		44.6 d
$_{27}$Co	58	57.935 755		70.8 d
	59	58.933 198	100	
	60	59.933 820		5.27 yr
$_{28}$Ni	58	57.935 347	68.3	
	60	59.930 789	26.1	
	61	60.931 059	1.1	
	62	61.928 346	3.6	
	64	63.927 968	0.9	
$_{29}$Cu	63	62.929 599	69.2	
	64	63.929 766		12.7 h
	65	64.927 792	30.8	
$_{30}$Zn	64	63.929 145	48.6	
	66	65.926 035	27.9	
	67	66.927 129	4.1	
	68	67.924 846	18.8	
	70	69.925 325	0.6	
$_{31}$Ga	69	68.925 581	60.1	
	71	70.924 701	39.9	
$_{32}$Ge	70	69.924 250	20.5	
	72	71.922 080	27.4	
	73	72.923 464	7.8	
	74	73.921 179	36.5	
	76	75.921 403	7.8	

ELEMENT	A	ATOMIC MASS (u)	NATURAL ABUNDANCE (%)	HALF-LIFE (IF UNSTABLE)
$_{33}$As	74	73.923 930		17.8 d
	75	74.921 596	100	
$_{34}$Se	74	73.922 477	0.9	
	76	75.919 207	9.0	
	77	76.919 908	7.6	
	78	77.917 304	23.5	
	80	79.916 520	49.8	
	82	81.916 709	9.2	
$_{35}$Br	79	78.918 336	50.7	
	80	79.918 528		17.7 min
	81	80.916 290	49.3	
$_{36}$Kr	78	77.920 397	0.35	
	80	79.916 375	2.25	
	82	81.913 483	11.6	
	83	82.914 134	11.5	
	84	83.911 506	57.0	
	86	85.910 614	17.3	
$_{37}$Rb	85	84.911 800	72.2	
	87	86.909 184	27.8	4.9×10^{10} yr
$_{38}$Sr	84	83.913 428	0.6	
	86	85.909 273	9.8	
	87	86.908 890	7.0	
	88	87.905 625	82.6	
$_{39}$Y	89	88.905 856	100	
$_{40}$Zr	90	89.904 708	51.5	
	91	90.905 644	11.2	
	92	91.905 039	17.1	
	94	93.906 319	17.4	
	96	95.908 272	2.8	
$_{41}$Nb	93	92.906 378	100	
$_{42}$Mo	92	91.906 809	14.8	
	94	93.905 086	9.3	
	95	94.905 838	15.9	
	96	95.904 675	16.7	
	97	96.906 018	9.6	
	98	97.905 405	24.1	
	100	99.907 473	9.6	
$_{43}$Tc	99	98.906 252		2.1×10^{5} yr
$_{44}$Ru	96	95.907 596	5.5	
	98	97.905 287	1.9	
	99	98.905 937	12.7	
	100	99.904 217	12.6	
	101	100.905 581	17.0	
	102	101.904 347	31.6	
	104	103.905 422	18.7	
$_{45}$Rh	103	102.905 503	100	
$_{46}$Pd	102	101.905 609	1.0	
	104	103.904 026	11.0	
	105	104.905 075	22.2	

ELEMENT	A	ATOMIC MASS (u)	NATURAL ABUNDANCE (%)	HALF-LIFE (IF UNSTABLE)
	106	105.903 475	27.3	
	108	107.903 894	26.7	
	110	109.905 169	11.8	
$_{47}Ag$	107	106.905 095	51.8	
	108	107.905 956		2.41 min
	109	108.904 754	48.2	
$_{48}Cd$	106	105.906 461	1.3	
	108	107.904 186	0.9	
	110	109.903 007	12.5	
	111	110.904 182	12.8	
	112	111.902 761	24.1	
	113	112.904 401	12.2	9×10^{15} yr
	114	113.903 361	28.7	
	116	115.904 758	7.5	
$_{49}In$	113	112.904 056	4.3	
	115	114.903 875	95.7	5×10^{14} yr
$_{50}Sn$	112	111.904 823	1.0	
	114	113.902 781	0.7	
	115	114.903 344	0.4	
	116	115.901 743	14.7	
	117	116.902 954	7.7	
	118	117.901 607	24.3	
	119	118.903 310	8.6	
	120	119.902 199	32.4	
	122	121.903 440	4.6	
	124	123.905 271	5.6	
$_{51}Sb$	121	120.903 824	57.3	
	123	122.904 222	42.7	
$_{52}Te$	120	119.904 021	0.1	
	122	121.903 055	2.5	
	123	122.904 278	0.9	$\sim 1.2 \times 10^{13}$ yr
	124	123.902 825	4.6	
	125	124.904 435	7.0	
	126	125.903 310	18.7	
	128	127.904 464	31.7	
	130	129.906 229	34.5	
$_{53}I$	127	126.904 477	100	
	131	130.906 119		8.0 d
$_{54}Xe$	124	123.906 12	0.1	
	126	125.904 281	0.1	
	128	127.903 531	1.9	
	129	128.904 780	26.4	
	130	129.903 509	4.1	
	131	130.905 076	21.2	
	132	131.904 148	26.9	
	134	133.905 395	10.4	
	136	135.907 219	8.9	
$_{55}Cs$	133	132.905 433	100	
$_{56}Ba$	130	129.906 277	0.1	
	132	131.905 042	0.1	

ELEMENT	A	ATOMIC MASS (u)	NATURAL ABUNDANCE (%)	HALF-LIFE (IF UNSTABLE)
	134	133.904 490	2.4	
	135	134.905 668	6.6	
	136	135.904 556	7.9	
	137	136.905 816	11.2	
	138	137.905 236	71.7	
$_{57}$La	138	137.907 114	0.1	1×10^{11} yr
	139	138.906 355	99.9	
$_{58}$Ce	136	135.907 14	0.2	
	138	137.905 996	0.2	
	140	139.905 442	88.5	
	142	141.909 249	11.1	5×10^{16} yr
$_{59}$Pr	141	140.907 657	100	
$_{60}$Nd	142	141.907 731	27.2	
	143	142.909 823	12.2	
	144	143.910 096	23.8	2.1×10^{15} yr
	145	144.912 582	8.3	$>10^{17}$ yr
	146	145.913 126	17.2	
	148	147.916 901	5.7	
	150	149.920 900	5.6	
$_{61}$Pm	147	146.915 148		2.6 yr
$_{62}$Sm	144	143.912 009	3.1	
	147	146.914 907	15.1	1.1×10^{11} yr
	148	147.914 832	11.3	8×10^{15} yr
	149	148.917 193	13.9	$>10^{16}$ yr
	150	149.917 285	7.4	
	152	151.919 741	26.7	
	154	153.922 218	22.6	
$_{63}$Eu	151	150.919 860	47.9	
	153	152.921 243	52.1	
$_{64}$Gd	152	151.919 803	0.2	1.1×10^{14} yr
	154	153.920 876	2.1	
	155	154.922 629	14.8	
	156	155.922 130	20.6	
	157	156.923 967	15.7	
	158	157.924 111	24.8	
	160	159.927 061	21.8	
$_{65}$Tb	159	158.925 350	100	
$_{66}$Dy	156	155.924 287	0.1	$>1 \times 10^{18}$ yr
	158	157.924 412	0.1	
	160	159.925 203	2.3	
	161	160.926 939	19.0	
	162	161.926 805	25.5	
	163	162.928 737	24.9	
	164	163.929 183	28.1	
$_{67}$Ho	165	164.930 332	100	
$_{68}$Er	162	161.928 787	0.1	
	164	163.929 211	1.6	
	166	165.930 305	33.4	
	167	166.932 061	22.9	

ELEMENT	A	ATOMIC MASS (u)	NATURAL ABUNDANCE (%)	HALF-LIFE (IF UNSTABLE)
	168	167.932 383	27.1	
	170	169.935 476	14.9	
$_{69}$Tm	169	168.934 225	100	
$_{70}$Yb	168	167.933 908	0.1	
	170	169.934 774	3.2	
	171	170.936 338	14.4	
	172	171.936 393	21.9	
	173	172.938 222	16.2	
	174	173.938 873	31.6	
	176	175.942 576	12.6	
$_{71}$Lu	175	174.940 785	97.4	
	176	175.942 694	2.6	2.9×10^{10} yr
$_{72}$Hf	174	173.940 065	0.2	2.0×10^{15} yr
	176	175.941 420	5.2	
	177	176.943 233	18.6	
	178	177.943 710	27.1	
	179	178.945 827	13.7	
	180	179.946 561	35.2	
$_{73}$Ta	180	179.947 489	0.01	$>1.6 \times 10^{13}$ yr
	181	180.948 014	99.99	
$_{74}$W	180	179.946 727	0.1	
	182	181.948 225	26.3	
	183	182.950 245	14.3	
	184	183.950 953	30.7	
	186	185.954 377	28.6	
$_{75}$Re	185	184.952 977	37.4	
	187	186.955 765	62.6	5×10^{10} yr
$_{76}$Os	184	183.952 514	0.02	
	186	185.953 852	1.6	2×10^{15} yr
	187	186.955 762	1.6	
	188	187.955 850	13.3	
	189	188.958 156	16.1	
	190	189.958 455	26.4	
	192	191.961 487	41.0	
$_{77}$Ir	191	190.960 603	37.3	
	193	192.962 942	62.7	
$_{78}$Pt	190	189.959 937	0.01	6.1×10^{11} yr
	192	191.961 049	0.79	
	194	193.962 679	32.9	
	195	194.964 785	33.8	
	196	195.964 947	25.3	
	198	197.967 879	7.2	
$_{79}$Au	197	196.966 560	100	
$_{80}$Hg	196	195.965 812	0.2	
	198	197.966 760	10.0	
	199	198.968 269	16.8	
	200	199.968 316	23.1	
	201	200.970 293	13.2	
	202	201.970 632	29.8	
	204	203.973 481	6.9	

ELEMENT	A	ATOMIC MASS (u)	NATURAL ABUNDANCE (%)	HALF-LIFE (IF UNSTABLE)
$_{81}$Tl	203	202.972 336	29.5	
	205	204.974 410	70.5	
$_{82}$Pb	204	203.973 037	1.4	1.4×10^{17} yr
	206	205.974 455	24.1	
	207	206.975 885	22.1	
	208	207.976 641	52.4	
	210	209.984 178		22.3 yr
$_{83}$Bi	209	208.980 388	100	$>2 \times 10^{18}$ yr
	212	211.991 267		60.6 min
$_{84}$Po	214	213.995 191		0.16 ms
$_{85}$At	215	214.998 646		0.10 ms
$_{86}$Rn	222	222.017 574		3.824 d
$_{88}$Ra	226	226.025 406		1.60×10^{3} yr
$_{90}$Th	230	230.033 131		7.7×10^{4} yr
	232	232.038 054	100	1.4×10^{10} yr
	233	233.041 580		22.2 min
$_{91}$Pa	233	233.040 244		27 d
$_{92}$U	233	233.039 629		1.6×10^{5} yr
	234	234.040 947		2.4×10^{5} yr
	235	235.043 925	0.72	7.04×10^{8} yr
	238	238.050 786	99.28	4.47×10^{9} yr
$_{93}$Np	237	237.048 169		2.14×10^{6} yr
	239	239.052 932		2.4 d
$_{94}$Pu	239	239.052 158		2.4×10^{4} yr
$_{105}$Ha	262	262.113 84		?

Appendix III
References

This listing contains references for collateral and extended reading on topics treated in the textbook. After the list of items of a general nature, applicable for the entire study, references are given for specific topics, identified by chapters within this textbook.

General introductory modern-physics textbooks on a comparable level

- Beiser, A. *Perspectives of Modern Physics.* 2d ed. New York: McGraw-Hill, 1973.
- Blanchard, C. H., Burnett, C. R., Stoner, R. G., and Weber, R. L. *Introduction to Modern Physics.* 2d ed. Englewood Cliffs, N.J.: Prentice-Hall, 1969.
- Ford, K. W. *Classical and Modern Physics,* vol. 3. New York: John Wiley & Sons, 1974.
- Gautreau, R., and Savin, W. *Theory and Problems of Modern Physics.* Schaum's Outline Series. New York: McGraw-Hill, 1978. A book on solving problems, including many worked-out solutions.
- Semat, H., and Albright, J. *Introduction to Atomic and Nuclear Physics.* 5th ed. New York: Holt, Rinehart, and Winston, 1972.
- Tipler, P. A. *Modern Physics.* 2d ed. New York: Worth, 1977.
- Wehr, M. R., Richards, J. A., and Adair, T. W., III. *Physics of the Atom.* 3d ed. Reading, Mass.: Addison-Wesley, 1978.

General introductory modern-physics textbooks on a more advanced level

- Born, M. *Atomic Physics.* 8th rev. ed. New York: Hafner, 1969. A classic.
- Eisberg, R., and Resnick, R. *Quantum Physics of Atoms, Molecules, Solids, Nuclei, and Particles.* New York: John Wiley & Sons, 1974.
- Norwood, J., Jr. *Twentieth-Century Physics.* Englewood Cliffs, N.J.: Prentice-Hall, 1976.
- Richtmeyer, F. K., Kennard, E. H., and Cooper, J. N. *Introduction to Modern Physics.* 6th ed. New York: McGraw-Hill, 1969. For many years, a standard intermediate-level textbook; many detailed references to original papers.

Other general references

- Boorse, H. A., and Motz, L., eds. *The World of the Atom.* In two volumes. New York: Basic Books, 1966. Actual texts of landmark documents in the history of atomic physics. Each of the original papers is preceded by a brief biographical sketch of the author and a commentary on the historical and scientific setting of the discovery.
- Bork, A. M., and Aarons, A. B. "Collateral Reading for Physics Courses." *American Journal of Physics* 35 (February 1967): 71–78. Extensive list of references on physics and related fields (art, history, philosophy, sociology, and so on).
- Chen, M., ed. *University of California, Berkeley, Physics Problems with Solutions.* Englewood Cliffs, N.J.: Prentice-Hall, 1974. Includes detailed worked-out solutions to problems in atomic, quantum, and statistical physics.
- Gamow, G. *Mister Tompkins in Paperback.* New York: Cambridge University Press, 1967. Whimsical, nontechnical writings on relativity, quantum theory, and atomic physics, including the adventures of Mr. Cyril George Henry Tompkins in a universe in which the speed of light (c) is small, the

gravitational constant (G) is large, and Planck's constant (h) is large, by a master writer for nonscientific audiences.

- Heisenberg, W. *Physics and Beyond* (translated from the German by A. J. Pomerans). New York: Harper & Row, 1971. Partly a biographical account by one of the chief participants of the development of twentieth-century physics, relating physics to philosophy and society.
- Kevles, D. J. *The Physicists: The History of a Scientific Community.* New York: Alfred A. Knopf, 1978. The story of the development over the past century, particularly in the United States, of physics, not only as a scientific enterprise, but also as a social, cultural, and sociological phenomenon.
- *Physics Today,* a monthly published by the American Institute of Physics, 335 East 45 Street, New York, N.Y. 10017, includes summary articles and announcements on current research findings.
- *Reviews of Modern Physics,* published by the American Institute of Physics, 335 East 45 Street, New York, N.Y. 10017, publishes most recent and authoritative listing of fundamental physical constants (typically February issue, on a biennial basis).
- *Scientific American,* a monthly with clearly written, beautifully illustrated articles of a relatively nontechnical nature on modern developments in science, including all important advances in physics, published by Scientific American, 415 Madison Avenue, New York, N.Y. 10017. A comprehensive set of OFFPRINTS of individual articles in past issues are available.
- Trigg, G. L. *Landmark Experiments in Twentieth-Century Physics.* New York: Crane, Russak, 1975. Sixteen critical experiments in modern physics and their significance are treated in some detail.

Special relativity (Chapters 2 and 3)

- Brecker, K. "A Guide for the Perplexed." *Nature* 278 (March 15, 1979): 215–18. Collection of direct and indirect quotations by and about Einstein, as a tribute on the hundredth anniversary of his birth (March 14, 1879).
- Einstein, A. *Meaning of Relativity.* 5th ed. Princeton, N.J.: Princeton University Press, 1956. A general introduction by the founder.
- French, A. P. *Special Relativity.* New York: W. W. Norton, 1968. An introductory, comprehensive treatment with applications to a variety of topics in modern physics.
- French, A. P., ed. *Einstein: A Centenary Volume.* Cambridge, Mass.: Harvard University Press, 1979. Collection of essays on and by Einstein, translations of some of his fundamental scientific papers, photographs; commentary on the man, his scientific accomplishments, his views on nonscientific topics. A veritable potpourri of Einsteiniana.
- Jammer, M. *Concepts of Space: The History of Theories of Space in Physics.* 2d ed. Cambridge, Mass.: Harvard University Press, 1969. A fundamental and philosophical consideration of space, including relativity theory.
- Marder, L. *Time and the Space Traveller.* Philadelphia: University of Pennsylvania Press, 1974. Strong emphasis on the twin paradox.
- Mermin, N. D. *Space and Time in Special Relativity.* New York: McGraw-Hill, 1968.
- Purcell, E. M. *Electricity and Magnetism.* New York: McGraw-Hill, 1965. An intermediate-level textbook in electricity and magnetism using special relativity throughout.

- Resnick, R. *Introduction to Special Relativity.* New York: John Wiley & Sons, 1968.
- Taylor, E. F., and Wheeler, J. A. *Space-Time Physics.* San Franscisco: W. H. Freeman, 1966. A rigorous, elementary treatment emphasizing the four-dimensional character of space-time.

Quantum and atomic physics (Chapters 3, 4, 5, 6, and 7)

- Cagnac, B., and Pebay-Peyroula, J.-C. *Modern Atomic Physics: Fundamental Principles* (translated by J. S. Deech). New York: Halsted Press, 1975. See especially the treatment of angular momentum and magnetic moments, Chapters 8–12.
- Christy, R. W., and Pytte, A. *The Structure of Matter: An Introduction to Modern Physics.* Menlo Park, Calif.: Benjamin-Cummings Publishing Co., 1965. Clear treatment of molecular and many-particle systems, Chapters 7–15 and 19–26.
- Cline, B. L. *The Questioners: Physicists and the Quantum Theory.* New York: Thomas Y. Crowell, 1965. The personalities and events, with fascinating anecdotes, during the development of the quantum theory.
- Cropper, W. F. *The Quantum Physicists and an Introduction to Their Physics.* New York: Oxford University Press, 1970. A historical account of the principal figures and developments in quantum theory, coupled with careful exposition of the theory.
- Eisberg, R. M. *Applied Mathematical Physics with Programmable Pocket Calculators.* New York: McGraw-Hill, 1977. Chapter 8 tells how to find solutions to the Schrödinger equation by numerical methods.
- Gamow, G. *Thirty Years That Shook Physics: The Story of Quantum Theory.* Garden City, N.Y.: Doubleday, 1966. A fascinating, nontechnical narrative.
- Heisenberg, W. *Physical Principles of the Quantum Theory.* New York: Dover, 1930. The matrix mechanics formulation of quantum theory and the uncertainty principle by the orginator.
- Heitler, W. *Elementary Wave Mechanics with Applications to Quantum Chemistry.* 2d ed. New York: Oxford University Press, 1956. A particularly lucid account of quantum theory in a very slim volume.
- Hoffman, B. *The Strange Story of the Quantum.* New York: Dover, 1959. A nontechnical book with many amusing analogies.
- Jammer, M. *The Philosophy of Quantum Mechanics: The Interpretation of Quantum Mechanics in Historical Perspective.* New York: John Wiley & Sons, 1974. Comprehensive treatment of the philosophical foundations of quantum theory.
- Margenau, H., and Murphy, G. M. *The Mathematics of Physics and Chemistry.* Huntington, N.Y.: R. E. Krieger Publishing Co., 1976. See Chapter 11 for a concise postulational treatment of quantum theory.
- Sands, M., ed. *The Feynman Lectures on Physics,* Vol. 3. Reading, Mass.: Addison-Wesley, 1965.
- Schilpp, P. A., ed. *Albert Einstein: Philosopher-Scientist.* New York, Harper & Row, 1959. See especially Chapter 7, "Discussion with Einstein on Epistemological Problems in Atomic Physics," by Niels Bohr. Although Einstein was confident that the quantum theory worked, he was uncomfortable with its intrinsic emphasis on uncertainties and probabilities; he devised

gedanken (thought) experiments intended to refute the uncertainty principle. As the Einstein-Bohr conversations show, all such attempts failed.
- Wilmott, J. C. *Atomic Physics.* New York: John Wiley & Sons, 1975.

Statistical physics and quantum devices (Chapters 8 and 9)

- Ashcroft, N. W., and Mermin, N. D. *Solid State Physics.* New York: Holt, Rinehart and Winston, 1976. Somewhat sophisticated; excellent three-dimensional diagrams.
- Bertolini, G., and Coche, A. *Semiconductor Detectors.* New York: John Wiley & Sons (Interscience), 1968. Nuclear detectors using semiconducting materials and the spectroscopy of nuclear radiation.
- Finkelstein, R. J. *Thermodynamics and Statistical Physics.* San Francisco: W. H. Freeman, 1970.
- Fowles, G. R. *Introduction to Modern Optics.* 2d ed. New York: Holt, Rinehart and Winston, 1975. See especially Chapter 9, on lasers.
- Françon, M. *Holography.* New York: Academic Press, 1974.
- Harrison, W. A. *Electronic Structure and Properties of Solids.* San Francisco: W. H. Freeman, 1979.
- Holden, A. *The Nature of Solids.* New York: Columbia University Press, 1968. A lucid account of elementary properties of semiconducting devices.
- Kittel, C. *Introduction to Solid State Physics.* 5th ed. New York, John Wiley & Sons, 1976. The standard in its field; also available in abbreviated form as *Elementary Solid State Physics.*
- Kock, W. E. *Lasers and Holography.* Garden City, N.Y.: Doubleday, 1969.
- Lengyel, B. A. *Lasers.* 2d ed. New York: John Wiley & Sons, 1971. A general introduction, with the principal features of the most common types of lasers described.
- Lynton, E. A. *Superconductivity.* 3d ed. New York: Halsted Press, 1971. A general introduction to superconductivity and a summary of the phenomenon.
- Omar, M. A. *Elementary Solid State Physics.* Reading, Mass.: Addison-Wesley, 1975.
- Reif, F. *Fundamentals of Statistical and Thermal Physics.* New York: McGraw-Hill, 1965. Statistical physics treated rigorously but from an elementary point of view.
- Siegman, A. E. *Introduction to Lasers and Masers.* New York: McGraw-Hill, 1971.
- Troup, G. *Masers and Lasers.* 2d ed. New York: John Wiley & Sons, 1963. *Maser* is the acronym for microwave amplification by the stimulated emission of radiation.

Nuclear physics (Chapters 10 and 11)

- *Atomic Radiation.* Camden, N.J.: RCA Service Company, 1962. Units of measurement for ionizing radiation; biological hazards and safety measures.
- Bishop, A. S. *Project Sherwood: The U.S. Program in Controlled Fusion.* Garden City, N.Y.: Doubleday, 1960. Fundamental principles of nuclear fusion and the control of this process.
- Bromley, D. A., ed. "Detectors in Nuclear Science." *Nuclear Instruments and Methods* 162 (1979): 1–738. Comprehensive review articles on all types of nuclear detectors.

- Enge, H. A. *Introduction to Nuclear Physics.* Reading, Mass.: Addison-Wesley, 1966.
- Evans, R. D. *The Atomic Nucleus.* New York: McGraw-Hill, 1955. Moderately advanced and comprehensive, especially for low-energy nuclear physics.
- Goldsmith, M., and Shaw, E. *Europe's Giant Accelerator.* New York: Crane, Russak, 1977. The story of the design, development, and construction of the 400-GeV proton synchrotron at CERN (Center for Nuclear Research), the multinational European cooperative high-energy research establishment near Geneva, Switzerland.
- Hill, R. D. *Tracking Down Particles.* Menlo Park, Calif.: Benjamin-Cummings, 1963. For the nonscientist; covers broad range of topics, with special concentration on accelerating machines and parity nonconservation.
- Kaplan, I. *Nuclear Physics.* Reading, Mass.: Addison-Wesley, 1966.
- Lapp, R. E., and Andrews, H. L. *Nuclear Radiation Physics.* 4th ed. Englewood Cliffs, N.J.: Prentice-Hall, 1972. Strong concentration on nuclear radiations and their detection.
- Livesey, D. L. *Atomic and Nuclear Physics.* Waltham, Mass.: Blaisdell, 1966.
- Livingston, M. S. *Particle Accelerators: A Brief History.* Cambridge, Mass.: Harvard University Press, 1969. Relatively nontechnical account of the development of high-energy accelerating machines.
- Onseph. P. J. *Introduction to Nuclear Radiation Detectors.* Plenum Press, 1975. Discussion of all principal types of detectors.
- Shutt, R. P. *Bubble and Spark Chambers.* New York: Academic Press, 1967.
- Vandenbosch, R., and Huizenga, J. R. *Nuclear Fission.* New York: Academic Press, 1973.
- Wapstra, A. H. and Bos, K. *Atomic Data and Nuclear Data Tables, 19,* 177 (1977), New York: Academic Press. Authoritative source for nuclide masses.

Elementary particles (Chapter 12)

For the most up-to-date accounts, articles in *Scientific American* are especially recommended.

- Capra, F. *The Tao of Physics: An Exploration of the Parallels between Modern Physics and Eastern Mysticism.* New York: Random House, 1975. A highly speculative attempt to show that modern physics, and especially elementary particle physics, has counterparts in Eastern thought.
- Ford, K. *The World of Elementary Particles.* New York: Blaisdell, 1963.
- Frisch, D. H., and Thorndike, A. M. *Elementary Particles.* Princeton, N.J.: Van Nostrand, 1964.
- Gell-Mann, M. and Ne'eman, Y. *The Eight-Fold Way.* Menlo Park, Calif.: Benjamin-Cummings, 1964.
- Longo, M. J. *Fundamentals of Elementary Particle Physics.* New York: McGraw-Hill, 1973.
- Tassie, L. J. *The Physics of Elementary Particles.* New York: John Wiley & Sons, 1973.
- Zukav, Gary. *The Dancing Wu Li Masters: An Overview of the New Physics.* New York: William Morrow, 1979. Covers quantum physics and relativity theory from their inceptions to the present for people with no mathematical or technical background.

Answers to Problems

Chapter 2

2–3. (*a*) 1150 ft/s; (*b*) +100 ft/s; (*c*) S'

2–4. (*a*) mv; (*b*) (1/2) mv^2; (*c*) zero; (*d*) (1/4) mv^2

2–5. (*a*) 0.20; (*b*) 4.74×10^4 m/s

2–6. $x' = 1.1 \times 10^6$ m; $t' = 3.8$ ms

2–8. 0.186c parallel to semimajor axis

2–10. Speed c at 60° to x-axis

2–11. +0.998c

2–12. (*a*) 0.80 c; (*b*) 0.929 c

2–13. (*a*) 3 m/s; (*b*) 5×10^{-15}%

2–14. (*a*) 0.693c along +x-axis; (*b*) 0.555c

2–15. (*a*) 0.936c; (*b*) 0.798c

2–16. 2.1×10^{-5} s (particle 1) and 3.6×10^{-6} s (particle 2)

2–17. 0.9998c parallel to length of rod

2–18. 5.3×10^{12} m

2–19. Toward the east

2–20. $\Delta t/\sqrt{1-(v/c)^2}$

2–21. 1.7×10^{-14} m

2–22. (*a*) $2l/c$; (*b*) $\dfrac{2l/c}{\sqrt{1-(v/c)^2}}$; (*c*) $T_0/\sqrt{1-(v/c)^2}$

2–23. 1.5×10^{-4} Å

2–24. (*a*) 2.0; (*b*) 0.10

2–26. (*a*) 7.0 mm; (*b*) 595 days

2–27. (*a*) 0.60c; (*b*) 12.8 ft; (*c*) No

2–28. (*a*) 0.99c; (*b*) 10c

2–30. 19,700 Å in *infrared* region

2–31. (*a*) Yes; (*b*) Yes; (*c*) No; (*d*) Yes

2–32. No

2–33. (*a*) 0.58c

2–35. (*a*) 12.5×10^9 yr; (*b*) 1.1×10^{10} 1-yr; (*c*) 20×10^9 yr

2–36. 1.3×10^{-8} s

2–37. 25 yr

2–38. (*a*) $(1 - 3.34 \times 10^{-12})(1.283 \times 10^5$ s);
(*b*) $(1 - 1.191 \times 10^{-12})(1.283 \times 10^5$ s);
(*c*) $(1 - 3.34 \times 10^{-12})(1.283 \times 10^5$ s);
(*d*) 2.75×10^{-7} s

2–39. (*a*) 250 yr; (*b*) 150 yr; (*c*) 333 yr; (*d*) 267 yr

2–40. (*b*) Only inertial system is that moving at velocity (c/2) along +x-axis for which Events 1 and 2 are simultaneous.

Chapter 3

3–1. 0.87c

3–3. 502

3–4. 0.14c

3–5. (*a*) m_0v and $\frac{1}{2}m_0v^2$; (*b*) zero and $\frac{1}{4}m_0v^2$

3–6. 2.6×10^{-15} m

3–7. −0.68c

3–8. 8.3×10^4

3–10. 0.999998c

3–11. 20 keV/c

3–12. $\sqrt{3}\, m_0c$

3–13. (*a*) 29 cm; (*b*) 10.3 m

3–14. 1.5 GeV

3–15. 0.95c

3–16. 0.98c

3–17. (*a*) 0.866c; (*b*) 0.51 MeV

3–18. 1.19 mass of proton

3–20. 1981 GeV/c in direction of proton velocity

3–22. 3.3×10^{-8} %

3–23. $2\,E_k/c^2$

3–24. 5.7×10^{-10}

3–25. 5.6×10^{19} J for spaceship alone

3–26. (*a*) 10^4 eV; (*b*) 99.5 MeV; (*c*) 5.9 GeV

3–27. (*a*) 6.02×10^{-13} kg; (*b*) 0.563 eV/molecule

3–28. (*a*) decreases; (*b*) 2.9×10^{-9} kg

3–29. (*a*) 0 °C, ice and water; (*b*) 7.5×10^{-15}

3–30. $6\ m_0c^2$

3–32. (*a*) $-E_0/\sqrt{3}\ c$; (*b*) $2\ E_0/\sqrt{3}$

3–33. 13.2 GeV

3–34. 65 MeV/*c* and 65 MeV

3–35. (*a*) $-0.454c$; (*b*) $0.949c$

3–36. $0.919c$ and $-0.253c$

3–37. $0.56c$ along bisector

3–38. (*a*) one photon with momentum 393.5 MeV/*c* parallel to incident pion and energy 393.5 MeV, the other with momentum 11.5 MeV/*c* antiparallel to incident pion and energy 11.5 MeV; (*b*) one photon with momentum 22.3 MeV/*c* perpendicular to incident pion and energy 22.3 MeV, the other photon with momentum 383 MeV/*c* in direction 3.3° to incident pion and energy 383 MeV.

3–40. 3.00×10^8 m/s

3–42. (*a*) 1.6 MW; (*b*) 0.26 MW

Chapter 4

4–1. 6.0 mJ and 2.0×10^{-11} kg · m/s

4–2. $P\lambda A/4\pi R^2 hc$

4–3. 5.0 eV

4–4. $2hc/\lambda_0$

4–5. 4000 Å to 6526 Å

4–6. (*a*) 4.74×10^{14} wavecrests/s $= 4.74 \times 10^{14}$ Hz (*b*) 3.2×10^{15} photons/s

4–7. (*a*) 6.8×10^{-30} J/s $= 4.2 \times 10^{-11}$ eV/s; (*b*) 1.1×10^{11} s $\approx 3.5 \times 10^3$ years, compared with 10^{-9} s experimental

4–8. (*a*) 2.0×10^4 photons/s; (*b*) 3.2×10^{-18} Å

4–9. 1.6×10^6 m $\approx$ 1000 mi

4–10. $\lambda_{max} = 4\lambda$

4–11. $7.90 \times 10^{-5}\ T$

4–12. (*a*) 4.82 eV; (*b*) 7.58 eV and 1.63×10^6 m/s

4–13. (*a*) 4.17×10^{-15} J · s/C compared with 4.14×10^{-15} J · s/C; (*b*) 2.10 eV

4–14. (*a*) 1.2×10^3 m/s; (*b*) $0.97c$

4–15. (*a*) 1.8×10^{-5} m/s; (*b*) $1 - (3.0 \times 10^{-14})$

4–16. (*a*) 3.1×10^{-8} s; (*b*) information is communicated by the electric field, which travels at the speed of light, 3.0×10^8 m/s.

4–17. (*a*) 0.10 kg; (*b*) 1.6×10^4 m/s = 9.6 mi/s; (*c*) 2.5×10^{11} m; (*d*) 2.8×10^{-5}

4–18. Three to two

4–19. 1.0 keV/*c*

4–21. (*a*) $E_\gamma = 0.1$ MeV and $p_\gamma = 0.1$ MeV/*c* northeast

4–22. 1.048 Å

4–23. 2.6×10^{-5} Å

4–25. (*a*) 2.0×10^{-7} Å; (*b*) 3.4×10^2 m/s

4–27. 0.0205 Å

4–29. (*c*) $\frac{2}{3}\ E_0$

4–30. 2.51 MeV

4–31. 67.5 MeV

4–33. (*a*) $5\ E_0/c$; (*b*) $5\ E_0$; (*c*) 11.5°

4–34. (*a*) 620 Å; (*b*) 5.45 Å; (*c*) 1.22×10^{-2} Å

4–35. (*b*) 3.0 cm (positron) and 0.70 cm (electron)

4–40. (*a*) no; (*b*) yes

4–41. water (9.38 cm), aluminum (0.917 cm), lead (0.00690 cm)

4–42. (*a*) 0.046 cm; (*b*) $I/I_0 = 4.73 \times 10^{-14}$

4–43. Lead (0.0057 cm), aluminum (0.55 cm)

Chapter 5

5–1. 3.91×10^{-6} eV

5–3. 0.50 Å

5–5. (*a*) 1.23 Å; (*b*) 3.70×10^{-2} Å; (*c*) 1.24×10^{-4} Å

5–6. (*a*) 1.1×10^{-3} m; (*b*) 3.9×10^{-10} m; (*c*) 3.6×10^{-6} eV; (*d*) 6.5×10^3 Å

5–7. (*a*) 150 V/m; (*b*) 6.0×10^{-6} eV; (*c*) Because of the difficulty in maintaining a 6.0×10^{-6} eV electron beam.

5–8. (*a*) 1.55×10^{-3} eV; (*b*) 1.63 Å

5–9. 60° and 180°

5–10. (*a*) 3.95×10^3 eV; (*b*) 15.3 eV; (*c*) 8.3×10^{-3} eV

5–11. (*a*) 4.5 Å; (*b*) 4

5–12. (*a*) 0.053 Å; (*b*) 5.3 Å

5-13. (*a*) Because the magnetic field deflects the electrons, the diffraction pattern is distorted; (*b*) No effect, because photons are unaffected by magnetic field.

5-14. 8.16×10^{-2} eV

5-15. 1.2×10^{9} eV

5-16. 1.63 Å (1st order), 0.81 Å (2nd order), 0.54 Å (3rd order), . . .

5-17. (*a*) 8.7×10^{-2} Å; (*b*) 0.59 cm, 1.2 cm, 1.8 cm

5-18. 2.0 Å, 1.4 Å

5-19. (*a*) 1.73 Å and 4.21×10^{6} m/s; (*b*) 1.67 Å and 4.36×10^{6} m/s; (*c*) 33.3°

5-20. (*a*) 1.92 Å

5-21. (*a*) 3.0×10^{32} photons/s; (*b*) 2.4×10^{21} photons/m^2·s

5-22. (*a*) 6.4×10^{21} photons/m^2·s; (*b*) laser photon flux constant, radio flux divergent.

5-23. (*a*) 240 photons; (*b*) 240 photons; (*c*) within angle of 0.018°

5-24. (*a*) 1.5×10^{18} photons/s; (*b*) 0.60 W/m^2

5-25. 1.5 eV

5-26. 1.24×10^{3} eV/*c*

5-28. 7×10^{-2} m

5-29. 1.7 mm/s

5-30. 410 bounces

5-31. (*a*) 5.5×10^{-2} Å; (*b*) 180

5-32. (*b*) $\Delta\theta = \infty$

5-33. (*a*) $mv\Delta v$; (*b*) $m\Delta v_x$

5-34. (*a*) *c*; (*b*) 1.06*c*; (*c*) 21.7*c*

5-35. Four

5-36. (*a*) 1.4 MeV; (*b*) 1.2×10^{9} eV

5-37. 6.2×10^{3} eV

5-38. (*a*) 2; (*b*) 1; (*c*) $n = 3.3 \times 10^{31}$

5-39. (*a*) 1.2×10^{5} eV; (*b*) 1.2×10^{5} eV; (*c*) any of first 56 states

5-40. 9 times ground-state energy

5-42. (*a*) 3.7×10^{8} eV; (*b*) 3.1×10^{53} eV

5-43. (*a*) infinite; (*b*) infinite; (*c*) infinite

5-44. (*a*) $E_I/9$

5-46. (*a*) $n(5.3 \times 10^{-30})$ m/s and $n^2(1.8 \times 10^{-43})$ eV (*b*) 1.2×10^{28}; (*c*) 5.3×10^{-30} m/s

5-47. (*a*) Square well: $(2n + 1)\, h^2/8mL^2$; Harmonic oscillator: *hf*; Inverse square: $E_I\,(2n + 1)/n^2(n + 1)^2$

5-48. (*b*) Zero

5-50. (*a*) larger than E_1

5-51. (*b*) 0.408 eV, 1.63 eV, 3.67 eV

Chapter 6

6-1. (*a*) 6.1×10^{-5} Å; (*b*) 4.1×10^{-4} Å; (*c*) wavelength smaller than separation distance

6-3. 4.8×10^{-14} m

6-4. (*a*) 3.1×10^{-2}; (*b*) 9.6×10^{-4}; (*c*) 9.7×10^{5}

6-5. (*a*) 25 min^{-1}; (*b*) 25 min^{-1}; (*c*) 2.9 min^{-1}; (*d*) 200 min^{-1}

6-6. (*a*) $r = 9.1 \times 10^{-14}$ m; (*b*) $r = 6(r_g + r_p)$, no nuclear reaction; (*c*) r (inner electron) $\approx 6.7 \times 10^{-13}$ m $= 12r$, proton within inner electron orbit

6-7. $(\sigma/\pi r^2) = 1.6 \times 10^{8}$; scattering to total target area $= 4.6 \times 10^{-5}$, which depends on thickness of target.

6-8. 2.2 keV

6-9. (*a*) 2.1×10^{-15} m; (*b*) 5.1×10^{-15} m

6-10. 1.46, 1.55, and 1.57 compared with 1.88 of Equation 6-7.

6-11. 9.7 cm

6-12. 0.1 mm

6-14. 3.26 eV

6-15. 122 eV

6-16. 12.1 eV, 1.89 eV, and 10.2 eV

6-18. From $n_u = 6$ to $n_l = 4$

6-19. Total electric charge, total number of electrons, total energy, and total linear momentum.

6-20. (*a*) 100 Å; (*b*) 0.01 eV

6-21. (*a*) $a = -1$, $b = -1$, $c = 2$; (*b*) $R = 0.529$ Å

6-22. (*a*) 6; (*b*) 2200

6-23. (*a*) $f_1 = 6.581 \times 10^{15}$ Hz, $f_2 = 0.823 \times 10^{15}$ Hz, and $\nu_{2\to1} = 2.468 \times 10^{15}$ Hz; (*b*) $f_{100} = 6.581 \times 10^{9}$ Hz, $f_{101} = 6.387 \times 10^{9}$ Hz, and $\nu_{101\to100} = 6.485 \times 10^{9}$ Hz; (*c*) $f_{1000} = 6.581 \times 10^{6}$ Hz, $f_{1001} = 6.561 \times 10^{6}$ Hz, and $\nu_{1001\to1000} = 6.572 \times 10^{6}$ Hz

6-24. 1×10^{5} K

6-25. (*a*) 1.5×10^{4} eV

6-26. (*a*) 10.21 eV; (*b*) 10.21 eV/*c*; (*c*) 10.21 eV/*c*; (*d*) 6×10^{-8} eV

6-27. (*a*) 9.3×10^{-8} eV; (*b*) average thermal energy = 3.9×10^{-2} eV

6-28. (*a*) $\hbar$; (*b*) $4\hbar$; (*c*) $\hbar$ and $4\hbar$

6-30. (*a*) $4 \to 2$; (*b*) H-atom $\hbar$, singly ionized helium atom $2\hbar$

6-31. (*a*) 7.45×10^4 eV; (*b*) 0.222 Å; (*c*) X-ray

6-32. 1.03×10^3 Å

6-33. (*a*) 3.385×10^3 eV/*c* and 6.106×10^{-3} eV; (*b*) 1.02 eV

6-34. (*b*) 0.12 Å

6-35. (*a*) 4.1×10^{-7} eV; (*b*) 5×10^{-5} Å and 4×10^{-8}

6-36. (*a*) 0.9337*c*

6-38. (*a*) 6561.14 Å; (*b*) 6564.67 Å

6-39. (*a*) 35.31 Å; (*b*) 2.84×10^{-3} Å

6-41. (*a*) 6.8029 eV; (*b*) 2430 Å

6-44. 0.040

6-45. 3.2

Chapter 7

7-1. 0.53 Å

7-3. $4S \to 2P$, $4D \to 2P$, $4P \to 2S$

7-4. 2P state

7-5. (*a*) $md^2/2$; (*b*) $E_j = j(j+1)\hbar^2/md^2$, $E_0 = 0$, $E_1 = 2\hbar^2/md^2$, $E_2 = 6\hbar^2/md^2$, $E_3 = 12\hbar^2/md^2$; (*c*) $\nu_{1\text{-}3} = 5\hbar/\pi md^2$, $\nu_{0\text{-}2} = 3\hbar/\pi md^2$

7-7. (*a*) $\sqrt{2}\hbar$; (*b*) $(\sqrt{6} - \sqrt{2})\hbar$

7-8. K^{+8}

7-10. $5S \to 4P$ or 3P, $4P \to 3D$ or 4S or 3S, $4S \to 3P$, $3P \to 3S$, $3D \to 3P$; transitions in visible range (1.6 to 3.2 eV) $5S \to 3P$, $3D \to 3P$, $3P \to 3S$

7-11. 4.1 eV

7-12. 223.6 eV

7-13. (*b*) $\geq 5 \times 10^5$

7-14. 4.54×10^{11} Hz

7-15. (*a*) $\approx 10^{20}$; (*b*) 10^{-20} rad

7-16. (*a*) 1.2×10^{-33} J·s; (*b*) 20

7-17. (*a*) 0 or $\sqrt{2}\hbar$; 0 or $\pm\hbar$

7-18. (*a*) $\sqrt{20}\hbar$; (*b*) 0, $\pm\hbar$, $\pm 2\hbar$, $\pm 3\hbar$, $\pm 4\hbar$

7-19. $\lambda = 4226.90$ Å, 4226.73 Å, and 4226.56 Å

7-20. 1.759×10^{11} C/kg

7-24. (*a*) 3.3×10^{25} s^{-1}; (*b*) 300*c*

7-30. (*a*) $P_{3/2} \to S_{1/2}$ has larger energy difference than $P_{1/2} \to S_{1/2}$; (*b*) Because $P_{3/2}$ level has 4 states and $P_{1/2}$ level only 2 states, Intensity ($P_{3/2} \to S_{1/2}$) = 2 × Intensity ($P_{1/2} \to S_{1/2}$)

7-31. (*a*) 1.3×10^{-5} eV; (*b*) 4.5×10^{-6} eV

7-34. One

7-35. (*a*) Two; (*b*) 0.38 cm

7-36. 119

7-37. $_{16}$S

7-38. (*a*) $\sqrt{3}\hbar/2$; (*b*) $\sqrt{3}\hbar/2$

7-39. (*a*) (1,0,0,1/2), (1,0,0, − 1/2), (2,0,0, − 1/2); (*b*) (2,0,0,1/2) and (2,1,0, − 1/2)

7-40. (*b*) $3^2P_{1/2}$

7-41. Ne and Hg

7-42. (*a*) 84.5 keV and 0.147 Å; (*b*) 1.47 keV and 8.44 Å

7-44. 8.00 keV and 8.58 keV

Chapter 8

8-1. 2.78 Å

8-2. ≈ 0.9

8-4. (*a*) −5.18 eV; (*b*) Yes

8-5. (*a*) 1.5 eV; (*b*) −4.9 eV; (*c*) −3.4 eV

8-7. Radius of Na^+ is less than radius of Ne; radius of Cl^- is greater than radius of Ar.

8-10. (*a*) 0.50; (*b*) 4.1×10^{-73}

8-11. (*a*) 1; (*b*) 0.881; (*c*) 0.119

8-12. One

8-14. (*c*) $\epsilon < \frac{kT}{2}$; (*d*) $\epsilon > \frac{kT}{2}$

8-16. 0.637

8-17. 1.98×10^4 K

8-20. 5.38×10^{-2} eV

8-21. 10^{-81}

8-22. (*b*) 1.4×10^3 K

8-23. 1.79×10^{-2} eV

8-24. (*b*) 0.134%

8–25. 3.87×10^{13} Hz

8–26. (*a*) 0.030 cal/gm • C°; (*b*) 0.15 cal/gm • C°; (*c*) 0.088 cal/gm • C°

8–27. Beryllium

8–28. (*a*) 0.056 cal/gm • C°; (*b*) 1.8×10^{-4} cal/gm • C°

8–29. (*a*) $p_y = p_z = 0$, $p_x = 3.3 \times 10^{-33}$ kg • m/s, $\epsilon = 7.3 \times 10^{-30}$ J for state (1,0,0) and $p_x = p_y = 3.3 \times 10^{-33}$ kg • m/s, $p_z = 0$, $\epsilon = 1.0 \times 10^{-29}$ J for state (1,1,0); (*b*) 2.7×10^{-30} J; (*c*) $\Delta\epsilon = 3.7 \times 10^{-32}$ J

8–30. 2.1×10^{15}

8–32. (*a*) 7.11×10^{12} Hz; (*b*) 341 K

8–33. 503

8–37. 1.1 mm

8–40. 1.26×10^{5} K

8–41. 1.55 K

8–43. 0.0282 eV

8–44. 5.5×10^{4} K

8–45. 0.98

Chapter 9

9–7. $\sim 10^{-8}$

9–8. 6.4×10^{-3} eV

9–9. 53 μm

9–10. (*a*) phosphorus, arsenic, antimony; (*b*) aluminum, gallium, indium

9–11. (*a*) *p* type; (*b*) antimony

9–12. (*a*) 6.4 Å; (*b*) 8.5 Å

9–13. (*b*) 107; (*c*) 0.76

9–14. 0.66 eV

9–15. (*a*) 1.1 V; *p* side

9–17. (*a*) 3×10^{12} W; 140 km

9–18. (*a*) 1.13×10^{4} Å; (*b*) 2.4

9–19. (*a*) 10^{10}/cm^3; (*b*) 2×10^{-13}

9–20. (*a*) 2×10^{6}; (*b*) 3×10^{3}

9–22. (*a*) 200 J; (*b*) 1.3×10^{9} J/m^3; (*c*) 77 mg

9–23. 1.1×10^{-33}

9–24. (*a*) 5s → 4p; (*b*) 4s → 3p

9–25. (*b*) 1×10^{14} s^{-1}

9–26. 2.9×10^{-7} s

9–29. 4.3×10^{-4}%

9–31. 8.79×10^{6}

9–32. 4.33×10^{-12} Hz

9–33. (*a*) 6.10×10^{-3}; (*b*) 6.14×10^{-3}

9–38. (*a*) 0.8 μm

Chapter 10

10–3. 21.3 MHz

10–4. 1.12 MeV

10–6. $\Delta m = 2 \times 10^{-7}$ u

10–9. (*a*) ^{42}Ca, ^{43}Ca, ^{44}Ca, ^{46}Ca, ^{48}Ca; (*b*) $^{39}_{19}$K, $^{38}_{18}$Ar, $^{37}_{17}$Cl, $^{36}_{16}$S; (*c*) $^{40}_{18}$Ar

10–10. $^{2}_{1}$H, $^{6}_{3}$Li, $^{10}_{5}$B, $^{14}_{7}$N

10–11. $\lambda = 1.44$ fermi

10–12. 216 nucleons

10–13. (*a*) $\rho \approx 3 \times 10^{17}$ kg/m^3; (*b*) 12 km

10–14. (*b*) 1.24 GeV, 1.24 GeV, 0.61 GeV

10–15. (*a*) $U = -kZe^2/R$ for $r \geq R$, and $U = \frac{1}{2}(kZe^2/R^3)\,r^2 - \frac{3}{2}\,kZe^2/R$ for $r \leq R$; (*b*) $\frac{3}{2}\,kZe^2/R$

10–16. (*a*) 190 MeV; (*b*) 53 MeV

10–17. (*a*) 20.6 MeV to remove first neutron, 5.5 MeV then to remove first proton, and 2.2 MeV to separate deuteron

10–19. (*a*) α decay; (*b*) 92 keV

10–20. (*a*) 18.72 MeV

10–21. (*a*) 11.23 MeV, 15.96 MeV, 1.94 MeV; (*b*) 13.12 MeV, 18.72 MeV, 4.95 MeV

10–22. (*a*) 0.000840 u; (*b*) 0.00297 u; (*c*) 0.00213 u compared with 0.002128 u (App. II)

10–23. 8.4%

10–26. 3.61×10^{-4}

10–27. 0.277 hr

10–28. (*a*) 61.6 μg; (*b*) 15.0 μCi

10–29. 6×10^{9} yr

10–31. 0.317 Å

10–32. (*a*) 0.16 MeV; (*b*) yes

10–33. (*b*) 0.78 MeV

10-34. (*a*) EC, β^+; (*b*) β^-; (*c*) β^-; (*d*) EC, β^+; (*e*) β^-, EC, β^+

10-35. (*a*) 0.863 MeV; (*b*) 0.863 MeV/*c*; (*c*) 57 eV

10-36. (*a*) 13.37 MeV; (*b*) 13.37 MeV

10-38. (*a*) 7; (*b*) 4

10-39. 51.7 MeV

10-40. Antineutrino with zero energy

10-41. 1.3×10^9 yr

Chapter 11

11-1. (*a*) β^-; (*b*) EC or β^+; (*c*) β^-; (*d*) EC or β^+; (*e*) EC or β^+

11-2. (*a*) 8.64×10^{-15} m; (*b*) 2.1×10^{-15} m

11-3. (*a*) stable $^{13}_{6}$C; (*b*) unstable $^{12}_{7}$N; (*c*) stable $^{13}_{6}$C; (*d*) stable $^{15}_{7}$N

11-4. (*a*) stable $^{20}_{9}$F; (*b*) stable $^{19}_{10}$Ne; (*c*) $^{19}_{8}$O decaying by β^-; (*d*) stable $^{22}_{11}$Na

11-5. $^{13}_{6}$C

11-8. Photon

11-9. Electron beam

11-10. (*a*) 2.226 MeV; (*b*) 3.338 MeV; (*c*) 6.645 MeV

11-11. (*a*) 1.876 MeV; (*b*) 13.09 MeV

11-12. 3.028 MeV

11-13. 9.99 MeV and 7.85 MeV

11-16. 0.87 MeV

11-17. $\sigma^{2/3}$

11-19. 3.1×10^{16}

11-20. (*a*) -1.261 MeV

11-21. (*a*) 13.2 MeV; (*b*) 6.0 MeV; (*c*) 2.2 MeV; (*d*) 2.8 MeV; (*e*) 3.4 MeV

11-22. (*a*) 1.513 MeV; (*b*) 7.115 MeV; (*c*) 7.787 MeV; (*d*) 8.833 MeV

11-23. (*a*) 272 MeV

11-25. 0.3 MeV

11-26. (*b*) 28%

11-28. 7.6×10^9 K

11-30. Less

11-31. (*a*) 4.5×10^{23} eV; (*b*) 6.6×10^{29} eV; (*c*) 5.6×10^{32} eV

11-32. (*a*) 280 fermi

11-33. 0.24 mm

11-34. (*a*) 2.9×10^5; (*b*) 4.6×10^{-14} C; (*c*) 0.23 V

11-36. (*a*) 276 V; (*b*) 253 V; (*c*) 174 V

11-37. 1.18 MeV, 0.16 MeV, 0.97 MeV, 1.33 MeV, 0.31 MeV, 1.12 MeV

11-38. (*a*) upward; (*b*) positron; (*c*) 15 MeV

11-40. (*a*) parabola; (*b*) helix

11-41. (*a*) 0.34 mm; (*b*) 33 m; (*c*) 14 mm; (*d*) 36 m

11-42. 2.5×10^5 m/s

11-43. (*a*) 40 MeV; (*b*) 2.5 MeV/nucleon

11-45. (*a*) 30.5 MHz; (*b*) 3.1×10^7 m/s; (*c*) 96.6 MeV/*c*; (*d*) 0.16 m

11-46. (*b*) 2; (*c*) 0.5 MeV

11-47. (*a*) 3.26×10^{-10}; (*b*) 3 in; (*c*) 1.6×10^{14} s^{-1}

11-48. (*a*) 1.67 T; (*b*) 2.1×10^{-5} s

11-49. (*a*) 2.0 MeV; (*b*) 1.0 MeV

11-50. (*a*) 8.73×10^{-6} s; (*b*) 8.73×10^{-6} s; (*c*) 4.84×10^{-4}

11-53. (*a*) 29 GeV; (*b*) 15 GeV

Chapter 12

12-1. (*b*) virtual proton and virtual photon

12-2. (*a*) $\approx 6 \times 10^{-22}$ s; (*b*) 90 fermi

12-3. (*a*) W^-; (*b*) W^-

12-5. (*a*) 2×10^{-8} eV; (*b*) 1×10^{-10} eV

12-6. (*a*) 1.5 fermi, (*b*) 0.6 fermi, (*c*) 1.5×10^{-4} fermi

12-7. (*a*) neutral pion; (*b*) photon; (*c*) neutrino

12-8. π^0_v and p_v

12-9. Two

12-11. 1.98 GeV

12-12. (*a*) K^0 (0.420 GeV), Λ^0 (0.34 GeV); (*b*) 1.27 GeV; (*c*) $\tau(K^0) = 1.67 \times 10^{-10}$ s, $\tau(\Lambda^0) = 2.71 \times 10^{-10}$ s; (*d*) $\tau_0(K^0) = 0.91 \times 10^{-10}$ s; $\tau_0(\Lambda^0) = 2.1 \times 10^{-10}$ s

12-16. (*b*) 2.68 GeV (Ω^-), 0.663 GeV (Ξ^-), zero (Σ^-)

12-17. (*a*) K^+; (*b*) $\bar{K}^0$; (*c*) K^+; (*d*) $\bar{K}^0$

12-19. Because they all decay by the strong interaction.

12-21. (*a*) 190 MeV

12-22. Δ^{++} and Δ^+

12-23. All reactions can occur, but the third and fourth reactions require minimum incident kinetic energies of 3 MeV and 9 MeV, respectively.

12–24. Threshold photon energies: Δ^+ (340 MeV), Δ^0 (337 MeV), Δ^{++} (534 MeV), Δ^- (531 MeV)

12–26. (*a*) charge not conserved; (*b*) yes, through the weak interaction; (*c*) yes, through the electromagnetic interaction; (*d*) yes, through the strong interaction; (*e*) yes, through the strong interaction, if the center-of-mass energy exceeds 607 MeV.

12–27. Δ^{++} (uuu), Δ^+ (uud), Δ^0 (udd), Δ^- (ddd)

12–28. (*a*) $d\bar{u}$; (*b*) $u\bar{s}$; (*c*) $s\bar{d}$

12–29. (*a*) ssd; (*b*) sss; (*c*) $\bar{s}\bar{d}\bar{u}$

12–30. (*b*) p (uud), n (udd), Σ^- (sdd), Σ^0 (sdu), Σ^+ (suu), Λ^0 (sdu), Ξ^- (ssd), Ξ^0 (ssu)

Index